THE SURVEYING HANDBOOK

THE SURVEYING HANDBOOK

SECOND EDITION

edited by Russell C. Brinker and Roy Minnick

SPRINGER-SCIENCE+BUSINESS MEDIA, B.V.

Library of Congress Cataloging-in-Publication

The surveying handbook/editors, Russell C. Brinker and Roy Minnick--2nd ed.
    p.   cm.
   Includes bibliographical references and index.
   ISBN 978-1-4613-5858-9   ISBN 978-1-4615-2067-2 (eBook)
   DOI 10.1007/978-1-4615-2067-2
   1. Surveying. I. Brinker, Russell C. (Russell Charles). II. Minnick, Roy.
  TA555.887   1994
  526.9--dc20   91-5241
                                                   CIP

British Library Cataloguing in Publication Data available

Copyright © 1995 by Springer Science+Business Media Dordrecht
Originally published by Chapman & Hall in 1995
Softcover reprint of the hardcover 2nd edition 1995

Printed on acid-free paper.

Contents

10   Survey Drafting   *180*
Edward G. Zimmerman

31 Optical Tooling   *798*
James P. Reilly

32 Land Descriptions   *811*
Dennis J. Mouland

33   The Role of the Land Surveyor in Land Litigation: Pretrial   *836*

John Briscoe

Preface

The first edition of *The Surveying Handbook*, although a ground breaker, was widely accepted. However, surveying is a dynamic profession with each new development just one step ahead of the next, and updating became critical. In addition, the editors received constructive criticism about the first edition that needed to be addressed. So, while the objective of *The Handbook* remains intact, the logical evolution of the profession, along with the need to recognize constructive criticism, led to the second edition.

New chapters have been added on water boundaries, boundary law, and geodetic positioning satellites. The chapter on land data systems was rewritten to provide a dramatic updating of information, thus broadening the coverage of *The Handbook*. The same may be said for the state plane coordinate chapter. The material on public lands and construction surveying was reorganized as well. Appendices were added to tabulate some information that was buried in the earlier edition in several places. Numerous other changes were incorporated to help the handbook retain its profession-wide scope, one step beyond the scope of an upper-division college textbook. Along with the most sophisticated techniques and equipment, the reader can find information on techniques once popular and still important.

Four new authors are welcomed to the list of contributors: Grenville Barnes, R. B. Buckner, Donald A. Wilson, and Charles D. Ghilani.

The editors and publisher feel confident that a second edition of *The Surveying Handbook* meets the objectives of broad, thorough coverage and current information, while recognizing the valuable advice and suggestions of first edition users.

RUSSELL BRINKER
ROY MINNICK
February 25, 1993

Preface to the First Edition

THE SURVEYING HANDBOOK has been written to fill the need for a comprehensive volume on professional surveying. In the past, similar books have been filled primarily with tables more readily obtained from other sources, while several of the more recent versions concentrate on a single area of the profession and are published by the American Congress on Surveying and Mapping, American Society of Photogrammetry, and American Society of Civil Engineers. The 36 chapters in this volume were prepared by 35 contributors, generally based on their special fields.

Obviously, even the largest handbook could not cover every phase of surveying in complete depth. But sufficient material is given herein to provide surveyors and others with suitable information outside their specialty field. It can then be determined whether a full-sized special book on a subject area is needed. Some surveying equipment sales and repair shops stock a small number of textbooks. Customers have asked "Why can't I get just one volume to refresh and guide me instead of having to buy half a dozen books?" It is hoped this volume will eliminate that problem.

Based on advance publication interest, surveyors, civil, agricultural and other engineers, foresters, architects, archeologists, geologists, small home builders, realtors, title companies, and lawyers will find useful information in THE SURVEYING HANDBOOK.

Abundant figures and tables are included in this volume. References to textbooks, technical journals, and magazines will help readers find additional sources of specific information desired. Profuse footnotes have been used only in Chapter 31, The Role of the Surveyor in Land Litigation: Pretrial. At most chapter ends, superscript numbers refer to the list of References and Notes, thereby retaining a cleaner appearance and reducing awkward typesetting.

THE SURVEYING HANDBOOK is written in an easy-to-read style that avoids word repetition and other excess verbiage. A handbook is supposed to be practical and that has been the goal of Contributors and Editors. Many Contributors have written their own textbooks or parts thereof, and nearly all are frequent authors of technical papers.

Contacting Contributors residing in 19 different states by telephone and letters has been an unexpected challenge for the Editors who have been so heavily dependent upon the contributors' efforts and cooperation. Their expenditure of time and funds for the extremely small stipend paid handbook contributors by publishers is greatly appreciated.

In addition to the typical textbook chapters, special ones on Survey Drafting; Mining Surveys; Optical Tooling (Industrial Applications); Land Descriptions; Pre-Trial Preparation; Courtroom Techniques; Survey Business Management; Surveying Charges, Contracts, Liability; Land Information Systems; and Surveying Profession, Registration, Associations, are included.

This handbook is the result of the labors over the last $5\frac{1}{2}$ years. Although basic principles of survey measurement remain the same, technology and sources of information may change. Recently, NGS has published portions of NAD-83, several states have adopted new state plane coordinate systems, and data storage and retrieval methods at primary survey information sources have been modernized. The surveying profession is not static, but is constantly changing in response to modern technology.

RUSSELL BRINKER
ROY MINNICK

Contributors

GRENVILLE BARNES
Assistant Professor
Surveying and Mapping
    Program
Civil Engineering Department
University of Florida
Gainesville, FL

BRO. B. AUSTIN BARRY, F.S.C.
Professor of Civil Engineering
Manhattan College, Bronx, NY

RUSSELL C. BRINKER, P.E.
Adjunct Professor of Civil
    Engineering (Retired)
New Mexico State University
Las Cruces, NM

JOHN BRISCOE
Attorney at Law
San Francisco, CA

R. B. BUCKNER, Ph.D., L.S.
Surveying Program Chair
East Tennessee State University
Johnson City, TN

EARL F. BURKHOLDER, P.L.S., P.E.
Consulting Geodetic Engineer
Klamath Falls, OR

BOYD L. CARDON, R.L.S.
Professor of Mathematics
Ricks College, Rexburg, ID

FRANK T. CAREY, L.S., R.L.S.
Boundary Officer
California State Lands Commission
Instructor, Sacramento City Colege

KENNETH S. CURTIS, L.S.
Professor Emeritus of Surveying
    and Mapping
Purdue University
West Lafayette, IN

RICHARD L. ELGIN, Ph.D., L.S., P.E.
Elgin, Knowles & Senne, Inc.,
Elgin Surveying & Engineering, Inc.,
Rolla, MO

JACK B. EVETT, Ph.D., P.E., L.S.
Professor of Civil Engineering
The University of North Carolina at
    Charlotte
Charlotte, NC

ROBERT J. FISH, R.L.S.
Phoenix, AZ

CHARLES D. GHILANI, Ph.D.
Surveying Program Chair
Pennsylvania State University

DAVID W. GIBSON, L.S.
Associate Professor
Surveying and Mapping
   Program
Civil Engineering Department
University of Florida
Gainsville, FL

E. FRANKLIN HART, L.S., P.E.
Professor of Civil Engineering
   Technology
Bluefield State College
Bluefield, WV

LARRY D. HOTHEM
Manager
GPS Research and Applications
Geometronics Standards Section
USGS National Mapping Division
Reston, VA

ANDREW KELLIE, Ph.D., L.S.
Associate Professor of Engineering
   Technology
Murray State University
Murray, KY

DAVID R. KNOWLES, Ph.D., L.S., P.E.
Elgin, Knowles & Senne, Inc.
Professor of Civil Engineering
University of Arkansas
Fayetteville, AR

GERALD W. MAHUN
Assistant Professor of Civil
   Engineering
University of Wisconsin, Platteville
Platteville, WI

PORTER W. MCDONNELL, P.E., L.S.
Professor of Surveying
Metropolitan State College
Denver, CO

DAVID F. MEZERA, Ph.D., P.L.S., P.E.
Associate Professor of Civil and
   Environmental Engineering
University of Wisconsin, Madison
Madison, WI

ROY MINNICK, L.S., R.L.S.
Tidelands and Waterways
First American Title Company
Santa Ana, CA

DENNIS MOULAND, P.L.S.
Cadastral Consultants, Inc.
Glenwood, CO

CARLOS NAJERA, L.S., R.L.S.
Boundary Officer
California State Lands Commission
Instructor, Sacramento City College

CAPTAIN DONALD E. NORTRUP
National Oceanic and Atmospheric
   Administration
Norfolk, VA

JOHN S. PARRISH, L.S.
Chief
Branch of Cadastral Survey
Bureau of Land Management
Carson City, NV

JAMES P. REILLY, Ph.D.
Academic Department Head
Surveying Department
New Mexico State University
Las Cruces, NM

WALTER G. ROBILLARD, L.S., R.L.S.
Attorney at Law
Atlanta, GA

ROBERT J. SCHULTZ, P.E., P.L.S.
Professor of Civil Engineering
Oregon State University
Corvallis, OR

JOSEPH H. SENNE, Ph.D., P.E.
Elgin, Knowles & Senne, Inc.
Professor Emeritus of Civil
    Engineering
University of Missouri, Rolla
Rolla, MO

M. LOUIS SHAFER, L.S., R.L.S.
Chief of Surveys, District III
California Department of
    Transportation
Marysville, CA

F. HENRY SIPE, L.L.S. #1
Consulting Land Surveyor
Elkins, WV

BRYANT N. STURGESS, L.S., R.C.E.
Boundary Officer
California State Lands Commission

WAYNE VALENTINE, P.E., L.S.
Geometronics Group Leader
U.S. Forest Service
Missoula, MT

ELLIS R. VEATCH II, P.S.
Senior Trainer
Ashtech, Inc.
Sunnyvale, CA

DONALD A. WILSON, R.L.S., R.P.F.
Land Boundary Consultant
Newfields, NH

PAUL R. WOLF, Ph.D.
Department of Civil and
    Environmental Engineering
University of Wisconsin, Madison
Madison, WI

EDWARD G. ZIMMERMAN, L.S., R.L.S.
Senior Land Surveyor
Supervisor of Survey Training and
    Professional Development
California Department of Transportation
Sacramento, CA

1

Surveying Profession, Registration, and Associations

Walter G. Robillard

1-1. INTRODUCTION

This chapter provides information about the professional organizations and their role in professional surveying. Addresses of key organizations are included in that direct contact can be made to obtain further information.

1-2. OVERVIEW

Prior to formation of the American Congress of Surveying and Mapping (ACSM) in the 1930s, surveying was an important part of civil engineering and had an appropriate number of courses in college civil engineering curricula. The American Society of Civil Engineers (ASCE) was the primary sponsor of surveying technical papers and continues to include them in the monthly civil engineering magazine and periodically in a journal of surveying engineering. Recently, an engineering surveying manual prepared by the Committee on Engineering Surveying of the Surveying Engineering Division has been published and is available for purchase from the ASCE.

The ACSM, through its quarterly journal and bimonthly bulletins, provides excellent articles on all pertinent items that along with its semiannual national meetings make member-

ship essential. Other worthy publications include the *Point of Beginning (P.O.B.) Magazine* and *Professional Surveyor*.

When civil engineering professional registration was first legislated in the early part of the 20th century, the civil engineering license included surveying privileges. Gradually, separate licensing of surveyors became the law in most parts of the United States.

1-3. THE FUTURE

Challenges of the future—e.g., space exploration, oceanographic research, urban and land planning and development, ecology and the use and search for natural resources—are dependent on and interrelated to the fields of mapping, charting, geodesy, and surveying.

1-4. BACKGROUND OF SURVEYING AND MAPPING

Records of land surveys date back to the Babylonian era, 3000 or more years ago. Boundary stones were used during those times to mark property in the valleys of the Tigris and Euphrates Rivers.

Geological relics from about 3000 B.C. are still preserved, depicting certain physical features of ancient Babylonia. Town plans of Babylon survive that date back to 2000 B.C.

In ancient Egypt, the valley of the Nile River was flooded frequently, and boundary stones were often shifted or washed away. The Egyptians developed a system of surveying through which they were able to perpetuate the boundary and property lines of that rich area.

Some surveys of ancient times relate to those of today. During the construction of the Aswan Dam on the Nile River, surveyors established precise points for use as guides in cutting, moving, and reassembling the statues at Abu Simbel. This was necessary in order to preserve the beauty and harmony of the original design and construction, which in turn depended significantly on measurements made by the surveying techniques of ancient times.

In the second century of the Christian era, Ptolemy introduced and named the system of latitude and longitude. The Vinland map, which is thought to have been made about A.D. 1440, delineated Iceland, Greenland, and a land mass called Vinland that represented the North American mainland. In 1594, Mercator devised geometrically accurate map-projection systems.

In the United States, the public-land system of townships, ranges, and sections was developed in 1784. In 1803, Lewis and Clark explored and surveyed the country along the Missouri River and west to the Pacific Ocean. Hassler and Blunt led the way in coastal charting in the 1850s; the Powell, Fremont, Hayden, King, and Wheeler surveys of the 1860s opened the development of the American West. Significant developments in aerial photogrammetry, as applied to surveying and mapping, occurred in the 1920s and are still going on. The 1960s brought the beginning of manned space exploration, climaxed by the landing on the moon. The Surveyor I through VII series of satellites contributed much valuable data leading up to that highly successful moon landing. In the 1970s, exploration of space by the United States continued with orbiting surveys of Mars, and the spectacular landing of the space vehicle on that planet, followed by transmission of both photos and detailed data concerning the surface. Surveyors always have been closely identified with exploration and the growth in complexity and sophistication of the cultural development that follows exploration.

1-5. THE SURVEYING PROFESSION

Surveyors are licensed by each state, usually under the authority of a board of registration. The addresses of the boards are listed in Appendix 1.

Laws governing the practice of surveying are enacted at state level. With the exception of public-land surveys, there are almost no federal laws regulating survey practice and no federal license or registration.

Qualifications for surveyors vary from state to state, but generally a pattern of six years of prescribed experience and a 16-hour written examination are the requirements for registration. Many states use portions of examinations prepared by the National Council of Engineering Examiners supplemented by a portion prepared by the state to test on specific state laws.

1-6. SURVEYING LITERATURE

A substantial body of literature about surveying exists. Booksellers specializing in surveying are listed in Appendix 3.

1-7. SURVEYING EDUCATION

Surveying degrees are offered by only a few colleges in the United States. More common is the two-year program offered by community colleges.

2

Surveying Field Notes, Data Collectors

Russell C. Brinker

2-1. INTRODUCTION

Surveying is defined in the 1978 *ASCE Manual No. 34: Definitions of Surveying and Associated Terms* prepared by a joint committee of the ASCE and ACSM as

"(1) The science and art of making all essential measurements in space to determine the relative positions and points and/or physical and cultural details above, on, or beneath the earth's surface and to depict them in usable form, or to establish the position of points and/or details. Also, the actual making of a survey and recording and/or delineation of dimensions and details for subsequent use. (2) The acquiring and/or accumulation or qualitative information and quantitative data by observing, counting, classifying, and recording according to need."

Examples are traffic surveying and soil surveying.

Manually or electronically made field notes are necessary to document surveying results. In this chapter, basic principles of good notekeeping will be discussed, detailed suggestions listed, and simple examples given. Many special noteforms have been designed to fit the specific requirements of various federal, state, city, and county agencies, large companies, property surveyors, and other organizations. Some of these specialized noteforms are included in later chapters. No single style is universally accepted and termed the "standard," even for a job as common as differential leveling. Diverse field conditions, equipment and personnel, and special needs cannot be served by rigid arrangements—e.g., property surveys often require recorders to improvise different noteforms. Tables of some surveying terms, abbreviations, and symbols used in noteforms are presented at the end of this chapter. A short list of surveying textbooks and other references is also provided.

2-2. IMPORTANCE OF FIELD NOTES

Field notes are the only truly permanent and original records of work done on a project. Monuments and corners set or found may be moved or destroyed, and maps prepared from notes sometimes show incorrect distances, angles, and locations of details. Obviously, one notekeeping error can ruin the accuracy and credibility of the succeeding steps: computing and mapping. Written documents (deeds) can jumble numbers and directions, and computer operators have been known to introduce their own mistakes. Original field notes are, therefore, the court of last resort.

A notekeeper's job is often the key assignment in a surveying field party; hence the party chief, who presumably is the most experienced and competent member, often assumes that responsibility. Numerical data must be recorded, sketches drawn, descriptions prepared, and mental calculations quickly made (as in first-order three-wire precise leveling), while one or more people shout things as they move around. On property surveys, the party chief, while keeping notes, may roam along boundary lines to get information. In a two-member differential-leveling unit, the notekeeper is also the instrument operator.

Property survey notes introduced as key-exhibit evidence in court cases can be a critical factor in decisions affecting land transfers by future generations. Land values continue to increase, so accuracy and completeness of surveys and notes are vital. The cross-referenced notes in a land surveyor's files become the saleable "good will" of the business.

The investment worth of surveying notes depends on the time and cost to reproduce any field work, plus the loss caused by their unavailability if immediately needed. The name, address, and telephone number of the person who prepared the notes and company that owns the field book must be lettered in India ink on the outside and inside cover. If a reward will be paid for return of the book if lost, it should be stated.

Because of possible omissions and copying errors, only original notes may be admitted in court cases, since they are the "best" evidence. Copies must always be clearly identified as such. Measurements not recorded at the time they were made or entered later from memory—which is even worse than copying from a scratch bit of paper—are definitely unreliable.

Since the time and date of erasures are always questionable and possible cause for rejection of the notes, erasures must not be made *on recorded measurements*. Also, the originally recorded material may later be found useful and correct. A pencil line should be run through a wrong number without destroying its legibility and the correct value placed above or below the deleted number. Part or all of a page to be canceled should be voided by drawing diagonal lines across it, but without making any part illegible, and prominently marked VOID. Erasing a nonmeasured line for a topographic sketch while in the field may be justified.

2-3.   ESSENTIALS OF SUPERIOR NOTES

Five primary features are considered in evaluating field notes:

1. *Accuracy.* This is the most important factor in all surveying procedures, including notekeeping.
2. *Composition.* Noteforms suitable for each project, with column headings arranged in order of readings and sufficient space provided for sketches and descriptions without crowding, promote accuracy, completeness, and legibility.
3. *Completeness.* A single omitted measurement or detail can nullify an entire set of notes and delay computing or plotting. On projects far from the office, time and money are wasted when returning to the field for missing data. Before leaving the survey site, notes must be carefully reviewed for closure checks and possible overlooked items.
4. *Clarity.* Planning logical field procedures before leaving the office enables a notekeeper to record measurements, descriptions, and sketches without crowding. Mistakes and omissions become more obvious, which helps to eliminate costly office errors in computing and drafting.
5. *Legibility.* Notes must be decipherable and understandable by all users, including those who have not visited the survey area. Neat, efficient-appearing notes are more likely to represent professional-quality measurements and inspire confidence in the field data.

2-4. FIELD BOOKS

Field books used in professional work contain valuable information acquired at considerable cost; they must survive rough usage and difficult weather conditions and last indefinitely. Various types are available, but bound books —the longtime standards with sewed binding, hard stiff covers of leatherette, polyethylene, or covered cardboard, and 80 leaves—are generally selected.

Stapled, sewed, and spiral-bound books are not suitable for most professional work. Duplicating field books may be convenient for jobs requiring progressive transfer of notes from field to office. The original sheet can be detached while a copy is retained in the field book. The loose-leaf original pages are filed in special binders.

Loose-leaf books have both advantages and disadvantages. The advantages include (1) a flat working surface; (2) the capacity to separately file individual project notes, thereby facilitating indexing and referencing, instead of wasting a partly filled book; (3) removable pages for shuttling between field and office; (4) easy insertion of preprinted noteforms, tables, diagrams, formulas, and other useful material; (5) the ability to carry different rulings in the same book; and (6) lower overall cost because the cover can be reused. Disadvantages are possible loss of some loose sheets and having the project data divided between field and office.

2-5. TYPES AND STYLES OF NOTES

Surveying field notes can be divided into four basic types: tabulations, sketches, descriptions, and combinations. The combination method is most common because it fits so many overall needs.

One axiom applies to all four types: If doubtful about the need for certain data, include them and make a sketch. A supplementary proverb, "One picture [sketch] is worth 10,000 words," might well have been written to guide surveying notekeepers. Preprinted noteforms for particular groups of surveys often use arrangements comparable to those illustrated in this text.

Left- and right-hand field book pages are generally paired and share the same number. The left page is commonly ruled in six columns for tabulations, with notes and sketches on the right-hand page. Column headings proceed from left to right in the order readings are taken and minor calculations made. Figures 2-1 and 2-2 are basic notes presented for illustrative purposes *only* to show two different tabulation arrangements on the same page.[2]

In Figure 2-1, distances between hubs are recorded *between* the hub letters, names, or numbers. Measurements *to* a hub, in stations, are placed *opposite* the hub. A sketch on the right-hand page may help but not be necessary, so the notes could be tabulations only. For a simple example of traverse distances, angles, and bearings, everything could be put on a sketch along with other information if only single angles are measured.

Figure 2-2, a combination of type, demonstrates that it is easier to follow the "open" style differential-leveling notes having a $(+)$ sight and height of instrument (HI) on *one* line, followed by the $(-)$ sight and elevation on the *next* one, rather than the "closed" type, which puts all four values on the *same* line. This is especially helpful when a less experienced person uses a noteform to check something.

The project title can run across the tops of both pages or be confined to the left one. The upper right corner of the right-hand page, away from descriptions and sketches, is a good place for these standard items: date, weather, party, and equipment type with serial number.

1. *Date, time of day* (AM *or* PM), both *starting* and *finishing times* are necessary for record purposes. The number of hours spent in the field on a job may help to assess the precision attained, work delays, and other factors.

2. *Weather conditions* ranging from extremely high to ultralow temperatures, sunshine, fog,

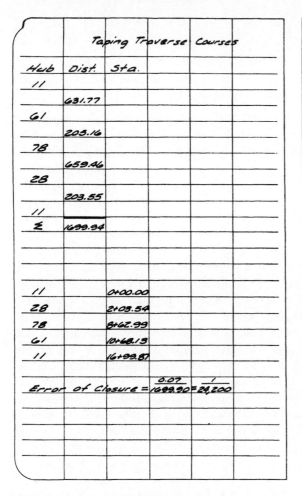

Figure 2-1. Taping traverse courses.

wind, dust, or rain and snowstorms can adversely affect surveying results. Equipment is designed to operate under deplorable conditions; humans are not, and precision suffers. Weather information is one source to consider when investigating necessary corrections and erratic closures.

3. *Party names* with initials and assignments are essential for present and future reference. Court cases and questions regarding survey details may require contacting the field personnel many years after a survey has been completed. Common job symbols are ⅄ for instrument operator, Ø for rodpersons, and N for notekeeper.

4. *Equipment make, model, type,* and *serial number,* along with adjustment condition, govern survey accuracy when properly handled by an experienced operator. Knowing which

instrument was employed may assist in isolating and correcting certain errors. For example, built-in errors of some electromagnetic distance-measuring instruments (EDMIs) and automatic level compensators are currently being investigated and corrections applied to various old surveys. However, unrecorded repairs and later adjustments, along with the lack of instrument identification on some projects, can prevent application of proper modifications.

Bench-mark descriptions should begin with the general location starting with state, county, town, or a familiar area if not otherwise covered by the job description or title. They then proceed through a recognizable feature, such as a street, bridge, building, etc., to a descrip-

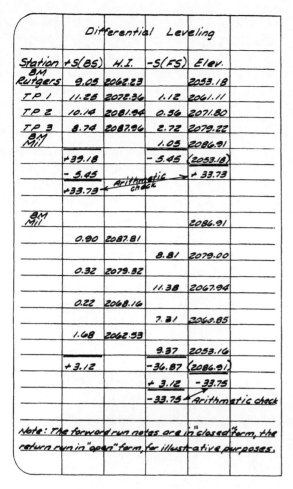

Differential Leveling			
Station	*+S(BS)*	*H.I.*	*-S(FS)*	*Elev.*

Station	*+S(BS)*	*H.I.*	*-S(FS)*	*Elev.*
BM Rutgers	9.05	2062.23		2053.18
TP 1	11.25	2072.36	1.12	2061.11
TP 2	10.14	2081.94	0.56	2071.80
TP 3	8.74	2087.96	2.72	2079.22
BM Mil			1.05	2086.91
	+39.18		-5.45	(2053.18)
	-5.45		*Arithmetic check*	+33.73
	+33.73			
BM Mil				2086.91
	0.90	2087.81		
			8.81	2079.00
	0.32	2079.32		
			11.38	2067.94
	0.22	2068.16		
			7.31	2060.85
	1.68	2062.53		
			9.37	2053.16
	+3.12		-36.87	(2086.91)
			+3.12	-33.75
			-33.75	*Arithmetic check*

Note: The forward run notes are in "closed" form, the return run in "open" form, for illustrative purposes.

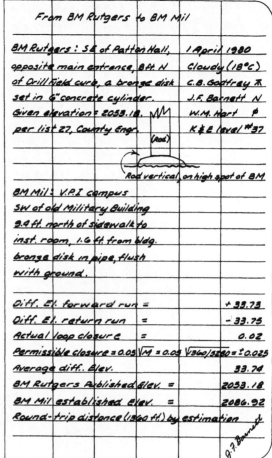

From BM Rutgers to BM Mil	
BM Rutgers: SE of Patton Hall,	*1 April 1980*
opposite main entrance, 8 ft. N	*Cloudy (18°C)*
of Drill field curb, a bronze disk	*C.B. Godfrey 🖉*
set in 6" concrete cylinder.	*J.F. Barnett N*
Given elevation = 2053.18,	*W.M. Hart φ*
per list 27, County Engr.	*K&E level #37*
Rod vertical, on high spot of BM	
BM Mil: V.P.I. campus	
SW of old Military Building	
9.4 ft. north of sidewalk to	
inst. room, 1.6 ft. from bldg.	
bronze disk in pipe, flush	
with ground.	
Diff. El. forward run =	*+33.73*
Diff. El. return run =	*-33.75*
Actual loop closure =	*0.02*
Permissible closure = 0.05√M = 0.05 √360/5280 = ±0.025	
Average diff. Elev.	*33.74*
BM Rutgers Published Elev. =	*2053.18*
BM Mil established Elev. =	*2086.92*
Round-trip distance (1360 ft.) by estimation	

Figure 2-2. Differential leveling (open and closed styles).

tion of the object or mark itself. When used later in the field book, it need only be referred to by page number and not described again.

Bench-mark names, such as Rock, Bridge, Hydrant, etc., provide a clue to the location and may reduce the number of ties required to find the point. On long lines, numbers in sequence are often preferable, but they can be subject to mistakes and do not give a recognition key for the mark.

Symbols and abbreviations save notekeeping time and space. There are standards for common items, but if unusual ones are required, they must be identified the first time employed.

A useful instrument for notekeepers is a moderately priced, small, dependable camera.

Photographs of found and set monuments, bench-mark locations, fence lines, and other items methodically described and referenced at the time they are taken, can eliminate or simplify lengthy lettered notes. The camera's position and aiming direction should be indicated by a symbol in the field book. All prints must be numbered, signed, and dated, then mounted in an album to be filed with the project field notes.

2-6. AUTOMATIC RECORDING

New *data collector* models are now available for use with *electronic distance-measuring instruments*, theodolites, and total-station equipment to automatically record distance and angle measurements electronically. Data collectors display and record measurements, in some cases by merely pushing a button. Reading and transcribing errors are eliminated in both field and office. Stored data are transferred from the collector to a field or office computer, then on to a printer that makes working plots and convenient page-wide printouts. Figure 2-3 shows the Lietz SDR2 electronic field book; Figure 2-4 defines the Lietz SDR2 flowchart from a recorder to the computer and other units in the assembly.

The K & E Vectron and Auto Ranger II provide a readout display for visual checking and transfer the panel measurements to a field computer, without manual input, for calculation and storage. The field computer can be interfaced with an office computer for printouts and permanent records.

The Wild TC1 recording attachment uses a magnetic tape cassette to store a complete block of measured data with built-in checks by touching a key. A cassette reader transfers the data to desk, mini, and large computers, and the information can then be transmitted from field to office by telephone.

The ABACUS' SDC71 ⓉⓂ Survey data collector provides all the power needed for survey applications, from easy, quick sequences to more complex projects. It is not tied to a single-brand total station, works well with most automated field equipment, and can be used independently as an electronic field book. The SDC71 allows the user to reduce source data to field coordinates and upload data to a variety of CAD systems.

Kern's Alphacord recording unit registers and stores automatically measured data that

Figure 2-3. Lietz SDR2 electronic field book. (Courtesy of the Lietz Company.)

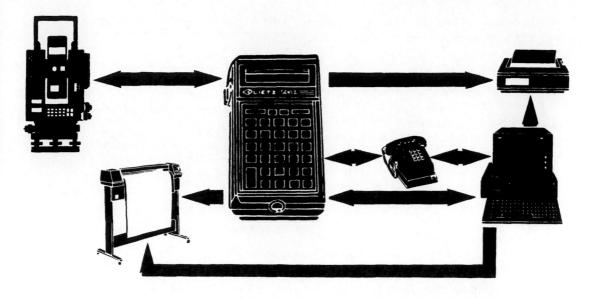

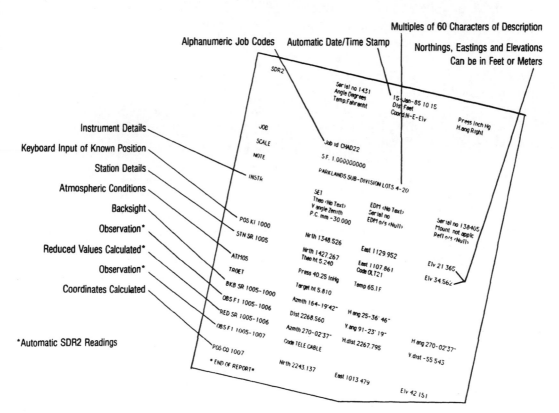

Figure 2-4. SDR2 flowchart. (Courtesy of the Lietz Company.)

along with keyed information is fed directly to a computer. While still in the field, the stored items can be relayed to a cassette tape recorder for later processing, thus making the unit's storage capacity available for repeated use.

The Magnavox 502 Georeceiver satellite surveyor automatically assures that valid data are properly recorded on magnetic cassette tape. Visual display permits manual recording of site position for the survey field notes.

All the data collectors briefly here have storage capacity for a normal day's operation. Data collectors are best suited for use with total-station instruments on projects providing many data to be passed on to a computer and other accessories. Among items to be considered are size, weight, power consumption, range of suitability for possible use with equipment from other manufacturers, storage media and capacity, manual and automatic data entry, clarity of display, ready access to repair facilities, initial and maintenance costs, and future upgrading feasibility.

Numerous changes and improvements have been made since a 1981 article, "Evaluating Data Collectors,"[3] was published in *P.O.B. Magazine.* The author tabulated six devices by name, manufacturer, physical characteristics, power supply, whether or not a stand-alone unit, communication, compatible equipment, data display, storage medium, memory size, availability, and price. The points discussed and evaluation checklist (see Table 2-1) are still pertinent.

A recent article, "Measuring the Productivity of Data Collectors and Total Stations" by Tom Donahue, evaluates four data-collection systems indoors at the Minneapolis Metro Dome. A Citation CI 450 top-mounted slope distance meter, representative of the equipment owned by a majority of surveying firms, with a Lietz TM 1A optical theodolite and standard field book were used for a base comparison.

Each system was timed and its performance measured over a typical traverse/topo project. Evaluations were also conducted to measure ease of use, flexibility, versatility, and overall performance. Six traverse points on the Metro Dome's field level and approximately 135 topo points on the field and lower deck areas were chosen. To more closely mirror field conditions, topo was shot from three different stations. The crew consisted of one equipment operator, two rod people and one person to set up stations.

Did the total stations and data collectors significantly increase productivity? The answer is

Table 2-1.   Data collectors evaluation checklist

Evaluation Checklist
1. Physical Characteristics (a) Dimensions of the unit (b) Total weight (c) Number of components in the system
2. Power Supply (a) Number and type of batteries required (b) Battery life per charge (c) Backup power supply
3. Interfacing the Data Collector to More Than One Type of Total Station
4. Method of Transmitting the Data to a Computer
5. Capacity of Data Collector
6. Type of Storage Medium (a) Solid-state (b) Tape
7. Features of Data Collector (a) Editing data (b) Computation (c) Checking data sequences (d) Type of data that can be entered
8. Type of Display (a) Light-emitting diode (LED) (b) Liquid crystal display (LCD)
9. Maintenance of Data Collector
10. Upgradability of the System

Source: Courtesy of *P.O.B. Magazine.*

yes and no. For radial surveys, with lots of data to be collected and where many shots can be taken from each traverse station, this type of equipment will more than pay back its original cost. However, where traversing only, or traversing with a small number of sideshots/station, an EDMI and field book would be almost as productive. The difference being that radial surveying is shot intensive, while traversing is almost set up and move. Consider the fact that it took the EDMI system approximately 40 to 50 seconds per shot to read and record topographic information versus 2.5 to 13 seconds for the total-station systems. All total-station/data collector systems finished the entire project in less than 1 hour and 40 minutes, while the EDMI/field book system took 3 hours and 45 minutes. That's 2.3 times slower than the slowest total station! Had the project consisted of the traverse stations only, the difference would have been minimal.

Other major findings of these evaluations were:

1. The weak link in the data collection chain is the transfer to a software package. This often involves coordination between different manufacturers and may change the field capabilities of a data collector. It is, therefore, imperative to evaluate this and other links in the system before purchasing.

2. For extensive radial surveying, a total station is two to three times faster than a mount-on EDM. This is mainly due to pointing time and one-button measurement.

3. For extensive radial surveying, a data collector is 35 to 50 percent faster than a field book. This assumes all the measured data can be electronically transferred to the collector. If some, or all, of the data need to be manually keyed in, then productivity is reduced to that of a field book.

4. The fastest total station had a coarse measurement mode, high-speed electronics, and wide beam width that made it ideal for radial stakeout and topographic surveys.

5. The best data collector was the only one which had built-in computation software and was also capable of calculating the traverse precision ratio in the field. Other built-in computation capabilities included resection, field stakeout, coordinate computation, and much more. In addition, readable field notes could be transferred directly to a printer without a computer.[4]

The advent of electronic recording has not diminished the need for highly competent notekeepers and field books. Since sketches, nonnumerical information, and descriptions must still be hand-prepared, the rapidly made measurements may increase a notekeeper's burdens on topographic and property surveys. But this responsibility is lessened somewhat by merely pointing, then just pushing buttons, instead of reading and recording. Cost of a data collector is an important factor for small surveying firms when considering how to enhance field and office equipment owned or contemplated for purchase.

Standard field books are readily accepted everywhere in court cases. Conversely, questions must be answered about magnetic tapes since they can be altered, erased by power mishaps or human error, lost, suffer deterioration, or make identification of the original versus a copy difficult. Also, electronic data collectors may present a possible hazard in underground surveys (see Chapter 29).

2-7. NOTEKEEPING POINTERS

Basic points, some previously mentioned, are listed as practical guides for notekeepers:

1. Letter the name, address, and telephone number of the field book's owner in India ink on the front and inside cover. State whether a reward will be paid for the return of a lost book.

2. Number all pages before first use of a field book. Left- and right-hand pages are paired and share the same number.

3. Employ the Reinhardt system of lettering for clarity, speed, and simplicity. Do not mix upper- and lowercase letters. Larger-size and uppercase lettering should be reserved for more important features.

4. Use a 3-H or harder pencil; keep it sharp, bear down.

5. Start a new day's work on a fresh page. For some projects in which large complicated sketches must be expanded, other considerations may be overriding.

6. Always record measurements immediately in the field book—not on scrap paper for later copying to improve appearance (a costly and dangerous act). "Rite in the Rain" field books now have waterproof paper specifically designed to accept field notes in wet or humid weather, even during rainstorms!

7. Carry a straightedge for ruling lines, a small protractor, and scales.

8. Make sketches to general proportion rather than exactly to scale or without advance planning. Keep in mind that preliminary estimates of the space required are often too small.

9. When in doubt, use sketches instead of tabulations.

10. Avoid crowding: It is one of the most common mistakes. If helpful, use several right-hand pages for descriptions and sketches to match a single left-hand page of tabulations, and vice versa.

11. If clarity is thereby improved, exaggerate details on sketches.

12. When possible, line up descriptions and sketches with corresponding data. For example, the beginning of a bench-mark description should be on the same line as its elevation.

13. Keep tabulated figures inside and off the column rulings with decimal points in line vertically.

14. Place a zero before the decimal point for numbers smaller than one—i.e., record 0.67 instead of just .67.

15. Record notes in an order that will facilitate office computations and mapping. For example, in stadia topography, number and read detail points in a clockwise rotation.

16. Letter measurements parallel with or perpendicular to sketch lines so they cannot be misunderstood. Machine-drawing type dimension lines are rarely used in surveying sketches.

17. Show the precision of measurements by recording significant zeros. Enter 2.60 instead of 2.6 if the reading was actually determined to hundredths.

18. Do not try to change a recording error by writing one number over another to transform a 3 to a 5, or a 7 to a 9.

19. Record what is read. Never "fudge" observations or closures. Surveying is an art and science. Art can stand retouching, science cannot.

20. Record aloud numbers given for recording. For a distance of 172.58, call out "one, seven, two, point five, eight" for verification.

21. To eliminate gross errors, make a mental estimate of all measurements before receiving and recording them.

22. Show essential computations made in the field so they can be checked later.

23. For compactness, employ conventional symbols and abbreviations. If not standard, identify them when first used or in a special table.

24. Use explanatory notes when they are pertinent.

25. Do not erase measured data—lines of a sketch can be deleted in the field.

26. Run a single line through an erroneous number and record the correct value above or below it.

27. To void a page, draw diagonal lines from opposite corners and letter VOID prominently without obscuring a number or any part of a sketch.

28. Letter COPY in large letters diagonally on copied notes but keep the lettering off a sketch or any numbers.

29. Run notes down the page except on route surveys where they progress upward to agree with sketches made while looking in the forward direction.

30. Review the notes, make all possible arithmetic checks, compute closures and error ratios, and record them before leaving the field. On large projects employing several parties, satisfactory closures indicate completed work and facilitate assignments for the next day.

31. Place a north arrow at and pointing to the top or left side of every page if possible since notes and drawings are read from the bottom or right side. A meridian arrow must be shown.

32. Title and index each project. Cross-reference every new job or continuation of a previous one by the client's organization, property owner, and description.

33. On all original notes, sign surname and initials in the lower right corner of the right-hand page. This is equivalent to signing a check and accepting responsibility for it.

2-8. ADDITIONAL BASIC NOTEFORMS

Additional noteforms covering basic and more advanced surveying operations are illustrated in Figures 2-5 through 2-16 (pp. 13–18) and in later chapters. They can serve as examples on

Differential Leveling (3-wire) with Foot Rod

BS(+)	Half-Interval	Mean(+)	FS(-)	Half-Interval	Mean(-)
(1) BM 29			TP 1		
8.266	161		3.491	171	
8.105	165	8.1037	3.320	168	3.3210
7.940	326		3.152	339	
(2) TP 1			TP 2		
6.574	216		4.623	204	
6.358	217	6.3577	4.419	200	4.4203
6.141	759	14.4614	4.219	743	7.7413
(3) TP 2			BM 30		
6.203	182		2.819	188	
6.021	178	6.0223	2.631	185	2.6320
5.843	1119	+20.4857	2.446	1116	-10.3733
		-10.3733			
		+10.1104 ft. (D.E.)			

The notekeeper must each time inform the instrument-man that the two half-intervals match acceptably within 3 (maybe 4) 1000 ths before the rod or the instrument is moved. If there is a bad matching, the notekeeper demands a complete re-reading. Cumulative summation of the half-interval columns enables cross-comparison at each set-up to maintain a cumulative balance of the sight distances, fore and aft, for the run.

From BM 29 to BM 30

3 April 1980

BM 29	98.461 (published)	Temp 18°C
D.E.	+10.110	A. Jones ⊼
BM 30	108.571 (newly establ.)	J. Spate N

C. Cohen R
Dietzgen Dumpy Level # 15
Dietzgen Phila. Rod # 19

BM 30 - Top of lower hinge of S.E. gate of 12' fence

BM 29 - Westerly end of door sill at Chapel entrance

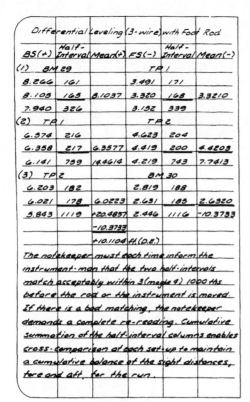

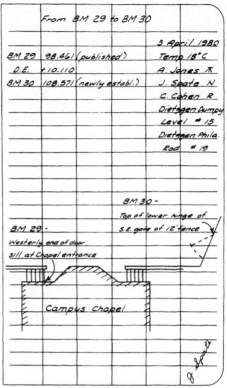

Figure 2-5. Differential leveling, three-wire.

Reciprocal Leveling

Sight	B.S.	F.S.	Sight	B.S.	F.S.
BM Stocker	1.467		BM Stocker		1.766
BM Monument		9.875			1.738
		9.897			1.753
		9.886			1.760
		9.892			1.758
		9.901			1.740
		9.888			1.748
		9.887			1.752
		9.885		Mean =	1.752
	Mean =	9.889	BM Monument	10.221	
D.E.		-8.422		+8.469	
Mean Diff. Elev.		-8.445			
El. BM Stocker		396.178	(Known from records)		
El. BM Monument		387.733	(Established)		

COPY

Across Baldwin Hills Arroyo
Stocker St., Los Angeles

BM Stocker: NE corner,	8 April 1980
Stocker and La Brea, 20 ft.	Fair, 20°C
from Stocker St. curb line	L.M. Wright ⊼
on ℄ of La Brea sidewalk,	A.R. Ferrell N
bronze plug set flush.	A.G. Thomas P
BM Monument: monument	Tilting K&E #17
in center of grass plot,	Phil. rod & target
Overhill Drive cul-de-sac,	# 67
an "X" chiseled in SW corner	
of base slab.	

Reciprocal procedure balances out the curvature and refraction error inherent in the unbalanced sights of a single set-up by having a long foresight for the first set-up and a long backsight for the second set-up.

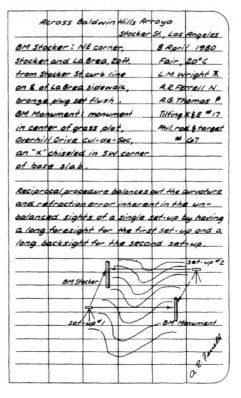

Figure 2-6. Reciprocal leveling.

		Profile		Across	
Station	B.S.(+)	H.I.	F.S. on TP(-)	F.S.(-)	Elev.
BM Rutgers	3.93	2057.11			2053.18
0+00.0				3.09	2054.02
+06.0				3.23	2053.88
+06.5				3.72	2053.39
+26.5				4.23	2052.88
+46.5				4.89	2052.22
+46.8				4.35	2052.76
0+50				4.6	2052.5
+52.5				4.84	2052.27
1+00				8.4	2048.7
+50				10.3	2046.8
2+00				12.4	2044.7
TP 1	1.34	2046.37	12.08		2045.03
2+50				3.3	2043.1
3+00				4.4	2042.0
+50				4.6	2041.8
4+00				5.2	2041.2
+50				5.7	2040.7
5+00				5.9	2040.5
+50				5.9	2040.5
6+00				5.5	2040.9
+50				5.7	2040.7
TP 2	6.88	2047.95	5.30		2041.07
7+00				7.1	2040.9

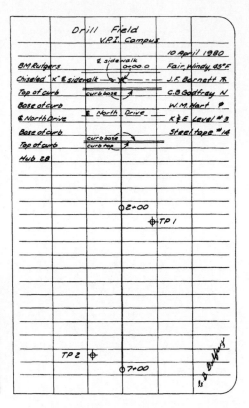

Figure 2-7.   Profile leveling across drill field.

Point	+S(BS)	H.I.	-S(FS)	Elev.	Cut	N*
BM Bridge	7.72	829.88	—	(822.16)	—	—
A0			5.0	824.9	9.9	1
A1			5.9	824.0	9.0	2
A2			6.8	823.1	8.1	2
A3			7.6	822.3	7.3	1
B0			4.2	825.7	10.7	2
B1			5.1	824.8	9.8	4
B2			7.6	822.3	7.3	4
B3			8.5	821.4	6.4	2
C0			3.0	826.9	11.9	1
C1			3.9	826.0	11.0	3
C2			5.2	824.7	9.7	4
C3			9.8	820.1	5.1	2
D1			2.1	827.8	12.8	2
D2			4.4	825.5	10.5	4
D3			10.1	819.8	4.8	2
E1			0.3	829.6	14.6	1
E2			3.8	826.1	11.1	2
E3			11.8	818.1	3.1	1
BM Bridge	7.71	(check)				

Borrow-Pit Leveling for Madison Factory Site.

* N = Number of rectangles the point touches.

$$Vol.\left(\tfrac{cu.}{yds.}\right) = \frac{area\ of\ rect.}{27}\left(\Sigma h_1 + 2\Sigma h_2 + 3\Sigma h_3 + 4\Sigma h_4\right)$$

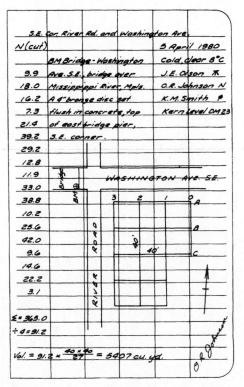

Figure 2-8.   Borrow-pit leveling.

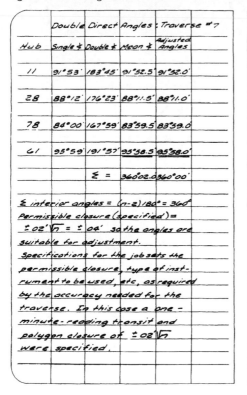

Figure 2-9 left page:

Closing The Horizon; Direction Instrument Method

Point Occ.	Point Obs.	Direction Reading	Angle	Adjusted Angle
61	78	8°24'		
61	11	104°22'	95°58'	95°58.7'
61	11	105°57'		
61	ALT	221°25'	115°28'	115°28.7'
61	ALT	223°57'		
61	78	12°29'	148°32'	148°32.6'
		Σ =	359°58'	360°00'
	Discrepancy =		0°02'	

Point Occ.	Obs.	Tel.	Direction Reading		Mean
61	78	D	109°58'54.0"		
	78	R	289°58'53.5"	53.8"	0°00'00.0"
61	11	D	205°57'47.6"		
	11	R	25°57'48.0"	47.8"	95°58'54.0"
61	ALT	D	321°26'24.6"		
	ALT	R	141°26'23.7"	24.2"	211°27'30.4"

Angles:
78 to 11   95°58'54.0"
11 to ALT   115°28'36.4"
ALT to 78   148°32'29.6"

Figure 2-9 right page:

This set was taken with a minute-reading transit, for familiarization and checking. Initial settings were at random. Since the horizon must close to 360°, the angles are adjusted assuming equal weights.

1 May 1980
Clear 75°F
C.A.Godfrey
Berger
Transit #9

This set was taken using a second-reading theodolite, again initial setting was at random. Readings taken clockwise on the point with telescope direct (D); at final point the telescope is reversed (R) and the points read counter-clockwise back to the initial point. Averaging the seconds gives the benefit of an additional value. Angles are found from the arc passed over.

C.A.Godfrey

Figure 2-9. Closing the horizon; direction instrument method.

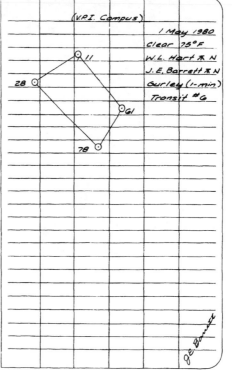

Figure 2-10 left page:

Double Direct Angles; Traverse #7

Hub	Single ∢	Double ∢	Mean ∢	Adjusted Angles
11	91°53'	183°45'	91°52.5'	91°52.0'
28	88°12'	176°23'	88°11.5'	88°11.0'
78	84°00'	167°59'	83°59.5'	83°59.0'
61	95°59'	191°57'	95°58.5'	95°58.0'
		Σ =	360°02.0'	360°00'

Σ interior angles = (n-2)180° = 360°
Permissible closure (specified) =
±02'√n = ±04' so the angles are
suitable for adjustment.
Specifications for the job sets the permissible closure, type of instrument to be used, etc, as required by the accuracy needed for the traverse. In this case a one-minute-reading transit and polygon closure of ±02'√n were specified.

Figure 2-10 right page:

(V.P.I. Campus)
1 May 1980
Clear 75°F
W.L. Hart ⊼ N
J.E.Barrett ⊼ N
Gurley (1-min)
Transit #6

J.E.Barrett

Figure 2-10. Double direct angles.

Balancing Traverse Angles and Computing

Hub	Direct Angle	Adjusted Angles ①	②	③
A	31°36'20"	31°35'40"		31°36'10"
B	150°35'40"	150°35'20"	150°35'20"	
C	67°48'00"	67°47'40"	67°47'30"	
D	154°26'20"		154°26'10"	
E	56°44'20"	56°44'00"	56°44'00"	
F	258°30'40"	258°30'00"	258°30'20"	258°30'30"
Σ	720°01'20"	720°00'00"	720°00'00"	720°00'00"

180°(n-2) = 720°00'00"
Closure = +0°01'20"

Given (or known)
azimuth AB =
N 55°26'00" E

Methods of adjustment

① Correct angles having short sights or poor observing conditions. For example, subtract 40" from angles A and F, or subtract 01'20" from an angle where only top of range pole was visible.

② Arbitrary but symmetrical distribution. Subtract 20" from angles B, C, E, and F.

③ Correction by pattern, useful in many-sided closed traverses.
Average corr. = 80"/6 = 13⅓" per angle. The successive difference column gives the correction; note the repetitive pattern in this column.

Bearings (Azimuths)

Method ③ Hub	Aver. Corr.	Rounded to 10"	Successive diff.
A	13⅓"	10"	10"
B	26⅔"	30"	20"
C	40"	40"	10"
D	53⅓"	50"	10"
E	66⅔"	70"	20"
F	80"	80"	10"

1 May 1980
C.N. Dunn

Bearing Calculation / Azimuth Calculation

Bearing Calculation	Azimuth Calculation
AB N 55°26'00" E +(SW)	AB 55°26'00"
B 150°35'20" +'	BA 235°26'00"
S 200°01'20" +	+B +150°35'20"
BC N 20°01'20" E+(SW)	BC 386°01'20" −360°
C 67°47'40" +'	CB 200°01'20"
S 93°49'00" +	+C + 67°47'40"
CD N 86°11'00" W−(SE)	CD 273°49'00"
D 154°26'20" +	DC 93°49'00"
DE S 68°15'20" W+(NE)	+D +154°26'20"
E 56°44'00" +	DE 248°15'20"
N 124°59'20" +	ED 68°15'20"
EF S 55°00'40" E−(NW)	+E + 56°44'00"
F 258°30'20" +	EF 124°59'20"
N 203°29'40" +	FE 304°59'20"
FA S 23°29'40" W+(NE)	+F 258°30'20" 203°
A 31°36'20" +	FA 562°29'40"
AB N 55°26'00" E +	AF 23°29'40"
(check)	+A + 31°36'20"
	AB 55°26'00" (check)

Figure 2-11. Balancing traverse angles and computing bearings (azimuths).

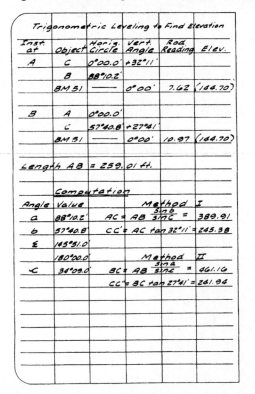

Trigonometric Leveling to Find Elevation

Inst. at	Object	Horiz. Circle	Vert. Angle	Rod Reading	Elev.
A	C	0°00.0'	+32°11'		
	B	88°10.2'			
	BM 51	—	0°00'	7.62	(144.70)
B	A	0°00.0'			
	C	57°40.6'	+27°41'		
	BM 51	—	0°00'	10.97	(144.70)

Length AB = 259.01 ft.

Computation

Angle	Value	
a	88°10.2'	
b	57°40.6'	
Σ	145°51.0'	
	180°00.0'	
C	34°09.0'	

Method I
$AC = AB \frac{\sin b}{\sin C} = 389.91$
$CC' = AC \tan 32°11' = 245.38$

Method II
$BC = AB \frac{\sin a}{\sin C} = 461.16$
$CC'' = BC \tan 27°41' = 241.94$

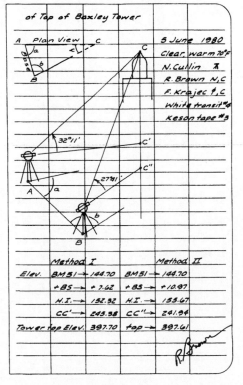

of Top of Baxley Tower

A Plan View C

5 June 1980
Clear, warm 70°F
N. Cuhlin ∏
R. Brown N,C
F. Krajec ♦,C
White transit #4
Keson tape #3

32°11'
27°41'

	Method I		Method II	
Elev.	BM 51 →	144.70	BM 51 →	144.70
	+BS →	7.62	+BS →	10.97
	H.I. →	152.32	H.I. →	155.67
	CC' →	245.38	CC'' →	241.94
Tower top Elev.	397.70		top →	397.61

R. Brown

Figure 2-12. Trigonometric leveling.

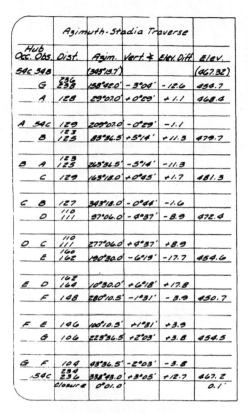

Azimuth-Stadia Traverse

Hub Occ. Obs.	Dist.	Azim.	Vert ✠	Elev. Diff	Elev.
54c 54B		(343°15.7')			(467.32)
G	236 238	158°42.0'	-3°06'	-12.6	454.7
A	128	29°07.0'	+0°29'	+1.1	468.4
A 54c	129	209°07.0'	-0°29'	-1.1	
B	123 125	83°36.5'	+5°14'	+11.3	479.7
B A	123 125	263°36.5'	-5°14'	-11.3	
C	129	163°18.0'	+0°45'	+1.7	481.3
C B	127	343°18.0'	-0°46'	-1.6	
D	110 111	97°06.0'	-4°37'	-8.9	472.4
D C	110 111	277°06.0'	+4°37'	+8.9	
E	160 162	190°30.0'	-6°19'	-17.7	454.6
E D	163 164	10°30.0'	+6°18'	+17.8	
F	148	280°10.5'	-1°31'	-3.9	450.7
F E	146	100°10.5'	+1°31'	+3.9	
G	106	225°36.5'	+2°03'	+3.8	454.5
G F	104	43°36.5'	-2°03'	-3.8	
54c	234 236	338°43.0'	+3°05'	+12.7	467.2
closure		0°01.0'			0.1'

Figure 2-13. Azimuth-stadia survey.

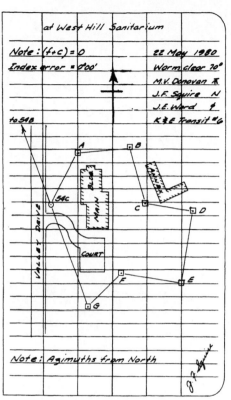

at West Hill Sanitarium

Note: (f+c) = 0   22 May 1980
Index error = 0°00'   Warm clear 70°
     M.V. Donovan ☊
     J.F. Squire N
     J.E. Ward ꝗ
to 54B   K&E Transit #6

Note: Azimuths from North

Topographic Details by Stadia

Point	Dist.	Azim.	Vert ✠	Diff in Elev.	Elev.
☊ @ 54c		(El. 467.32)		h.i. = 4.8	
54B		(343°15.7')	—		—
CB1	100	45°52'	+0°59'	+1.7	469.0
CB2	71	91°10'	+0°58'	+1.2	468.5
CB3	95 94	123°09'	-2°25'	-4.0	463.3
ℓ Dr. 4	39	196°37'	-3°15'	-2.2	465.1
ℓR 5	49	347°41'	-1°01'	-0.9	466.4
ℓR 6	102	334°07'	5.1 at 0°00'	-0.3	467.0
ℓ Ct. 7	106 105	137°51'	-4°03'	-7.8	459.5
ℓ Ct. 8	115 114	149°27'	-3°59'	-8.0	459.3
CP 9	108 107	341°15'	-3°16'	-6.1	461.2
A	128	29°07.0'	+0°29'	+1.1	468.4
54B		(343°15.7')	—	—	—
☊ @ A		(El. 468.4)		h.i. = 4.7	
54c	129	(209°07.0')	-0°29'	-1.1	467.3
CB 12	47	129°15'	3.8 at 0°00'	+0.9	469.3
CP 13	39	130°45'	+1°28'	+1.0	469.4
CP 14	60 58	58°07'	+5°53 on 3.7	+6.2	474.6
FL 15	89 88	32°00'	+6°37'	+10.7	479.1
FL 16	76	0°14'	+3°36'	+4.8	473.2
FC 17	83	337°02'	+1°11'	+1.7	470.1
CP 18	54	331°45'	4.4 at 0°00'	+0.3	468.7
CP 19	41	305°10'	-3°12'	-2.3	466.1
54c	—	(209°07.0')	—	—	—

Figure 2-14. Topographic details by stadia.

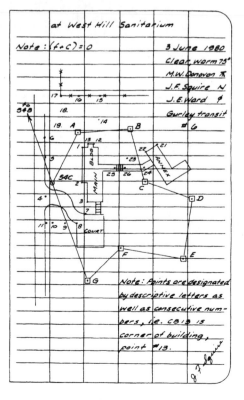

at West Hill Sanitarium

Note: (f+c) = 0   3 June 1980.
     Clear warm 75°
     M.W. Donovan ☊
     J.F. Squire N
     J.E. Ward ꝗ
to 54B   Gurley transit #6

Note: Points are designated by descriptive letters as well as consecutive numbers, i.e. CB13 is corner of building, point #13.

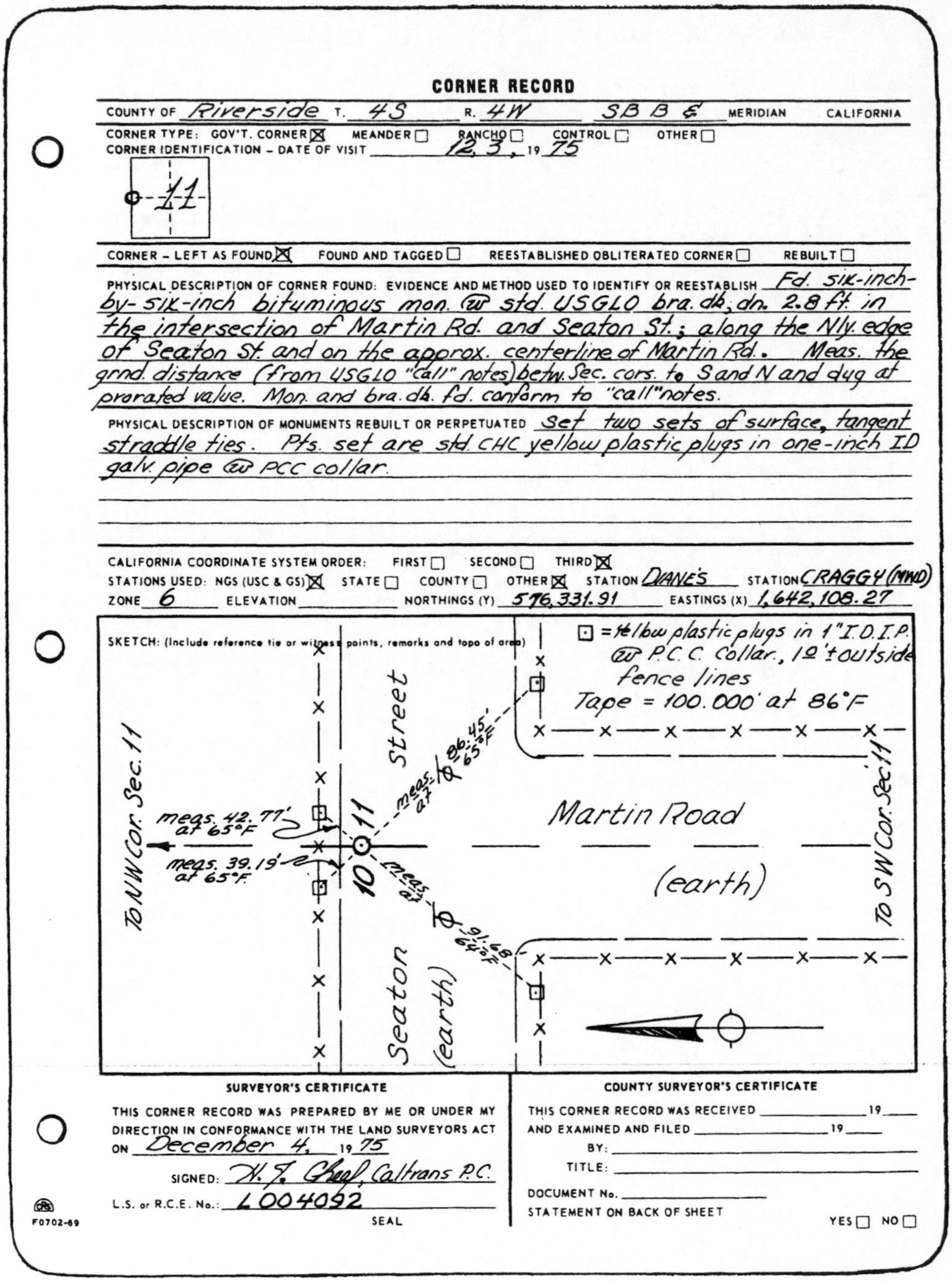

Figure 2-15. Corner record.

which more complicated and special survey forms are built, with individual preferences exercised. Preprinted and more advanced types used by various agencies and organizations are included in pertinent chapters. All figures in this chapter except 2-3 and 2-4 are excerpted from *Noteforms for Surveying Measurements*.

NOTES

1. *ASCE Manual No. 34: Definitions of Surveying and Associated Terms*. NY, NY 1984. ACSM and ASCE.
2. R. C. Brinker, B. A. Barry, and R. Minnick. 1981. *Noteforms for Surveying Measurements*, 2nd ed. Rancho Cordova: Landmark Enterprises, CA, pp. 12, 14.
3. T. Donahue. 1986. Measuring the productivity of data collectors and total stations. *P.O.B. Magazine 11*(2), 85.

REFERENCES

McLean, J. E., Jr. 1981 Choosing an EDM data collector. *P.O.B. Magazine* 7(1), 36.

Kavanagh, S. J., and G. Bird. 1992 *Surveying Principles and Applications*, 3rd ed. Englewood Cliffs, NJ: Prentice Hall.

Wolf, P. R. and R. C. Brinker. 1994 *Elementary Surveying*, 9th ed. New York: Harper Collins.

3

Measurement Errors

Bro. B. Austin Barry, FSC

3-1. INTRODUCTION

Surveying is the art and science of making measurements. The notion of "exact" measurement, of "perfect" result, of "accurate" work is quickly dispelled when trying to duplicate an angle or distance measurement or difference of elevation. It is also evident when different people make the same measurement.

3-2. READINGS

Generally, when reading any graduated scale, the final digit is estimated—an appraisal of the distance between fine-scale graduations, such as 6.27 in Figure 3-1. This can be the end of a 50-ft tape, with graduation in tenths and half-tenths also marked, or a rod reading taken for elevation of a point.

Note that 6.27 would be the estimate of most observers, not 6.26 or 6.28, though these figures would almost surely be estimated by some others. Obviously, if extra care is warranted, a scale with finer graduations—say, to thousandths—might be used and the readings made with a magnifier, probably to ten-thousandths of a foot. Such readings, if repeatedly made by an observer or observers,

might vary more widely in the last digit (estimated). Such readings made to ten-thousandths instead of hundredths are more accurate than those of Figure 3-1, and the apparently wider fluctuation in the last place is not nearly as serious.

3-3. REPEATED READINGS

Assume a series of observed readings using the fine graduations and magnifier that permits readings to ten-thousandths of a foot (see Table 3-1). If the readings had been taken to thousandths only, all would have been listed as 6.276; if to hundredths, 6.28; if to tenths only, 6.3 units. In this case, the best value attainable is the arithmetic mean or average of the set. It would be recorded as 6.27603 or perhaps 6.2760 units.

It is never possible to obtain absolutely correct fourth- or fifth-decimal-place figures in the example simply because the method of measuring is not sufficiently refined. An exact value does exist but cannot be identified. The objective is to get what may be termed a best available result by refined measurements and techniques of successive readings.

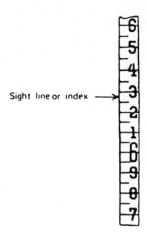

Sight line or index ⟶

Figure 3-1. Interpolation.

3-4. BEST VALUE

Because an average is the "best available value," it can be used, although without assurance that it is correct or incorrect. Since in a set of measurements the true answer is unknown, it must be concluded that the mean value—and, in fact, any of the 10 measured values in Table 3-1—contains an error. In this context, "error" means the difference between a measured and true (or correct) value. Note that this does not apply to a "count" of bolts, cans, cartons, etc.

Results differ, perhaps only slightly, but it means no measurement in a set can be selected as the correct figure or exact result. By examining the range of measured values, the worst ones can be eliminated and those retained that cluster close together. It is the

Table 3-1. Repeated readings

1	6.2763
2	6.2757
3	6.2761
4	6.2760
5	6.2761
6	6.2758
7	6.2760
8	6.2764
9	6.2759
10	6.2760
	Mean = 6.27603

mean or average of these more reliable bunched values that is logically accepted as a "best" value.

3-5. ACCURACY AND PRECISION

Accuracy is descriptive of exactness or trueness of a measurement, its correctness. *Precision*, on the other hand, describes the closeness to one another of several measurements for the quantity. It speaks of the measurer's care and acumen and the instrument quality. Precision is revealed only by repeating measurements and then observing discrepancies among the results and variations of each from the set's mean.

3-6. ERRORS IN MEASUREMENT

In surveying, many measurements of quantities are made. Each contains errors: *systematic* (*cumulative*) and *accidental* (random). It is never possible to find a correct or true value for the quantity being measured, as opposed to *counts* of chaining pins, plumb bobs, level rods, etc. However, a reliable value is obtained if systematic errors are corrected and accidental ones studied for sign and size.

3-7. SYSTEMATIC ERRORS

Systematic errors in a measurement are proportionate to some influencing cause. When evaluated for size and sign, they can be corrected and eliminated.

Example 3-1. A 100-ft steel tape standard at 68°F will be shorter when used at 28°F by an amount

$$E_t = kL \, \Delta t = (0.00000645)(100.000)(68 - 28)$$

$$= 0.0258 \text{ ft}$$

When used to lay out a 560-ft length on construction, the required distance is

$$560.000 + (5.60)(0.0258) = 560.000 + 0.1445$$
$$= 560.144 \text{ ft.}$$

For the same temperature conditions, a taped measurement between two fixed hubs on the job that is reported as 346.842 ft must be corrected by

$$E_t = (0.00000645)(346.842)$$
$$\times (68 - 28) = 0.0895 \text{ ft}$$

The corrected field length is 346.842 − 0.0895 = 346.753 ft.

Example 3-2. A reading on a distant level rod is affected by earth curvature and atmospheric refraction $E_{CR} = 0.574$ ft (M^2), where M is miles. If the reading seen from 325 ft is 6.354, the corrected rod result equals $6.354 - 0.574(325/5280)^2 = 6.352$ ft.

Example 3-3. If a velocity meter or pressure gauge is calibrated and found to consistently read 10% high, all readings can be corrected by 10%. In a similar way, the vertical circle index error of a transit can be applied to each vertical angle reading as a correction.

Systematic errors also can be corrected by compensation procedures or devices.

Example 3-4. In extending line AB on the ground to a point C by setting up a transit on B, backsighting to A, and plunging the telescope to set C, a maladjustment of the transit might place C to the left of its proper location. By repeating the procedure, starting with inverted telescope, point C will fall to the line's right. Correct placement of point C is midway between the two.

Example 3-5. In differential leveling, an instrument whose line of collimation is not parallel to the bubble-tube axis will give correct results if the foresight and backsight distances are kept equal, thus compensating in each pair of sightings for instrument error, as well as curvature, and refraction.

Example 3-6. In using electronic distance-measuring instruments (EDMIs), the velocity of light is affected by air temperature, atmospheric pressure, and vapor pressure, so distances must be corrected by a calculated sum. To correct directly for length errors, it is possible to modulate the EDMI circuitry for meteorological and environmental conditions.

For any measurement to give a true value, all systematic errors must be identified, analyzed, and corrected. Every source should be examined, since systematic errors can be natural, personal, or instrumental. Until all are isolated and corrected, accidental-error theory has no application. All that follows assumes systematic errors have been eliminated by proper corrections. However, note that evaluating and applying a correction still leave room for accidental errors.

3-8. ACCIDENTAL ERRORS

Accidental errors in measurements are random in nature, probably small rather than large, and equally liable to be plus as minus. They do not accumulate, but are partly compensating in nature. Logically, by repeating a measurement and calculating the mean (average) of several measurements, a safer and better value for the quantity is secured.

3-9. ERRORS VERSUS VARIATIONS

Because accidental errors are random and unpredictable, they cannot be evaluated or quantified; thus, corrections to counteract them are indeterminate. Making successive measurements, however, and comparing the results disclose differences in values. Studying the mean of a set of several values and their variations v from the mean indicates the reliability of each value. It is logical to assume unseen and unknowable errors x behave like visible and understandable variations v, so the mean of many measurements should be close to the quantity's correct value. Reliance is placed on variations to judge the mean value's nearness to truth.

3-10.   DISTRIBUTION OF ACCIDENTAL ERRORS

A large set of measurements of a quantity can be represented in a bar graph called a *histogram* (Figure 3-2). Connecting the bar tops by a faired curve, a frequency distribution curve, permits visual representation of the measurements and their variation from the average or mean. Observation shows that

1.   Small variations from the mean value occur more frequently than large ones.
2.   Positive and negative variations of the same size are about equal in frequency, rendering their distribution symmetrical about a mean value.
3.   Very large variations seldom occur.

These three characteristics can be seen in Figure 3-2, where variations are plotted. The three

are also characteristics of accidental errors, for the very existence of variations is explainable only by the presence of accidental errors in measurements. Therefore, it is not only convenient but also permissible to speak almost interchangeably of variations and errors.

The *normal* or *Gaussian* distribution is the most important of many possible distributions, since it has a wide range of practical applications. It is sometimes called the bell-shaped distribution, which typifies measurement distributions in practice. The histogram and frequency curve of Figure 3-2 are symmetric and shaped like a bell, thus that indicating the set of measurements is a normal distribution. The following mathematical model adequately describes such a distribution:

$$y = \frac{1}{\sqrt{(2\pi)}\sigma} \cdot e^{-(x-u)^2 / 2\sigma^2} \qquad (3\text{-}1)$$

The plot of this is called the *normal distribution curve;* if the height of the curve is standardized so that the area underneath it is equal to unity, then the graph is called a probability curve (Figure 3-3).

Focusing on the three obvious variation traits, Figure 3-4 depicts a set of several measurements of a quantity—say, the distance taped between two monuments. The true but unknowable length is indicated as "true value," with an error x_i, and the measurement's mean is shown with its variation v_i. The average error, while unknown, is shown as

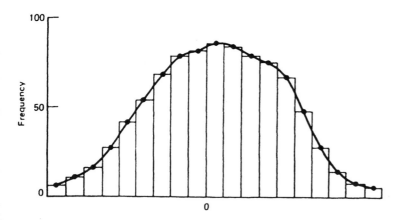

Figure 3-2. Histogram with superimposed frequency distribution curve.

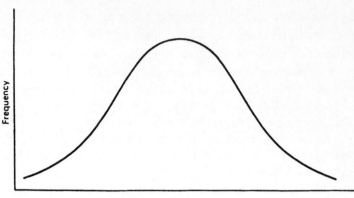

Figure 3-3. Normal distribution curve.

$\Sigma x/n$. Although surveyors and engineers can never equate a variation from the mean v_i with an error x_i, the pattern of occurrence of errors (invisible, unknown) is reasonably assumed to follow that of variations (visible, knowable). It is seen that

$$\underset{\text{(arithmetic mean)}}{\overline{X}} = \underset{\text{(true value)}}{\overline{X}_0}$$
$$-\underset{\text{(mean error)}}{\frac{\Sigma x}{n}} \quad (3\text{-}2)$$

and also

$$\underset{\text{(any variation)}}{v_i} = \underset{\text{(corresponding error)}}{x_i}$$
$$-\underset{\text{(mean error)}}{\frac{\Sigma x}{n}} \quad (3\text{-}3)$$

When n is larger, it signifies that the mean measured value $\overline{X}$ is closer to the true value (mean of the population). It is also apparent that variations then become more nearly equal

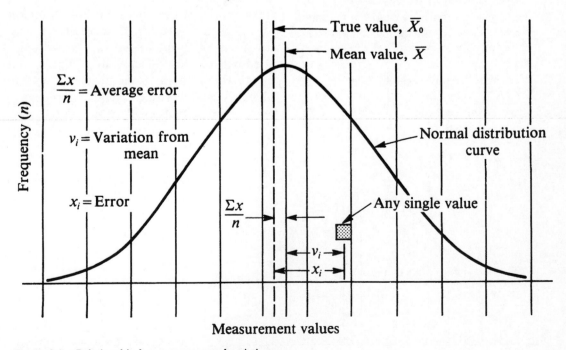

Figure 3-4. Relationship between error and variation.

to errors. Though the true magnitude of a measured quantity is never determinable, it can be ascertained as closely as required by taking enough measurements. If the range of variations narrows to a small value, the range of errors also constricts, rendering the mean value $\bar{X}$ predictably close to the true one $\bar{X}_0$. This hypothesis enables observers to speak confidently of true value and true error.

3-11. STANDARD DEVIATION

A practical indicator used to describe the reliability or worth of a set of repeated measurements is the *standard deviation*, defined as

$$\sigma_s = \pm \sqrt{\frac{\Sigma v^2}{(n-1)}} \qquad (3\text{-}4)$$

Thus, if n measurements of a quantity are obtained, each made in the same manner, the mean value of the set can be employed and given a degree of acceptance by citing the set's standard deviation. Table 3-2 (see p. 25) illustrates the computation of precision for a set of

Table 3-2. Measure of precision for set A

Measured Value	Variation v	v^2
165.861	−0.003	0.000009
165.866	+0.002	0.000004
165.860	−0.004	0.000016
165.864	0.000	0.000000
165.863	−0.001	0.000001
165.865	+0.001	0.000001
165.864	0.000	0.000000
165.863	−0.001	0.000001
165.863	−0.001	0.000001
165.866	+0.002	0.000004
165.864		
Mean		$\Sigma v^2 = 0.000037$

$$\sigma_s = \pm \sqrt{\frac{0.000037}{(10-1)}} = \pm 0.002$$

Best value (mean) = 165.864
Measure of precision $\sigma_s = \pm 0.002$

10 measured values. Today, good hand calculators perform this task easily through the keys marked $\Sigma +$, $\bar{X}$, and σ.

Statistical theory, borne out by extended measurement observations, enables helpful

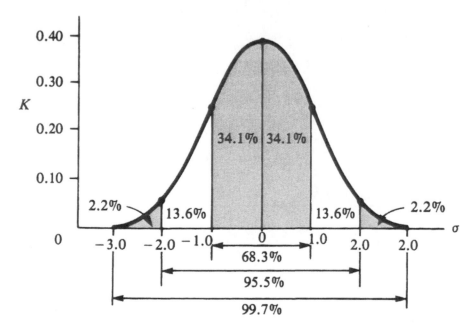

Figure 3-5. Characteristics of the normal distribution curve.

conclusions about a set of measurements. The normal distribution curve (see Figure 3-5) has the following characteristics:

1. The area beneath the entire curve is unity or 100% probability that all measurements will fall somewhere in the curve's range.
2. The area beneath the curve bounded by $\pm \sigma_s$ is 68.3%, and $\pm \sigma_s$ is the 68.3% "error."
3. The area beneath the curve bounded by a $\pm 2\sigma_s$ is 95.5%, i.e., $\pm 2\sigma_s$ represents the 95.5% "error."
4. The area beneath the curve bounded by $\pm 3\sigma_s$ is 99.7%, so $\pm 3\sigma_s$ defines the 99.7% "error."

Stated differently, another measurement should fall with a 68.3% probability within the $\pm \sigma$ bounds; a 95.5% probability between $\pm 2\sigma$; and a 99.7% probability inside $\pm 3\sigma$. These limits are referred to as a one-sigma or 68.3% confidence level, a two-sigma or 95.5% confidence level, and a three-sigma or 99.7% confidence level. Values used for other confidence levels are given in Table 3-3.

3-12. USES OF STANDARD DEVIATION

There is a practical use for standard deviation in comparing sets of measurements of a quantity.

Example 3-7. Comparing set A (Table 3-2) with set B, we obtain

$$\text{Set } A: 165.864 \pm 0.002$$
$$\text{Set } B: 165.867 \pm 0.006$$

The standard deviation for set A suggests it is a better one, although both have validity. Now assume the following two additional sets:

$$\text{Set } C: 165.862 \pm 0.007$$
$$\text{Set } D: 165.864 \pm 0.004$$

To find the weighted mean of all four sets, a weight is accorded each set proportional to the inverse square of its standard deviation. Of the four, it is clear the greatest confidence can be placed in set A, the least in set C.

Set	X	σ_s	Weight Factor	Weight	$Wt(X - 165.860)$
A	165.864	± 0.002	$(1/0.002)^2 = 250,000$	12.25	0.049
B	165.867	± 0.006	$(1/0.006)^2 = 27,778$	1.36	0.010
C	165.862	± 0.007	$(1/0.007)^2 = 20,408$	1.00	0.002
D	165.864	± 0.004	$(1/0.004)^2 = 62,500$	3.06	0.012
				17.67	0.073

Weighted mean = $165.860 + 0.073/17.67 = 165.860 + 0.004 = 165.864$

Table 3-3. Size of error in a single measurement of a set

Name of Error	Symbol	Value	Certainty (%)	Probability of Larger Error
Probable	E_p	$0.6745\sigma_s$	50	1 in 2
Standard deviation	σ_s	$1.0\sigma_2$	68.3	1 in 3
90% error	E_{90}	$1.6449\sigma_s$	90	1 in 10
Two-sigma or 95.5% error	$2\sigma_s$	$2\sigma_s$ or $3E_p$	95.5	1 in 20
Three-sigma or 99.7% error	$3\sigma_s$	$3\sigma_s$	99.7	1 in 370
Maximum*	E_{max}	$3.29\sigma_s$	99.9 +	1 in 1000

*Some authorities regard the 95.5% error as the "maximum error." Neither view is absolutely correct, since the theoretical maximum error is $\pm \infty$, which does not occur in practice. It is, then, a good practical decision to use the 95.5 or 99.9⁺% error as the "practical" maximum that is tolerable.

Another practical use for standard deviation is to determine whether one set of measurements is significantly different from another set. If the discrepancy between the means of the two sets is not more than twice a value called σ_{DIFF}, they can be accepted as measurements of the same quantity.

$$\sigma_{DIFF} = \sqrt{(\sigma_s)_A^2 + (\sigma_s)_B^2} \qquad (3\text{-}5)$$

Example 3-8. Suppose two sets are compared as follows:

Set E: 165.848 $\pm$ 0.006,

$$\sigma_{DIFF} = \sqrt{(0.006)^2 + (0.010)^2}$$

Set F: 165.867 $\pm$ 0.010, $\qquad 2\sigma_{DIFF} = 0.024$

$\overline{\quad 0.019 \quad}$ (difference)

Conclusion: Set E is not significantly different from set F, since 0.019 is not more than twice σ_{DIFF} ($= 0.024$). Therefore, sets E and F can be regarded as valid measurements of the same quantity, and inspection indicates they can be combined in the weighted mean procedure.

Set	X	σ_s	Weight Factor	Weight	$W_i(X - 165.860)$
A	165.864	± 0.002	250,000	25.0	0.100
B	165.867	± 0.006	27,778	2.8	0.020
C	165.862	± 0.007	20,408	2.0	0.004
D	165.864	± 0.004	62,500	6.2	0.025
E	165.848	± 0.006	27,778	2.8	-0.034
F	165.867	± 0.010	10,000	1.0	0.007
				$\overline{39.8}$	$\overline{0.122}$

Weighted mean = 165.860 + 0.122/39.8 = 165.860 + 0.003 = 165.863

3-13. VARIANCE AND STANDARD DEVIATION

Variance is another measure of scatter among measured values in a set of measurements. Preferred by some users, it is simply the square of the standard deviation; thus,

$$\text{Variance} = \sigma_s^2 = \frac{\Sigma v^2}{(n-1)} \qquad (3\text{-}6)$$

Comparing the four sets in Section 3-12, for instance, would show the following:

Set	Variance V		
A	0.000004	or	1/250,000
B	0.000036	or	1/27,800
C	0.000049	or	1/20,400
D	0.000016	or	1/62,500

Other measures exist, but are not covered here:

1. Standard error of the standard deviation = $\pm \sigma_s / \sqrt{2n}$.
2. Standard error of the variance = $\pm \sigma_s^2 \sqrt{2n}$.
3. Standard error of coefficient of variation = $V / \sqrt{2n}$.
4. Standard error of the median = $1.25 \sigma_s / \sqrt{n}$.

3-14. USE OF STANDARD SPECIFICATIONS FOR PROCEDURE

In measurements of any kind, reliance is placed on an established procedure that has been used repeatedly, many hundreds of times, to establish the validity of results. Thus, when standard specifications for a task are followed, only a limited number of measurements is needed to be certain the standard deviation from a shortened set is acceptable. Experience shows a larger number of measurements does not give a greatly different result for the mean or average value of the set.

Confidence inspired by following a fixed procedure or specifications also applies to instruments used in making measurements. A specific EDMI is advertised to give distances accurately to $\pm(5 \text{ ppm} + 4 \text{ cm})$ because many thousands of test observations were made by the manufacturer. Thus, any measurement with this instrument by experienced personnel can be accorded the same accuracy. For instance, a length measured as 1543.02 m has an error value of $\pm(0.0077 + 0.04) = \pm 0.048$ m. This is properly regarded as the standard deviation of the measurement. The measurement precision can be stated as $\pm 0.048/1543$ or 1 in 32,100.

Example 3-9.   The Kern DM502 accuracy has improved from $\pm(5 \text{ mm} + 5 \times 10^{-6} \text{ D})$ to $\pm(3 \text{ mm} + 5 \times 10^{-6} \text{ D})$ by recent technical advances. These limits are the one-sigma (68.3%) values obtained by analysis of many measurements. Using this instrument in the prescribed manner assures it is part of the large family or population of measurements already made and thus able to share in that established reliability.

3-15.   DISTRIBUTION OF ACCIDENTAL ERRORS

Virtually all surveying measurement errors conform to a pattern called normal distribution. The theoretically perfect normal distribution (normal probability curve) shown in Figure 3-3 is the plot of equation

$$y = (h/\sqrt{\pi})e^{-h^2 x^2} \qquad (3\text{-}7)$$

the familiar bell-shaped curve. It is symmetrical, with flatness or peakedness dependent on error sizes (variations from true value).

Natural phenomena and surveying measurements follow the same law of normal

Gaussian distribution for heights of 17-year-olds in a school system; weights of apples gathered from a single tree; weights of babies at birth; and repeated distance, angle, or level measurements. Other distribution patterns result from imposed influences and are mostly Poisson distributions. Examples are the arrival of ships or trains, traffic grouping on a street, incidence of storms, telephone demand, and road accidents. Plotting Poisson distributions or predicting results is not possible by the methods used here, which depend on normal (natural, uninfluenced) distribution.

To study a large set of surveying measurements, plot them as a histogram (bar graph) and superimpose a curve connecting the tops of the bars (see Figure 3-2). If this curve looks bell-shaped, its normally distributed results establish confidence that the rules of probability are fulfilled.

3-16.   PLOTTING THE NORMAL PROBABILITY CURVE

To facilitate comparison, the *normal probability curve* can be plotted at a scale consistent with the histogram. The normal distribution equation is rendered in the form

$$y = \frac{KnI}{\sigma_s} \qquad (3\text{-}8)$$

where I is the class interval, and the following values of K are used:

x	K
$\bar{x}$ (mean)	0.39894
$\bar{x} \pm 0.5\sigma_s$	0.35206
$\bar{x} \pm 1.0\sigma_s$	0.24197
$\bar{x} \pm 1.5\sigma_s$	0.12953
$\bar{x} \pm 2.0\sigma_s$	0.05399
$\bar{x} \pm 2.5\sigma_s$	0.01753
$\bar{x} \pm 3.0\sigma_s$	0.00443
$\bar{x} \pm \infty$	0.00000

(If individual values are used, $I = 1$; if grouped by 2s or 5s, etc., $I = 2$ or 5, etc.)

Example 3-10.   In testing an automatic level instrument, the marker recorded a set of

439 rod readings, with the mean $\overline{X} = 6.5782$ and $\sigma_s = \pm 0.00304$ ft. The ordinates for the superimposed normal distribution curve are

x	y
$\bar{x}$ (mean)	57.5
$\bar{x} \pm 0.5\sigma_s$	50.8
$\bar{x} \pm 1.0\sigma_s$	34.9
$\bar{x} \pm 1.5\sigma_s$	18.7
$\bar{x} \pm 2.0\sigma_s$	7.8
$\bar{x} \pm 2.5\sigma_s$	2.5
$\bar{x} \pm 3.0\sigma_s$	0.6
$\bar{x} \pm \infty$	0.0

The normal curve superimposed on the histogram (Figure 3-6) shows a quite good fit.

$\overline{X}$. It pertains to this sample, not the population as a whole. A sample, or set, representing the whole population distribution will have a mean value $\overline{X}$, but not one equal to the true mean of the whole population. Nor will another sample, or another, etc. Every set likewise has its own standard deviation, not that of the entire population, and unlike other sets (Figure 3-7). It is clear also that an average of the means of several sets will get closer to the population mean. The logic of this is that working with the whole population progressively by sets, the average of all means must ultimately equal the population mean. It follows, therefore, that if the sample is larger, the mean of the sample will more likely approach the true value of the population mean.

3-17. MEANING OF STANDARD DEVIATION

The σ_s indicates that any next measurement introduced in this set should, with 68.3% certainty, fall within the $\pm \sigma_s$ range of the mean

3-18. MEANING OF STANDARD ERROR

The standard deviation of any set or sample of n items yields $\sigma_m = \pm \sigma_s / \sqrt{n}$, which is the

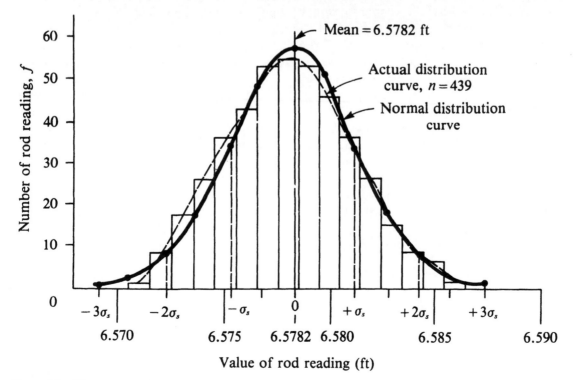

Figure 3-6. Histogram of 439 rod readings with actual distribution curve and normal distribution curve plotted.

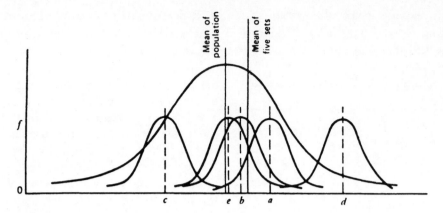

Figure 3-7.   Relationship of samples to population: *a* through *e* are the means of sets of individual measurements.

standard error of the mean. This is a measure of the sample validity. It signifies that with 68.3% confidence the set's mean lies within $\pm \sigma_m$ of the true value of the population mean, or with 95.5% confidence it lies within $\pm 2\sigma_m$, etc.

If several sample sets of measurements are taken from the same population, the sample means vary somewhat from sample to sample and will themselves form a sampling distribution (of $\overline{X}$). The standard deviation of the group means is less than the σ_s of any sample. To find the standard deviation of the population of measurements of a particular quantity, work from one set, assuming that the population would, if entirely covered by samples, be found to have a mean value near that of the one sample, and a standard deviation of the means of all the similar samples repeated enough to cover the entire population. Recalling that the variance of the set is σ_s^2, then the variance of the means of sets is

$$\text{Variance } (\overline{X}) = \text{variance } (\Sigma x_1/n)$$

$$= 1/n^2 \text{ variance } (\Sigma x_i)$$

But variance $(x_i) = \sigma^2$ for all i

$$\text{Variance } (\overline{X}) = (1/n^2)(\sigma^2 + \sigma^2 + \cdots + \sigma^2)$$

$$= n\sigma^2/n^2 = \sigma^2/n$$

Thus,

$$\sigma_m = \frac{\sigma_s}{\sqrt{n}}$$

Although nature's distribution of apple sizes or men's weights may yield a mean value and standard deviation showing the whole population's scatter, surveying measurements are different. Surveying measurements do not exist until they are made; they cannot include all possible measurements of an angle, length, or difference of elevation. Therefore, from one sample, use the mean and standard error as an indication of correct value—e.g., the mean of the population of measurements if all possible measurements were made. The σ_m is an indicator of the standard deviation of the population, all based on a reasonable number of measurements in a single set.

Example 3-11.   Set *A* of Table 3-2 would yield $\sigma_m = \pm(0.002/\sqrt{10}) = 0.0006$ ft. With 68.3% assurance, the true value is within ± 0.0006 of 165.864 $(= \overline{X})$, or between 165.863 and 165.865 ft.

Example 3-12.   In the 439 measurements of Example 3-10, the mean $\overline{X}$ is 6.5782 and $\sigma_s = \pm 0.00304$. Then

$$\sigma_m = \pm 0.00304/\sqrt{439}$$

the "maximum" error (Table 3-3) is $3.29\sigma_m$ or ± 0.000477, which gives a 99% confidence that the true value lies between 6.5777 and 6.5787 ft. Such a large number of measurements would be unlikely except in testing a procedure or developing a new instrument.

Example 3-13. From records of some 30 surveys of similar precision, a single mean value of difference of elevation is sought, along with a measure of its precision (accuracy). The following values are grouped randomly into six sets, size five, for study:

	1	2	3	4	5	6	The 30-Set
	3.717	3.622	3.651	3.775	3.611	3.697	
	3.621	3.594	3.632	3.583	3.622	3.564	
	3.753	3.609	3.661	3.656	3.527	3.595	
	3.558	3.695	3.524	3.633	3.648	3.616	
	3.675	3.659	3.623	3.577	3.706	3.639	
$\overline{X}$	3.665	3.636	3.618	3.645	3.623	3.622	3.635
σ_s	0.077	0.041	0.055	0.080	0.065	0.050	0.060
σ_m	0.034	0.018	0.025	0.036	0.029	0.022	0.011

Range bracketing the population mean with a 68.3% confidence:

	1	2	3	4	5	6	The 30-Set
	3.631	3.618	3.593	3.609	3.594	3.600	3.624
	to	to	to	to	to	to	to
	3.699	3.654	3.643	3.681	3.652	3.644	3.646

Any one of the six X and σ_m values is representative of the mean and standard deviation of the whole population. For further comparison, the $\overline{X}$ and σ_m values are shown for the 30 measurements regarded as a single sample set; it is seen that each of the six smaller sets has a range that brackets the $\overline{X}_{30}$ ($= 3.635$). Further, each set's mean is a valid contender for the population mean, which can, of course, never be known for sure. This example also demonstrates that a small sample can and sometimes must be used, but a larger sample gives a more refined result.

3-19. PLOTTING THE NORMAL DISTRIBUTION IN OTHER FORMS

The shape of a normal curve (bell-shaped) depends on the standard deviation σ_s, which spreads out the curve when it is larger. Whatever the mean and standard deviation, however, one in three observations will lie beyond one standard deviation from the mean, one in 20 beyond two, etc. Drawing the cumulative distribution curve in another form makes

some things clearer. This is done by plotting "percentage-smaller-than" against values of the measurement.

Example 3-14. A set of 16 angle measurements, tallied in ascending order, (Table 3-4), is indicated by percentages calculated on the basis of $(n + 1)$ for reasons to be explained later.

$n = 16$, calculated mean $= 134°37'19.21''$
$n + 1 = 17$, $\sigma_s = \pm 03.48''$
Mean minus $\sigma_s = 134°37'15.73''$
Mean plus $\sigma_s = 134°37'22.69''$

This small array of numbers is grouped in ascending order with class width of $02''$ and a histogram plotted. Connecting the tops to form a frequency distribution curve (Figure 3-8) shows it is not bell-shaped or satisfactory.

The 16 results are better portrayed on a cumulative *frequency distribution curve* (the S-curve, Figure 3-9). This curve can be held to virtually a straight line between the 15.8% and 84.2% values (the $\pm \sigma_s$ limits) and made to pass through the plotted points in a "best-fit"

Table 3-4.   Set of 16 angle measurements

X	f	f	%	Class Width	Class Mean	f
134°37′13.8″	1	1	5.9			
14.8″	1	2	11.8	13.5–15.5″	14.5″	2
15.8″	1	3	17.6	15.5–17.5″	16.5″	4
16.1″	1	4	23.5	17.5–19.5″	18.5″	3
16.8″	1	5	29.5	19.5–21.5″	20.5″	4
17.4″	1	6	35.3	21.5–23.5″	22.5″	1
17.5″	1	7	41.2	23.5–25.5″	24.5″	1
19.1″	1	8	47.1	25.5–27.5″	26.5″	1
19.4″	1	9	52.9			
20.3″	1	10	58.8			
20.8″	1	11	64.7			
20.9″	1	12	70.6			
21.4″	1	13	76.5			
23.4″	1	14	82.4			
24.0″	1	15	88.2			
26.0″	1	16	94.1			

manner. Then the points outside this range, usually spoken of as the 15 to 85% range, will tail off to form an S-curve. Observing the curve at 50% shows the mean value, which should verify that previously calculated. The 15 and 85% points are values marking the 68.3% limits of certainty.

A still better way to plot a set of values is on arithmetic probability paper designed to plot any normal or *Gaussian* distribution as a straight line. There are two ways to use such paper for a meaningful plot and examine the scatter:

1. Plot the individual points, percentage-smaller-than versus the actual values; draw a best-fit straight line, and read $\bar{X}$ (mean) at 50% and the limits of σ_s at the 15.8 and 84.2% lines.

2. Calculate the mean and σ_s values, draw a straight line through them, and finally plot individual values to see how well they fit.

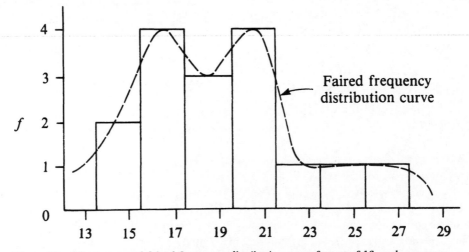

Figure 3-8.   Histogram and faired frequency distribution curve for set of 16 angle measurements.

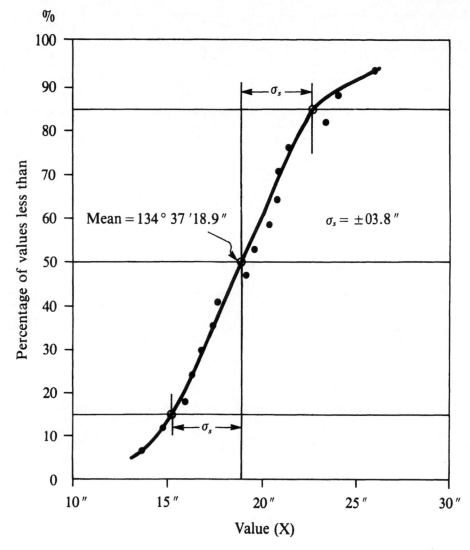

Figure 3-9. Cumulative frequency distribution curve for set of 16 angle measurements.

Example 3-15. The 16 angle values are plotted (Figure 3-10) by method a and the curve drawn as a best-fit line. Percentages are calculated using $(n + 1)$ in the denominator to plot the end point(s) of the curve. Comparison of the angle measurements with those of Example 3-14 shows these results. From the plot on arithmetic probability paper,

Mean = 134°37'19.2", $\sigma_s = 04.0''$

From calculation,

Mean = 134°37'19.21", $\sigma_s = 03.5''$

The plotted points are seen to fit a faired straight line decently well, so they can be visually adjudged to conform with normal distribution and be accepted for any further statistical treatment. The scaled outlines of Figure 3-10 can be photocopied and used for arithmetic probability paper.

3-20. PROPAGATION OF ACCIDENTAL ERRORS

Basic to the combined effect of accidental errors is their tendency to cancel themselves

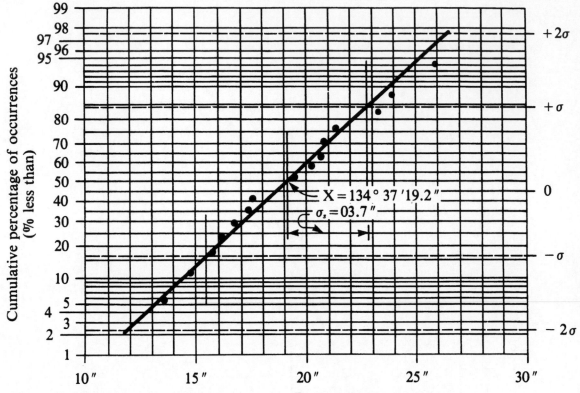

Figure 3-10. Plot of 16 angle measurements to find mean $\bar{X}$ and standard deviation σ_s of the set.

out (as likely to be $+$ as $-$) and their propensity to cluster about the mean—more frequently small, seldom large. This gives rise to the law of compensation

$$E_{\text{total}} = \pm\sqrt{E_1^2 + E_2^2 + E_3^2 + E_4^2 + \cdots + E_n^2} \quad (3\text{-}9)$$

Example 2-16. If a 100-ft steel tape is corrected for temperature to $\pm5°F$, a random error of ±0.0032 ft results. Comparison with a standard tape may expose an error of ±0.002 ft. When applying tension within 2 lb, there is a possible error of ±0.0013 ft. Therefore, total effective error for each tapelength from these three sources is

$$E_{\text{total}} = \pm\sqrt{0.0032^2 + 0.002^2 + 0.0013^2}$$
$$= \pm0.0040 \text{ ft}$$

3-20-1. Addition

When adding measured quantities having known errors, the error of the sum is

$$E_s = \pm\sqrt{e_1^2 + e_2^2 + e_3^2 + \cdots + e_n^2} \quad (3\text{-}10)$$

Example 3-17. If the steel tape of the previous example is known to have a ±0.0040-ft error each time it is used, the error in taping 1600 ft would be

$$E_s = \pm\sqrt{16(0.0040)^2} = \pm4(0.0040)$$
$$= \pm0.016 \text{ ft}$$

Example 3-18. The error in the sum of two measured angles is computed similarly.

Angle $AOB = 15°32'18.9'' \pm 05''$
Angle $BOC = 67°17'45.0'' \pm 15''$
Angle $AOC = 82°50'03.9'' \pm 15.8''$

$$(= \pm\sqrt{05^2 + 15^2})$$

3-20-2. Subtraction

The same rule applies as in addition, readily understood by considering subtraction as negative addition: similar to Equation (3-10).

Example 3-19. If a known line AB is 1867.857 ± 0.018 ft long and a segment AP on the line is taped to be 195.009 ± 0.010 ft, then the remaining segment *PB* is

$$1672.848 \pm (\sqrt{0.018^2 + 0.010^2} = \pm 0.021)\ \text{ft}$$

3-20-3. Multiplication

When multiplying two or more measured quantities having accidental errors, this general equation for relative errors applies

$$E_{\text{product}} = \pm A \cdot B \sqrt{\left(\frac{E_A}{A}\right)^2 + \left(\frac{E_B}{B}\right)^2} \quad (3\text{-}11)$$

Example 3-20. A field measured as 160.881 ± 0.026 ft long by 75.007 ± 0.001 ft wide has an error in its area (12,067.20 ft²) of

$$\pm 12,067.20 \sqrt{\left(\frac{0.026}{160.881}\right)^2 + \left(\frac{0.011}{75.007}\right)^2}$$

$$= \pm 2.63\ \text{ft}^2$$

3-20-4. Division

The error in a quotient of two measured quantities is

$$E_{\text{quotient}} = \pm \frac{A}{B} \sqrt{\left(\frac{E_A}{A}\right)^2 + \left(\frac{E_B}{B}\right)^2} \quad (3\text{-}12)$$

Example 3-21. If the area of the rectangular plot is somehow known to be 49,650 ± 10 ft² and the width dimension measured several times is found to be 175.62 ± 0.46 ft, the calculated length dimension is 282.72 ± 0.74 ft, since the error is

$$\pm 282.72 \sqrt{\left(\frac{10}{49,650}\right)^2 + \left(\frac{0.46}{175.62}\right)^2}$$

$$= \pm 0.74\ \text{ft}$$

3-20-5. Other Operations

The volume of a rectangular tank or bin whose three dimensions are measured has an error

$$E_{\text{volume}} = \pm L \cdot W \cdot H$$

$$\times \sqrt{\left(\frac{E_L}{L}\right)^2 + \left(\frac{E_W}{W}\right)^2 + \left(\frac{E_H}{H}\right)^2} \quad (3\text{-}13)$$

A cube, if all three sides are measured, will

side *A* is measured and the other dimensions are regarded as equal, the volume error is

$$E_{\text{volume}} = \pm A^3 \sqrt{\left(\frac{E_A}{A}\right)^2 \times 3}$$

$$= \pm A^2 E_A \sqrt{3} \quad (3\text{-}14)$$

The general form for error in raising to a power any quantity containing an error is

$$E_{\text{power}} = \pm A^n \sqrt{n \left(\frac{E_A}{A}\right)^2}$$

$$= \pm E_A \cdot A^{(n-1)} \sqrt{n} \quad (3\text{-}15)$$

The volume of a measured sphere, because only one dimension is measured, follows the power rule just given.

$$\text{Volume of sphere} = \tfrac{4}{3}\pi r^3$$

$$E_{\text{sphere}} = \pm (\text{vol}) \sqrt{n \left(\frac{E_r}{r}\right)^2}$$

$$= \pm \frac{4\sqrt{3}}{3} \pi r^2 E_r \quad (3\text{-}16)$$

Example 3-22. If a sphere's radius is measured as 10.00 ± 0.08 ft, the calculated volume is 4188.8 ft³ and the error will be

$$E_{\text{vol}} = \pm \frac{4\pi}{3} 10^3 \sqrt{3 \left(\frac{0.08}{10}\right)^2} = \pm 58.0\ \text{ft}^3$$

The error in volume of a cylindrical tank can be found by first analyzing the area of the circular end, then working with it, the length, and the errors in both.

Example 3-23. A cylindrical tank of diameter 10.00 ± 0.02 ft is 30.00 ± 0.04 ft long. End area is $\tfrac{1}{4}\pi D^2 = 78.540$ ft² and volume is 78.540 × 30.00 = 2356.2 ft³.

$$E_{\text{area}} = \pm \tfrac{1}{4}\pi 10^2 \sqrt{2 \left(\frac{0.02}{10}\right)^2} = \pm 0.222\ \text{ft}^2$$

$$E_{\text{vol}} = \pm (30.00)(78.540)$$

$$\times \sqrt{\left(\frac{0.222}{78.540}\right)^2 + \left(\frac{0.04}{30.00}\right)^2}$$

The area of a right triangle whose altitude and leg are both measured has an area error computed similarly.

Example 3-24. The sides of a right triangle are measured as 100.000 ± 0.021 and 35.000 ± 0.012 ft. Area is $\frac{1}{2}ab = 1750.00$ ft^2.

$$E_{\text{area}} = \pm\frac{1}{2}ab\sqrt{2\left(\frac{E_a}{a}\right)^2 + \left(\frac{E_b}{b}\right)^2}$$

$$= \pm1750\sqrt{\left(\frac{0.021}{100}\right)^2 + \left(\frac{0.012}{35}\right)^2}$$

$$= \pm0.70 \text{ ft}^2 \qquad (3\text{-}17)$$

3-21. AREA OF A TRAVERSE

The error in area of a closed traverse can be found similarly, using the *relative accuracy* concept. If the traversing procedure prescribed for, say, a 1/5000 accuracy is followed, the area is obtained by multiplying total latitude by meridian distances. If we assume that a traverse area already obtained is 1,062,323 ft^2 and the sum of the latitudes equals 1867.812 ft, the meridian distances total

$$1,062,323/1867.812 = 568.753 \text{ ft}$$

The error ascribable to each factor is

$$E_{\text{latitude}} = \pm1,967,812/5,000 = \pm0.374 \text{ ft}$$
$$E_{\text{meridian}} = \pm568.763/5,000 = \pm0.114 \text{ ft}$$

$$E_{\text{area}} = \pm1,062,323\sqrt{\left(\frac{0.374}{1,868}\right)^2 + \left(\frac{0.114}{569}\right)^2}$$

$$= \pm300.9 \text{ ft}^2$$

It may be noted that this is the same as

$$E_{\text{area}} = \pm(\text{area})\sqrt{\left(\frac{1}{5000}\right)^2 + \left(\frac{1}{5000}\right)^2}$$

$$= \pm300.5 \text{ ft}^2$$

or, more simply,

$$E_{\text{area}} = \pm(\text{area})(1/5000)\sqrt{2} = \pm300.5 \text{ ft}^2$$

Had the traverse been run to 1/10,000 accuracy,

$$E_{\text{area}} = \pm(\text{area})(1/10,000)\sqrt{2} = \pm150.2 \text{ ft}^2$$

Common traverse relationships between linear and area accuracies are given in Table 3-5.

3-22. ERRORS AND WEIGHTS

Results obtained from different measurements of the same quality can be combined to find a *weighted mean* by assigning proportionately greater weights to measurements, or measurement sets, that have smaller standard deviations σ_s or smaller standard errors σ_m. Weights should be inversely proportional to the square of the sigma quantities.

Example 3-25. Several sets of linear measurement are listed. It is desired to find the combined weighted mean.

Table 3-5. Summary table of traverse area error

Order/Class	Linear Accuracy	Area Accuracy
First	1/100,000	$(1/100,000)\sqrt{2} = 1/70,700$
Second/I	1/50,000	$(1/50,000)\sqrt{2} = 1/35,300$
Second/II	1/20,000	$(1/20,000)\sqrt{2} = 1/14,100$
Third/I	1/10,000	$(1/10,000)\sqrt{2} = 1/7070$
Third/II	1/5000	$(1/5000)\sqrt{2} = 1/3530$

Set	X	σ_s	Weight Ratio	Weight w	$w \cdot (X - 387)$
A	387.071	±0.025	$(1/0.025)^2 = 1,600$	5.38	0.382
B	387.126	±0.042	$(1/0.042)^2 = 567$	1.91	0.240
C	387.080	±0.019	$(1/0.019)^2 = 2,770$	9.32	0.746
D	387.112	±0.058	$(1/0.058)^2 = 297$	1.00	0.112
				17.61	1.480

Weighted mean = 387.000 + 1.480/17.61 = 387.084 ft

Example 3-26. A weighted mean value of elevation difference between two bench marks can be calculated if a set of level runs by one party is combined with that of another party, each set being the result of several runs having a calculated standard deviation.

Set	X	σ_s	Weight Ratio	w	$w(X - 166)$
A	167.212	±0.182	$(1/0.182)^2 = 30.190$	1.000	1.212
B	166.978	±0.071	$(1/0.071)^2 = 198.373$	6.571	6.380
				7.571	7.592

Weighted mean = 166.000 + 7.592/7.571 = 167.003 ft.
(The unweighted mean would be 167.095 ft.)

Example 3-27g. In differential leveling, one party made a run using second-order/class I methods, another followed the same route using third-order/class II methods. The weighted mean of the two results is found through the published relative accuracy values.

Set	Difference of Elevation (m)	Order/ Class	Relative Accuracy	Weight Ratio	w	$w(X - 41)$
Q	41.0962	2nd/I	±1.0 mm$\sqrt{K}$	$\left(\dfrac{1}{1.0K}\right)^2 = \dfrac{1}{K}$	1.69	0.1626
R	41.1076	3rd/II	±1.3 mm$\sqrt{K}$	$\left(\dfrac{1}{1.3K}\right)^2 = \dfrac{0.59}{K}$	1.00	0.1076
					2.69	0.2702

Weighted mean = 41.0000 + 0.2707/2.69 = 41.1004 m. (The unweighted mean would be 41.1019 m.)

Example 3-28. A bench mark is to be established using three different level runs following varied procedures and contrasting specifications. A weighted mean is required, with weights fixed inversely proportional to the square of the published relative errors. Distances in kilometers are shown.

Run	Elevation m	k	Order/Class	Relative Error e	Weight Ratio $1/e^2$
A	46.1672	1.5	2nd/I	± 1.0 mm$\sqrt{1.5}$ = ± 1.225	0.667
B	46.2107	2.0	3rd	± 2.0 mm$\sqrt{2.0}$ = ± 2.828	0.125
C	46.1810	1.5	2nd/II	± 1.3 mm$\sqrt{1.5}$ = ± 1.592	0.395

Run	Weight	Weight (Elevation $-$ 46)
A	5.336	0.8922
B	1.000	0.2107
C	3.160	0.5720
	9.496	1.6759

Weighted mean = $46.0000 + 1.6759/9.496 = 46.1765$ m

3-23. CORRECTIONS

Corrections for measured quantities should be inversely proportional to their weights, directly proportional to the squares of accidental errors or standard deviations or standard errors.

Example 3-29. Angles of a triangle were measured with different instruments. Lacking a better guide, errors assigned to each point are the nominal capabilities of different instruments used at these points.

Point		Instrument Type	E	E^2	Correction Index	Correction	Adjusted X
A	70°13'50"	10-sec	$\pm 10''$	100	1	$1/41(40) = 01''$	70°13'49"
B	58°45'20"	20-sec	$\pm 20''$	400	4	$4/41(40) = 04''$	58°45'16"
C	51°01'30"	1-min	$\pm 60''$	3600	36	$36/41(40) = 35''$	51°00'55"
	180°00'40"				41		180°00'00"

Example 3-30. A line *AB* is measured by first-order methods, and then again in two segments *AC* and *CB* by third-order methods. Adjusted values are needed for all three measured distances. (The value of *AB* must obviously equal the sum of *AC* plus *CB*, finally.)

Line	X	Precision Fraction	e*	$1/e^2$	w	w(X − 2104)
AC	1416.912					
CB	687.901					
	2104.813	1/5000	0.421	5.643	1	0.813
AB	2104.697	1/20,000	0.105	90.299	16	11.152
Diff =	0.116				17	11.965

*(1/5000)(2104.813) = 0.421
(1/20,000)(2104.697) = 0.105
Weighted mean = 2104 + 11.965/17 = 2104.704

Since (*AC* + *BC*) must be adjusted downward by (2104.813 − 2104.704 = 0.109) to equal this weighted mean, the two segments (proportionately) are

$$AC = 1416.912 - (1417/2105)(0.109)$$
$$= 1416.839$$

$$CB = 687.901 - (688/2105)(0.109)$$
$$= 687.865$$

Sum = 2104.704   (check)

The weight ratios could be obtained directly from the expressions of precision; thus,

Constant × e^2	Weight Ratio (= $1/e^2$)
Inverse of $(1/5000)^2$ = 40.0 × 10^{-9}	25 × 10^6 or 1
Inverse of $(1/20,000)^2$ = 2.5 × 10^{-9}	400 × 10^6 or 16

Another way to arrive at these new values is as follows, noting that adjustments are made in inverse ratio to the weights:

Segment	e	$1/e^2$	w	Adjustment Ratio	Adjustment*
AC + CB	0.421	5.643	1	1/1 = 1.0000	0.109
AB	0.105	90.70	16	1/16 = 1.0625	0.007
				1.0625	0.116

*Adjustment calculations:
    (1/1.0625)(0.116) = 0.109
(0.0625/1.0625)(0.116) = 0.007
                Sum = 0.116

3-24. SIGNIFICANT FIGURES IN MEASUREMENTS

3-24-1. Exact Versus Doubtful Figures

When a measurement is made, all digits in the result are exact if they are obtained by counting or finding a point that lies between two markers. Digits are doubtful when they result from estimating. *Significant figures* include all exact digits, plus a doubtful one. Several rules and conventions will be given.

3-24-2. Use of Zero

A zero is not significant when it serves merely to place the decimal.

1. 0.00584 contains three significant figures, seen more clearly if written as 5.84×10^{-3}.

2. 34,000 mi has two significant figures, unless it is clearly intended that the value is exact. Writing it 3.4×10^4 is not common, but is better usage (scientific form), or as 34×10^3 (engineering form).

When used otherwise, a zero is significant.

3. 4.6007, 9,1030, or 4100.0 each has five significant figures.

4. 0.076130 also has five significant figures and could well be written 7.6130×10^{-2} or 76.130×10^{-3}.

3-24-3. Rules of Thumb for Significant Figures

The following are reasonable rules and conventions:

1. Unless some precision indicator is affixed—e.g., standard deviation—the usual interpretation for the last (doubtful) digit is plus or minus one-half a unit in the last column. Thus a measured length of 81.713 means the range of uncertainty extends from 81.7125 to 81.7135 units.

2. Although use of only one doubtful figure in the final result is anticipated, it is desirable to use two doubtful digits throughout the calculation and round off only at the end.

3. Adding and subtracting several measured values limit the result to show no more than the least valid item. The following examples are obvious:

6.27	56.17	
4.3	11.036	367.796
13.876	79.3015	−28.7
24.4	146.51	339.1

4. In multiplying or dividing, the result must not be credited with more significant digits than appear in the term with the smallest number of significant figures, as shown here:

$$6.7153 \times 4.67 = 31.4 \quad (\text{not } 31.360451)$$

$$(86.85 \times 10^4)^2$$
$$= 754.3 \times 10^9 \quad (\text{not } 754.29225 \times 10^9)$$
$$850.436/4.56 = 186 \quad (\text{not } 186.499123)$$

However, 8 and 9 are almost two-digit numbers, and occasionally an extra digit in the product is warranted; thus,

$$9.703 \times 4.07062 = 39.497$$
$$(\text{instead of just } 39.50)$$

5. Calculators sometimes convey a false sense of precision, so care must be taken to cut back and properly round off the final result, like this

$$\tfrac{1}{2}(87.645 \times 8.6305)$$
$$= 756.42 \text{ or } 756.420 \quad (\text{not } 756.4201725)$$

3-24-4. Rounding Off

When dropping excess digits, raise the last one and retain it if the discarded quantity is greater than one-half, or leave it unchanged if the discarded quantity is smaller than one-half; thus,

4.796 becomes 4.80 or 4.8 or 5

8.512 becomes 8.51 or 8.5 or 9

If the quantity to be discarded is exactly 5, round off the preceding digit to the nearest even value; thus,

> 10.675 becomes 10.68 or 10.7
> 10.685 becomes 10.68 or 10.7
> 10.695 becomes 10.70 or 10.7
> 10.705 becomes 10.70 or 10.7
> 10.6749 becomes 10.67,   and
> 10.6751 becomes 10.68

3-24-5. Using Exact Values

The procedures described apply to quantities resulting from measurements. If exact values are implied in a statement—e.g., a 2000-ft radius curve—the number of significant digits is not limited. This and others are stated and discussed here.

A field of 89,102.6 ft^2 is properly converted to acres using the following exact conversion factor:

89,102.6/43,560

$$= 2.04551 \text{ acres} \quad (\text{but not } 2.04551423)$$

If a 2000-ft radius is specified for a circular curve, this sets the degree of curve D at

$$\frac{5729.577951}{2000 \text{ (exact)}} = 2.864788976°$$

$$\text{or} \quad 2.8647890 \text{ to 8 digits}$$

This comes from a ratio in the circle

$$\frac{D \text{ (degree of curve)}}{360°} = \frac{100 \text{ ft}}{2\pi R \text{ ft}}$$

$$\text{or} \quad R = 5729.577951/D$$

With a good calculator, the value of π is given to eight, nine, or 10 valid digits, and the 2000 ft is implied to be equipped with an endless row of exact digits (zeros). Preserving these digits is necessary to find the length of a circular curve. Assuming an intersection angle of $43°47'34'' = 43.79278°$ and a 2000-ft radius, we obtain

$$L_c = 43.79278/2.864789$$

$$= 15 + 28.656 \quad (\text{stationing designation})$$

This requires seven digits, minimally, and inexactness can readily occur if the two angular values are carelessly truncated early on.

REFERENCES

BARRY, B. A. 1978 *Errors in Practical Measurement in Science, Engineering, and Technology*. Rancho Cordova, CA: Landmark Enterprises.

BUCKNER, R. B. 1983 *Surveying Measurements and Their Analysis*. Rancho Cordova, CA: Landmark Enterprises.

CRANDALL, K. C., and R. W. SEABLOOM. 1970 *Engineering Fundamentals in Measurements, Probability, Statistics, and Dimensions*. New York: McGraw-Hill.

4

Linear Measurements

Kenneth S. Curtis

4-1. INTRODUCTION

Surveyors are fundamentally concerned with the measurement of horizontal and vertical distances and angles and, more recently, with direct positioning. These, then, are used in various combinations in traversing, triangulation, trilateration, mixed-mode operations, mapping, layout staking, leveling, etc.

Linear distance measurement can be achieved by (1) *direct comparison measurement* with a tape, either fully supported on the ground or suspended in catenary; (2) *optical distance-measurement methods* by remote angular observation on a variable- or fixed-base length held horizontal or vertical, such as in tacheometry, stadia, or subtensing; and (3) *electromagnetic distance instruments* utilizing the travel time of radio or light waves converted to distance. This chapter covers direct tape measurement and optical distance measurement. Chapter 5 discusses electromagnetic distance-measuring instruments (EDMIs).

To a surveyor, the word distance usually refers to the horizontal length between two points projected onto a horizontal plane. Many measuring devices yield slope distances, which must be converted to horizontal. Maps and land areas are based on horizontal measurements or dimensions. Whether to measure a distance by pacing, taping, stadia, or with a

highly accurate EDMI depends on the accuracy required, and this, in turn, depends on the purpose of the measurement. Only a person thoroughly familiar with all types of measuring techniques can choose the optimum and most cost-effective procedure.

4-2. UNITS OF LINEAR MEASUREMENT

Several methods are used to measure distances. They range from rather inaccurate estimates to very precise instrumental procedures. Most early measurement units were derived from physical dimensions associated with parts of the human body. For example, the cubit, digit, palm, hand, span, foot, yard, pace, and fathom can be traced to human anatomy. Many others, such as the rod, pole, perch, chain, furlong, mile, and league, are extensions of these basic units (Table 4-1). Three barleycorns laid lengthwise equaled one inch, 12 of them equaled one foot. Many old units have been discarded in favor of the basic ones, foot and meter. Much of the world has now converted to the meter-decimal system (SI units) as illustrated in Table 4-2. Numerous English-related countries, such as the United States, remain slow to completely convert from the foot (English) system.

Table 4-1. Units of length

Unit	Inches (in.)	Feet (ft)	Yards (yd)	Rods (rd)	Chains (ch)	Meters (m)
1 inch	1	0.08333	0.02778	0.00505	0.00126	0.02540
1 foot	12	1	0.3333	0.0606	0.01515	0.3048
1 yard	36	3	1	0.1818	0.04545	0.9144
1 rod	198	16.5	5.5	1	0.25	5.0292
1 chain	792	66	22	4	1	20.1158
1 mile	63,360	5280	1760	320	80	1609.35
1 meter	39.37	3.281	1.094	0.199	0.04971	1

The primary standard of length in the United States, the National Prototype Meter 27, a 90% platinum and 10% iridium bar, is housed at the National Bureau of Standards (NBS) in Gaithersburg, MD. It is identical in form and material with the International Prototype Meter deposited at the International Bureau of Weights and Measures at Sevres, France, and also with other national prototype meters distributed in 1889, in accordance with a treaty known as the Convention of the Meter, dated May 20, 1875. The meter was originally conceived as 1/10,000,000 part of a meridional quadrant of the earth.

In 1960, the official definition of the meter was redefined as a length equal to 1,650,763.73 wavelengths in a vacuum of the radiation of the orange-red light of a krypton-86 atom. The International Prototype bar was abrogated in favor of a natural and indestructible standard thought to have an accuracy adequate for metrology's modern needs. However, in 1983, the meter was redefined again as the distance traveled by light in a vacuum during 1/229,792,458 sec. It is claimed that this new definition allows the meter to be defined 10 times more accurately and achieves the goal of using time, the most accurate basic measurement, to define length.

In the United States, since 1893, the yard has been defined in terms of the meter by the following relations: 1 yd = 3600/3937 m or 1 m equals 39.37 in. exactly. This legal ratio is used to define the "U.S. Survey Foot." In 1959, after several years of discussion, the United Kingdom and United States agreed to establish a new uniform relationship between the yard and meter as

1 yard = 0.9144 meter exactly

or 1 foot = 0.3048 meter   (international foot)

or 1 inch = 25.4 millimeter

Since the new value of the yard is smaller by two parts per million than the 1893 yard, only in large-scale geodetic survey data is the difference important.

The land-surveying profession is uniquely sensitive to metric system usage because many problems can arise in converting recorded

Table 4-2. Metric units of length

Unit	Micrometers (μm)	Millimeters (mm)	Centimeters (cm)	Decimeters (dm)	Meters (m)	Kilometers (km)
1 micrometer	1	0.001	0.0001			
1 millimeter	1000	1	0.1	0.01	0.001	
1 centimeter	10,000	10	1	0.1	0.01	0.00001
1 decimeter	100,000	100	10	1	0.1	0.0001
1 meter	1,000,000	1000	100	10	1	0.001
1 kilometer			100,000	10,000	1000	1

measurements from nonmetric units to SI units. It may appear quite simple: Just use a 30-m tape instead of a 100-ft tape. However, over several hundred years, land descriptions have been recorded in the English system of units or the Spanish vara (Table 4-3), and future generations may never completely get away from these historical units.

For a fee, the metrology division of the NBS provides calibration of line standards of length and measuring tapes. Using Invar base-line tapes and highly accurate electronic distance-measuring instruments, the National Geodetic Survey (NGS) has established nearly 200 calibration base lines (CBLs) across the United States, providing a means to detect constant and scale errors in measuring instruments. A typical CBL layout consists of four monuments located at 0, 150, 430, and 1400 m, relatively positioned with accuracies approaching one part per million.

4-3. DIRECT COMPARISON DISTANCE-MEASUREMENT METHODS

4-3-1. Pacing

Rough estimates of distances can be made by eye, based on experience in observing commonly used dimensions, such as 100 ft, 100 yd, or a city block. Distances can also be scaled from using a variety of map measures, maps, or aerial photographs.

A better approximation is obtained by walking the distance with a natural or artificial *pace*. The length of a human pace varies and few can develop a 3-ft artificial place to mea-

Table 4-3. Units used in land measurements

1 Gunter's chain (100 links) = 66 ft = 4 rods or poles or perches
80 Gunter's chains = 1 mile
1 vara = 32.993 in. in Mexico, 33 in. in California, and 33$\frac{1}{2}$ in. in Texas

sure distances, without creating fatigue. Therefore, it is best to determine your natural pace by walking over a course of known length —300 to 900 ft—several times, to standardize the pace.

Many factors can affect pace length, including slope and roughness of the terrain, shoe weight and clothing type, pacing speed, fatigue, and age. In addition, it is difficult to keep count of the steps. Sometimes, strides (two paces) are counted using a notched stick or mechanical tally register. A *pedometer* strapped to the pacer will automatically record the distance covered in miles after it has been adjusted to the wearer's pace. A similar instrument, called a *passometer*, automatically records the number of paces.

With a little practice, a good pacer can attain results within 1% of the true distance (1/100). No special equipment is required for its many practical applications, one of which is detection of blunders that can occur in taping or in other more accurate distance-measuring procedures.

4-3-2. Odometer or Measuring Wheel

The *odometer* is a device similar to the distance recorder in an automobile speedometer. It is attached to a wheel of known circumference and rolled over the distance to be measured. Results obtained depend on the topography and surface irregularities, but on level smooth ground may yield 1/200 accuracy. Measuring wheels serve as rough checks on more accurate measurements and can be useful in reconnaissance and preliminary surveys. Wheels of 2, 4, or 6 ft in circumference are most popular. They include precision totalizers or counters, which can be reset; one model has a battery-powered electronic totalizer and an LCD counter (Figure 4-1). Some measuring wheels can be attached to a vehicle with a rear-mounted hitch, allowing longer distances to be measured while moving at speeds of up to 8 mph.

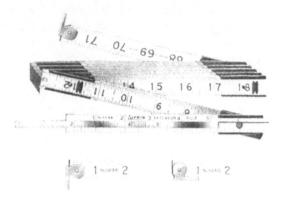

Figure 4-1. Measuring wheel. (Courtesy of Rolatape Corporation.)

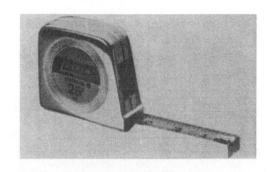

Figure 4-2. Folding rule and short power hardware tape. (Courtesy of Lufkin and the Cooper Group.)

4-3-3. Folding Rules and Hardware Tapes

Folding rules and short power tapes, sometimes referred to as hardware tapes, are variously made, differently graduated (Figure 4-2), and used on all types of building construction sites. A rule, being stiff, can be held in any position desired whereas a tape is flexible and generally needs two people for measuring. Some tapes are graduated in feet, inches, and eighth-inch (or sixteenth-inch), or feet, inches, and decimals. Except on building construction, surveyors generally ignore inches and work in feet and decimals (tenths and hundredths). Since some tapes are also graduated in metric units, surveyors need to carefully check the units before using any tape.

4-3-4. Woven or Fiberglass Tapes

The flexibility of *woven* or *fiberglass tapes* makes them extremely effective under many conditions where steel tapes are impractical.

However, due to moisture and temperature, all woven tapes are liable to shrink or stretch and frequently should be compared with steel tapes to determine their accuracy and actual measuring length. Woven tapes, usually 50 to 150 ft long, are a combination of dacron fibers and coatings that have the stability of fiberglass and flexible strength of polyester. They feature high dielectric strength for safety on construction sites near high-tension circuits. One maker of woven nonmetallic tapes reinforces the first 9 in. of line with green plastic—green indicating nonconductivity. Short tapes are normally enclosed in a case. Some cloth tapes have fine metal strands of wire woven lengthwise into their fabric and are truly metallic tapes. They should not be used around electrical units.

A relatively new fiberglass tape, made of thousands of strands of glass fibers coated with polyvinyl chloride, is flexible, strong, nonconductive, and will not need a temperature correction (Figure 4-3). Under normal use, with tension lower than 5 lb, a correction to compensate for elasticity is seldom required. When greater tension is applied, some small corrections are needed; e.g., 0.02 in. per 3 ft at 11 lb and 0.04 in. per 3 ft at 22 lb. These tapes are available in lengths of 50 to 300 ft in a metal case or an on open-type reel. They are practical for locating details in mapping or checking reference distances.

4-3-5. Steel and Invar Tapes

The most common tapes used in surveying practice are steel ribbons (*band chains*) of constant cross section, usually varying in width from $\frac{1}{4}$ to $\frac{3}{8}$ in. and in thickness from 0.008 to 0.025 in. Normal lengths are 100, 200, or 300 ft. Metric tapes of the same thickness and width usually are 30, 50, 60, or 100 m long (Figure 4-4).

Graduations and identifying numbers are stamped either on soft (babbitt) metal previously embossed at the tape divisions or etched in the tape metal. Riveted, heavy-plated brass end-clips or rings provide a place to attach leather thongs, tension handles, or hooks to allow one person to make measurements unassisted (Figure 4-3). Most steel tapes come on reels and are stored on them. If a reel proves awkward, the tape can be removed from it and, when not in use, wound up into 5 ft-loops

to form a figure 8, and then "thrown" into a circle about 8 in. in diameter. The common 100-ft steel tape weighs from $1\frac{1}{2}$ to $2\frac{1}{2}$ lb, depending on thickness and width. If a steel tape gets wet, it should be wiped dry with a cloth and again with an oily cloth. Steel tapes are quite rugged, but if tightened with kinks in them, they break rather easily.

Tapes are marked in many ways to satisfy user desires. For example, some tapes have the last foot of each end divided into decimals, but others have an extra subdivided foot added to the zero end. Tapes with an extra foot are called *add tapes;* those without an extra foot are termed *cut tapes.* The latter type is becoming extinct because the subtraction required for each measurement is a possible source of error. Several variations are available such as divisions subdivided through their entire length. Others have zero points about $\frac{1}{2}$ ft from the end, or both end points at the outer edges of the end loops instead of being on a line itself. Before using them, surveyors must be completely familiar with the divisions and markings of all tapes.

Steel tapes expand or contract due to changes in temperature. Nickel-steel alloy tapes, known as *Invar* (which has a coefficient of thermal expansion about $\frac{1}{30}$ that of steel), *Lovar,* and *Minvar,* are used in high-precision surveying on geodetic base lines and as a standard of comparison for other working

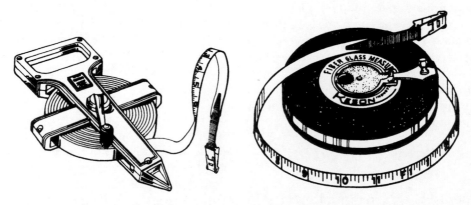

Figure 4-3. Fiberglass measuring tapes (in an open reel case and a metal case). (Courtesy of Keson Industries, Inc.)

 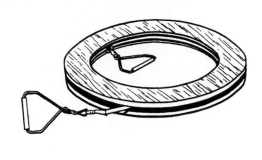

Figure 4-4.   Steel tape on open reel and Invar tape in wooden case. (Courtesy of Lufkin and the Cooper Group.)

tapes. They are almost always wound on an oak plywood reel (Figure 4-4). The nickel-steel alloy tapes are relatively insensitive to temperature, but the metal is soft, somewhat unstable, easily broken, and their cost is perhaps 10 times that of ordinary tapes.

Some tape equipment companies offer the option of a graduated thermometer scale, which corresponds to the tape contraction and expansion, as a variable terminal mark of the tape. The distance measured then depends on the prevailing temperature. Also, there are separate 6-in. wooden rules graduated with temperature corrections for 50- and 100-ft steel tapes.

Another handy device is the *topographic trailer tape* (Figure 4-5), used in conjunction with an *Abney hand level* to obtain horizontal distances by measuring along slopes. The tape, approximately $2\frac{1}{2}$ ch long, is basically 2 ch plus a distance on the trailer equal to the number of graduations indicated by the topographic arc reading. The total length thus measured on a slope equals a horizontal dis-

tance of 2 ch (132 ft). If the same procedure is carried out by reading the tape's reverse side, the distance is 1 ch (66 ft). It is a perfect tape for surveyors, foresters, and mappers to get slope corrections.

A device of historical importance in the United States is the *Gunter's chain* (Figure 4-6), which had extensive use in land surveying and the public land surveys during the 1700s and 1800s. This basic chain was 66 ft long and divided into 100 parts or links. Each link was equal to 0.66 ft or 7.92 in. and made of heavy

Figure 4-5.   Topographic trailer tape. (Courtesy of Keufel & Esser Co.)

Figure 4-6.   Gunter's chain. (Courtesy of Keuffel & Esser Co.)

wires connected by loops and three connecting rings with end handles. Intermediate tags of various design identified every tenth link. The handles had a length adjustment feature to compensate for chain lengthening due to wear on the 600 to 800 connecting wearing surfaces. Distances were recorded in chains and links or chains and decimals; e.g., 20 ch 12.4 lk or 20.124 ch. The Gunter's chain, 66 ft long, therefore, was $\frac{1}{80}$ of a mile (4 rods). Ten chains square is equal to 43,560 ft^2 or 1 acre —a very useful system.

Subsequently, an engineer's chain of 100 ft with 100 1-ft links was developed. Chains were replaced by development of the 100-ft steel tape. However, the chain unit remains a fundamental part of land-surveying practice, and a steel tape graduated in chain units is available. The term *chaining* continues to be used interchangeably with *taping*, even though a tape is used in the measurement.

4-3-6. Taping Accessories

To measure distances accurately with a tape, a number of so-called accessories are necessary or desirable. Although wooden stakes and tacks probably provide greater accuracy, *chaining pins* or *taping arrows* generally are used to mark the tape ends or intermediate points on the ground. They are also helpful as tallies to count the number of tape lengths in a given line. The pins, made of heavy steel wire, are usually 14 in. long, pointed at one end with a round loop at the other, and brightly painted with alternate red and white bands. A standard set consists of 11 pins on a steel ring or in a leather *quiver*, which can be attached to a surveyor's belt. After 1000 ft, the rear tapeperson is holding 10 pins if standard procedures are followed.

Since steel tapes are calibrated to measure correctly when under a definite tension, for precise measurements a *tension handle* or *spring balance* is attached to one end of the tape and a desired tension or pull applied. They are also used to counteract the effect of *sag* when measuring without a fully supported tape. The usual spring balance reads up to 30 lb in $\frac{1}{2}$-lb increments or 15 kg in $\frac{1}{4}$-kg calibrations. Without a tension handle, tapepersons have to estimate the proper pull.

Since tension must always be applied to a tape, especially at intermediate points, wrapping it around one hand is not recommended. Instead, a tape *clamping handle* should be used, permitting tension to be applied by a scissors-type grip, which does not slip or damage the tape. Without using a clamp, the tape could be slightly bent and kinked. Kinks, once introduced, cannot be entirely straightened out and create weak spots where future fracture will likely occur.

Plumb bobs are employed to place the tape directly over a point when the tape must be suspended above it. The commonly employed plumb bob is a fine quality, accurately centered brass bob that comes in various sizes, varying in weight from 6 to 18 oz, with a fine hardened steel point. All bobs provide for attaching a cord to the top; some carry replacement steel points in the bob. For convenience, plumb bobs are normally carried in a sewn leather sheath and sometimes fastened to a *gammon reel*, which provides instant rewind of the plumb-bob string, up and down adjustment, and an accurate sighting target (Figure 4-7).

Range poles are used to mark ground point locations and the direction of a line on which taping must proceed. They normally are 1 to $1\frac{1}{4}$ in. in diameter, either round, octagonal, or with deep corrugations to diffuse surface glare. Usually, they are 6 to 8 ft long or in multiple sections totaling perhaps 12 ft and equipped with a steel pointed shoe and shank. Made of wood, metal, or fiberglass, they are alternatively painted red and white in 1-ft, or 50-cm, sections and can be used for rough measurements. They are not javelins and should not be used to loosen rocks or stakes. Some equipment companies offer a short tripod to support range poles.

When striving for high accuracy, *pocket thermometers* are used to obtain air temperature and, it is hoped, provide an adequate estimate

Figure 4-7. Plumb bob fastened to a gammon reel. (Courtesy of Lietz Co.)

of the tape temperature during measurement. A common type is 5 in. long with a scale reading to 2°F (from −30°F to +120°F), carried in a protective metal case with pocket clip. There are also *tape thermometers* available that can be easily fastened directly to the tape and should ensure a more accurate tape temperature. Several years ago, a company offered a *tape temperature corrector*, a thermometer mounted on the standards of a transit. It had a scale in decimals of a foot for temperature corrections to be applied to a 100-ft steel tape. After reading the thermometer, the tapeperson can measure long or short as indicated.

Tape-repair kits contain sleeve splices, a combination hand punch and splicing tool, and eyelet rivets to facilitate field repairs of broken tapes. A simple and rapid method for emergency repairs is also available, consisting of sheet-metal sleeves coated with solder and flux to be fitted over the broken ends and hammered down tightly. Then, using heat from a match, the tape is securely fastened together. Repaired tapes should be used only on rough work.

To keep the tape ends at equal elevations when measuring over rough or sloping terrain, a simple *hand level* is used. It consists of a bubble mounted on a metal sighting tube and reflected by a 45° mirror or prism into the tube so the bubble can be observed at the same time as the terrain. If the slope angle or percentage of grade is desired, an Abney-type hand level or clinometer with a graduated vertical arc attached to its side is available (Figure 4-8).

Other taping accessories include detachable tape-end hooks with serrated face grips; tape-end leather thongs, tape rings or tape handles; and various size reels sometimes sold separately from the tapes.

4-3-7. Taping Procedures

Taping techniques and procedures vary because of differences that exist in kinds of tapes available, terrain traversed, project requirements, long-established practices, and personal preference. Regardless of the methods used, surveyors must be masters of their operations and fully understand the consequences of different techniques. Tapes are employed for two fundamental measurements: (1) to measure the distance between two existing physical points; or (2) to lay out and mark a distance called for in plans, specifications, land descriptions, or plats.

As pointed out earlier, a horizontal distance is the desired result. There are three basic methods of taping: (1) *horizontal taping*, (2) *slope taping*, and (3) *dynamic taping* (Figure 4-9). In horizontal taping, the tape is held

Figure 4-8. Abney-type hand level and clinometer. (Courtesy of Keuffel & Esser Co.)

horizontally and the end or intermediate points transferred to the ground or other surface. In slope taping, the tape length, or a portion thereof, is marked on the inclined surface supporting it, the slope determined, and the corresponding horizontal distance computed. In dynamic taping, a sloping taped distance from a transit or theodolite spindle is measured along with the vertical angle, and the horizontal distance subsequently calculated.

In all operations, careful attention must be given to proper tape support, proper alignment, use of correct tension, skill in handling plumb bobs and placement of pins, accounting for temperature variations, and any other factors that might affect accuracy of the result.

4-3-8. Taping over Level Ground

If we assume a level line void of tall grass and underbrush, the 100-ft tape can be laid on the ground, thus fully supported throughout. Under these conditions, the proper tension, generally applied, is about 10 lb. If the ground is irregular or contains obstructions, the tape may have to be suspended in catenary, plumb bobs used, and more tension applied.

To ensure a measurement is kept on line, range poles are placed at the terminal points and sometimes at intermediate ones, depending on terrain and visibility. The rear tapeperson keeps the head tapeperson on line by eye using hand signals or voice communication. If more accuracy is desired, a transit or theodolite is used to maintain alignment.

Taping can be accomplished by two people: a rear tapeperson and head tapeperson. The head tapeperson carries the zero end of the tape forward, aligns it, and after a 10-lb pull is applied, marks the distance by sticking a taping pin at right angles to the tape but at approximately a 45° angle with the ground. This permits, if necessary, more exact plumbing over the point with a transit or taping bob. As the measurement progresses, the taping pins are collected by the rear tapeperson and used as tallies to keep track of the full tape lengths. Careful attention should also be given to marking the zero and terminal points on a tape, which may be graduated in different ways. This is particularly important for fractional tape lengths.

When the end of a line is reached, the head tapeperson stops and the rear tapeperson moves up to the last pin set. The tape is moved until a full foot mark is opposite the pin and

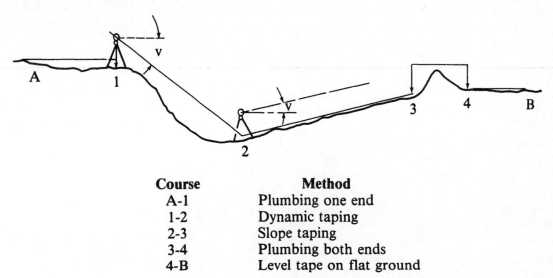

Course	Method
A-1	Plumbing one end
1-2	Dynamic taping
2-3	Slope taping
3-4	Plumbing both ends
4-B	Level tape on flat ground

Figure 4-9. Horizontal, slope, and dynamic taping. (Adapted from photo, courtesy of Lufkin and the Cooper group.)

the terminal point falls within the end-foot length, usually subdivided into tenths and perhaps hundredths. The rear tapeperson notes the foot mark number held, the head tapeperson's reading is subtracted, if using a tape having the first foot subdivided (a *subtract* or *cut tape*), or added if employing a tape with an additional subdivided foot (an *add tape*). Most taping errors or mistakes are made in measuring the fractional tape lengths, or in keeping a correct tally of full tape lengths (Figure 4-10).

4-3-9. Horizontal Taping on Sloping or Uneven Ground

In measuring on sloping or uneven ground, it is standard practice to hold the tape horizontal and use a plumb bob at one or both ends. More tension, usually 20 to 25 lb, must be applied in order to obtain a 100-ft horizontal distance. Plumbing the tape above 5 ft is difficult, and wind can make accurate work impossible. Bracing both forearms tightly against the body reduces swaying and jerking the tape.

When a full 100-ft measurement becomes impossible, it is divided into subsections of shorter lengths totaling 100 ft. Referred to as *breaking tape*, this is most important on steeper sloping ground. A hand level removes the guesswork from estimating whether the tape ends are at the same elevation. *Taping downhill is preferable*, since it is easier to set a forward pin with the plumb bob than to keep the rear end plumbed over a set pin with tension on the tape.

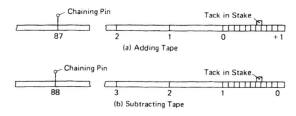

Figure 4-10. Reading partial tape lengths. (Courtesy of Brinker/Wolf.)

4-3-10. Slope Taping

Instead of breaking tape every few feet on a steep but uniform slope, it may be desirable to tape along the slope, determine the slope angle with an Abney level or transit or the elevation difference, and then compute the horizontal distance (see Section 4-3-12). Considerable practice is required for field personnel to do accurate taping in hilly or rolling terrain.

4-3-11. Dynamic Taping

Dynamic taping is similar to slope taping but is done from the transit horizontal axis (Figure 4-10) and is sometimes referred to as the *transit-and-floating-tape* technique. It is best accomplished with a fully graduated tape, sometimes on distances up to 200 or 300 ft. A transit is set up over the beginning mark and the first taping point established on line ahead. The head and rear tapepersons combine to measure and record the slope distance from the new taping point to the transit axis. A vertical angle is read from the transit and the horizontal distance calculated. The instrument is moved ahead to the first taping point, a second one set forward on line, and the procedure repeated. A person holding at midtape should use a tape clamp to avoid kinking or bending the tape. This method is surprisingly fast, accurate, and permits measurements to be made across typical obstacles.

By exercising strict attention to some basic concepts, surveyors can achieve a high degree of precision in taping. Several of the techniques noted may be necessary to accurately measure a distance of only several hundred feet. Another good practice is to measure each distance twice—forward and back—perhaps with a different tape.

4-3-12. Systematic Errors in Taping

The total error present in a measurement made in the field is equal to the algebraic sum of all random (accidental) errors and all systematic errors contained therein. Every attempt must be made to identify any systematic errors and apply corrections to nullify them. Systematic errors in taping are caused by the following conditions that may exist during measurement:

1. Tape not its nominal length.
2. Temperature of tape during measurement is not that at which it was standardized.
3. Tension (pull) applied to the tape is not the same as that when standardized.
4. Measurements were made along slopes instead of in a horizontal plane.
5. Tape was not fully supported throughout but in catenary (suspended with sag).

To achieve a prescribed relative accuracy in taping measurements, the raw observed data have to be corrected to get the true or best length, since measurements are seldom made under ideal conditions; adjustments must be made for pull, temperature, and mode of support.

Incorrect Length of Tape

Although tapes are precisely manufactured, they become worn, kinked, stretched, or improperly spliced after breaks and should be checked periodically against a standard. This can be accomplished in several ways. Most surveying offices either have a special tape to be used only for checking or standardizing other tapes, or maintain permanent marks 100.00 ft apart to check working tapes.

For higher-precision work or to maintain a standard tape, the NBS will, for a fee, issue a certificate for a submitted tape, giving its length to the nearest 0.001 ft at 68°F (20°C) for any specific tension and support conditions. Also, various other governmental agencies maintain the capability to standardize tapes as a service to the public. The NGS has established a number of base lines around the country where tapes and EDMIs can be calibrated by surveyors.

Applying corrections caused by incorrect length of tape is a simple matter but should be

carefully considered. Assume that the actual length of a 100-ft tape is 99.98 ft and a distance between fixed points measured with this tape was recorded as 1322.78 ft. Since each full tapelength was short by 0.02 ft, the correct length is

$$(100.00 - 0.02) \times 13.2278 = 1322.78 - 0.26$$
$$= 1322.52 \text{ ft}$$

However, if a certain distance is to be established, such as in staking out, with a tape known to be too short, the reverse is true, so add the 0.26 ft—i.e., lay out a length of 1323.04 ft. In approaching this problem of incorrect length of tape, Table 4-4 is useful in making the corrections.

Correction for Temperature

The coefficient of thermal expansion of the steel used in common tapes is 0.00000645 per unit length per 1°F. If the length of a tape is known at some standardized temperature, T_0 (NBS uses 68°F or 20°C), the correction can be obtained from

$$C_t = 0.00000645 \, (T_f - T_o) \, L \qquad (4\text{-}1)$$

As an example, if a steel tape known to be 100.00 ft long at a standardized temperature of 68°F is to be used in the field at 43°F (T_f), the actual length of the tape is 0.0097 ft short. This causes a change in length of about 0.01 ft for each change in temperature of 15°F, which is frequently used in approximate calculations. An astounding example is the change in temperature of 75°F between measurements in the summer and winter in some areas. The length of a 100-ft steel tape changes 0.05 ft due to this temperature difference, equivalent to a discrepancy of 2.6 ft in a mile. Errors in taping due to temperature are frequently overlooked

Table 4-4.   Corrections for incorrect tape length

When Tape Is	To Lay Out a Distance	To Measure a Line Between Fixed Points
Too Long	Subtract	Add
Too Short	Add	Subtract

by inexperienced surveyors, but obviously they cannot be disregarded. In SI units the coefficient of thermal expansion of steel is 0.0000116 per unit length per degree C. The corrections can be plus or minus in sign.

Tape temperatures are difficult to measure, particularly on partly cloudy summer days. *Tape thermometers* clamped on near the ends do not contribute to their sag, and so are most reliable. Some tapes have a terminal graduation that varies with temperature.

Corrections for Tension or Pull

Steel tapes are standardized at some specified tension and, being elastic, change length due to variations in the tension applied. An ordinary 100-ft steel tape stretches only about 0.01 ft for an increase in tension of 15 lb. If *spring balances* are used to maintain the prescribed pull, errors caused by tension variations are negligible. Without a spring balance, the tension applied usually varies either above or below the standard and can be considered an accidental error and disregarded in all but precise measurements. *Inexperienced tapepersons are likely to apply tension lower than the standardized figure.*

The formula for tension correction is derived using the modulus of elasticity E, which is the ratio of unit stress to unit strain. The total correction for elongation C_P of a tape length L is calculated from the expression

$$C_P = \frac{(P_f - P_o)L}{AE} \qquad (4\text{-}2)$$

in which P_f is the applied field pull in pounds, P_o the standard tension, A the tape cross-sectional area, and E the modulus of elasticity of the tape—taken as 29,000,000 lb per square in. for steel. The cross-sectional area can be calculated from the tape length, its weight, and the specific weight of steel.

Corrections for Slope

In reducing slope measurements to their horizontal lengths—required in slope and dynamic taping—a correction must be applied to the measured distance equal to the difference between the hypotenuse s and side d of a right triangle having its vertical side equal to h.

When the difference in elevation of the two points h is measured, the correction for slope C_h can be obtained from the formula

$$C_h = \frac{h^2}{2s} + \frac{h^4}{8s^3} \qquad (4\text{-}3)$$

Usually, the first term in the equation is sufficient. If the vertical angle α is measured along with the slope distance, the following formula applies:

$$C_h = s(1 - \cos \alpha) - s \text{ vers } \alpha \qquad (4\text{-}4)$$

This correction has a negative sign since the hypotenuse is always longer than the other side.

Corrections for Sag

A steel tape suspended and supported only at the end points takes the form of a *catenary curve*. Obviously, the horizontal distance between its ends will be shorter than for a tape fully supported throughout its entire length. The difference between the curve and chord lengths is the sag correction and always has a negative sign. Sag is related to weight per unit length and the applied tension. For a tape supported at its midpoint, the total effect of sag in the two spans is considerably smaller. More intermediate supports further reduce the sag to zero when fully supported.

The following formula is used to compute the sag correction C_s:

$$C_s = \frac{w^2 L^3}{24 P^2} = \frac{W^2 L}{24 P^2} \qquad (4\text{-}5)$$

where w is weight of tape in pounds per foot, L the unsupported length between supports, $W = wL$ the total weight of tape between supports, and P the total applied tension in pounds.

These formulas show that greater tension is required for a tape not supported throughout. In fact, there is a theoretical normal tension for each tape that increases its length exactly by the shortening due to sag. This pull can be determined practically using an actual mock-up or by theoretically employing the following formula in a trial-and-error method:

$$P_f = \frac{0.204W\sqrt{AE}}{\sqrt{P_f - P_o}} \qquad (4\text{-}6)$$

Sources of Error and Mistakes in Taping

In addition to the major sources of systematic error in taping, three other conditions must also be carefully monitored: (1) faulty alignment, which has an adverse effect on accuracy—thus, a tape held 1.5 ft off line in 100 ft causes an error of 0.01 ft in distance; (2) taping pins must be set in proper position; and (3) sag is difficult to evaluate when a strong wind blows on an unsupported tape.

Several possible common mistakes made in taping and recording cannot be tolerated and include (1) faulty tallying, (2) misreading the tape graduations, (3) improper plumbing, (4) reversing or misunderstanding the calls in recording numbers, or dropping or adding one foot, and (5) mistaking the end mark.

Taping continues to be a basic operation in many aspects of surveying because it is fast and easy for measuring relatively short distances.

4-4. OPTICAL DISTANCE-MEASUREMENT METHODS

4-4-1. General Introduction

The alternative to using ground or catenary taping or electromagnetic methods of distance measurement is to use optical methods. Although some surveyors only relate optical methods to conventional stadia procedure, there is a large family of instruments, methods, and procedures generally classified as optical distance measurement.

Several of these methods were popular in the past. Now, with the advent of electromag-

netic distance technology and its versatility, most indirect optical methods—with the possible exception of stadia tacheometry—have been relegated to the status of little-used substitutes. They have, nevertheless, played an important role in attempts to develop distance-measuring equipment that would be rapid although not as accurate. Some had only limited practical application, others were very costly, a few quite complicated, and a number had great potential but surveyors did not recognize their capabilities.

Electronic distance instruments are generally rather expensive and not useful over the range of distances used in optical measurement. Therefore, tapes, EDMIs, and optical instruments complement rather than displace one another.

The term *tacheometry* or *tachymetry* means "rapid" measurement. Actually, any measurement made rapidly could be considered tacheometric but general practice is to include only optical measurements by stadia, subtense bars, etc. Thus, exceedingly fast EDMIs, covered in Chapter 5, are not included in this term.

Only general coverage is attempted in this text; however, two British books published in 1970 are completely dedicated, in great detail, to optical distance measurement. Smith groups various optical devices into the following categories[1]:

1. Instruments on the rangefinder principle
   (a) Fixed-base
   (b) Fixed-angle
2. Theodolites
   (a) Conventional stadia tacheometry
   (b) Tangent tacheometry
   (c) Wedge attachments to theodolites
3. Self-reducing tacheometers
   (a) Using vertical staves-diagram tacheometers
   (b) Using horizontal staves-double-image tacheometers
4. Subtense bar
5. Planetable alidades
6. Miscellaneous items

As will be apparent later in a discussion of principles, optical distance measurement really only combines a fixed quantity with a variable one. All optical instruments involve an angle and a base, which are either fixed or variable; also, the base may be horizontal or vertical. Smith summarizes the possibilities in a convenient form (Table 4-5).

4-4-2. Stadia Tacheometry

Stadia tacheometry is the commonly known procedure that utilizes two supplementary horizontal (stadia) lines placed at equal distances above and below the central horizontal line in an instrument's telescope. Usually, they are short lines to differentiate them from the longer main horizontal one. A graduated vertical rod is sighted and the intercept between the stadia lines read. Modern internal-focusing instruments have a fixed distance between the stadia lines so when the telescope is horizontal and the rod vertical, distance D from the instrument center to the rod equals 100 times the stadia intercept S as in Figure 4-11.

$$D = 100S \qquad (4\text{-}7)$$

In some instruments, the stadia multiplier k is 333.

It is not always possible to keep the telescope horizontal; more commonly, the line of sight is inclined. The stadia intercept S then must be multiplied by the cosine of the verti-

cal angle α to make AB perpendicular to D (see Figure 4-12).

$$D = 100AB \qquad (4\text{-}8)$$
$$D = 100S \cos \alpha \qquad (4\text{-}9)$$

and

$$H = D \cos \alpha = S\,100\,(\cos^2\alpha) \qquad (4\text{-}10)$$
$$V = D \sin \alpha = S\,100\,(\sin \alpha \cos \alpha)$$
$$= S\,100\,(\tfrac{1}{2} \sin 2\alpha) \qquad (4\text{-}11)$$

These formulas are referred to as the *inclined-sight stadia formulas* and form the basis for several stadia reduction (computing) devices.

It should be noted that some older instruments have external-focusing telescopes and the distance X (see Figure 4-13) is proportional to the stadia intercepts—from the external principal focus of the lens. To obtain D, the distances $f + c$ must be added. Their sum is usually about 1 ft and must be applied when using the older external-focusing instruments. It is almost zero and is ignored for the newer internal-focusing type.

4-4-3. Stadia Reduction Devices

The inclined-sight stadia Equations (4-10) and (4-11) must be solved many times when the *stadia method* is used to locate features in transit-stadia and planetable mapping. The desired vertical and horizontal distances V and

Table 4-5. Optical distance-measuring devices

Section	Form	Base	Angle	Base	Resulting Distance
1	Fixed-base rangefinder	Fixed	Variable	Horiz.	Slope
2	Fixed-angle rangefinder	Variable	Fixed	Horiz.	Slope/Horiz.
3	Stadia tacheometry	Variable	Fixed	Vertical	Slope
4	Tangent tacheometry	Fixed	Variable	Vertical	Slope
5	Wedge attachment	Variable	Fixed	Horiz.	Slope/Horiz.
6	Diagram tacheometers	Variable	Fixed	Vertical	Horiz.
7	Double-image tacheometers	Variable	Fixed	Horiz.	Horiz.
8	Subtense bar	Fixed	Variable	Horiz.	Horiz.
9	Planetable alidades	Variable	Fixed	Vertical	Horiz.

Source: Courtesy of J. R. Smith. ''Optical Distance Measurement,'' Granada Publishing Ltd., England, 1970.

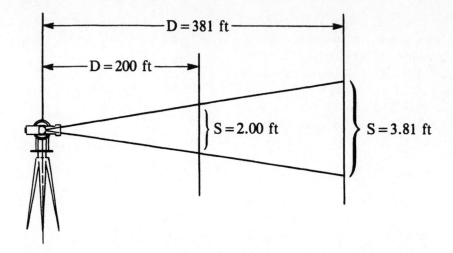

Figure 4-11.   Stadia with level sight. (Adapted from photo, courtesy of Keuffel & Esser Co.)

H come from the equations. Although not true with modern hand-held programmable calculators, using the formulas in direct application was slow and tedious. Therefore, many reduction devices were designed to facilitate the computing. These include (1) stadia tables (Tables 4-6 and 4-7), (2) diagrams, (3) slide rules, (4) *stadia circles*, (5) *Beaman arcs*, (6) self-reducing curved stadia lines, and (7) cam-operated movable reticle lines. Instruments using curved stadia lines and cams will be discussed later under diagram tacheometers.

Stadia tables were included in surveying textbooks for years. They assume a stadia in-

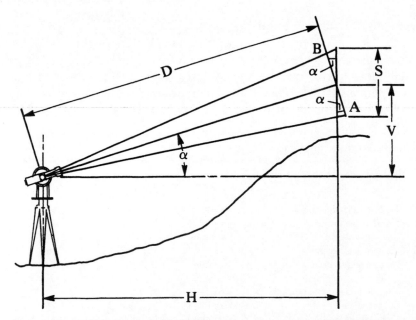

Figure 4-12.   Inclined-sight stadia. (Adapted from photo, courtesy of Keuffel & Esser Co.)

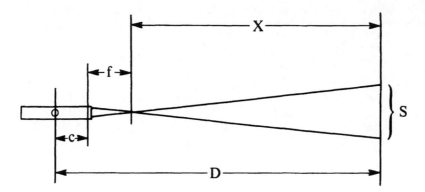

Figure 4-13. Geometry of external-focusing telescope. (Adapted from photo, courtesy of Keuffel & Esser Co.)

Table 4-6. Horizontal corrections for stadia intercept 1.00 ft

Vert. Angle	Horiz. Cor. for 1.00 ft	Vert. Angle	Horiz. Cor. for 1.00 ft	Vert. Angle	Horiz. Cor. for 1.00 ft
0°00′		5°36′		8°02′	
	0.0 ft		1.0 ft		2.0 ft
1°17′		5°53′		8°14′	
	0.1 ft		1.1 ft		2.1 ft
2°13′		6°09′		8°26′	
	0.2 ft		1.2 ft		2.2 ft
2°52′		6°25′		8°38′	
	0.3 ft		1.3 ft		2.3 ft
3°23′		6°40′		8°49′	
	0.4 ft		1.4 ft		2.4 ft
3°51′		6°55′		9°00′	
	0.5 ft		1.5 ft		2.5 ft
4°15′		7°09′		9°11′	
	0.6 ft		1.6 ft		2.6 ft
4°37′		7°23′		9°22′	
	0.7 ft		1.7 ft		2.7 ft
4°58′		7°36′		9°33′	
	0.8 ft		1.8 ft		2.8 ft
5°17′		7°49′		9°43′	
	0.9 ft		1.9 ft		2.9 ft
5°36′		8°02′		9°53′	
					3.0 ft
				10°03′	

Results are correct to the nearest foot at 100 ft and to the nearest 1/10 ft at 100 ft, etc.

With a slide rule, multiply the stadia intercept by the tabular value and subtract the product from the horizontal distance.

Example: vertical angle, 4°22′; stadia intercept, 3.58 ft.

$$\text{Corrected horiz. dist.} = 358 - (3.58 \times 0.6) = 356 \text{ ft}$$

Table 4-7 gives the vertical heights for a stadia intercept of 1.00 ft. With a slide rule, multiply the stadia intercept by the tabular value.

Example: vertical angle, 4°22′; stadia intercept, 3.58 ft.

$$\text{Vertical height} = 3.58 \times 7.59 = 27.2 \text{ ft}$$

Table 4-7. Vertical heights for stadia intercept 1.00 ft

Min.	0°	1°	2°	3°	4°	5°	6°	7°	8°	9°
0	0.00	1.74	3.49	5.23	6.96	8.68	10.40	12.10	13.78	15.45
2	0.06	1.80	3.55	5.28	7.02	8.74	10.45	12.15	13.84	15.51
4	0.12	1.86	3.60	5.34	7.07	8.80	10.51	12.21	13.89	15.56
6	0.17	1.92	3.66	5.40	7.13	8.85	10.57	12.27	13.95	15.62
8	0.23	1.98	3.72	5.46	7.19	8.91	10.62	12.32	14.01	15.67
10	0.29	2.04	3.78	5.52	7.25	8.97	10.68	12.38	14.06	15.73
12	0.35	2.09	3.84	5.57	7.30	9.03	10.74	12.43	14.12	15.78
14	0.41	2.15	3.89	5.63	7.36	9.08	10.79	12.49	14.17	15.84
16	0.47	2.21	3.95	5.69	7.42	9.14	10.85	12.55	14.23	15.89
18	0.52	2.27	4.01	5.75	7.48	9.20	10.91	12.60	14.28	15.95
20	0.58	2.33	4.07	5.80	7.53	9.25	10.96	12.66	14.34	16.00
22	0.64	2.38	4.13	5.86	7.59	9.31	11.02	12.72	14.40	16.06
24	0.70	2.44	4.18	5.92	7.65	9.37	11.08	12.77	14.45	16.11
26	0.76	2.50	4.24	5.98	7.71	9.43	11.13	12.83	14.51	16.17
28	0.81	2.56	4.30	6.04	7.76	9.48	11.19	12.88	14.56	16.22
30	0.87	2.62	4.36	6.09	7.82	9.54	11.25	12.94	14.62	16.28
32	0.93	2.67	4.42	6.15	7.88	9.60	11.30	13.00	14.67	16.33
34	0.99	2.73	4.47	6.21	7.94	9.65	11.36	13.05	14.73	16.39
36	1.05	2.79	4.53	6.27	7.99	9.71	11.42	13.11	14.79	16.44
38	1.11	2.85	4.59	6.32	8.05	9.77	11.47	13.17	14.84	16.50
40	1.16	2.91	4.65	6.38	8.11	9.83	11.53	13.22	14.90	16.55
42	1.22	2.97	4.71	6.44	8.17	9.88	11.59	13.28	14.95	16.61
44	1.28	3.02	4.76	6.50	8.22	9.94	11.64	13.33	15.01	16.66
46	1.34	3.08	4.82	6.56	8.28	10.00	11.70	13.39	15.06	16.72
48	1.40	3.14	4.88	6.61	8.34	10.05	11.76	13.45	15.12	16.77
50	1.45	3.20	4.94	6.67	8.40	10.11	11.81	13.50	15.17	16.83
52	1.51	3.26	4.99	6.73	8.45	10.17	11.87	13.56	15.23	16.88
54	1.57	3.31	5.05	6.79	8.51	10.22	11.93	13.61	15.28	16.94
56	1.63	3.37	5.11	6.84	8.57	10.28	11.98	13.67	15.34	16.99
58	1.69	3.43	5.17	6.90	8.63	10.34	12.04	13.73	15.40	17.05
60	1.74	3.49	5.23	6.96	8.68	10.40	12.10	13.78	15.45	17.10

tercept of 1 ft and list values in parentheses in Equations (4-10) and (4-11) for various values of vertical angle α. To compute H and V, appropriate tabular figures are obtained for α and multiplied by S. Some tables list corrections that are multiplied by S and subtracted from $100S$ to get values of H.

The sole purpose of the 10-in. Kissam slide rule or Cox (circular) stadia rule is to obtain H, V, and/or horizontal corrections by using appropriate S and α values.

A stadia circle consists of two special H and V scales attached to planetable alidades and some transits, on request, which permits the observer to read the same values obtainable from stadia tables instead of the vertical angle. These numbers must then be multiplied by S. Since spacing of the graduations is irregular, a vernier is not needed. A stadia arc is essential in planetable mapping.

To avoid minus readings, the V-index reading on level sights is 50 on most arcs, so 50 must be subtracted from each reading to obtain the true multiplier. The H multiplier normally is near 100. The Beaman arc is the same as a stadia circle, except the horizontal scale H yields a percentage correction that must be subtracted from $100S$. The vertical scale V is the same as on a stadia circle. All the reduction devices were developed to ease the burden and make computing the great volume of stadia side-shots simpler. Self-reducing tacheometers having curved stadia lines or cams are also based on the inclined-sight stadia formulas.

4-4-4. Accuracy of Stadia Measurements

The ordinary level rod graduated to hundredths of a foot can be used in stadia work for maximum sight lengths of 300 to 400 ft. For longer sights, special stadia rods with graduations to 0.1 ft, yd, or m are easier to read and more satisfactory. Observational technique is an individually developed procedure to ensure efficient data measurement, recording, and rod movement.

For most large-scale mapping and contour intervals, horizontal distances to the nearest foot and elevations nearest 0.1 ft suffice, so stadia easily meets this requirement. Stadia-read distances are normally in the 1/300 to 1/500 range, although by carefully calibrating the stadia interval factor F/I, using a target attachment, and making repeated measurements, accuracies as high as 1/1000 to 1/2000 have been recorded. For measuring longer distances, where the full stadia interval does

not fall on a typical rod, a half- or quarter-interval can be read and used to obtain distance. This is particularly useful for small-scale mapping with a planetable.

4-4-5. Diagram Tacheometers

The diagram tacheometer (or self-reducing tacheometer) utilizes a diagram of *reduction curves*, projected into the field of view, and allows horizontal distance H and height difference V to be read directly as intercepts on a graduated vertical rod (Figures 4-14 and 4-15). These reduction curves, based on the inclined-sight stadia formulas, on a diagram plate rotate when the telescope is tilted by turning a series of planet gears. The horizontal multiplication constant is always 100, but there are four height-reading curves, depending on the vertical angle involved. Multiplication factors of 10, 20, 50, and 100 are used.

Rod intercepts are read separately from a base zero-curve line. The instrument is a standard theodolite with this added facility for making stadia reductions. A standard leveling rod can be used, but a special tacheometric staff with a double wedge-shaped mark at 1 m (or 4 ft) above the base allows the zero curve to be set accurately. A staff with an extendable

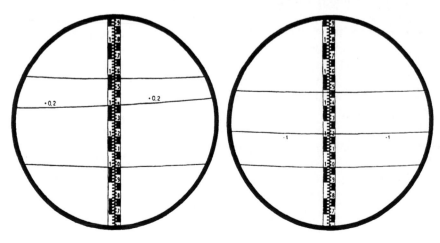

Distance: 57.2 m
Diff. of elevation: $+0.2 \cdot 40.1 = +8.02$ m

Distance: 48.5 m
Diff. of elevation: $-1 \cdot 21.7 = -21.7$ m

Figure 4-14. Reduction tacheometer with vertical staff. (Courtesy of Wild Heerbrugg Instruments, Inc.)

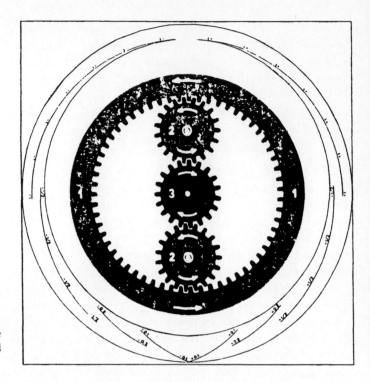

Figure 4-15. Planet gears and diagram plate with reduction curves. (Courtesy of Wild Heerbrugg Instruments, Inc.)

leg is available for sighting at a definite height of instrument.

Another self-reducing tacheometer similar to the diagram instrument has a fixed and movable reticle along with a cam (Figure 4-16). In this instrument, the lines are straight but the cam is based on the inclined-sight stadia formulas. This mechanical reduction system has a fixed reticle containing a vertical and horizontal line. A second horizontal line is on a movable reticle. As the telescope inclination is changed, a cam geared to the horizontal axis raises or lowers the movable reticle, so a rod intercept between two horizontal lines yields the horizontal distance. After switching with a knurled ring (to change the cam configuration), the rod intercept supplies a vertical distance. Again, the instrument is a standard theodolite with an added reduction mechanism.

4-4-6.   Planetable Alidades

Planetable alidades invariably contain some type of stadia reduction system. As examples, the Keuffel & Esser standard alidade uses a stadia circle with H and V scales; the newer self-indexing model has a separate optical scale-reading eyepiece to view the three scales, including the typical elevation angle (Figure 4-17). Berger and Gurley alidades use the closely related Beaman arc. Modern Swiss-made Wild and Kern alidades employ the diagram tacheometric self-reducing method with curved-reduction lines. A horizontal multiplier of 100 and three multipliers for difference in elevation (20, 50, and 100) are appropriately marked on the reticle and read directly.

4-4-7.   Optical Wedge Attachments

A distance-measuring wedge is a simple theodolite accessory. When attached to the telescope objective housing, a slope distance reading is taken on a horizontal staff set up at the target point. A counterweight screwed tightly to the eyepiece end holds the telescope in balance. Besides the fixed-angle wedge, a parallel-plate micrometer functions similarly to that used in precise leveling, except it measures in a horizontal rather than vertical plane.

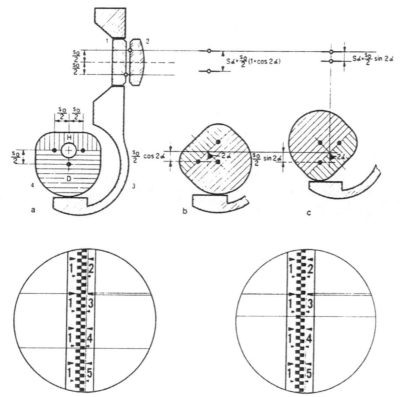

Horizontal distance reading 15.6 m
The switching ring is set at D

Height difference reading 6.4 m
The switching ring is set at Δ H
Sighting height on the rod 1.30 m

Figure 4-16. Mechanical reduction system employing a cam-operated movable reticule with straight lines. (Courtesy of Kern Instruments, Inc.)

Of the two optical elements, the wedge supplies a fixed-angle δ (normally 34'22.6") and the parallel-plate micrometer aids in resolution of the fine-reading part of the variable base b (Figure 4-18). Therefore, a slope distance and the required horizontal distance are obtained by using a measured vertical angle for reduction.

In optical distance measurement, a distance D is derived from the parallactic angle intercepting a staff length d. The glass wedge covers only the middle section of the objective lens. Therefore, it deflects only those rays that actually pass through the wedge, while the others go straight through. When sighting a horizontal staff, two images are seen displaced

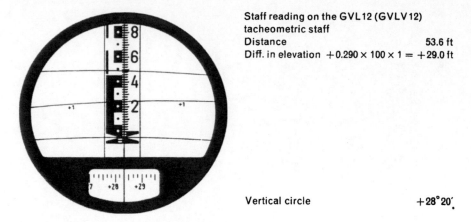

Staff reading on the GVL12 (GVLV12)
tacheometric staff
Distance   53.6 ft
Diff. in elevation +0.290 × 100 × 1 = +29.0 ft

Vertical circle   +28°20′.

Figure 4-17. Modern planetable alidade with reduction curves. (Courtesy of Wild Heerbrugg Instruments, Inc.)

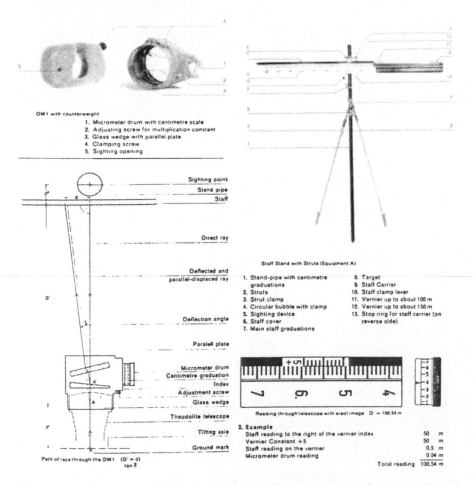

Figure 4-18. Distance-measuring wedge attachment with horizontal staff. (Courtesy of Wild Heerbrugg Instruments, Inc.)

in relation to each other by a similar d. This variable spread represents $1/100$ of the slope distance D (multiplication constant $= 100$).

The lower half of a horizontal staff consists of the main graduations (lines with 1-cm intervals); the upper half has two verniers with graduation intervals of 0.9 cm. The inner vernier is used to measure distances from 10 to 100 m and the outer one (with the distinguishing sign of $+5$) for distances from 60 to 150 m. The $+5$ indicates an added constant of 50 m. The accuracy obtainable with the wedge attachment falls in the range of $1/5000$ to $1/10,000$—much better than with stadia tacheometry. A wedge attachment is employed mainly in measuring traverse sides. The time involved setting up the horizontal staff rules out its use in multiple side-shot detailing.

4-4-8. Double-Image Tacheometers

These are similar in operation to wedges except the effect of a vertical angle is automatically eliminated. A "fixed" angle δ is obtained by an optical device consisting of two rotating wedges, which together with a micrometer system (using a rhombic prism) reduces slope distance to the horizontal ($D \cos \alpha$). The micrometer is used only to resolve the fine reading part of the distance. The same horizontal staff is employed with the double-image tacheometer.

Some double-image tacheometers are equipped with a device to set the rotating wedges so their effect is eliminated when the telescope is horizontal, and the deflection increases proportionally to sin α for telescope inclinations. Then the staff reading gives $D \sin \alpha$ corresponding to the difference in elevation between the instrument and staff. Maximum range is about 150 m and accuracies are comparable to the wedge attachments. Their principal advantage is the automatic reduction of readings.

4-4-9. Subtense Tacheometry

The subtense method essentially consists of accurately measuring the variable angle subtended by a fixed horizontal distance or base (Figure 4-19). Generally, the standard subtense bar is 2 m long and consists of a metal tube with a target at each end controlled by an Invar wire, so it is unaffected by temperature changes. The bar is mounted horizontally on a tripod and its small triangular targets can be internally illuminated for night operation. A small optical sight ensures that the bar is perpendicular to the sight line.

 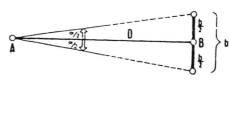

Figure 4-19. Measuring with a subtense bar. (Courtesy of Wild Heerbrugg Instruments, Inc.)

The distance D is derived from the following formula:

$$D = b/2 \cot \alpha/2 \qquad (4\text{-}12)$$

Since base length b is exactly 2 m, distance D totally depends on the horizontal angle measurement α. To obtain accuracies of 1/5000, sights must be limited to 400 to 500 ft and the angle measured to $+1''$ of arc with a precise 1-sec theodolite. Several sets of angles should be read to ensure this $1''$ accuracy. Tables furnished by subtense-bar makers reduce or eliminate computations.

The method's principal advantage over taping lies in its ability to measure over rough terrain, across gullies and wide streams. Furthermore, no slope correction is necessary, since the horizontal angle subtended by the bar is independent of the inclination of the sight line; therefore, the horizontal distance is obtained directly. Since introduction of electronic distance-measuring instruments, the subtense method is no longer able to compete.

4-4-10. Tangential Tacheometry

A variation on stadia tacheometry is called tangent tacheometry, sometimes described as vertical subtense. Two targets are set a known distance apart on a vertical rod. Then by measuring the subtended angle *and* vertical angle, the required horizontal distance and difference in height can be computed.

4-4-11. Rangefinders

Rangefinders are fixed-base or fixed-angle types (Figure 4-20). From a fixed-base b and fixed angle at A, the system's optics are manipulated to vary angle at $B(\alpha)$ until the partial images of Y seen through A and B are coincident in the instrument's field of view. Distance $AY(L)$ is then a direct function of variable α and constant b with angle A normally arranged as a right angle.

In fixed-angle rangefinders, angle B of α is fixed and distance b is varied by sliding a prism unit at B along the bar until the two partial images of Y seen through A and B are coincident. Both fixed-base and fixed-angle rangefinders provide the slope distance, but the vertical angle must be used to get a horizontal equivalent.

Wild makes a small hand-held rangefinder, TMO, with a fixed 25-cm base, and a larger tripod-mounted coincidence rangefinder, TM2, with a long 80-cm base. Depending upon range, measuring accuracies range from 1/10 to 1/150. The TM2 has a measuring range from 300 to 5000 m. Their distinct advantage over other methods is that inaccessible dis-

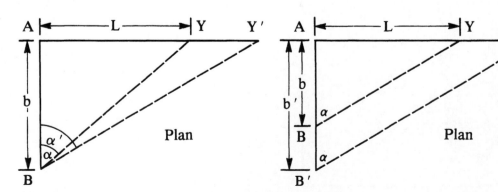

Figure 4-20.   Geometry of fixed-base and fixed-angle rangefinders. (Courtesy of R. J. Smith, "Optical Distance Measurement," Granada Publications Ltd., England, 1970.)

tances can be measured and a rodperson is not required.

REFERENCES

ALLAN, A. L. 1977 An alidade for use with electro-optical distance measurement. *Survey Review* (July).

American Congress on Surveying and Mapping 1978. *Metric practice guide for surveying and mapping*, 12 pages, A.C.S.M., 210 Little Falls Street, Falls Church, Virginia 22046.

ASPLUND, E. C. 1971 A review of self reducing, double image tacheometers. *Survey Review* (Jan.).

BAUER, S. A. 1949 The use of geodetic control in surveying practice. *Surveying and Mapping 9*(3), 187–191. Falls Church, VA.

BERRY, D. W. 1954 Tacheometry with the Ewing stadialtimeter. *Surveying and Mapping 14*, 479–483. Falls Church, VA.

BIESHEUVEL, H. 1955 The double image telemeter. *Empire Survey Review 13*(98), 160–164.

BOLTER, W. H. 1956 The self-indexing alidade. *Surveying and Mapping 16*, 157–163. Falls Church, VA.

BRINKER, D. M. 1971 Modern taping practice versus electronic distance measuring. *ACSM Fall Convention Papers*, pp. 354–356. Falls Church, VA.

COLCORD, J. E. 1971 Tacheometry in survey engineering, *Journal of the Surveying and Mapping Division*, A.S.C.E., *97*(SU1), 39–52. New York, NY.

COLCORD, J. E., and CHICK, F. H. 1968 Slope taping. *ASCE Journal of the Surveying and Mapping Division 94*(SU2), 137–148. New York, NY.

CURTIS, K. S. 1976 *Optical distance measurement.* Indiana Society of Professional Land Surveyors (ISPLS) Surveying Publication Series No. 7. Indianapolis, IN.

DEUMLICH, F. 1982 *Surveying instruments.* (transl.). Berlin/New York: de Gruyter.

DRACUP, J. F., FRONCZEK, C. J., and TOMLINSON, R. W. 1977 (Revised by Spofford, P. R. 1982). Establishment of calibration base lines. *NOAA Technical Memorandum NOS NGS-8*, NOAA, (August). Rockville, MD.

EMERY, S. A. 1954 Nomography applied to land surveying. *Surveying and Mapping 14*(1), 59–63. Falls Church, VA.

FRONCZEK, C. J. 1977 (Reprinted with corrections 1980). Use of calibration base lines. *NOAA Technical Memorandum*, NOS NGS-10, NOAA, (Dec.). Rockville, MD.

GOLLEY, B. J., and SNEDDON, J. 1974 Investigation of dynamic taping. *Journal of the Surveying and Mapping Division, ASCE 100*(SU2), 115–122.

HARRINGTON, E. L. 1955 The stepping method of stadia. *Surveying and Mapping 15*(4), 460. Falls Church, VA.

HIDNERT, P., and KIRBY, R. K. 1953 New method for determining linear thermal expansion of invar geodetic surveying tapes. *J. Research NBS 50* 179, RP2407.

HODGES, D. J., and GREENWOOD, J. B. 1971 *Optical distance measurement.* Butterworth and Co., England.

Instruction Manual, *Self-reducing engineer's tacheometer*, KI-RA, Kern & Co., Ltd., Aaru, Switzerland.

JUDSON, L. V. 1956 *Calibration of line standards of length and measuring tapes at the national bureau of standards;* National Bureau of Standards Circular No. 572.

KENNGOTT, R. 1980 EDM calibration baseline at Westchester Community College, Valhalla, New York, *ACSM Paper Proceedings*, Fall Technical Meeting, Niagara Falls, New York, (Oct.), pp. LS-2-F-1 to 8.

KISSAM, P. 1961 Stadia will cut your costs. *Engineering Graphics 1*(5), 6.

KISSAM, P., et al. 1940 *Horizontal control surveys to supplement the fundamental net;* A.S.C.E. Manual No. 20. New York, NY.

LOVE, J., and SPENCER, J. 1981 Explanation and availability of calibration base line data. *ACSM Paper Proceedings*, Fall Technical Meeting, San Francisco, California, (Sept.), pp. 209–217.

Lufkin Rule Company. 1972. *Taping techniques for engineers and surveyors*, Apex, North Carolina.

MALTBY, C. S. 1954 Alidade development in the geological survey. *Surveying and Mapping 14*, 173–184.

MELTON, G. D. 1969 Devices for distance measurement—The Wild RDS. *Proceedings of the Second Annual Land Surveyors' Conference*, University of Kentucky, pp. 54–60.

MILLER, C. H., and ODUM, J. K. 1983 Calculator program for reducing alidade or transit stadia traverse data. *Surveying and Mapping 43*(4), 393–398. Falls Church, VA.

MOFFITT, F. H. 1975 Calibration of EDM's for precision measurement. *Surveying and Mapping 35*(2), 147–154. Falls Church, VA.

MOFFITT, F. H. 1958 The tape comparator at the University of California. *Surveying and Mapping 18*(4), 441–444. Falls Church, VA.

MORSE, E. D. 1956 The practical application of close tolerance in survey work. *Report of 5th Annual Texas Surveyors Association Short Course*, pp. 59–69.

MORSE, E. D. 1951 Concerning reliable survey measurements. *Surveying and Mapping 11*(1), 14–21. Falls Church, VA.

MUSSETTER, W. 1953 Stadia characteristics of the internal focusing telescope. *Surveying and Mapping 13*(1), 15–19. Falls Church, VA.

MUSSETTER, W. 1956 Tacheometric surveying. *Surveying and Mapping 16*(Part I), 137–156; (Part II), 473–487. Falls Church, VA.

O'QUINN, C. A. 1976 Florida's electronic distance measuring equipment base lines. *ACSM Paper Proceedings*, 36th Annual Meeting, Washington, DC, (Feb.), pp. 118–127.

POLING, A. C. 1965 A taped base line and automatic meteorological reading instruments for the calibration of electronic distance measuring instruments. *International Hydrographic Review 42*(2), 173–184. Monaco.

RICK, J. O. 1975 ACSM student chapter offers tape comparison service to surveyors. *Surveying and Mapping 25*(3), 239–243. Falls Church, VA.

SAASTAMOINEN, J. 1959 Tacheometers and their use in surveying. *Canadian Surveyor 14*(9), 444–453.

SHEA, H. J. 1946 Second-order taping through use of taping bucks. *Surveying and Mapping 6*(2), 103–109. Falls Church, VA.

SHEPARD, C. D. 1932 Nomographic chart for steel tapes. *Civil Engineering Magazine 2*(7), 440–442. ASCE, New York, NY.

SMIRNOFF, M. V. 1952 The use of the subtense bar. *Surveying and Mapping 12*(4), 390–392. Falls Church, VA.

SMITH, J. R. 1970 *Optical distance measurement.* Granada Publishing Ltd., England.

STANLEY, D. R. 1952 The microptic alidade. *Surveying and Mapping 12*, 25–26. Falls Church, VA.

TURPIN, R. D. 1954 A study of the use of the Wild telemeter DM1 for a closed traverse. *Surveying and Mapping 14*, 471–477. Falls Church, VA.

WAGNER-SMITH, R. W. 1961 Errors in measuring distances by offsets. *Surveying and Mapping 21*(1), 73–77. Falls Church, VA.

WOLF, P. R., WILDER, B., and MAHUN, G. 1978 An evaluation of accuracies and applications of tacheometry. *Surveying and Mapping 38*(3), 231–244. Falls Church, VA.

5

Linear Measurements: EDM Instruments

Porter W. McDonnell

5-1. INTRODUCTION

Electronic measurement is a modern method of precise and rapid determination of slope (line-of-sight) distances. Some electronic distance-measuring instruments (EDMIs) are designed to reduce the slope distances to horizontal distances.

EDMIs can be used over water and from one high point to another—e.g., over buildings, etc. They have maximum ranges varying from 500 to 64,000 m, or 0.3 to 40 mi. The introduction of electronic distance measuring has simplified traversing work: The selection of traverse stations can be made without concern for the feasibility of taping the sides. The tedious, slow process of clearing the entire traverse line and proceeding 100 ft at a time for the measurement is eliminated. In some cases, electronic traversing now takes the place of triangulation for horizontal control over large areas. Construction layout work can be done from a single central station (radial line stakeout), and short-range EDMIs may be incorporated into theodolite designs to create all-purpose instruments usually known as "total stations."

The many instruments available differ in detail, but nearly all depend on the precision of a quartz crystal oscillator and determination of distance by measuring a "phase shift." (Two recent models, not covered here,[1] make use of a "transit time technique.") An instrument is set up at one end of the distance to be measured. It transmits a beam of infrared light *or* microwave, which serves as a carrier for the waves used for measurement. The beam is received at the other end of the distance by a reflector, when using an infrared beam, or by another electronic instrument if employing a microwave beam. In either case, the beam is returned to the master instrument. A *reflector* consists of one or more "corner prisms" (Figure 5-1) or, for short distances, just a molded scotch light or bicycle reflector. The reflectors are designed to return the light beam, even if they are only approximately pointed toward the source. For the microwave beam, the remote instrument is similar to the master instrument and thus more than a passive reflector. Light sources other than infrared (visible lasers, incandescent light) have also been used.

Several infrared instruments are shown in Figures 5-2, 5-3, and 5-4, and a microwave instrument is seen in Figure 5-5. An EDMI "modulates" the light or microwave beam to pulsate at each of several different frequencies and an increased amplitude. These *pattern* frequencies permit the comparison of several

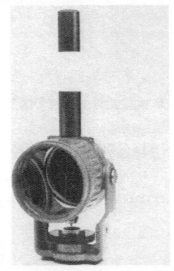

Figure 5-1. Single- and multiple-prism reflectors used to receive and return light beams in electronic distance measurements. They are shown in tilting mounts. (Courtesy of the Lietz Co.)

phase shifts of returning beams. An on-board computer deduces distance from these data—i.e., by a phase comparison of the returning beam with an internal branch of the outgoing beam.

5-2. PRINCIPLES

Sections 5-3 through 5-6 present a discussion of the principles of electronic distance-measuring instruments, based on a paper by Gort.[2] It refers to the HP 3800 instrument for

Figure 5-2. RED 2A distance-measuring unit yoke mounted on a theodolite, and the RED 2L mounted in a tribrach. (Courtesy of the Lietz Co.)

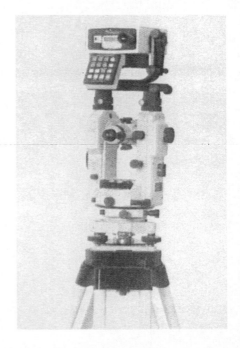

Figure 5-3. Stinger infrared EDM instrument. (Courtesy of IR Industries, Inc.)

which Gort had the design responsibility. Hewlett-Packard no longer makes EDMIs, but the following explanation illustrates the function of current models as well as those of the recent past.

5-3. THE APPLICATION OF MODULATION

In general, the measurement of any distance is accomplished by comparing it to a multiple of a calibrated distance, e.g., by using a 100-ft tape. In electronic distance meters, the same comparison principle is used: The calibrated distance is the wavelength of the modulation on a carrier (light or microwave). In the HP 3800, the effective wavelength is a precise 20 ft, which is related to the modulation frequency by

$$\lambda = v/f \qquad (5\text{-}1)$$

where

λ = modulation wavelength
v = velocity of light
f = modulation frequency

Figure 5-4. Kern DM 503 mounted on an electronic theodolite. The unit may be used on conventional theodolites as well. It features transmitting optics *above* the telescope and receiving optics *below* it. (Courtesy of Kern Instruments, Inc.)

Figure 5-5. Microwave-type EDM instrument, the Tellumat CMW20. (Courtesy of Teludist, Inc.)

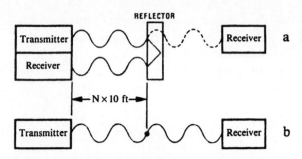

Figure 5-6. Distance being measured is a multiple of 10 ft; the wavelength is 20 ft. (Courtesy of Hewlett-Packard Co.)

Suppose the distance to be measured is an exact multiple n of 10 ft as shown in Figure 5-6a. The total optical path, however, will be $2n \times 10$ ft, which is shown by folding out the reflector-to-receiver path (Figure 5-6b). As the total path is $2n \times 10$ ft, the total phase delay will be $n \times 360°$. (Each 20-ft wavelength represents a full 360° phase delay.) The phase difference between a transmitted beam and received beam is also $n \times 360°$, which cannot be distinguished from a 0° phase difference. Figure 5-6b shows the sine wave reaching the receiver at the same point in a cycle as when it was transmitted.

In general, the distance to be measured may be expressed as $n \times 10 + d$ ft. Figure 5-7 shows the total optical path by folding out the returned beam for clarity. The total phase delay ϕ between transmitted and received signal becomes

$$\phi = n \times 360° + \Delta\phi$$

in which $\Delta\phi$ equals the phase delay due to the distance d. As $n \times 360°$ is equivalent to 0° for

a phase meter, the angle can be measured and will represent d according to the relation

$$d = \left(\frac{\Delta\phi}{360} \times 10\right) \text{ft}$$

In order to find the number n of 10-ft multiples, a 200-ft modulation wavelength is used next. This results in another ambiguity, of multiples of 100 ft. Of course, the procedure can be repeated with a 2000-ft wavelength to resolve this ambiguity, and so on.

5-4. THE INDEX OF REFRACTION OF AIR

The accuracy of distance measurements depends, among other things, on the calibration of the measurement unit. Since the modulation wavelength is used as a measurement unit in electronic distance meters, this wavelength has to be accurately established. Equation (5-1) shows that the accuracy in λ depends on v and f. The velocity of light in air may be expressed as

$$v = \frac{c}{n} \tag{5-2}$$

where c = vacuum velocity of light and n = index of refraction of air.

As c is a universal constant, only n has to be determined in order to find v. The index n is a function of the wavelength used, and the

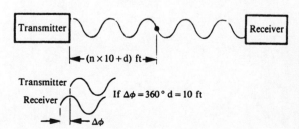

Figure 5-7. Distance being measured is *not* a multiple of 10 ft; the returning signal is out of phase with the transmitted beam. (Courtesy of Hewlett-Packard Co.)

density and composition of the air. Thus, it is affected by the atmospheric pressure and temperature at the time of measuring. As the accuracy of a distance measurement depends on determination of this index, the question arises: How accurately must the pressure and temperature be known? Figure 5-8a shows the relationship between the error in a distance measurement and inaccuracy of the temperature. Figure 5-8b displays the error as a function of pressure measurement inaccuracies. These graphs, used together, demonstrate that a 10° high estimate of temperature combined with a 1-in. low-pressure estimate causes a distance error of 15 ppm, or 1 part in 67,000—sufficiently accurate for most survey work.

In the case of microwave instruments, the humidity also is a factor (see Section 5-7).

5-5. DECADE MODULATION TECHNIQUE

As mentioned previously, a 360° phase delay equals 10 ft in the phase meter at the highest modulation frequency (24.5 MHz). The HP 3800, however, uses four different modulation frequencies in order to measure a full 10,000 ft without ambiguity. The four modulation frequencies are related in decade steps and yield phase-meter constants, which are 360° = 10 ft at one end of the scale and 360° = 10,000 ft at the other end. The fraction of 10 ft is determined first, using the 24.5 MHz modulation; the fraction of 100 ft is then determined by the 2.45 MHz modulation; the fraction of 1000 ft by applying 245 kHz modulation; and the fraction of 10,000 ft by using 24.5 kHz modulation. Table 5-1 shows this in compact form and indicates how the ambiguity is resolved for a distance of 6258.31 ft.

With respect to multiples of 10,000 ft, a readout can be ambiguous. The range of the HP 3800 is specified as 10,000 ft for good viewing conditions in the daytime, but it is possible to measure longer distances under favorable atmospheric conditions. The instrument would not distinguish between 6258.31 and 16,258.31; in fact, it is limited to a six-digit display.

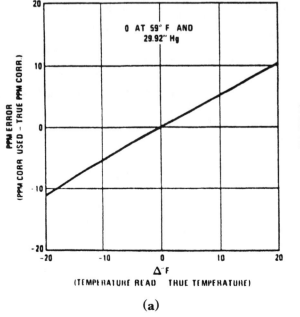

(a)

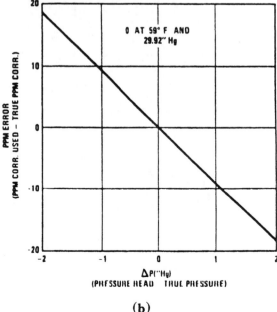

(b)

Figure 5-8(a) and (b). Distance errors in ppm caused by using incorrect temperature and atmospheric pressure. (Courtesy of Hewlett-Packard Co.)

Table 5-1. Decade modulation for HP 3800

Modulation Frequency	Phase-Meter Constant	Distance Incl. Ambiguity	HP 3800 Readout
24.5 MHz	$360° = 10$ ft	$n_1 \times 10 + 8.31$ ft	xxx8.31
2.45 MHz	$360° = 100$ ft	$n_2 \times 100 + 58.31$ ft	xx58.31
245 kHz	$360° = 1000$ ft	$n_3 \times 1000 + 258.31$ ft	x258.31
24.5 kHz	$360° = 10,000$ ft	$n_4 \times 10,000 + 6258.31$ ft	6258.31

5-6. ENVIRONMENTAL CORRECTION

It was evident from the description of the decade modulation technique and readout system that the effective modulation wavelength must be kept accurately at 20 ft or its decade multiples. If a fixed modulation frequency is used, the wavelength would vary in accordance with changes in the index of refraction of air. The HP 3800 modulation frequencies can be slightly varied to compensate for index changes and thus keep the effective wavelength an accurate 20 ft. The master oscillator in the frequency generator is set manually by a control on the power unit. Figure 5-9 gives a correction to be dialed in as a function of temperature and pressure. This should be

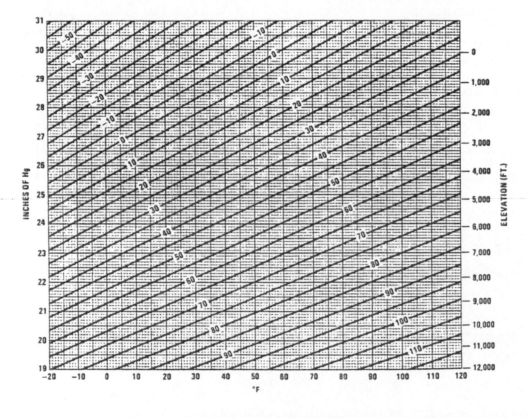

Figure 5-9. A ppm correction chart for atmospheric conditions. (Courtesy of Hewlett-Packard Co.)

done before the distance measurement, as it calibrates, in effect, the instrument's distance unit. Newer instruments permit the simple keying in of temperature and pressure, eliminating the need for a diagram or graph.

5-7. LIGHT BEAM VERSUS MICROWAVE

In general, the use of microwave beams offers the advantages of penetration through fog or rain and, usually, a longer range. Two instruments—transmitting and receiving—are required, as opposed to a single instrument and reflecting prism. The two units also provide a speech link. Microwave instruments are considerably more sensitive to humidity than light-beam models. At normal temperature, an error of 1 part in 100,000 will be introduced by a 2.7°F (or 1.5°C) error in the difference between wet- and dry-bulb thermometer readings. This effect increases on warmer than normal days. However, corrections for meteorological conditions (temperature, pressure, and humidity) are easily made. Microwave beams are wider than light beams and can present difficulties in underground surveys, indoor work, or measurements made close to a water surface. Narrower beams have been used in recent years. A microwave instrument appears in Figure 5-5.

Light-wave instruments are much more common. In a 1984 article by McDonnel in *P.O.B. Magazine*, all EDMIs were tabulated listing 47 models.[3] Only two models use microwaves. Of the remaining 45 instruments, two employ helium neon lasers as a light source and 43 use infrared. In addition, all total-station instruments are of the infrared type. For short distances, such as a half-mile, atmospheric conditions are not as important and approximate values of temperature and pressure are often sufficient (see Sections 5-4 and 5-6). Humidity is not a factor. Most light-wave models are limited to shorter ranges than mi-

crowave devices, but the HeNe laser instruments will measure up to 25,000 m using nine prisms (the K & E Ranger V-A) or 64,000 m to a bank of 30 prisms (K & E Rangemaster III).

5-8. INSTRUMENTAL AND REFLECTOR ERRORS

The National Geodetic Survey (NGS) recommends that electronic distance-measuring instruments be checked occasionally. For this purpose, they have assisted in establishing a number of calibration base lines in each state. Their specifications for these lines state that each one should be about a mile long and have intervisible monuments at 500 ft, 1150 ft, 2600 ft, and the ends.

Such a base line can be used to detect an instrument's frequency shift. An improperly tuned instrument may have a frequency error, making it comparable to a tape of incorrect length. In addition, it is possible to make a frequency check by employing a frequency counter, available from the manufacturer or other electronics supplier. Such a check should be made at regular intervals if high-order work or surveys with long lines are being performed.

The corner cube prisms used with light-wave instruments (see Figure 5-1) have a so-called "effective center." The location of this center is not geometrically obvious because it involves the fact that light travels more slowly through glass than air. The effective center will be behind the prism itself and generally not directly over the station to which the measurement is desired. Thus, there is a "reflector constant," sometimes amounting to 3 or 4 cm, to be subtracted from the measurement. The amount is accurately determined by the manufacturer. It may be partially or totally offset by advancing the electrical center of the transmitting unit during manufacture. In the case of a particular instrument, e.g., there is a correction of +320 mm at the transmitting unit. If it is used with a reflector that has a constant of 40 mm, a combined correction of 280 mm is required. This correction may be dialed into the unit and automatically added by an internal computer before a distance is displayed. Field determination of the constants is possible but not usually necessary. A discussion of this topic and errors caused by untilted or misaligned prisms is provided in a technical report by Kivioja and Oren.[4]

5-9.   SLOPE CORRECTION

EDM instruments all measure slope (line-of-sight) distances only. Reduction of inclined measurements to horizontal distance is accomplished in several ways, sometimes by a built-in microprocessor. The *vertical* projection of the slope distance (a difference in elevation) may also be determined. The latter process is a modern variation of trigonometric leveling.

A simple right-triangle reduction may be applied if the distance is short or the precision needed is modest. The effects of earth curvature and atmospheric refraction become important for longer distances. For short ones, the horizontal component is

$$S \cos \alpha \quad \text{or} \quad S \sin Z \quad (5\text{-}3)$$

where S is the slope distance, α the vertical angle, and Z the zenith angle. If the elevations of both ends of the measurement are known, the elevation difference h will serve as a basis for slope reduction. The horizontal distance is

$$S - \frac{h^2}{2S} - \frac{h^4}{8S^3} \quad (5\text{-}4)$$

in which the last term is usually negligible.

Consideration must be given to the mounting position of an EDM unit and the design of the prisms-and-target combination. For example, an EDMI may be telescope-mounted, causing its electrical center to rotate forward

or backward about the horizontal axis of the telescope for an inclined sight. A special "telescope-mount target" (Figure 5-10) may be used to allow the prism reflector to be positioned at the right distance above the painted target and tilted at a similar angle. The vertical or zenith angle is measured to the painted target and the slope distance read to the prism. An optical sight (*collimator*) is shown above the prism housing in the figure. The horizontal distance is calculated by Equation (5-3). The target in Figure 5-11 provides a vertical pole with threads on top for mounting a prism and an adjustable painted target directly below it. It is intended for use with a yoke-mounted EDM unit, such as those in Figures 5-2 and 5-3, which have their own tilting axes (vertically above that of the theodolite). Slope reduction again involves the simple Equation (5-3) in the preceding paragraph.

The target intended for use with a yoke-mounted EDMI (Figure 5-11) is much less expensive than the telescope-mount model (Fig-

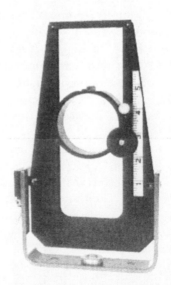

Figure 5-10. Prism holder and target designed for use with an EDM instrument mounted on the telescope of a theodolite. The prism holder is set above the target at the same offset as exists between EDMI and telescope. (Courtesy of SECO Manufacturing Co., Inc.)

Figure 5-11. Target pole threaded on top for a prism holder, designed for use with a yoke-mounted EDM instrument. The target is set below the prism at the same vertical offset as exists between the axes of the yoke and the telescope. (Courtesy of SECO Manufacturing Co., Inc.)

ure 5-10). It is easy to program a pocket calculator to make a slope reduction in which a telescope-mounted EDMI is used with the less expensive, nontilting target. The formula is

$$S \cos \alpha - d \sin \alpha \qquad (5\text{-}5)$$

where d is the distance from the theodolite axis to the top-mounted EDMI axis and α the vertical angle with its proper sign. It is sufficiently accurate to place the painted target below the prism at the same distance d. Theoretically, it should be at distance $d \sec \alpha$.

There is practically no distance error if a telescope-mounted EDMI reads an inclined distance to a prism, and then the theodolite measures the slope *to the prism* rather than a painted target below it. If $S = 100.00$ ft, $\alpha = 5°17'$, and $d = 0.50$ ft, the error in horizontal distance will be 0.001 ft (too small to measure), and it decreases for longer measurements. In Equation (5-5), vertical angle α will be differ-

ent and the last term, $-d \sin \alpha$, will not be needed. Thus, Equation (5-3) may be used.

Instruments such as the Kern DM503 in Figure 5-4 transmit the beam *above* the telescope barrel and receive the returned beam *below* it. They require a tall rectangular prism. The telescope is pointed to the middle of the prism.

Curvature of the earth and refraction of the atmosphere must be considered when measuring long lines. In 1977, Gort described an example involving a slope measurement of 1085.276 m (about two-thirds of a mile).[5] The zenith angle was $78°11'42''$. Ignoring the curvature and refraction correction would have caused a distance error of 1 part in 30,000. The effect on the computed difference in elevation would have been about 7.5 cm (or 0.25 ft). Gort was reporting on the development of the Hewlett-Packard 3820A Total Station instrument (no longer in production), which contained a microprocessor for computing the corrections. The equations used by the on-board computer were

$$\text{Horizontal distance} = S(\sin Z - E_1 \cos Z) \quad (5\text{-}6)$$

$$\text{Vertical distance} = S(\cos Z + E \sin Z) \quad (5\text{-}7)$$

in which

$$E_1 = \frac{0.929 S \sin Z}{6,372,000}$$

and

$$E = \frac{0.429 S \sin Z}{6,372,000}$$

In the latter terms, 6,372,000 is the earth's radius. For the case discussed, the horizontal distance is 1062.287 m, the vertical distance 222.103 m. For trigonometric leveling over short distances, disregarding curvature and refraction, the vertical distance would, of course, be $S \cos Z$.

If we assume the same zenith angle, but a slope distance of only 305.000 m (about 1000 ft), the horizontal distance is 298.549 m without a curvature and refraction correction, and 298.546 m with it. The vertical distance is 62.397 m by right-triangle trigonometry and 62.403 with the correction.

Equation (5-6) uses the instrument elevation as the datum. Thus, if the sight is long and steeply inclined, as in the first example, a reciprocal observation (from the other end) will give a different answer. The two answers should agree if each is reduced to a common datum. The distance found in the example, 1062.287, would be 1062.324 at the datum of the higher station. Each can be reduced to sea level by the following factor:

$$\frac{6,372,000}{6,372,000 + H}$$

where H is the station elevation in meters. Equations (5-6) and (5-7) use an average coefficient of refraction (0.071).

5-10. ACCURACY SPECIFICATIONS

Equipment manufacturers usually list accuracy as a standard deviation or mean square error (nearly equivalent concepts). The specification given is a two-part quantity: a constant uncertainty (independent of distance) and a parts-per-million term (proportionate to distance). Figure 5-12 depicts this graphically for a typical instrument and also shows the claimed ranges to single and triple prisms. The stated accuracy is $\pm(5 \text{ mm} + 5 \text{ ppm})$ for the Topcon GTS-2. The 90% error, equal to 1.64 times the standard deviation, is also plotted, indicating an "allowable error" for which the instrument could be considered suitable.

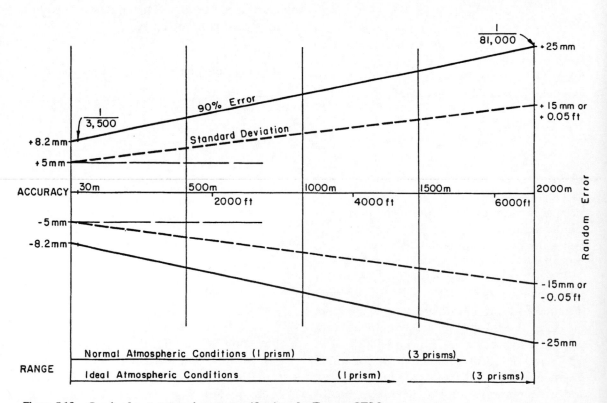

Figure 5-12. Graph of accuracy and range specifications for Topcon GTS-2.

For short measurements, the constant part of the random error is significant and exceeds the errors of ordinary taping. The figure shows that at 30 m nearly all the error is the constant type and the 90% error is 1 : 3,500. At 2000 m (a maximum range), two-thirds of the error is the proportionate kind, and the total error is 1 : 81,000. Thus, as in angle measurements, short traverse sides should be avoided.

Figure 5-12 may be said to present a "worst case." The various sources of error are complex and do not necessarily add together in the manner shown. Periodicity and calibration errors are parts of the first term (5 mm) and are systematic for a given instrument measuring a particular distance, but will vary with other lengths. Principal factors in the second term (5 ppm) are *noise* due to heat shimmer, which is random, and crystal oscillator error that is systematic. Some references show the first and second terms in the accuracy statement to propagate as two random errors. At 2000 m, the maximum range in Figure 5-12, the standard deviation would be

$$\sigma = \sqrt{5^2 + \left[\frac{5(2,000,000 \text{ mm})}{1,000,000} \right]^2} = \pm 11.2 \text{ mm}$$

instead of ± 15 mm.

The random errors discussed here are in addition to those caused by inaccurate estimates of temperature and atmospheric pressure, and incorrect centering of instrument and target. Systematic errors can be caused by improper calibration, inexact pointing on the reflector (for a peak return signal), and assuming an incorrect prism constant.

5-11. TOTAL-STATION INSTRUMENTS

Hewlett-Parkard invented the name Total Station to promote its Model 3810A, which sensed vertical angles electronically, as well as distances, and automatically applied sines and cosines to generate elevations and horizontal distances. The name caught on, either because such an exciting new type of instrument seemed to deserve a special new name, or surveyors forgot that the old term *tacheometer* or *tachymeter* would be appropriate (as would electronic tacheometer). Gradually, the name Total Station was applied to instruments that combined an electronic distance-measuring instrument and a theodolite, but had no electronic angle sensing or automatic data reduction. (Pentax simply called this an "EDM theodolite.") With its various meanings, the term Total Station became as common as dumpy level and began appearing without capital letters.

In 1983, *P.O.B. Magazine* recognized the popularity of the name total station but saw a need to define and name three separate types, as follows:

> *Manual total station.* Both distance measuring and angle measuring make use of the same telescope optics (coaxial). Slope reduction of distances is done by optically reading the vertical angle (or zenith angle) and keying it into an on-board calculator or any pocket calculator (see Figure 5-13).

> *Semi-automatic total station.* Contains a vertical angle sensor for automatic slope reduction of distances (without keying in the slope angle). Horizontal angles are read optically.

> *Automatic total station.* Both horizontal and vertical angles are read electronically for use with slope distances in a data collector or internal computer.[6]

Note that a theodolite with a mount-on EDM unit was *not* classified as a total station, except in the case of an electronic (digitized) theodolite, in which a modular total station was the design commitment, as in Figure 5-4.

Manual total stations are theodolites with built-in EDM units. Typically, these neat compact instruments weigh less than traditional

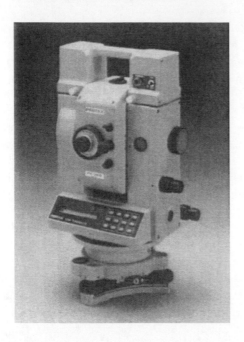

Figure 5-13. Manual total station, combining an EDM instrument, theodolite, and preprogrammed calculator in a single instrument. After zenith angle is keyed in, display shows slope distance, horizontal distance, and vertical distance in turn. Instrument offers tracking mode, stakeout mode, and distance-averaging mode. Optional cordless control unit is available. (Courtesy of Pentax Corp.)

American vernier transits. The model in Figure 5-13, and some of its competitors, have a detachable carrying handle containing the battery.

Vertical angle sensors (as on semiautomatic models) may function only in the telescope-direct position and over a limited range, such as ±40°. Thus, it may be necessary to override the sensor at times.

While the instruments are "total" in many respects, *it is important to note that a solar observation will damage the EDM unit unless an objective filter or Roelofs prism protects the optics.*

The manual and semiautomatic total stations represent an amazing advance in convenience and portability in the short history of the popular short-range electronic distance-measuring instruments. (The HP 3800 was introduced in 1971.) Even more dramatic and

revolutionary, however, are the automatic total stations. The Zeiss Elta 2, e.g., permits "free stationing." The instrument is set up on a new station (position unknown). A random combination of direction measurements or combined direction and distance measurements is used for up to five targets. The target coordinates are read from memory and the station coordinates are calculated, using a least-squares adjustment if there is redundant data. Also displayed is the standard error of the coordinates and adjusted scale factor of the system. The Elta 2 can perform a similar operation to obtain the height of instrument (HI) above datum using a distance and slope angle to one or up to five bench marks. The measurements, of course, are sensed electronically and need not be keyed in manually.

The Kern E1 (Figure 5-4) and E2 contain tilt sensors and thus are able to apply a correction to measured horizontal angles, if the instrument is not level. This is also true on the Geodimeters 140, 420, and 440, some Jena models, and the HP 3820. The Kern E2 and HP 3820 will even display the amount of inclination, so a surveyor can level the instrument exactly, if preferred, and they also correct for the effects of an inclined horizontal axis.

The Kern models, Geodimeter 140, the MK-III, and Omni I are designed to make full use of the powerful HP 41CX calculator. Values generated by the total station can be transferred directly to the calculator and stored for computation. The Kern remote receiver (on the sight rod) is also depending on this calculator to determine required orthogonal offsets in stakeout work.

Automatic total stations are frequently used with data collectors. Generally, data collectors are electronic supplements to the conventional field book, permitting a convenient interface with a computer and plotter, and remote transmission of data by telephone, using an acoustic modem (see Figure 5-14).

Because the horizontal circle is read electronically, the instrument can display the angle in degrees or grads, or subtract the

Figure 5-14. Data collector or electronic field book. Geodat 126 uses HP-41CX calculator as "administrative system" (display, keyboard, calculating capacity, and memory module) and adds HP Interface Loop, RS-232C interface, and additional memory for storing measurement data, file handling, and special programs. (Courtesy of Geodimeter.)

angle from 360° to display a counterclockwise angle. It is important to note that the smallest angular unit displayed (1″, e.g.) is not a manufacturer's claim of accuracy. One manufacturer will, at extra cost, modify its instrument for greater accuracy and yet not change the smallest displayed unit.

The combination instruments (manual, semiautomatic, and automatic total stations) are revolutionizing field procedures for all kinds of surveying.

NOTES

1. P. W. McDonnell. 1985. Total station survey. *P.O.B. Magazine 10* (6) (Aug.–Sept.), 20.
2. P. W. McDonnell. 1984. EDM instruments survey. *P.O.B. Magazine 9* (3) (Feb.–March), 17.

3. L. A. Kivioja, and W. A. Oren. 1981. A new correction to EDM slope distances in applications of untilted corner reflectors, effects of misaligned reflectors and determination of reflector constants. Final Technical Report CE-G-81-1, Purdue University, School of Civil Engineering, West Lafayette, IN.
4. A. F. Gort. 1977. The Hewlett-Packard 3820A electronic total station. Proceedings of the Fall Convention, ACSM. Washington DC.
5. P. W. McDonnell. 1983. Total station survey. *P.O.B. Magazine 8* (6) (Aug.–Sept.), 16.

REFERENCE

Bird, R. G. 1989. *EDM Traverses, Measurement, Computation, Adjustment.* New York: John Wiley & Sons.

6

Angle Measurement: Transits and Theodolites

Edward G. Zimmerman

6-1. INTRODUCTION

Three dimensions or combinations thereof must be measured to locate an object with reference to a known position: (1) horizontal length, (2) difference in height (elevation), and (3) angular direction. This chapter discusses the design and uses of surveyors' transits and theodolites to measure horizontal and vertical angles.

6-2. ANGULAR DEFINITION

An angle is defined as the difference in direction between two convergent lines. A *horizontal* angle is formed by the directions to two objects in a horizontal plane, or by lines of intersection in the horizontal plane with the vertical plane containing the objects (Figure 6-1). In surveying, one of the directions that forms a *vertical* angle is usually either (1) the direction of the vertical (*zenith*) (hence, the angle is termed the *zenith distance*) or (2) the line of the vertical plane in which the angle lies with the horizontal plane, therefore called an *angle of elevation* (+), or *angle of depression* (−) (Figure 6-2).

6-3. UNITS OF ANGULAR MEASUREMENT

The *sexagesimal system* uses angular notation in increments of 60 by dividing the circle into 360 deg; degrees into 60 min; and minutes into 60 sec. Therefore, a complete circle contains 360°, 21,600′, and 1,296,000″. This angular system is employed almost exclusively by surveyors, engineers, and navigators in the United States, as well as extensively in other parts of the world.

The *centesimal system* of angular measurement is based on a circle of 400 increments or *grads* (400^g); 100 centesimal minutes (100^c) per grad; and each centesimal minute split into 100 centesimal records (100^{cc}). The ease of addition and subtraction expressed in centesimal form leads to a decimal notation. Thus, $210^g 71^c 84^{cc}$ is noted as 210.7184^g. This method is used widely throughout Europe.

The *mil system* divides a circle into 6400 increments or *mils*. An angle of 1 mil subtends an arc of approximately 0.98 unit on a circle of 1000-unit radius.

A *radian* is the angle subtended at the center of a circle by an arc equal in length to the circle's radius. It is equal to $360°/2\pi$ or approximately 57°17′44.8″. Table 6-1 lists conversions between the four systems described.

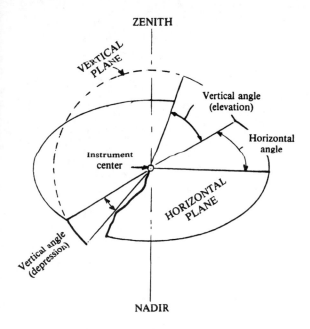

ZENITH

VERTICAL PLANE

Vertical angle (elevation)

Horizontal angle

Instrument center

HORIZONTAL PLANE

Vertical angle (depression)

NADIR

Figure 6-1. Horizontal and vertical angles and respective reference planes.

6-4. BEARINGS AND AZIMUTHS

Directions of lines being surveyed must be determined and tied to a fixed line of known direction commonly defined as a reference meridian. A meridian is either a real survey line or an imaginary reference line to which all courses of a survey are angularly related.

Four basic classifications of meridians commonly used by surveyors are as follows:

1. *Astronomic meridian.* A line on the earth's surface having the same astronomic longitude at every point (see Figure 6-3).
2. *Magnetic meridian.* The vertical plane in which a freely suspended magnetized needle, under no transient artificial magnetic disturbance, will come to rest.
3. *Grid meridian.* A line through a point parallel to the central meridian or *y*-axis of a rectangular coordinate system.
4. *Assumed meridian.* An arbitrarily chosen line with a directional value assigned by the observer.

The direction of a line is the horizontal angle from a reference meridian and can be expressed as an azimuth or bearing.

6-4-1. Bearings

The bearing angle of a line is measured from the north or south terminus of a reference meridian to the east or west, giving a reading always smaller than 90°. Bearings can be astronomic, geodetic, grid, assumed, computed, forward, backward, and in property surveys, record, or deed bearings. In Figure 6-4, the bearing of a line *OA* in the northeast quadrant is measured clockwise from the meridian. Thus, its bearing angle noted is N 45° E. Likewise, the bearings are measured counterclockwise, so for line *OB* it is S 37°43' W. Line *OD* in the northwest quadrant has a bearing angle of N 47°25' W.

6-4-2. Azimuths

Azimuths are angles measured clockwise from any reference meridian and range from 0 to 360° (Figure 6-4). They do not require letters to identify their quadrant. For example, the azimuth of line *OA* is 45°; line *OB*, 123°17', line *OC*, 217°43'; and line *OD*, 312°35'. Figure 6-4 also shows that azimuths can be calculated readily from bearings, and vice versa. In plane

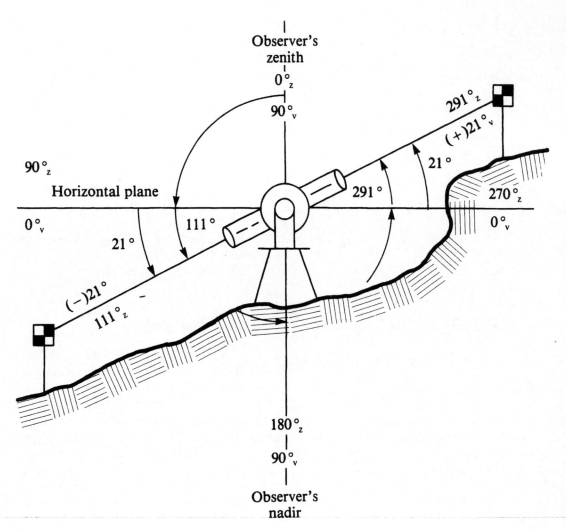

Figure 6-2. Vertical and zenith angles.

Table 6-1. Comparison of various angular definitions.

	Degrees	Grads	Mils	Radians
1 deg =	1	1.11111	17.77778	0.017453
1 grad =	0.9	1	16.0	0.015708
1 mil =	0.05625	0.0625	1	0.0009875
1 rad =	57.29578	63.66198	1018.59164	1

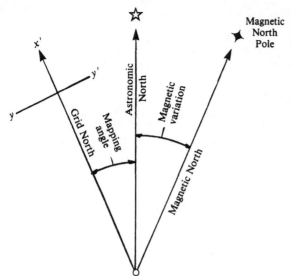

Figure 6-3. Comparison of meridians.

surveying and navigation, aximuths usually are measured from north, but in geodesy and the military services they are referenced to south.

Every line has two directions that differ by 180°. In Figure 6-5, the forward bearing from point K to point L is N 26° E, and the back bearing is S 26° W; likewise, the forward azimuth of line KL is $26°\,AZ_N$, and the back azimuth is $205°\,AZ_N$.

6-5. OPERATIONAL HORIZONTAL ANGLES

Three types of horizontal angles, shown in Figure 6-6, are defined as follows:

1. *Interior angles* are measured clockwise or counterclockwise between two adjacent lines of a closed polygonal figure.
2. *Deflection angles*, right or left, are measured from an extension of the preceding course and the "ahead" line. It must be noted whether the deflection angle is right (R) or left (L).
3. *Angles to the right* are turned from the back line in a clockwise or right-hand direction to the "ahead" line.

Angles are normally measured with a surveyor's transit or theodolite, but they can also be obtained with a sextant, compass, alidade, planetable, or tape. This chapter considers only angular measurements by transit or theodolite.

6-6. HISTORY AND DEFINITION OF SURVEYOR'S TRANSIT AND THEODOLITE

A transit is the surveying instrument having a horizontal circle divided into degrees, minutes, seconds, or other units of circular measurement, and has an alidade that can be reversed in its support without being removed. It is equipped with a vertical circle or arc. Transits are used to measure horizontal and vertical angles, differences in elevation, and horizontal distances. Modern transits may vary in appearance or construction from earlier counterparts, but their principles of operation and use are comparable.

The surveyor's transit probably originated in England during the 16th century. Reference was made in an early engineering essay to a "Topographical Instrument" that appears to be an ancestor of the modern transit. The first American-made transit was most likely constructed by William Young of Philadelphia, in 1831. However, it has been reported that Edmund Draper also constructed a transit in Philadelphia in the same year.

Theodolites are precision surveying instruments consisting of an alidade with a telescope and an accurately graduated circle, and equipped with the necessary levels and optical-reading circles. The glass horizontal and vertical circles, optical-reading system, and all mechanical parts are enclosed in an alidade section along with three leveling screws contained in a detachable base or tribrach. The convenience and inherent precision of theodolites have greatly expanded their use,

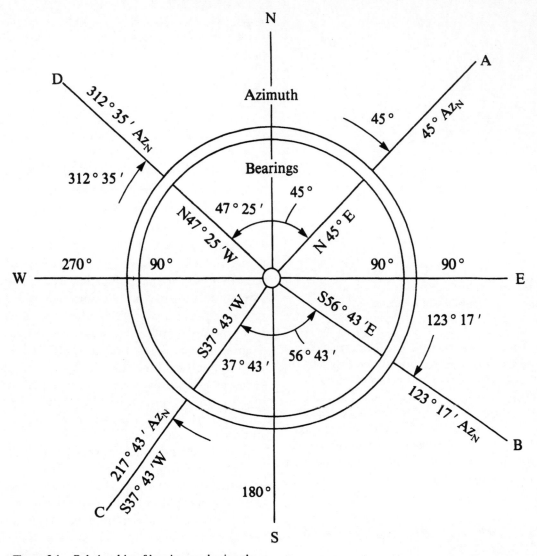

Figure 6-4. Relationship of bearings and azimuths.

and on most types of surveys, they now largely replace transits.

The first optically read theodolite was produced in the early 1900s. Dr. Heinrich Wild designed the then-revolutionary instrument to replace the cumbersome, awkwardly read transit-type theodolites. This new instrument allowed an operator to simultaneously observe both sides of a horizontal circle by means of an optical micrometer, and to determine the arithmetical mean to within a few seconds.

Wild's first optically read theodolite, known as the Wild TH-1, became available in 1923

and was the ancestor of the modern Wild T-2 theodolite. In 1935, Wild also designed an instrument manufactured by the Kern Company. This theodolite was christened the Kern DK-2 and became the forerunner for the Kern Company's present line of optical theodolites.

The pioneer theodolites, as well as their modern counterparts, share the following basic features: (1) They are compact, lightweight, and easy to operate; (2) are shock-, weather-, and dustproof; (3) have high pointing and reading accuracy; and (4) use glass circles and precise graduations that permit small instru-

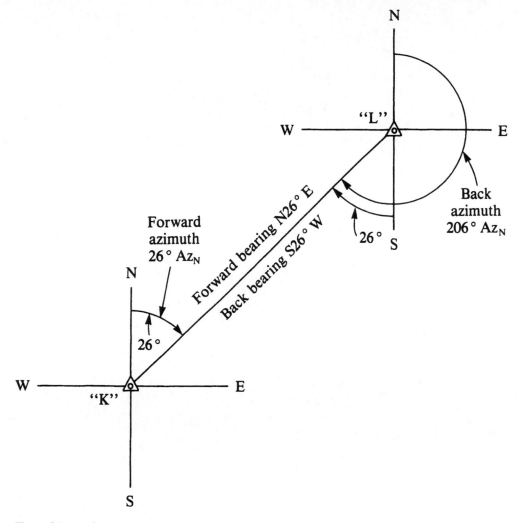

Figure 6-5. Back or forward bearings and azimuths.

ments to be used in triangulation work. A wide variety of theodolites is available for modern use, most of them based on Wild's original design.

6-7. ANGLE-DISTANCE RELATIONSHIP

Surveyors must know several relationships between an angular value and its corresponding subtended distance. Memorizing two simple trigonometric functions, $\sin 1' = \tan 1' = 0.00029$ (approx.) and $\sin 1° = \tan 1° = 0.0175$

(approx.), permits a quick manual or hand calculator check on angle-distance relationships.

$1''$ of arc

  $= 1$ ft at 40 mi or 0.5 m at 100 km (approx.)

  $1'$ of arc $= 1$ in. at 340 ft (approx.)

$1'$ of arc

  $= 0.03$ ft at 100 ft, or 3 cm at 100 m (approx.)

Surveyors must strive to maintain a balance in precision for angular and linear measurements. In Table 6-2, if distances in a survey are

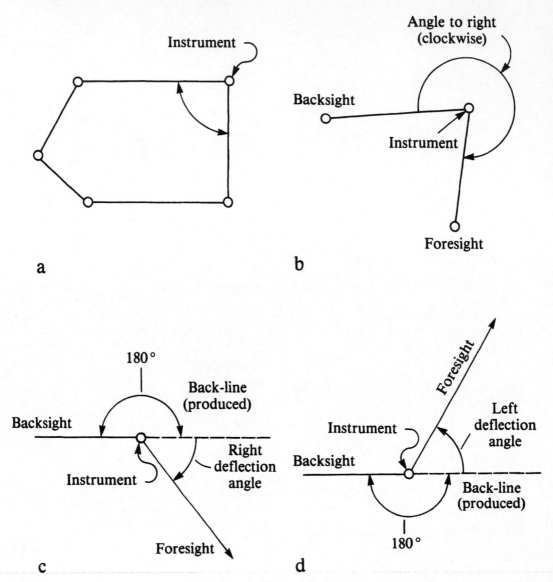

Figure 6-6. Different types of horizontal angles. (a) Interior angle. (b) Angle to right. (c) Deflection angle right. (d) Deflection angle left.

Table 6-2. Comparison of angular and linear errors.

Standard Error of Angular Measurement	Linear Error in		Accuracy Ratio
	1000 ft	300 m	
05′	1.454	0.436	1/688
01′	0.291	0.087	1/3440
30″	0.145	0.044	1/6880
20″	0.097	0.029	1/10,300
10″	0.048	0.015	1/20,600
05″	0.024	0.007	1/41,200
01″	0.005	0.001	1/206,000

to be measured with a relative precision of 1 part in 20,000, the angular error should be limited to 10″ or smaller.

6-8. TRANSIT

6-8-1. Surveyor's Transit

A transit can be called "the universal surveying instrument" because of its many uses. Surveyors and engineers employ transits to (1) measure or lay out horizontal angles and directions; (2) determine vertical angles, differences in elevation (leveling), and distances indirectly (tacheometry); and (3) prolong straight lines. The horizontal and vertical angles read have their vertex at the "instrument center," a point located at the intersection of the instrument's vertical and horizontal axes, as shown in Figure 6-7.

When properly setup and leveled over a survey point, the instrument's axis is identical to a vertical line between the point and its zenith. The horizontal axis is, of course, perpendicular to its vertical axis. Thus, the line of sight (pointing of the telescope) can be rotated simultaneously in both horizontal and vertical planes while maintaining a correct geometric relationship. The line of sight is used to turn a vertical or horizontal angle and measure both with one set of pointings.

6-8-2. Telescope

A transit telescope closely resembles that of a level, as described in Chapter 7. Most level and transit telescopes are equipped for an erecting image, but to obtain superior optical qualities, some high-precision instruments have an inverting eyepiece with fewer lenses. Modern transit telescopes are the internal-focusing

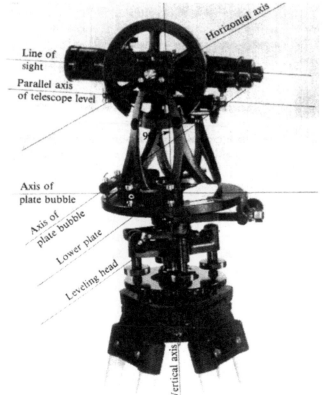

Figure 6-7. Geometry of the transit. (Courtesy of Teledyne-Gurley Co.)

type, generally have a 20- to 30-diameter magnifying power, and are equipped with stadia cross wires (see Figure 6-8 for typical cross-wire patterns). Stadia measurements with transits are discussed in Chapter 22.

A telescope is mounted on the transit's horizontal axis by axles resting in bearings on top of standards that are integral with the upper plate (see Figure 6-9). Adjustable bearings ensure a truly horizontal axis, so the telescope can be rotated 360° in an accurate vertical plane. The telescope is locked in any vertical position by a telescope clamp screw and fine pointing made by turning the telescope tangent screw. A vertical circle connected to the horizontal axis is read by means of the vertical vernier attached to one of the standards. A telescope level tube with a sensitive vial is fastened to the telescope's underside allowing the transit to also be used as a level.

6-8-3. Upper Plate

The alidade of a transit consists of a horizontal circular plate attached to a vertical spindle that allows the upper plate to rotate about the instrument's vertical axis. Two adjustable level vials, one parallel to the telescope and the other at right angles to it, are connected to the plates along with two verniers designated as A and B, set 180° apart. The level vials are also perpendicular to the vertical axis, bringing it to vertical when the vials are leveled.

6-8-4. Lower Plate

A lower plate (Figure 6-7) mounted on a hollow (outer) *spindle* has a graduated horizontal circle on its upper face. The upper plate or inner spindle perfectly fits the outer spindle of the lower plate, which rests in the tapered bore of the leveling head. With the exception of two observation windows, in which verniers and graduated horizontal circle meet and are viewed, the lower plate is completely covered by the upper plate. A clamp on the lower plate locks the inner and outer spindles so the upper and lower plates can be turned about the outer spindle as a single unit. This rotation is controlled by the lower clamp and tangent screw.

6-8-5. Leveling Head

The leveling head is a two-piece structure consisting of a base plate and collar that screws onto the tripod head. A vertical socket, which accepts the lower-plate outer spindle, is built into the base plate. Integral with the socket are four "spider" arms located 90° apart, accepting the leveling screws. The vertical socket-leveling screw unit is attached to the base plate by a half ball-and-socket joint held in place by a sliding plate beneath it. When the leveling screws are tightened, the sliding plate is pulled up against the underside of the base to hold the transit in a horizontal position. If the screws are loosened slightly, the transit head can be shifted a small distance horizontally on the base plate. Thus, a fine lateral adjustment is achieved when centering the instrument over a survey point.

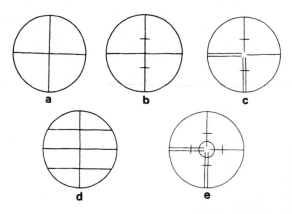

Figure 6-8. Cross-wire patterns. (a) Standard pattern on less precise transits. (b) Upper and lower wires for stadia; shorter to avoid confusion. (c) Double wires allow centering rather than covering sighted object at distance. (d) Extended stadia lines. (e) For direct solar observation; diameter of circle is 15'45".

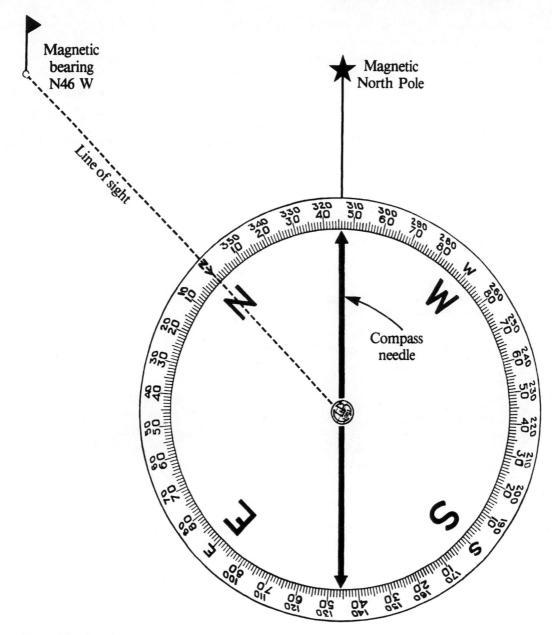

Figure 6-9.　Transit compass rose.

Leveling the instrument by lengthening one set of screws and shortening the other rotates the instrument's vertical axis about the ball-and-socket joint. A short chain with a hook for the plumb-bob string is attached to the lower end of the spindle.

As the transit is leveled, the spindle axis becomes a continuation of the plumb line and places the instrument center directly over the survey point. Some transits are equipped with an optical plummet to pass a line of sight through a prism downward along the vertical axis of the instrument. When the transit is leveled, the plummet's line of sight is vertical, so the instrument can be centered over a survey point without using a plumb bob.

Summary

A brief review of the functions of the transit's clamps and tangent screws follows:

1. Vertical rotation of the telescope is controlled by the vertical motion clamp and vertical tangent screw, generally mounted on the right-hand standard of the transit.

2. The upper plate clamp locks the upper and lower circles together; the upper tangent screw permits a small differential rotation between the two plates.

3. A lower plate clamp locks it to the leveling head; the lower tangent screw rotates the lower plate in small increments relative to the leveling head.

4. If the upper plate clamp is locked and the lower one unlocked, the upper and lower plates rotate as a unit, thereby enabling the sight line to be pointed at an object with a preselected angular value set on the plates.

5. With the lower clamp locked and the upper clamp loose, the upper plate can be rotated about the lower one to set a desired angular value. By locking the upper clamp, an exact reading or setting is attained by turning the upper tangent screw.

6-8-6. The Compass

The compass needle is made of magnetized steel and equipped with a cup-type jeweled center bearing to support the needle on a sharply pointed steel pivot post, thus allowing the needle to rotate freely in response to the earth's magnetic field. The needle assembly is contained within a glass-covered compass box attached to the upper plate. The steel mounting post is identical to the vertical axis of the instrument.

A lifter raises the needle off the pivot when not in use. If allowed to rest on the pivot while the instrument is being transported, the compass needle bearing will be quickly worn or damaged. When released, the needle rotates until it is aligned with the magnetic north (or south) pole.

The compass circle or "rose" directly beneath the needle is graduated, as shown in Figure 6-9. Note that east and west are reversed in the compass circle. The needle does not move during an observation; it remains stationary, pointing at the magnetic pole, while the circle (and line of sight) rotate beneath the needle. Thus, it provides the sight-line direction without need to switch the E and W letters when recording a magnetic bearing.

On transits with a compass circle fixed to the upper plate, cardinal N and S points of the circle are in the same vertical plane as the sight line; directions observed are therefore magnetic. Other transits are equipped with a movable compass circle, so a known or assumed declination can be set by rotating the circle. Directions are then referred to true north (or true south in the southern hemisphere).

At best, the expected accuracy of a compass bearing is no better than about 1°. For additional techniques and precautions in using a compass, see Chapter 21.

6-8-7. Tripod

Most surveying instruments are mounted on a three-legged stand known as a tripod. It consists of two parts: (1) the upper component or tripod head and (2) a set of three legs.

The tripod head has a male thread, usually $3\frac{1}{2}$ in. in diameter with eight threads per inch, on which a transit is secured. Smaller or older instruments may have a 3-in.-diameter thread size. Most theodolites employ a different mounting system that has a special $\frac{5}{8}$-in.-diameter bolt with 11 threads per inch. Transit and theodolite tripods are not interchangeable; however, the ensuing discussion applies to transits and theodolites, as well as tripods for levels, sighting targets, and other surveying instruments.

Tripod legs are attached by adjustable tension hinges, permitting the legs to swing on a line radial to the tripod head's center. Legs may be fixed in length or adjustable, solid construction or split. A nonfixed-leg tripod is preferable because it is simpler to level the plates when setting up the instrument over a fixed point. Tripods are generally $4\frac{1}{2}$ to 5 ft tall. Special extension legs are available that raise the telescope height to 8 or 10 ft for sights over brush or other obstructions. All tripods have pointed metal feet, which are pushed into soft ground by stepping on a metal spur fixed at the lower end of each foot.

To assure the best results from a surveying instrument, a stable tripod setup is required. Always plant tripod feet firmly in the ground to form an angle of approximately 60° between the leg and ground, with the telescope at eye-level height, if possible. When setting up on a sloping surface, one tripod leg should be pointed uphill to provide a stable instrument. On a hard surface, tripod feet should be set in a crack or depression, or anchored to each other by a chain loop or wooden framework. This prevents the legs from sliding outward and collapsing the tripod.

Attention must be given to proper tripod maintenance. Bolts attaching the metal feet to legs must be tightened periodically to maintain rigidity. Leg hinges should be tight enough to almost hold an unsupported leg in a horizontal position. If too tight, tripod hinges may bind then unexpectedly release, ruining an accurate setup.

6-8-8. Graduated Circles

Transits are equipped with *horizontal* and *vertical circles*, generally constructed of brass, bronze, or aluminum. During manufacture, graduation lines and numbers are engraved on the circles by machine and filled with black or white paint for visibility.

Vertical circles are commonly divided into quadrants of 90° each, with the number of degrees increasing to 90° in both directions from a 0° mark. The 0° mark of each set of quadrants coincides with the instrument's horizontal plane. The circle is divided into half-degree increments with every 10° division line numbered (Figure 6-10).

Horizontal circles are graduated either in half-, one-third, or one-quarter-degree increments, with every 10° division line numbered. Different circles available are graduated from (1) 0 to 360° in a clockwise direction; (2) 0 to 360° in both directions; or (3) 0 to 360° in a clockwise direction with four 0 to 90° quadrants superimposed, as shown in Figure 6-11. Most surveyors prefer the 0 to 360° in both directions style.

6-8-9. Verniers

Verniers are additional scales used to more accurately read transit circles. One vernier attached to the left-hand transit standard is used in conjunction with the vertical circle. Two other verniers are integrated with the upper plate, 180° apart, for reading the horizontal circle. A vernier enables readings of the circle much closer than the smallest circle division allows. Vernier principles are thoroughly discussed in Chapter 7 and apply to a transit vernier.

Before using a vernier, determine the value of the smallest interval on the adjacent graduated horizontal or vertical circle. Verniers are designed to subdivide the scale unit into an equal number of vernier divisions. The angular value of the smallest vernier division is its *least count*. For example, in Figure 6-12b, the smallest vernier division is 30″; hence, it is classified as a 30-sec least count vernier. Most verniers are 1′, 30″, or 20″ least count. Examples of various verniers are given in Figure

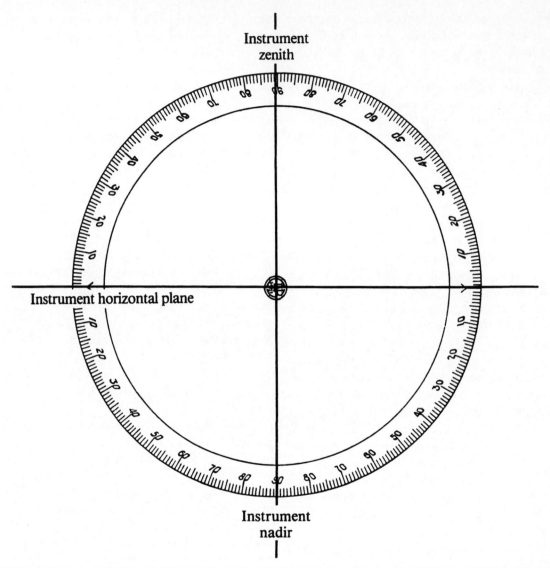

Figure 6-10. Vertical circle for transit.

6-12; note that all verniers can be read in either direction. This type of vernier is a *double-direct* vernier.

To determine (measure) a horizontal angle, read the horizontal circle as close as its least count permits with the index (zero) mark of the vernier. Next, read the value of the vernier line that is coincident with any division line on the circle; only one vernier division line will exactly align with a circle division line. Com-

bine the two values read to determine the final value. For example, in Figure 6-12b, when we read left to right, the vernier's index mark falls between 130°00′ and 130°20′ on the circle, to provide a "rough" reading of 130°10′. Coincidence occurs at 09′30″ on the vernier. Combining 130° and 09′30″ equals 130°09′30″, the final reading.

The vernier must always be read in the direction that the angle is being turned. Proper

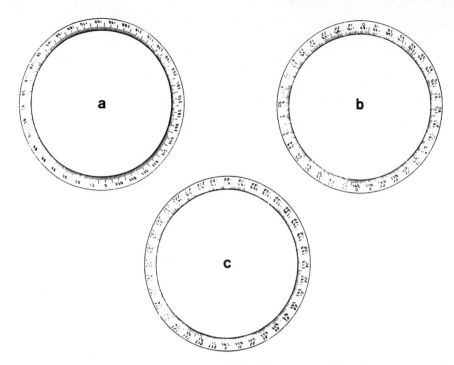

Figure 6-11. Horizontal circles for transits. (a) Graduated 0 to 360°, clockwise. (b) Graduated 0 to 360°, both directions. (c) Graduated 0 to 360°, clockwise, quadrants superimposed.

reading direction is assured by matching slopes of the numbers on the circle and vernier.

A magnifying glass must always be used when reading a vernier. An experienced operator using a glass can consistently estimate to one-half the vernier's least count. To avoid parallax error, observers must always position their eyes directly above the matching lines. Looking at division lines on either side of and immediately adjacent to the index mark is helpful to verify that a selected line is the correct choice. If two sets of vernier and scale lines seem to almost coincide, so a symmetrical pattern exists, interpolating a middle number is reasonable.

6-8-10.   Transit Operation

To remove a transit from its carrying case, hold either the leveling head or both standards. *Do not lift* by grasping the telescope.

While holding the transit in both hands (with all motion clamps loose), attach the instrument to a securely positioned tripod. Tighten the threads enough to prevent slack, but do not bind the leveling head on the tripod by overtightening. The transit/tripod unit can be carried by folding together the legs and placing it on your shoulder. Keep the tripod feet forward and the weight of the unit comfortably balanced. Indoors or in bushy terrain, the tripod should be carried under your arms in a horizontal position with the transit forward, giving you more control in avoiding obstacles.

6-8-11.   Instrument Setup

On reaching a setup location, extend the legs and center the tripod/transit by eye over the survey point. Attach the plumb-bob string to a hook under the leveling head, using a

GRADUATED 30 MINUTES READING TO ONE MINUTE
DOUBLE DIRECT VERNIER

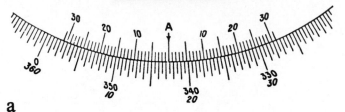

a

Reading L to R = 17° 25′
R to L = 342° 35′

GRADUATED 20 MINUTES READING TO 30 SECONDS
DOUBLE DIRECT VERNIER

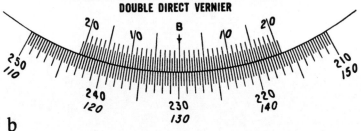

b

Reading L to R = 130° 09′ 30″
R to L = 229° 50′ 30″

GRADUATED 15 MINUTES READING TO 20 SECONDS
DOUBLE DIRECT VERNIER

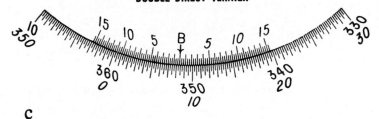

c

Reading L to R = 08° 24′ 20″
R to L = 351° 35′ 40″

Figure 6-12.   Transit verniers.

slipknot to create a loop of string. Slide the slipknot up or down to adjust the line length so the plumb bob is approximately $\frac{3}{8}$ in. above the survey point. Bring the plumb bob into rough alignment over the point by moving one tripod leg at a time in a radial direction. The leveling plate is now brought to roughly horizontal by shifting an appropriate tripod leg in a circumferential direction. If an adjustable leg tripod is being used, combine lengthening or shortening the legs, with movement, to rough level and center.

Align the plate levels over the opposite leveling screws and turn each screw of a pair in opposite directions (lengthening one, shortening the other) with the thumb and forefinger of each hand. *Do not overtighten;* keep both screws in light contact with the leveling plate. A bubble will travel in the same direction as rotation of the left thumb. After both bubbles have been centered, slightly loosen a screw of each set to ease pressure on the leveling plate and shift the head laterally to bring the plumb bob exactly over the survey station. Retighten the screws, check plate bubbles for level, and relevel if necessary. Rotate the upper plate 180° and recheck for level. If the bubbles move off center by more than one division, an adjustment is necessary. See Chapter 8 for checking and adjustment procedures.

A transit equipped with an optical plummet is set up in a similar fashion. The instrument is roughly centered with a plumb bob and leveled. The plumb bob is removed and final centering achieved with the special cross-slide tripod head on optically plumbed transits. After optically centering the instrument, it is necessary to relevel. The plummet's line of sight is also the vertical axis of a transit, so the instrument must be recentered after releveling. An optical-plummet transit is faster to set up because the time needed to adjust and damp the plumb bob's swing is eliminated. It has a more accurate means of centering, particularly on a windy day.

When transit operations have been completed, center the leveling head, equalize lengths of the leveling screws, and lightly tighten all clamps prior to "boxing" or putting away the instrument. At this time, check the transit to make sure it is clean and relatively dust-free. Be certain a cap is placed over the objective lens and the instrument fits securely in the case. If undue pressure is required to close the carrier, check to see whether the transit is correctly positioned. Any moisture on the instrument should be dried prior to putting it away.

6-8-12. Measuring a Horizontal Angle with a Transit

A horizontal angle is measured by first setting the plates to read 0°00', then backsighting along the reference line from which the angle is to be measured. After the plates are "zeroed" using the upper motion clamp, upper tangent screw, and vernier, the line of sight is brought onto the backsight point by turning the lower clamp and lower tangent screw. The upper clamp is loosened and the telescope rotated independently of the circle until the line of sight is on the foresight target (see Figure 6-13). The angular reading is obtained from the circle and vernier and recorded.

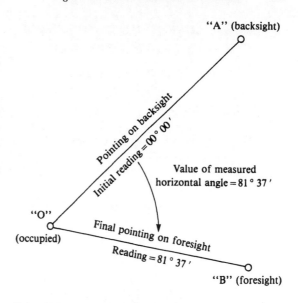

Figure 6-13. Single measurement of a horizontal angle.

A step-by-step procedure follows:

1. Loosen both clamps and bring the 0° circle mark roughly opposite the vernier index mark. This can be accomplished quickly by rotating the lower plate with fingertips pressed against its bottom surface.

2. Tighten upper clamp and bring the zero (0°00') mark of the circle into precise alignment with the vernier index, using the upper tangent screw. When upper clamp is tight, circle and vernier (upper plate) are locked together as one rotating unit.

3. With lower clamp still loose, point telescope (rotating upper and lower plate unit by hand) to backsight.

4. Tighten lower clamp and use lower tangent screw to align the vertical cross wire with backsight.

5. Loosen upper clamp and rotate upper plate until the telescope is roughly aligned with foresight.

6. Tighten upper clamp and finish pointing by aligning vertical cross wire on the foresight by using upper tangent screw.

7. Using a magnifying glass, read and record angular value indicated by vernier.

6-8-13. Laying out a Horizontal Angle with the Transit

In Figure 6-14, line *IC* will be laid out using angular reference to line *IA* employing the following procedure:

1. Set up transit at *I* and zero plates.
2. Point telescope along line *IA* and sight backsight *A*. Achieve fine pointing with lower tangent screw.
3. Loosen upper clamp, rotate upper circle by hand, and find the preselected angle.
4. Tighten upper clamp and perfect reading with upper tangent screw.

The line of sight is now in the required direction and point *C* can be established on line *IC*.

6-8-14. Measuring Horizontal Angles by Repetition

Repeated measurements of an angle increase accuracy over that obtained from a single measurement. To measure a horizontal angle by repetition, obtain an initial reading with a transit or repeating theodolite, as discussed in Section 6-8-12. Then, continue measuring as follows:

1. Read and record value for the initial angle, as noted in Section 6-8-12.

2. Loosen lower clamp, plunge (transit) the telescope, rotate upper/lower plate unit, and point to backsight.
3. Tighten lower clamp and perfect backsight, pointing with the lower tangent screw. The telescope is now inverted and aligned on backsight, with the initial angle reading remaining set on the horizontal circle.
4. Loosen upper clamp, rotate upper plate, and point at foresight.
5. Tighten upper clamp and complete foresight pointing, using the upper tangent screw.
6. A second angular measurement is accumulated on the plates and read as the sum of the first and second angle. Divide the sum by 2 (or the number of repetitions) to determine average value of the angle.

Repetitions are continued until the required measurement accuracy is met. Repeated sightings are fashioned in even-numbered sets with the telescope plunged on alternate observations. The initial and final readings of a set are made with the same precision. The mean of a set (final reading/number of repetitions) has a precision exceeding that afforded by the vernier least count and scale graduations. Assume an angle 63°21'21" is measured with a 30" transit. A single observation can be read correctly to within 30" or 63°21'30" (possible error limit ± 15"). Measured twice, the observed reading

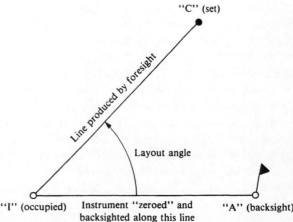

Figure 6-14. Layout horizontal angle by single sighting.

"I" (occupied)   Instrument "zeroed" and backsighted along this line   "A" (backsight)

(to $\pm 15''$) is 126°42′30″. Divided by 2, the average is 63°21′15″, correct to one-half the vernier's least count and has an error limit of $\pm 07''$. Measured four times, the 253°25′24″ reading averaging 63°21′22″ is correct to within one-quarter of the least count, with a possible error of $\pm 03''$.

As computed, the expected accuracy of a measurement is in direct proportion to the number of observations. However, factors including possible eccentricity in instrument centers, errors in plate graduations, instrument wear, and random errors associated with reading, setting, and pointing an instrument fix a practical limit on the number of repetitions at about six or eight. Beyond this number, there is little or no appreciable increase in accuracy.

Systematic instrument errors result because, in actual practice, geometrical relationships of a transit or theodolite (Section 6-8-1) cannot be exactly maintained. When operations with repeating instruments include direct and inverted pointings, systematic errors of the same magnitude occur in opposite directions, thus largely canceling them out. Also, repeating an angular measurement provides a check for, and exposes, reading blunders.

In summary, measuring angles by repetition (1) improves accuracy, (2) compensates for systematic errors, and (3) eliminates blunders.

6-8-15. Laying Out Angles by Repetition

Establishing an angle with accuracy greater than can be expected from a single pointing is accomplished by adaptations of methods detailed in the preceding section. In Figure 6-15, *IA* is an existing reference line, *IB* the line to be established.

An instrument is set up at *I*, plates zeroed, and the sight line directed to a backsight on *A*. Turn specified angle *AIB* as accurately as the least count permits, fixing *IB'*. Temporary point *B'* is set at the required horizontal distance *IB*. Angle *AIB'* is then measured by

repetition enough times to attain a desired accuracy (Section 6-8-14). Compare the value obtained for angle *AIB'* with that of angle *AIB*. Any difference is angle *BIB'*, the angular correction needed to locate *B* within specified limits. Usually, the correction angle is too small to be laid off with a transit, so a direct offset is measured along an arc of a circle formed by radius *IB*. A convenient method of accomplishing the offset computation is to multiply length *IB* by angle *BIB'* converted to radian measure. For example (see Figure 6-15), to set *B* 725 ft from *I* on a line 47°28′ right of line *AI* to an angular accuracy of $\pm 05''$ using a 1-min transit:

1. Set up instrument at *I*, level, zero plates, and backsight on *A*.

2. Turn off 47°28′ right and establish *B'* at 725 ft.

3. Measure angle *AIB* six times by repetition and read total measurement of 284°46′.

4. The mean of total angle (284°46′/6) is 47°27′40″, accurate to $\pm 05''$.

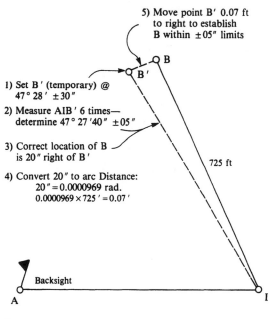

Figure 6-15. Layout horizontal angle by repetition.

5. Comparing angle *AIB'* with angle *AIB* discloses a difference of 20″ to the left.

6. Compute the correction arc: 20″ (= 0.00009696 rad) × 725 ft = 0.07 ft.

6-8-16. Extending a Straight Line

Often, it is required to prolong a straight line forward from an existing point. Set up a transit or theodolite at the end of the line and locate a new point on the line ahead. As shown in Figure 6-16, straight line *HI* must be extended to *J*. However, *J* is well beyond the range of visibility, but line *IJ* can be established from a setup on *I*. The transit is backsighted to *H*, the upper and lower plates clamped, telescope plunged, a new point *I* sighted, and a point *I* set on the extended line. This procedure is repeated until the line reaches *J*. It is more accurate to plunge the telescope than turn 180° with the horizontal circle.

If the telescope is misaligned or horizontal axis not level, it will be necessary to use the

"double-centering" method of prolonging a line as shown in Figure 6-17. To establish *C*, the instrument is set at *B*, the upper and lower plates locked, and a backsight taken to *A*. The telescope is plunged and a temporary mark made at *C*. Then, with the telescope remaining inverted, the alidade is rotated 180°, again backsighted on *A*, the telescope plunged, and a second mark *C'* established. The two foresights will have equal and opposite errors if the instrument is not in perfect adjustment. Therefore, the correct location of *C* will lie midway between C_1 and C_2.

6-8-17. Establishing Line Beyond an Obstruction

Figure 6-18a depicts a simple method to extend a line around an obstruction. This procedure should be used only on low-accuracy surveys. For greater accuracy, the method in Figure 6-18b is employed. A 90° angle is turned at *A* to establish *A'* at a suitable offset from the survey line. *B'* is set at 90° and the same offset distance as *AA'* creating line *A'B'* parallel with *AB*. Line *A'B'* is then extended by setting *C'* and *D'*, which are now occupied to establish *C* and *D* at 90° and offset distance *BB'*. The survey distance is obtained from the total length of line segments *AB*, *B'C'*, and *CD*. To ensure reliability, 90° angles at *A*, *B*, *C'*, *D'* must be accurately turned. It is also important to make lines *AB* and *C'D'*, and offset distance *AA'* as long as practical.

Another method of bypassing obstacles is shown in Figure 6-19c. The transit is set up at *B* and deflection angle α, no larger than needed to clear the obstruction, is turned to locate *C*. Point *C* is occupied and an angle 2α deflected in the opposite direction from the first angle. *D* is established on the resulting prolongation of *AB* at the same distance as *BC*. Occupying *D* and turning a deflection angle *x* in the same direction as angle *B* produce line *ABD* ahead to *E*. It is necessary

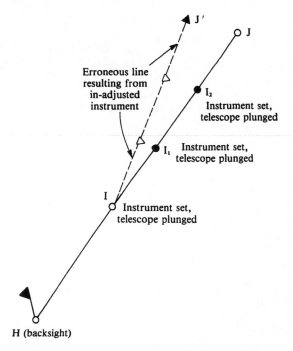

Figure 6-16. Extend straight line, single plunge.

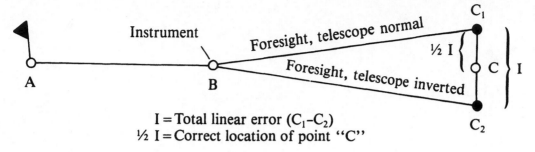

I = Total linear error (C_1–C_2)
½ I = Correct location of point "C"

Figure 6-17.   Double centering.

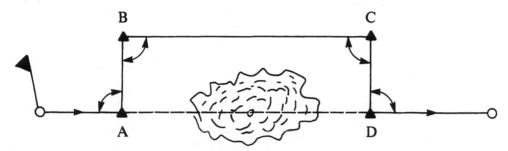

Turn 90° angles @ A, B, C, D
Make distance AB and CD equal

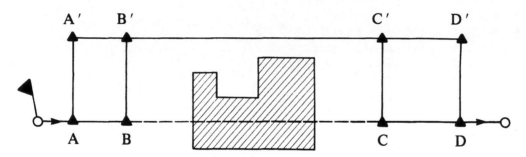

Turn 90° angles @ A, B, C ', D '

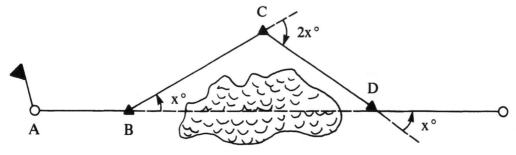

Turn x° deflection (left) @ B; turn 2 x° deflection (right)
@ C; establish D @ a distance equal to B–C; turn x° deflection
(left) @ D.

Figure 6-18.   Extending straight lines beyond obstructions.

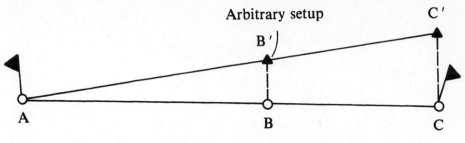

Lateral shift B–B′ = C–C′ × A–B/A–C

Figure 6-19. Wiggling in.

to calculate distance *BD*. Lengths *AB* and *CD* are equal so a convenient equation to use is

$$BD = 2 \times BX \times \cos \alpha \qquad (6\text{-}1)$$

6-8-18.　Wiggling In

It may be necessary to set up on a line between two points not intervisible, but both discernible, from an intervening setup. Estimate the alignment by eye and set up the transit. Backsight to the far point and plunge the telescope to check on the nearer one (Figure 6-19). Measure the resulting length *CC′* and use it to estimate the lateral shift required to bring the instrument on line. This operation may have to be repeated a few times until the final shift is within the instrument's sliding head limit. When the correct position is attained, check the alignment by double centering the transit between terminal points of the line.

6-8-19.　Vertical Angles

Before measuring a vertical angle, check for index error by carefully leveling the instrument and centering bubble in telescope vial with the vertical clamp and tangent screw. Check proper adjustment by rotating the upper plate 180°. If the bubble remains centered, the instrument is in correct adjustment. Read vertical circle with telescope bubble centered; if its reading is not 0°00′, an index error exists, so all measured vertical angles must be cor-

rected. For example, if a +00°03′ index error is found, it has to be subtracted from all altitude (+) angles or added to depression (−) angles.

To begin readings, place the horizontal cross wire exactly on the point being observed by using the vertical clamp and tangent screw. The angle is read on the vertical circle and vernier, then any index error is applied.

For a transit equipped with a 360° vertical circle, vertical angles are observed with the telescope in both direct and inverted positions. The mean of angles read is correct, free of index error.

Some modern transits have a movable vertical vernier, which is controlled by a separate tangent screw and level bubble. The vernier is referenced to horizontal when the bubble is centered and, therefore, minimizes any index error. With transits of this type, the vernier bubble is leveled prior to reading vertical angles.

6-8-20.　Sources of Error

Detailed definitions of errors affecting measurements made with transits and theodolites are discussed in Chapter 3. Classifications investigated here are (1) instrumental errors, (2) personal errors, and (3) natural errors.

Instrument Errors

Proper functioning of transits and theodolites depends on a precise geometrical relationship between instrument components. Although accurately manufactured and assembled, instruments will not achieve and maintain this relationship without provisions for adjustment. If an instrument is in proper adjustment, geometry is maintained within acceptable limits. However, when maladjustment is present, unacceptable errors may result unless certain operational procedures are followed. Sources of instrumental errors are as follows:

1. *Eccentricity of verniers*. If the difference between A and B vernier readings is not exactly 180° and constant around the entire horizontal circle, the verniers are offset. In this case, use only vernier A or the mean of both verniers. If the difference varies for circle positions, eccentricity of instrument centers is indicated. To compensate for this condition, observe angles at several other positions on the circle, mean the A and B vernier readings, and determine averages of observed values.

2. *Imperfect graduations*. Only high-precision measurements are affected by this condition. If angular measurements are distributed around the circle and verniers meaned, this error can be reduced to a minimum.

3. *Plate bubble adjustments*. The vertical axis of an instrument will not be truly vertical if maladjusted plate bubbles are used to level an instrument. Any inclination from vertical creates a variable error in both horizontal and vertical measurements that cannot be equalized by direct and inverted observations. A possible compensation is to center plate bubbles and rotate the upper plate 180°. Observe the length of bubble run-out and bring each bubble back one-half that distance toward the center by manipulating the leveling screws. This will bring back the vertical axis to near vertical. Do not move leveling screws until all angular measurements of a set have been completed.

4. *Line of sight not perpendicular to horizontal axis*. When an instrument's telescope having this error is plunged, the sight line describes a cone of error, whose center is the intersec-

tion of sight line and horizontal axis. The error is most apparent in measuring a horizontal angle on a backsight at a markedly different angle of elevation than for the foresight, or when plunging the telescope to prolong a line or turn deflection angles. All these errors can be minimized by meaning equal numbers of readings with the telescope direct and reversed.

5. *Horizontal axis not perpendicular to vertical axis*. If this error is present, the line of sight will describe an inclined plane when the telescope is plunged. A horizontal angle measured with the backsight and foresight at different elevations will be erroneous. This error is similar to point 4 but it can be controlled by direct and plunged telescope pointings.

6. *Line of sight not parallel with axis of telescope level vial*. This condition creates an error when an instrument is used for spirit leveling. It is eliminated by balancing lengths for fore- and backsights. An error is also introduced when observing vertical angles, but compensated for by meaning equal numbers of direct and inverted sightings.

Summary

A brief review follows:

(a) Instrument errors from inadjustment of plate bubbles or the horizontal axis affect vertical angles and increase in magnitude as vertical angles get larger.

(b) Instrument error misaligning the sight line is greatest when plunging the telescope. No error results if the telescope is not plunged when measuring a horizontal angle between back- and foresight points at the same elevation.

(c) All instrument errors are systematic but can be decreased to an acceptable level by meaning equal numbers of readings with the telescope in normal and inverted positions. Half the readings are too large, half too small. Averaging the sum gives the correct angle.

Systematic errors are kept to a minimum by keeping instruments in correct adjustment.

Surveyors can perform certain adjustments in the field, as outlined in Chapter 8. However, if an instrument is damaged by being dropped or from an accidental blow, it should be sent to a professional repair facility.

Personal Errors

These errors result from limitations of human eyesight and judgment and are considered to be accidental. Examples are described in the following list:

1. *Not centering the instrument over an occupied point.* This affects all horizontal angles measured there. Error magnitude varies inversely with the lengths of courses observed. A transit set 0.04 ft off center to measure an angle between sides 200 ft long causes an error of approximately 1 min. For 2000-ft sides, the error is about 6 sec. Thus, although reasonable care should be taken in instrument centering over a station, spending extra time in positioning the transit perfectly is unnecessary, particularly on long sights.

2. *Not sighting directly on a point.* This error has the same effect mentioned in the previous point. A pointing 0.10 ft off a target 350 ft away produces an angular error of approximately 1 min. The same pointing error for a mark 2000 ft distant is roughly 10 sec. Greater care must be exercised on shorter sight distances, and a narrower object (plumb-bob string or pencil point) used.

3. *Misreading vernier.* An accidental error occurs when the observer does not use a magnifying glass, or reads the scale and vernier graduations in a nonradial direction. An experienced instrumentperson can correctly estimate to one-half the vernier's least count.

4. *Improper focusing (parallax).* Care must be taken to sharply focus the cross wire and objective-lens images. Horizontal and vertical angles suffer in accuracy when improper focus causes parallax.

5. *Level bubbles not centered.* Plate bubbles should be checked frequently during operations, but not releveled during a measurement set (as they are in differential, profile, and other leveling projects). If an instrument is accidentally bumped, relevel and begin the interrupted operation again.

6. *Displacement of tripod.* Survey personnel must exercise care when walking around an instrument. If set up on soft ground, it can easily be displaced by one step near a tripod foot. The instrument can also be disturbed if contacted by loose clothing or a carelessly carried tool. If moved, reset it and repeat the work in progress when the disaster occurred.

To summarize, personal errors are accidental and cannot be entirely eliminated, only reduced in number and magnitude through proper techniques. They constitute the major factor in angular measurement inaccuracy. The prime sources of personal errors are caused by failing to exactly read and set vernier and micrometer scales and in not making perfect pointings on targets.

Natural Errors

Natural errors are defined as those created by the following:

1. Poor visibility resulting from rain, snow, fog, or blowing dust.

2. Sudden temperature changes, causing uneven expansion or contraction of instruments or tripods.

3. Unequal refraction deforming the line of sight or inducing a shimmering effect (heat waves) that make it difficult to accurately observe targets.

4. Settlement of tripod feet on hot pavement, or soft and soggy ground.

5. Gusty or high-velocity winds that vibrate or displace an instrument, move plumb-bob strings, and make sighting procedures difficult.

In summary, natural errors generally are not enough to affect work of ordinary precision. To lessen their effect in higher-order, certain steps can be taken, such as reducing temperature changes and refraction problems by shielding the transit from the sun with an umbrella, or performing work at night. When surveying in soft or swampy areas, support tripod legs on long wooden stakes driven into any unstable ground. Always discontinue work when weather conditions become unreasonably severe.

6-9. OPTICAL THEODOLITES

6-9-1. Descriptions

Two types of theodolites are available: double-center and directional. Both share certain features, but each has unique operating principles. Differences in them will be discussed.

Compared with transits, theodolites are compact and generally weigh only about 10 lb. Their vertical axis is cylindrical and rotates on precision ball bearings. The horizontal and vertical circles are made of glass and have precisely etched graduation lines and numerals. Optional models have graduations in degrees, grads, or mils. The circles, optical-reading system, and mechanical parts are totally enclosed within a weather- and dust-proof housing. All circle readings and bubble position checks can be made from the eyepiece end of the telescope, thereby eliminating unnecessary movement around the instrument.

Telescopes are usually short, fully transitable, and equipped with a large objective lens; they contain a glass reticle having an exactly engraved set of cross lines, and have internal focusing to provide sharp views even at relatively short ranges. The alidade can be detached from its mounting or tribrach.

Three screws supporting the tribrach are used to level the instrument in concert with a circular (*bull's-eye*) bubble on the tribrach, and a single level is mounted on the alidade. The vertical circle is equipped with either an indexing level bubble or automatic compensator to establish a horizontal reference plane minimizing "index error."

Angles are read through an optical system consisting of a microscope and series of prisms. An adjustable mirror on the outside of the instrument housing reflects light into the reading system; a battery-powered light provides illumination for night work.

6-9-2. Repeating Theodolites

This type of theodolite is constructed with a double vertical axis similar to a transit, cylindrical rather than conical as in a transit. Repeating theodolites are equipped with upper and lower circle clamps and tangent screws and can be used to measure angles by repetition.

Generally, horizontal and vertical circles are graduated in 20-sec or 1-min increments, both circles having the same least count. Some theodolites read directly to 6, 10, or 30 sec. Most horizontal circles are divided from 0 to 360° in a clockwise direction. Vertical circles are also graduated from 0 to 360°, 0° corresponding to the instrument's zenith. With the telescope level, in normal position, a zenith angle of 90° is read; in an inverted position, the zenith angle is 270°. Figures 6-20 and 6-21 show typical repeating theodolites. Each of these models is equipped with automatic compensators to correctly orient the vertical circle; both have a telescope magnification of $30 \times$ and $30''/2$ mm plate-bubble sensitivity. Both types use optical plummets.

The Wild T-1 (Figure 6-20) can be read directly to $6''$ with estimation to $3''$ for both circles. Figure 6-20 shows the field of view observed through the T-1's reading microscope. Zenith angles appear in the upper window (V); horizontal angles are seen in the lower window (Hz); and the micrometer readings viewed in the middle opening. To read an angle, rotate the micrometer knob (located on the right standard) and center a circle graduation line between the double index marks. Direct numerical reading of the micrometer setting then is found in the right-side window.

Figure 6-20. T-1 repeating theodolite. (Courtesy of Wild Heerbrugg Instruments, Inc.)

The angle depicted in Figure 6-20 is

Upper window (zenith angle)	87°
+ Right window (micrometer setting)	27′09″
Final reading (to nearest 03″) =	87°27′09″

Both circles appear simultaneously in the reading microscope, so the Hz circle is yellow, the V circle white.

The Kern K1-S (Figure 6-21) is identical in basic operation to the Wild T-1 except for its circle-reading method. Figure 6-21 presents

Figure 6-21. Kern K1-S repeating theodolite. (Courtesy of Kern Instruments, Inc.)

the K-1's reading microscope field of view. Upper window (V) displays a zenith angle, middle opening (H) the clockwise horizontal angle, and bottom window (arrow to left) exhibits the counterclockwise horizontal angle. Observed angles are direct scale observations, read to 0.5 min with estimation to 0.1 min. Thus, the readings in Figure 6-22 are

Zenith	78°35.7′
Horiz. (clockwise)	68°21.8′
Horiz. (counterclockwise)	291°38.2″

Basic angular measurement operations with a repeating theodolite are identical to those for a transit. Sections 6-8-14 through 6-8-16 are also applicable to repeating theodolites. On older theodolites, a spirit level attached to the vertical circle must be centered prior to reading a zenith angle. Modern theodolites are equipped with an automatic compensator to minimize inclination of the vertical axis in zenith angle measurement.

6-9-3. Directional Theodolites

A directional theodolite is not equipped with a lower motion. It is constructed with a single vertical axis and cannot accumulate angles. It does, however, have a horizontal circle positioning drive to coarsely orient the horizontal circle in any desired position. A directional-type theodolite with plate-bubble sensitivity of, generally, 20″/2 mm division is more precise than repeating theodolites.

Directions, rather than angles, are read. After sighting on a point, the line direction from instrument to object is noted; when a pointing is taken on the next mark, the difference in directions between them is the included angle. Optical-reading systems of direction instruments permit an observer to simultaneously view the circle at diametrically opposite positions, thus compensating for any circle eccentricities.

Theodolites shown in Figures 6-22 and 6-23 are typical of directional theodolites. Both are

Horizontal circle
reading—105.8224$_g$

Vertical circle
reading—94° 12′ 44″

Figure 6-22. Wild T-2 directional theodolite. (Courtesy of Wild Heerbrugg Instruments, Inc.)

equipped with a micrometer scale that provides horizontal and vertical circle readings directly to 1″ (with estimation to 0.1″), automatic compensators for vertical circle orientation, and optical plummets.

Figure 6-22 shows the field of view through the reading microscope of a Wild T-2 theodolite. The upper window exhibits vertical lines above and below a thin horizontal line, representing simultaneous readings on opposite sides of the horizontal or vertical circle. Rotation of the micrometer knob brings a set of lines into coincidence and moves a pointer in the middle window to display circle readings directly to 10 min. Single minutes and seconds estimated to the nearest 0.1 sec are obtained through the bottom window. The reading in Figure 6-22 is

Middle window (after coincidence)	94°10′
+ Lower window (after coincidence)	02′44″
Final reading	94°12′44″

Choice of a vertical circle (white) or horizontal circle (yellow) is determined by the position of a selector knob, located on the instrument's right standard.

A different format is used for reading the Kern DKM-2AE. Figure 6-23 illustrates the view seen in its reading microscope. The upper window displays a vertical circle reading in degrees (large number) and nearest 10 min (framed by cursor). The vertical (V) and horizontal (H) coincidence scales are located directly below the top window. Each contains a pair of double index lines superimposed over a pair of single lines regulated by the micrometer control knob. To read an angle, obtain coincidence of the selected circle by rotating the micrometer knob and centering the single line within the double index marks. Degrees and tens of minutes are read in the upper window: the large upper window for vertical angles, the large lower window for horizontal angles. The direction in Figure 6-23 is read as follows:

Coincidence achieved	
Top window (vertical)	85°30′
+ Bottom window (sec/min)	35′14″
Zenith angle	85°35′14″

When observing horizontal angles, use the third window from the top (H) to achieve coincidence and read the horizontal angle immediately below.

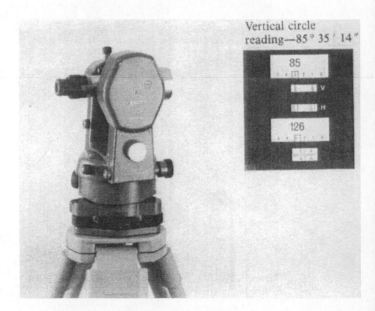

Figure 6-23. Kern DKM 2-A directional theodolite. (Courtesy of Kern Instruments, Inc.)

6-9-4. Electronic Theodolites

Recent developments in electronics have been incorporated into surveying instrument manufacturing, and several digital theodolites are now on the market. They are designed to automatically read, record, and display horizontal and vertical angles. Their basic design and operation is similar to those previously discussed; however, these electronic models can display results digitally or store data directly in an electronic data recorder for later retrieval, computing and plotting by an office computing system.

Figure 6-24 shows the Wild T2000, a newly developed electronic theodolite. When equipped with a Wild Distomat EDMI, the T2000 becomes a total station, capable of measuring and displaying horizontal and vertical angles, horizontal distance, and elevation difference. An on-board computer reduces slope distances and corrects horizontal distances for earth curvature and refraction. Coordinates for a currently occupied station are calculated if starting values are entered for the initial traverse point.

T2000 uses a dynamic angle-measuring system, allowing a full scan of the rotating circle during each measurement. Graduations around the entire circle are considered during each observation, eliminating the effect of graduation errors and any residual eccentricity of the circle and axis. Internal precision of the measuring system is about $\pm 0.2''$. However, the manufacturer states that when atmospheric and pointing errors are considered,

Figure 6-24. Wild T-2000 electric theodolite equipped with an EDMI. (Courtesy of Wild Heerbrugg Instruments, Inc.)

the standard deviation for a mean normal and inverted pointing is $\pm 0.5''$.

6-9-5. Theodolite Setup

Roughly level and center the tripod over the point to be occupied. Lift the theodolite from its case by the standards or, if equipped, the lifting handle. Place the instrument on the tripod and attach it by firmly tightening the tribrach attachment bolt. The instrument can be centered like a transit (horizontal location by plumb bob, fine centering by optical plummet), but the following method using an adjustable-leg tripod is faster.

After selecting a setup point, place one tripod leg about 2 ft beyond it. Next, grasp the other two legs and, while looking through the optical plumb, position legs so the point occupied is visible in the plummet eyepiece. Push legs firmly into the ground and, while looking through the plummet, bring the cross wires over the setup point by adjusting the level screws. Next, roughly center the bull's-eye level by changing the tripod leg lengths to approximately center and level the instrument over a point. The setup is completed by carefully leveling the theodolite and then precisely positioning it over the mark by lateral shifting on the tripod head. During final centering of the instrument, do not rotate it; if the tripod head is not level, rotation will take a theodolite out of level.

Theodolites are equipped with three leveling screws, a single plate level, and a circular bull's-eye level used mainly for coarse setup purposes. Final leveling is done with the screws and plate bubble in the following four steps:

1. Rotate instrument and align the plate-bubble axis with two leveling screws.
2. Center plate bubble by adjusting the screws, rotate the instrument 90°, and use the third screw to recenter bubble.
3. Repeat those procedures, then reverse to make a final leveling check.
4. If the alidade bubble moves more than two divisions upon reversal, it must be readjusted.

6-9-6. Horizontal Angles with Directional Instruments

As noted in Section 6-9-3, a directional theodolite reads directions or "positions" on its horizontal circle. The difference in directions to two points is the angle included between them.

The following methods are used with the Wild T-2 theodolite, but can also be employed with other directional instruments. Figure 6-25 is a diagram of the measurement procedure for a theodolite set up at Q. With the horizontal clamp loose, make a rough pointing to K, tighten clamp, and perfect pointing with the horizontal tangent screw. Coincidence is achieved with the micrometer knob and line direction QK observed as $12°31'16''$. Next, loosen the horizontal clamp, rotate instrument, and point roughly on L. Tighten horizontal clamp, and using the horizontal tangent screw perfect pointing and observe line direction QL as $76°11'39''$. Subtracting QL from QK gives an angle $63°40'23''$.

Note that no attempt was made to set the horizontal circle on zero, although the theodolite has a control to move it to a predetermined approximate position. This control is very coarse and does not permit fine settings. Trying to make exact settings of the "seconds" portion of any position is not recommended.

Measuring Horizontal Angles by Repetition

Repetitive direction measurement requires each line to be observed with the telescope in direct and reversed positions. The directions are meaned and the results used to calculate an angle. A complete set of direct and reverse observations to a point is termed a *position*. An example set of field notes for directions to points H, I, and J, with a Wild T-2 theodolite setup at K is shown in Figure 6-26.

The following measurement procedure refers to those noted:

1. Loosen horizontal clamp and point at H, the left-most station of the set, designated

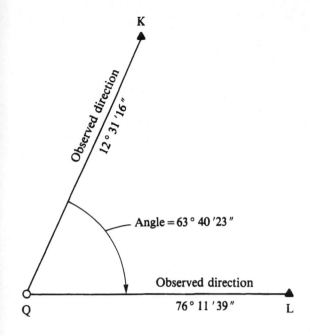

K

Observed direction
12° 31′16″

Angle = 63° 40′23″

Observed direction
76° 11′39″

Q

L

Figure 6-25. Horizontal angle with direction
instrument. Single measurement.

the initial point. Initializing on *H* permits directions to be read and recorded in a clockwise sequence.

2. Lock the horizontal clamp and use the circle-drive knob to set approximately 00°00′10″ on the circle.

3. Perfect pointing and read the directions to *H* (00°00′34″). Loosen horizontal clamp and point on *I*; tighten clamp and finish pointing.

4. Read and record the direction to *I* (202°21′53″).

5. Use the same routine to observe and record the direction to *J* in completing the position's first step.

6. Next, loosen the horizontal clamp, rotate alidade 180°, reverse the telescope, and point again to *J*.

7. Read and record the direction, which will differ by approximately 180° from the first reading.

8. With scope inverted, sight to *I* and *J* (counterclockwise), completing the first position or set of angles.

In this example, it was required to complete two positions. Distribute the readings uniformly around the circle by making a second initial position pointing at *H*, with approxi-

mately 90°20′20″ set on the circle. Perfect pointing, read and record the direction to *H* (90°22′29″). Complete a second position by measuring the remaining directions in the set with the telescope direct and reversed.

Always reduce notes prior to leaving all occupied stations by meaning direct and inverted observations for every pointing in the position, then determine the mean direction for each position. Compute all the meaned directions by subtracting the value of the meaned initial direction (*KH*) to get final directions. If any position varies from the mean of all positions by more than ±5″, reject it and reobserve that particular position.

6-9-7. Zenith Angles with a Theodolite

Unlike transits, theodolites are not equipped with a telescope level. One of two different mechanical arrangements is used to orient a theodolite's vertical circle to its zenith.

The first method, used on older instruments and more precise theodolites, is to attach a spirit level on the vertical circle. A separate tangent screw rotates the circle about its horizontal axis to center the bubble and

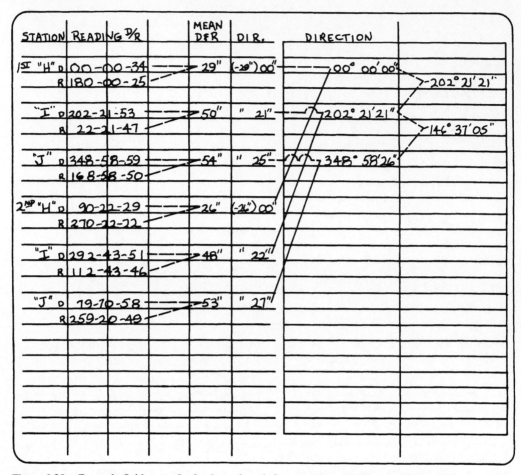

Figure 6-26. Example field notes for horizontal angle by repetition with direction instrument.

minimize index error. The general procedure in using this instrument type is to point the horizontal cross wire on an observed object, then center the bubble orienting the circle. Neglecting to center the bubble introduces an accidental error of unknown size into a measured angle.

The second indexing method utilizes an *automatic compensator* responding to the influence of gravity. With the theodolite properly leveled, the compensator is free to bring the vertical-circle index to its true position in much the same way an automatic level compensator functions. Automatic compensators are generally of two types: (1) *mechanical*, whereby a suspended pendulum controls prisms directing light rays of the optical reading system or

(2) an *optical* system, in which the optical path is reflected from the level surface of a liquid.

Zenith observations generally follow this routine:

1. Point instrument on object to be observed.
2. With the telescope in direct mode and vertical clamp loose, set horizontal cross wire on object, tighten vertical clamp, and perfect pointing with vertical tangent screw.
3. If the instrument is not equipped with an automatic compensator, index the vertical circle by centering bubble, then read and record the vertical angle.
4. Rotate alidade 180°, loosen vertical clamp, repoint on target with telescope inverted, and read and record the observed zenith angle.

5. A mean zenith angle is obtained by first adding the direct and reverse readings to obtain the algebraic difference between their sum and 360°, then dividing this difference by 2, and algebraically adding the result to the first (direct) series measurement.

6. The result is the zenith angle corrected for any residual index error.

If greater reliability is required, repeat the steps and determine mean values. The result of an individual set should agree with those of at least two sets within the following limits:

1″ theodolite ± 5″

20″ theodolite ± 12″

1′ theodolite ± 12″

If these limits are not met, read additional sets until the rejection limit is satisfied.

Zenith (Z) angles can readily be converted to vertical (V) angles. For example, 79°51′14″ Z is resolved:

$$(90°00′00″ - 79°51′14″) = +10°08′46″ \text{ V}$$

A 101°31′06″ Z equals

$$(90°00′00″ - 101°31′06″ V) \text{ or } -11°31′06″ V$$

Zenith angles from the inverted sightings are 273°16′47″ Z equals

$$(273°16′47″ - 270°00′00″) = +03°16′47″ \text{ V}$$

and 264°21′32″ Z equals

$$(264″21′32″ - 270°00′00″) = -05°38′28″ \text{ V}$$

Figure 6-2 illustrates the relationships.

6-9-8. Horizontal Angle Layout by Directional Theodolites

Layout of horizontal angles can be done in the following manner:

1. Set up, level, and center the instrument over a selected point; observe and record the direction to reference backsight.

2. Add the required angle to the reference direction to determine the direction of line being established.

3. Put this answer in the instrument, using the micrometer knob to set the desired single minutes and seconds.

4. Loosen the horizontal clamp and rotate alidade until degrees and tens of minutes are roughly located in the microscope.

5. Lock the horizontal clamp and finish setting, employing the horizontal tangent screw to bring appropriate division lines into coincidence.

6. Sight through the telescope to mark a point on the established line.

7. Loosen the horizontal clamp, rotate alidade 180°, invert telescope, and sight to newly set point.

8. Read and record the direction, loosen horizontal clamp, point on backsight, determine direction, and calculate the angle.

9. If the second angle differs by more than 10 sec from a first (layout) angle, repeat the entire procedure from a new start.

10. If a wide variance continues on additional repetitions, check the theodolite for maladjustment.

6-9-9. Forced Centering

Most modern theodolites are mounted in a detachable tribrach that permits the instrument to be quickly interchanged with an EDM reflector or sight pole without disturbing integrity of the tripod/tribrach setup. To take full advantage of the interchangeability and ‶forced-centering″ operation, a survey crew should be equipped with three or more tripod/tribrach sets and the necessary adapter hardware.

A list of steps in the forced-centering procedure follows:

1. On completing observations at a station, the theodolite is detached from its tribrach, leaving a tripod/tribrach unit centered and leveled over the station.

2. The theodolite is carried ahead to the next station and attached to a tripod/tribrach

from which the foresight target has been removed.

3. The rearmost or former backsight unit is picked up and carried forward to a new station to be observed, set up and a foresight target fastened.

Advantages of forced centering are obvious: Instead of three separate setups at every station (foresight, theodolite occupation, and backsight), only a single placement of the tripod/tribrach unit is necessary. Two opportunities for accidental setup errors have been eliminated.

6-9-10. Expected Accuracy of Theodolites

Results derived from testing, manufacturer's technical specifications, and conservative assumptions indicate that the accuracy of measurements made by experienced personnel, under favorable conditions and using instruments in good adjustment, are reasonably expected to be within the following limits.

For a 1-sec theodolite, most angles measured should have a probable error not more than

(a)	One position (1 direct, 1 reverse)	$\pm 4''$
(b)	Two positions	$\pm 3''$
(c)	Four positions	$\pm 2''$

For a 1-min theodolite, the maximum error in most angles measured by repetition is

(a)	Turned twice	$\pm 7''$
(b)	Four repetitions	$\pm 4''$
(c)	Six repetitions	$\pm 3''$
(d)	Twelve repetitions	$\pm 2''$

7

Leveling

Robert J. Schultz

7-1. INTRODUCTION

Leveling is a process to determine the vertical position of different points below, on, or above the ground. In surveying operations, vertical elevations and vertical control are generally derived independently of horizontal control. Some modern positioning devices, termed total stations, allow simultaneous determination of spatial coordinates. *Elevation* is the vertical distance above a well known datum or arbitrary reference surface. Elevations are helpful for the placement of a water drain line to provide free gravity flow, construction of a sports playing field, and among other applications, the vertical layout of a roadbed to allow a smooth flow of trucks and trains, which must ascend or descend sloping terrain.

Usually, elevation measurements are made above a specific reference surface, such as *mean sea level*. This surface may be defined as the position of the ocean if all currents and tides cease to exist. It is then projected under the land surface. It is a surface on which gravity measurements would all be the same value, and hence it may be called an equipotential gravity surface and, more specifically the *geoid*. The earth's gravity field decreases with distance above mean sea level. Scientific studies have located this surface by such varying techniques as continuous *tide gage* readings and calculations employing artificial variations in satellite orbital elements.

7-2. DEFINITIONS

Figure 7-1 demonstrates a basic vocabulary of words used in leveling literature. All surfaces shown are referenced to the physical plumb-bob line at a point, such as *A*. This line is the direction that the string of a free-hanging plumb bob takes in a still atmosphere. Conventional leveling equipment is constructed to place the telescope line of sight in a plane perpendicular to the plumb-bob line.

A *vertical line* follows the direction of gravity (plumb-bob line) through any point, such as *B*. If points *A* and *B* are several miles apart, curvature of the earth causes plumb-bob lines through *A* and *B* to converge. Because of curvature, mass density changes and hidden masses below the earth's surface, all vertical lines are not parallel, even at close spacing. Generally, however, gravity variations are small, and these lines can be considered parallel in most applications. Surveys performed under

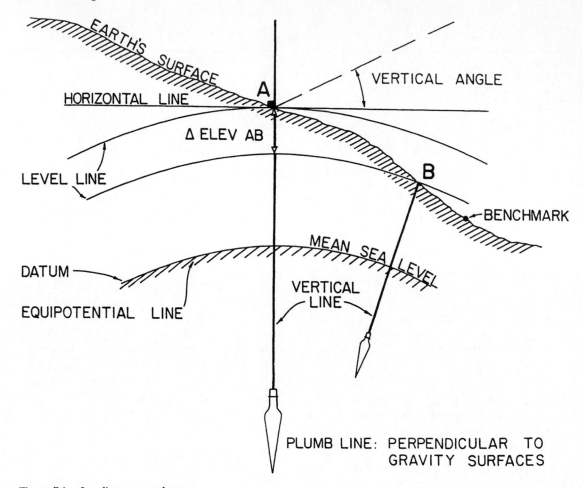

Figure 7-1.   Leveling nomenclature.

this assumption are considered to be *plane* surveys.

A *horizontal line* is perpendicular to the vertical line at a point under consideration. Spirit or pendulum-type levels make the line of sight horizontal, hence, the sight line can be rotated in the horizontal plane when an instrument is properly set up.

A *level line* is a line in a level surface. A level surface has all points perpendicular to the direction of gravity and hence is curved. To many people, horizontal and level surfaces are synonymous, but surveyors draw a clear distinction between them and attempt to measure vertical distances between level surfaces to obtain elevations.

The *difference in elevation* between two points *A* and *B* is expressed as vertical distance

through point *A* along a vertical line to the intersection of a level line or surface through point *B*. The elevation of point *B* in Figure 7-1 is found by subtracting the difference in elevation *AB* from *A* to give *B* a lower elevation. Points below mean sea level are considered negative quantities.

A *bench mark* (BM) is a permanent point of known elevation. It is located by arbitrary assignment of a fixed elevation or extension of vertical control. High-order field work based on a beginning reference surface, such as mean sea level, usually provides the basis for the network. The BM should be permanent, stable, and a recoverable object such as a brass cap, steel pipe, or man-made object listed with a description in some public agency's table of vertical-control data. The most commonly used reference datum in the United States is the *National Geoditic Vertical Datum of 1929,* formerly known as the *Sea Level Datum of 1929.*

7-3.   DIRECT LEVELING

Elevations of points are determined by direct and indirect means. Most vertical control for engineering construction is accomplished by *direct leveling.* In this method, elevation differences between a continuous short series of horizontal lines is determined by direct observations on graduated rods, using an instrument equipped with a sensitive spirit level, or a pendulum-type "automatic" level. Figure 7-2 illustrates this procedure.

First- and second-order vertical-control stations are established by the National Geodetic Survey (NGS) along lines at approximately 1-km spacings to form grids 50 to 100 km square. Other federal, state, city, and county organizations provide control of lower order. The U.S. Geological Survey prepares Control Survey Data maps at a scale of 1 : 250,000, showing major federal lines for use by local communities and surveyors. Those desiring information on first- and second-order NGS lines can write or call: The Director, NGS Information Center, NOAA, Rockville, MD 20852, to order data sheets listing adjusted mean sealevel bench marks included in the national network.

Many state highway offices also provide additional information about local vertical control. It is common to use these given data and run a line of levels to a construction job site, thus providing a temporary bench mark to control all other site elevations. Farmers wishing to drain their fields may assign an arbitrary elevation to a fixed point or rock outcrop in the area and refer all elevations to it.

7-4.   PROCEDURES IN DIFFERENTIAL LEVELING

From its written description, a beginning bench mark is located; then 10 to 100 m from the bench mark, the instrument is set up and leveled. A backsight (BS) to the bench mark is observed on a plumbed graduated rod, and the height of instrument (HI) determined by the formula HI = elev BM + BS. The BS may be negative if the rod is held in a mine survey against the tunnel roof. A foresight (FS) reading is taken on the rod held at any suitable point, thus creating a *temporary bench mark* (TBM) or *turning point* (TP). To eliminate systematic instrumental errors, the backsight and foresight should be approximately equal in length. Then elev TP (or TBM) = HI − FS. A series of such setups is taken until the final permanent bench mark is reached. The instrument can be set up and operated this way, by a skilled operator, under most topographic

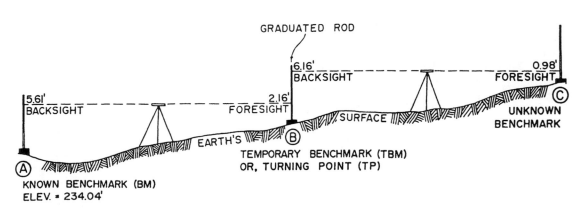

Figure 7-2.   Direct leveling.

conditions. Briefly stated, differential leveling consists of taking one backsight and one foresight of approximately equal length to achieve high-precision leveling.

In Figure 7-3, the height of a building is $B''B = AB' \tan \alpha + AB' \tan \beta = AB'(\tan \alpha + \tan \beta)$. Vertical angles above the horizon are positive ($+$), those below it are negative ($-$). Thus, elev B = elev A + HI + AB' tan β, where β is a negative angle and elev B'' = elev A + HI + AB' tan α, where α is a positive angle.

7-5. TRIGONOMETRIC LEVELING

An indirect technique for measuring elevation differences is to read the vertical angle and distance to a point using a clinometer, transit, or theodolite. For short distances, plane surveying principles are applied.

In Figure 7-3, the height of a building is $B''B = AB' \tan \alpha + AB' \tan \beta = AB'(\tan \alpha + \tan \beta)$. Vertical angles above the horizon are positive ($+$), those below it are negative ($-$). Thus, elev B = elev A + HI + AB' tan β, where β is a negative angle and elev B'' = elev A + HI + AB' tan α, where α is a positive angle.

In this procedure, precision of vertical angles α and β and linear measurement AB'

must be compatible to yield an accurate answer. Table 7-1 shows the relations between linear and angular errors, and Table 7-2 the precision of computed values.

In the right triangle $AB'B''$ (Figure 7-3), if α is 5° exactly and the horizontal distance AB' is 1000 ft exactly, the vertical distance $B'B''$ equals 87.49 ft. An error of 1 sec in the vertical angle measurement would have little effect on the height $B'B''$. However, an error of 1 min would cause errors of 0.15, 0.29, and 0.44 ft at 500, 1000, and 1500 ft, respectively. If the horizontal distance measurement AB' is also in error, the accuracy of the height determination $B'B''$ would decrease owing to a combination of the angular- and distance-measurement errors. In general, these errors are kept about equal to one another, thus forming a circle of error and creating a small plus or minus allowable tolerance in the actual height determination.

7-6. EARTH SHAPE CONSIDERATIONS

Over long distances, the effects of *earth's curvature* and *atmosphere refraction* must be considered in leveling, as shown in Figure 7-4. A properly set up and leveled instrument, in adjustment, has its line of sight perpendicular

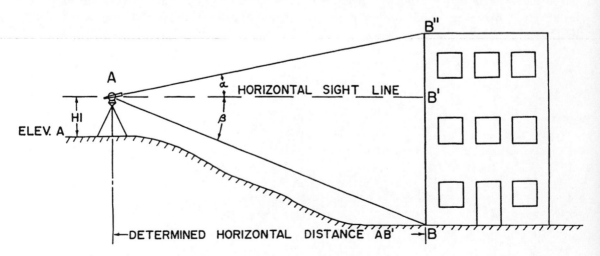

Figure 7-3. Elevation by vertical angles.

Table 7-1. Relation between linear and angular errors

Allowable Angular Error for Given Linear Precision			Allowable Linear Error for given Angular Precision				
Precision of Linear Measurements	Allowable Angular Error	Least Reading in Angular Measurements	Allowable Linear Error in				Ratio
			100′	500′	1000′	5000′	
$\frac{1}{500}$	6′53″	5′	0.145	0.727	1.454	7.272	$\frac{1}{688}$
$\frac{1}{1000}$	3′26″	1′	0.029	0.145	0.291	1.454	$\frac{1}{3440}$
$\frac{1}{5000}$	0′41″	30″	0.015	0.073	0.145	0.727	$\frac{1}{6880}$
$\frac{1}{10,000}$	0′21″	20″	0.010	0.049	0.097	0.485	$\frac{1}{10,300}$
$\frac{1}{50,000}$	0′04″	10″	0.005	0.024	0.049	0.242	$\frac{1}{20,600}$
$\frac{1}{100,000}$	0′02″	5″	0.002	0.012	0.024	0.121	$\frac{1}{41,200}$
$\frac{1}{1,000,000}$	0′00.2″	2″	0.001	0.005	0.010	0.048	$\frac{1}{103,100}$
		1″		0.002	0.005	0.024	$\frac{1}{206,300}$

to the plumb line and, except for the atmospheric refraction, the line of sight would lie in a horizontal plane. The earth's curved level surface departs from the horizontal by a distance c.

The normal ellipsoidal earth model has doubled curvature with independent radii in the *meridian* and *prime vertical*, which is at 90° to the meridian. If a spherical earth is assumed and low precision satisfactory, a single radius of approximately 20.9×10^6 ft can be assumed. A more precise radius depends on the observer's *latitude* and the sight-line *azimuth* of the observation.

A practical expression for curvature is $c = 0.667M^2$, where c is the earth's curvature in feet, and M the distance in miles. The coefficient 0.667 contains appropriate factors for geometry, unit conversion, and the earth's radius.

Due to a difference in density, an optical sight line passing through the atmosphere re-

fracts or bends back toward the earth. This refraction effect r is usually taken as one-seventh the effect of curvature and helps compensate for that factor. The refraction correction requires knowledge of temperature, pressure, and relative humidity, which are difficult to evaluate over long distances, so simplifying assumptions are generally used.

The combined effect of curvature and refraction is given by

$$(c + r) = h = 0.574M^2$$

The following shows that $(c + r)$ increases rapidly with distance:

	$(c + r)$ Effect				
Distance	200 ft	500 ft	1000 ft	1 mi	2 mi
h ft	0.001	0.005	0.021	0.574	2.296

Table 7-2.   Precision of computed values

Size of Angle and Function		Angular Error				
		1′	30″	20″	10″	5″
		Precision of Computed Value Using Sine or Cosine				
Sin 5° or cos 85°		$\frac{1}{300}$	$\frac{1}{600}$	$\frac{1}{900}$	$\frac{1}{1800}$	$\frac{1}{3600}$
10	80	$\frac{1}{610}$	$\frac{1}{1210}$	$\frac{1}{1820}$	$\frac{1}{3640}$	$\frac{1}{7280}$
20	70	$\frac{1}{1250}$	$\frac{1}{2500}$	$\frac{1}{3750}$	$\frac{1}{7500}$	$\frac{1}{15,000}$
30	60	$\frac{1}{1990}$	$\frac{1}{3970}$	$\frac{1}{5960}$	$\frac{1}{11,970}$	$\frac{1}{23,940}$
40	50	$\frac{1}{2890}$	$\frac{1}{5770}$	$\frac{1}{8660}$	$\frac{1}{17,310}$	$\frac{1}{34,620}$
50	40	$\frac{1}{4100}$	$\frac{1}{8190}$	$\frac{1}{12,290}$	$\frac{1}{24,580}$	$\frac{1}{49,160}$
60	30	$\frac{1}{5950}$	$\frac{1}{11,900}$	$\frac{1}{17,860}$	$\frac{1}{35,720}$	$\frac{1}{71,440}$
70	20	$\frac{1}{9450}$	$\frac{1}{18,900}$	$\frac{1}{28,330}$	$\frac{1}{56,670}$	$\frac{1}{113,340}$
80	10	$\frac{1}{19,500}$	$\frac{1}{39,000}$	$\frac{1}{58,500}$	$\frac{1}{117,000}$	$\frac{1}{234,000}$
		Precision of Computed Value Using Tan or Cot				
Tan or cot 5°		$\frac{1}{300}$	$\frac{1}{600}$	$\frac{1}{900}$	$\frac{1}{1790}$	$\frac{1}{3580}$
10		$\frac{1}{590}$	$\frac{1}{1180}$	$\frac{1}{1760}$	$\frac{1}{3530}$	$\frac{1}{7050}$
20		$\frac{1}{1100}$	$\frac{1}{2210}$	$\frac{1}{3310}$	$\frac{1}{6620}$	$\frac{1}{13,250}$
30		$\frac{1}{1490}$	$\frac{1}{2980}$	$\frac{1}{4470}$	$\frac{1}{8930}$	$\frac{1}{17,870}$
40		$\frac{1}{1690}$	$\frac{1}{3390}$	$\frac{1}{5080}$	$\frac{1}{10,160}$	$\frac{1}{20,320}$
45		$\frac{1}{1720}$	$\frac{1}{3440}$	$\frac{1}{5160}$	$\frac{1}{10,310}$	$\frac{1}{20,630}$
50		$\frac{1}{1690}$	$\frac{1}{3390}$	$\frac{1}{5080}$	$\frac{1}{10,160}$	$\frac{1}{20,320}$
60		$\frac{1}{1490}$	$\frac{1}{2980}$	$\frac{1}{4470}$	$\frac{1}{8930}$	$\frac{1}{17,870}$
70		$\frac{1}{1100}$	$\frac{1}{2210}$	$\frac{1}{3310}$	$\frac{1}{6620}$	$\frac{1}{13,250}$
80		$\frac{1}{590}$	$\frac{1}{1180}$	$\frac{1}{1760}$	$\frac{1}{3530}$	$\frac{1}{7050}$
85		$\frac{1}{300}$	$\frac{1}{600}$	$\frac{1}{900}$	$\frac{1}{1790}$	$\frac{1}{3580}$

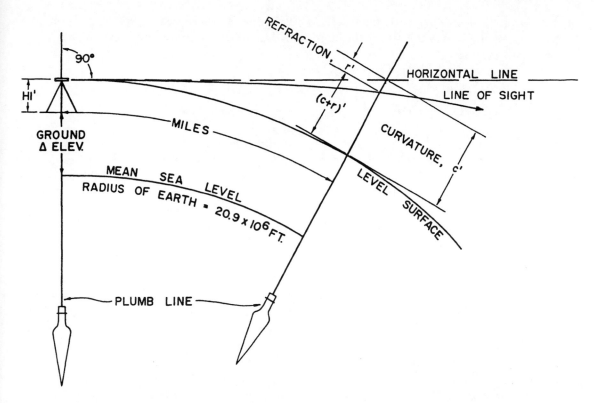

Figure 7-4.   Curvature and refraction.

This effect is generally neglected in construction surveys. In precise control surveys performed by the direct technique, backsights and foresights are closely balanced to reduce or cancel this systematic error, as shown in Figure 7-5, where $(c + r)_1$ cancels $(c + r)_2$ when $d_1 = d_2$.

Another way to compensate for this error is *reciprocal leveling*, where sightings must be taken across a gorge, canyon, or river. An instrument is set up on both sides of the obstacle, and the level rods are read simultaneously to cancel or reduce the errors. This is accomplished by meaning the results. An assumption

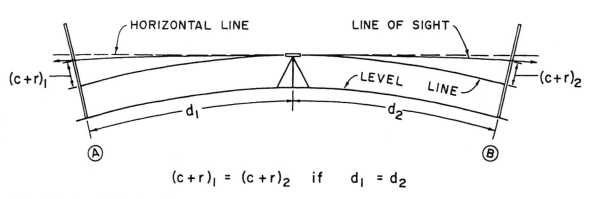

$$(c + r)_1 = (c + r)_2 \quad \text{if} \quad d_1 = d_2$$

Figure 7-5. $(c + r)$ Balancing effect.

made in reciprocal leveling is that refraction at both ends of the line is the same, although this is seldom the case.

7-7. INDIRECT LEVELING

In *trigometric leveling*, a similar curvature and refraction condition exists, as illustrated in Figure 7-6. The elevation of point C above A' is $BD + DE - BC$. For long sights, triangle $A'BD$ is still considered to be a right triangle and horizontal distance $A'D$ is taken as the level surface length AE. In current EDM practice, it is usually convenient to measure the slope distance $A'B$. Vertical distance BD is calculated as $A'D \tan \alpha$ and DE is the equivalent curvature and refraction correction $(c + r)$.

A theoretical way to overcome this $(c + r)$ correction is to simultaneously observe with two theodolites at A and C, thereby cancelling the systematic errors. Rarely are refraction conditions identical at both ends of a line, but averaging helps to distribute the error and is the generally accepted measuring technique for long-distance trigometric leveling.

7-8. OTHER LEVELING PROCEDURES

In addition to standard field surveying instruments, specialty items such as *barometers*, *lasers*, and the *global positioning system* (GPS) (Chapter 15) with receivers and antennas are available for unique leveling tasks. Air pressure changes with height, but precision surveying *altimeters* have been developed that use changes in barometric pressure to determine elevation differences from a base station with a precision of 2 ft. For small area surveys, a single altimeter may suffice for local photogrammetric control or rough spot-elevations, where easy access exists for vehicles. The procedure sets the altimeter to a known base elevation, and then readings are taken at desired locations. Finally, on returning to the base station, any difference in this reading from the original is distributed linearly around the loop.

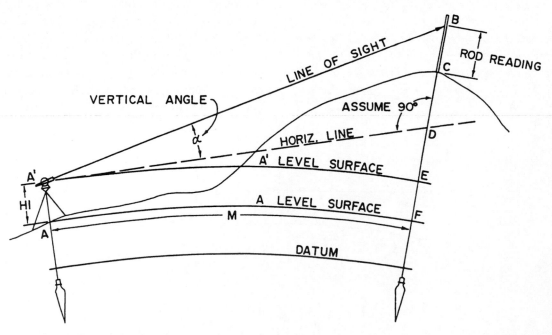

Figure 7-6. Trigonometric leveling.

An improved technique uses two altimeters. One is placed at the base bench mark and its changes noted, so time-dependent corrections can be applied to the roving instrument. Additional altimeters permit a leapfrogging technique to produce higher accuracy.

Construction site lasers have been developed that hold line for horizontal control but also provide vertical elevations. Alignment lasers can be placed in sewers, water lines, and construction trenches, and positioned on a slope. A target attached to the pipe end shows the low-power beam and permits pipe ends to be positioned on the proper design slope.

Another laser device can be leveled to rotate the beam over the area in a horizontal plane. Elevations to the nearest one-tenth of a foot on the building site are determined. A sliding light-sensing device attached to the side of the leveling rod can be moved up or down to receive the rotated laser light and show how far the rod's end is below the laser plane. Some systems allow an audio as well as a visual determination of the foresight. Systems of this type can be run by an equipment operator and do not require a two-person surveying crew.

The GPS consists of earth *satellites*, which transmit ultrastable signals with timing using cesium and rubidium atomic clocks, ground-based equipment to monitor the satellites, and a receiver to convert the signals into positions in a given reference frame. The ground equipment needed to use the system consists of an antenna, receiver, and computer that passively receives the signals from four or more satellites and computes a three-dimensional position.

7-9. INSTRUMENTATION

Leveling instruments are designed to have the sight line in or near the level surface; this is accomplished in a variety of ways by both simple and complex devices. For example, building inspectors may wish to check a con-

crete flat slab to determine if high and low points are within $\pm 1/8$ or $\pm 1/4$ in. of the mean surface elevation stated in job specifications. By flooding the slab with water, it is a simple task to measure the puddle depth with a carpenter's rule and sight across the water to estimate ridge heights. Early Roman builders used water leveling troughs with sight vanes to level aquaducts and tunnels—crude engineering, but projects were constructed and civil works accomplished with such elementary devices. Modern instrumentation improves the precision of layout and construction.

7-10. HAND LEVEL

A basic instrument used in construction today is the *hand level*. It consists of a spirit bubble and sighting horizontal wire in a telescope having zero or 2 × magnification. Figure 7-7 shows a spirit bubble vial.

Low-cost hand levels generally have a sighting chamber with no optics and only a horizontal cross wire and mirror to show the bubble image on the wire. Sights are considered level when the bubble is centered on the wire. Low-powered optics are often introduced into the hand level for work up to 50 or 75 ft. Unaided sights are generally used up to 30 ft for elevations to 0.1 ft, the normal specification figure for grading around building construction sites. Demands for more accuracy are met by typical engineers' levels, which include *dumpy* and *automatic levels*.

7-11. LEVELING VIALS

Level vials are rated by their *sensitivity*, the degree of tilt that moves the bubble through one 2-mm (or 0.01 ft or 0.1 in. in older instruments) division etched on the vial, or by the radius of curvature ground on the inside by their manufacturer. A short radius causes the bubble to be insensitive, but suitable. A hand-level bubble might have only one center grad-

GLASS VIAL
with circular curvature

PAINTED GRADUATIONS
at 2 mm spacings

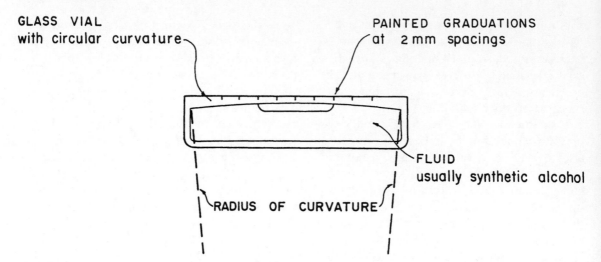

FLUID
usually synthetic alcohol

RADIUS OF CURVATURE

Figure 7-7.   Bubble vial.

uation, so the observer estimates when the bubble is centered. A more precise instrument, the dumpy level, has a 20″ of arc/2-mm divisions or 68-ft radius bubble sensitivity. For first-order instruments, the values can be 2″/2 mm at a 680-ft radius. Special construction techniques are employed to grind the curva-

ture into the glass vial at these high sensitivity values.

Two techniques are available to test bubble sensitivity. The laboratory procedure is to set the bubble into a level bubble trier, which is a slotted piece of wood hinged at one end and mounted on a precise screw. The vial is placed

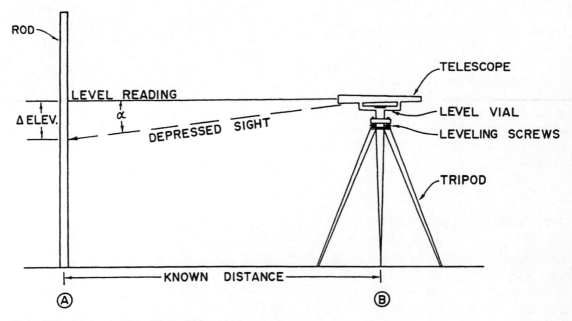

Figure 7-8.   Testing bubble vial sensitivity.

in the apparatus slot and raised or lowered by the screw. When the screw thread pitch is known, the number of turns of the screw to move the bubble through a known number of graduations tells the bubble sensitivity.

A field technique is shown in Figure 7-8. Here an engineer's level is set up some known distance from a leveling rod and a reading is taken. The level is then tilted through a known number of divisions on the bubble, and a new reading is taken on the rod. The sensitivity of the instrument s is given by

$$ s = \tan^{-1} \alpha \left[\frac{\Delta \text{elev}}{(AB)(\# \text{ of div})} \right] $$

where Δelev and AB are in the same units, and s is usually expressed in minutes or seconds of arc/2-mm division.

The fluid inside bubble vials is usually synthetic alcohol, which will not freeze at low temperatures. With the extremely precise instruments used for first-order work, such as a universal theodolite or precise level, the quantity of fluid in the bubble tube can be regu-lated by a special chamber attached to the vial. This ability to change the length of the bubble allows the instrument operator to compensate for temperature changes and use the most sensitive portion of the vial.

7-12. ENGINEERS' LEVELS

A series of engineers' levels with sensitive level bubbles, high telescope magnification, and fine-pitch leveling screws have evolved for construction and precise surveying work. Modifications include prisms instead of level bubbles, spherical ball-seat leveling surfaces rather than leveling screws, and tilting telescopes instead of fixed ones. Figure 7-8 shows a tripod, the leveling screws to tilt the instrument head so the leveling bubble or leveling device can be centered, and a telescope permitting long sights to be made.

A modern level is shown in Figure 7-9. Light rays enter the telescope *objective lens*, which has a special light-gathering coating on

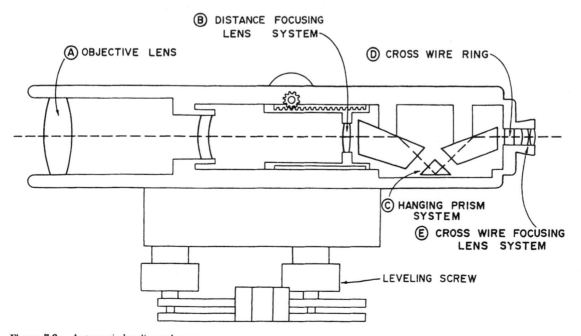

Figure 7-9. Automatic leveling telescope.

its outer surface, and pass through a distance-focusing movable lens that allows the image of a distance object to be in focus on the cross-wire ring. The pendulum system shown has two fixed prisms and a freely hanging third prism that passes the horizontal rays onto the cross-wire ring. To allow the pendulum system to hang free, a low-sensitivity spirit bubble is mounted on the instrument, and rough leveling of the instrument produces an "automatic leveling" of the sight line.

Different manufacturers have designed various systems to accomplish automatic leveling; instruments so constructed are called *automatic levels*. At the telescope's eyepiece end, a magnifying lens system exists enabling the instrument operator to simultaneously view the cross-wire ring and distant image. If focusing is not performed exactly, a condition known as *parallax* can exist and cause small reading errors by the observer.

In older telescopes, the cross wires were spider webs cemented to a ring. Newer instruments have lines etched on a glass reticle that sits in a metal holder. The holder can be repositioned to correct for systematic errors in the hanging prism system.

7-13. THE DUMPY LEVEL

The engineer's dumpy level is an older instrument that contains leveling screws, a spirit bubble, and a telescope (Figure 7-10). The four leveling screws are worked in opposite pairs to position the bubble at the vial center. The bubble follows the directional movement of the left thumb. Automatic levels generally have three leveling screws and the telescope elevation is changed slightly by raising or lowering one or two leveling screws. Four-screw instruments rotate around a spherical ball seat, and once leveled, they can be releveled to the original position if the bubble wanders away from the center position.

A telescope *line of sight* is the straight line through the center of the objective lens and intersection of the horizontal and vertical cross wires. Instrument design requires the line of sight to be made parallel to the axis of the level bubble vial—i.e., the line tangent to the radius of curvature of the vial at the marked center location when the instrument is exactly leveled. Because it is difficult to construct an instrument with a fixed telescope and bubble, each part can be shifted slightly so adjustments

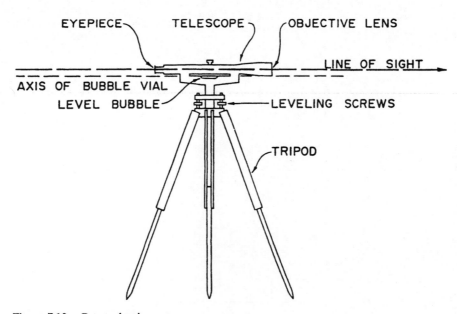

Figure 7-10.   Dumpy level.

to remove systematic instrumental error may be made in the shop or field. On construction instruments such as the old *wye level*, the dumpy and the automatics, their telescopes have approximately 32× magnification, and sensitivity of the bubble vials is about 20″/2-mm divisions.

added to the rod reading. This system can be very exact and repeated for a statistical determination of the difference to be added to the rod reading. Automatic instruments used in construction can be fitted with a removable optical micrometer and approach the accuracy of special, extremely precise, automatic levels used for first-order leveling work.

7-14. FIRST-ORDER LEVELS

For first-order work, a spirit-bubble-type instrument with greater magnification and bubble sensitivity has been constructed with a horizontal pin around which the telescope can rotate when driven by an extremely fine-pitch screw. This tilting level also has a reticle containing three horizontal wires, so the horizontal distance to the leveling rod can be read by stadia (see Chapters 6 and 22). To assist in obtaining an exact middle-wire reading, an *optical micrometer* is built onto the end of the telescope; this arrangement is shown in Figure 7-11. A *planoparallel lens* serves as the *objective lens* and can be tilted through a range equal to the 5- or 10-mm graduations on the leveling rod. An observer moves the micrometer to make the line of sight fall exactly on a rod graduation. The micrometer reading is then

7-15. LEVELING RODS

To determine the height of instrument above a bench mark, a graduated rod is held vertically on the point and a reading taken. For precise work, rod graduations are in the SI system (meters), whereas on construction and other work in the United States, markings are in decimal feet. When a hand level is used, a carpenter's rule graduated in feet and inches might be employed. Rods come in a variety of sizes and shapes, but the *Philadelphia* type has an advantage, since the rodperson can independently check instrument operator readings. These rods generally come in two pieces, which allow readings up to 12 or 13 ft to be taken and checked by setting a target on the graduations, which are shown to 1/100 of a foot (see Figure 7-12).

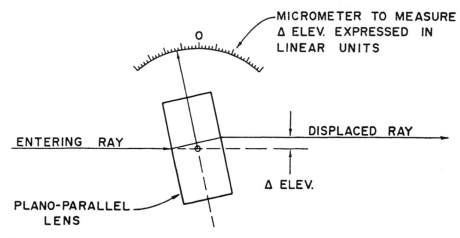

Figure 7-11. Optical micrometer.

The foot lengths are indicated by red numbers, and the 0.01-ft wide graduations are painted black on a white background. The intersection of the black and white portions of the graduations defines the value. As shown in the red 3-ft portion of the rod, the horizontal cross-wire reading is 3.64 ft. A vernier target can be slipped over the rod and readings taken to 0.001 ft with the direct-reading vernier for instrument sight distances of 200 ft or shorter. This target can be used on the lower portion of the rod, and a technique called *high rod*, employing a downward-reading scale on the back of the rod, permits the rodperson to verify high-rod readings with a properly set target and vernier on the rod's back side.

The metric-faced construction rods are usually color-coded orange for meter readings with centimeter spacings. The instrument operator reads the meters, decimeters, and centimeters directly from the rod. Then, if a fine reading is desired, the number of millimeters is estimated. A target with scale provides a millimeter check estimate.

On precise work, all readings are taken in the SI system. The rods are of special one-piece construction approximately 3 m long, which presents special shipping and carrying prob-lems. The rod face is made of Invar steel and held to the face under tension. A thermometer is usually provided so a temperature correction can be applied. The smallest graduations on the rod face are in centimeters or 5-mm values. The optical micrometers on first-order instruments allow repeatable readings to 0.1 mm, the equivalent of 0.0003 ft or 0.00003 in. Because of these small values, plumbing the rod is usually done with an attached bull's-eye level or special holder and special care given to placing the rod on a hard permanent surface. *Portable turning points* are carried into the field and used to provide a stable platform for the rods.

7-16. NOTEFORMS

Different noteforms are used for various standard direct-leveling procedures. When running a third-order control project, traverse (when suitable), around a construction site, a closed noteform is generally used (see Table 7-3 and Figure 2-2).

This noteform contains one *backsight* (+S) and one *foresight* (−S) per setup and illustrates differential leveling. The *height of instru-*

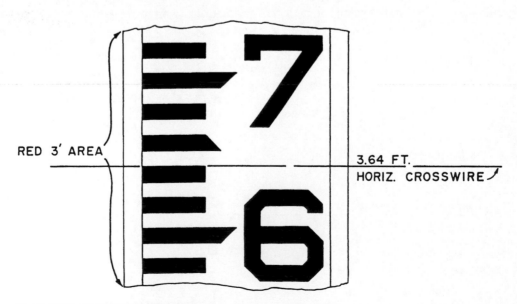

Figure 7-12.   Philadelphia rod face detail.

Table 7-3. Establishing height of green pin (closed noteform)

STA	+S	HI	−S	ELEV (ft)
BM 235	3.47	239.23	—	235.76
TP-1	9.10	246.62	1.71	237.52
TP-2	7.91	253.46	1.07	245.55
Green Pin	3.07	252.57	3.96	249.50
TP-3	0.13	245.59	7.11	245.46
TP-4	1.90	240.16	7.33	238.26
BM 235			4.41	235.75
	+ 25.58		− 25.59	− 235.76
			+ 25.58	− 0.01✔
			− 0.01✔	

Table 7-4. Left-hand page, partial set of three-wire notes (in mm)

STA	Thread Backsight	Mean	Thread Interval	Sum of Intervals
	1216			
209	1108	1108.3	108	215
	1001		107	
	3325			
	1326			
TP-1	1237	1237.0	89	178
	1148	2345.3	89	393
	7036			

ment (HI) is determined for each setup but not done in precise leveling where only the difference in elevation between two bench marks is desired. As shown, the Σ backsights − Σ foresights = final elevation − initial elevation. Accidental errors in leveling do not usually allow a circuit that starts and stops at the same point to generally close with the same value. This misclosure would have to be adjusted out by some technique consistent with the caliber of the work.

In first- or second-order control work, the notes include readings on a *precise level's* upper, middle, and lower wire, thereby allowing several checks and providing a stadia distance. One page of the field book is devoted to backsights, the other to foresights. The *stadia-interval factor* for the instrument must be known, and acceptable precomputed values for the maximum length of sight, maximum differences in lengths observed per setup, and maximum differences in cumulative distances for a section are computed and recorded by the notekeeper. A partial set of notes is shown in Table 7-4.

7-17. PROFILE LEVELING

When topographic conditions will not allow differential leveling or low precision is suffi-

cient to accomplish the job, *profile leveling* is used in combination with differential leveling. Side shots with no checks can be taken in determining the profile of a road and/or the elevation of a manhole. An example of profile notes is shown in Table 7-5.

A separate column of *intermediate foresights* (IFS) was listed, with *side shots* taken to stations 0 + 00, 1 + 00, 2 + 00 and 2 + 52 on Highland Street. A station in highway work is 100 ft on the ground and Sta 0 + 00 is the beginning point. Sta 1 + 00 is a point 100 ft away and usually on the street centerline. Manhole (MH) number 17 might be a grade shot taken to the top or to an invert in the manhole. A note or sketch on the right-hand page of the field book would clarify the situation. The nonchecked profile shots are recorded to only 0.1 ft, which is consistent with street subgrading work.

7-18. PRECISE LEVELING

A precise level without an optical micrometer was used for the notes in Table 7-4. The rod reading was estimated to the nearest millimeter at the three-wire positions and intervals between the upper-middle and middle-lower wires recorded in column four. A check is performed here, since these intervals should be the same, but they are usually allowed to

Table 7-5.   Highland Street profile

STA	+S	HI	−S	−IFS	ELEV (ft)
BM 235	3.16	238.92	—	—	235.76
TP-1	7.71	244.87	1.76		237.16
0 + 00				1.7	243.2
1 + 00				3.5	241.4
2 + 00				7.1	237.8
2 + 52				6.7	238.2
MH #17				10.6	234.3
TP-2	7.15	250.51	1.51		243.36
Green Pin			1.00		249.51
	+18.02		−4.27		−235.76
	−4.27				+13.75✔
	+13.75✔				

differ by up to 3 mm. Since the upper interval is greater by one unit than the lower one, the mean of 1108.3 is entered in column three. A check of this mean is obtained by adding the three thread backsights to 3325 and dividing by 3. The sum of the intervals is recorded for use with the right-hand page to ensure that the proper sight lengths have been maintained and balance properly. The final product from the note page is the *mean sum* of the backsights, 2345.3 mm, as shown. The difference in elevation is obtained by adding algebraically the mean sum of the foresights from the field book's right-hand page.

Many variations exist in noteforms and instrumentation. Optical micrometers require special notes, as do rods that have offset graduations on the left and right side to help improve the work by providing a statistical value and check against gross mistakes. Large governmental agencies and private firms generally use standard in-house noteforms written for computer reduction, if not already recorded in that format in an electronic field-data collector.

7-19. PROFILES IN HIGHWAY DESIGN

As noted earlier, the purpose of leveling is to locate objects in the vertical direction. Some standard leveling applications include (1) road profiles, (2) sewer and drainage design, (3) *borrow pits*, and (4) simple mapping.

A profile is a vertical section through the surface of the ground along any fixed line. In highway design work, it is important to know elevations along the proposed route centerline and plot them both along the centerline and at right angles to it. Profile levels are run over the proposed centerline. They are plotted in a form similar to that shown in Figure 7-13, and the road slope (grade or gradient) is selected by comparing centerline cuts and fills on trial grade lines.

The horizontal and vertical scales of a profile are generally different. The abscissa (*x*-axis) is usually noted in stations and the ordinate (*y*-axis) usually in feet above a datum. If the data are put into a computer graphics system and viewed on the computer screen, optimum road grades can be determined without a need for profile paper and design drafting boards.

After the proposed centerline is chosen, a designer will normally want to calculate the volumes of cut-and-fill materials required for the project. This information can be obtained in the field with a hand level or engineer's level and tape through a process known as *slope staking*. It can also be done photogrammetrically in a computer system that contains a *digital terrain model* at a fixed-grid spacing. In either case, stakes must be positioned in the field for construction control. See Chapter 24 for more details on noteforms.

7-20. LEVELING FOR WATER AND SEWER DESIGN

A similar but different application is in the design of a sewer or gravity-feed water system. The basic principle used is that water flows downhill, and a leveling process establishes the relative elevations of critical ground elevations. For a sewer line, a set of profile levels is taken along the proposed sewer location and a ground profile similar to Figure 7-13 drawn. Next, the survey engineer picks the critical depth and slope locations for manholes and prepares a set of profile drawings for contractor inspection and bidding. In constructing a sewer line, leveling is again used to locate the depth of trench cut and to place the pipe at its design elevations. In locating the manhole's pipe elevations, *inverts* are first established as

these occur at breaks in grade. Then batter boards are erected over the trench to contain horizontal and vertical control. To set the inside bottom of each piece of pipe at its correct elevation, workers *stringline* the pipe and measure vertically from the string. On larger jobs, a surveying laser is set up in the trench or pipe and the beam directed down the centerline of the proposed pipe location to targets attached to the pipe. Construction surveying consists of many applications where leveling plays a major role in the building process (see Chapter 24).

7-21. EARTHWORK QUANTITIES

An application of leveling to earthwork quantity calculations is in borrow-pit volumes (see Figures 5-8 and 7-14).

After it has been determined that suitable materials are to be removed from the ground for placement at another site, a surveyor lays out a horizontal grid of equal spacing, and

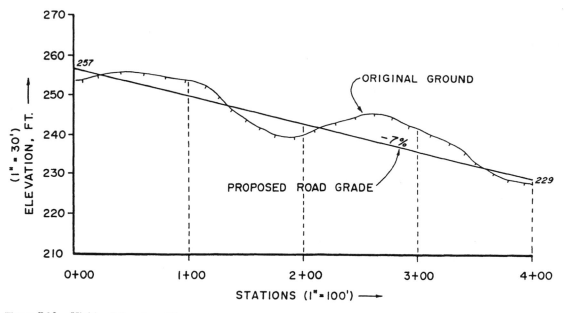

Figure 7-13. Highland Street profile.

numbers and letters the grid lines to reference the intersection points as 2-B, 3-C, etc. Level sights are taken on the grid corners and other points at breaks in the ground slopes, from a single instrument setup if possible, or by running a level circuit. Ground elevations are calculated. During and after excavation, whenever earthwork quantities are desired (perhaps for payment purposes), the intersection points are releveled and the volume of material removed from the regular prisms is determined by the following equation:

$$V = \ell \times w \, \frac{(\Sigma h_1 + 2\Sigma h_2 + 3\Sigma h_3 + 4\Sigma h_4)}{4 \times 27}$$

Here ℓ and w represent the individual length and width grid dimensions and the h's represent the difference in elevations at the number of prism corners to be counted in the calculations. In Figure 7-14, 2-B is the change in elevation of one corner of prism a, 2-C would be a corner common to prisms a and b, and 3-C would be a corner common to the four prisms a, b, c, and d. When the edge of excavation does not fall along the grid line, a wedge of earth might exist that should be included in the calculations. If the excavation is rough, or a better determination of irregular sides and bottom of the pits is needed, grid spacing should be reduced and volumes determined on smaller size prisms of earth.

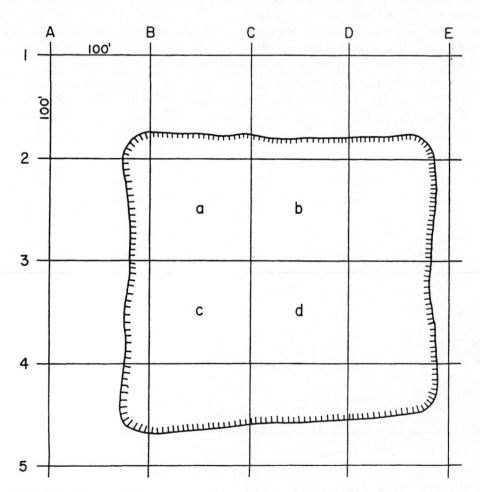

Figure 7-14.   Borrow pit.

7-22. MAPPING

In laying out a subdivision or a house lot, the survey engineer should map the topographic surface to best decide how the terrain can be used to enhance the house setting. This can be done by the grid technique for areas that are flat or small in size (see Figure 7-15). In this example, the ground elevations on the grid points have been established by leveling and a map drawn showing a series of contour lines that represent points of similar elevations. These contours present ground configurations and can be envisioned by the designer and owner to assist in house location.

7-23. ERRORS

As in all surveying work, when instruments are used, it is possible to make *blunders, systematic errors,* and *accidental errors* (see also Sections 3-5 to 3-10). Proper notekeeping and appropriate field procedures should eliminate the first two,

while multiple readings can reduce the third to a minimum. Examples of blunders in leveling include (1) using the wrong point for a bench mark, (2) reading the rod incorrectly, (3) reading on the stadia cross wire instead of the middle wire, and (4) transposing numbers in field books. Because of the repetitive nature of direct leveling, it is important that the instrument operator and notekeeper not slip into bad habits or shortcuts to speed up the work. When looking for a control bench mark from which to begin a survey, the descriptions are sometimes vague and the wrong point can be selected as the beginning monument. This can be checked by locating two known bench marks and leveling between them. In addition to checking an error in monument identification, this procedure will also disclose any differential settlement of one or both points.

Leveling rods that are not self-checking can cause reading errors when the horizontal cross wire is close to a foot or meter mark. The reading 1.92 ft can easily be cited as 2.92 ft because the instrument operator sees the large red 2 numeral by the 0.92 reading and may

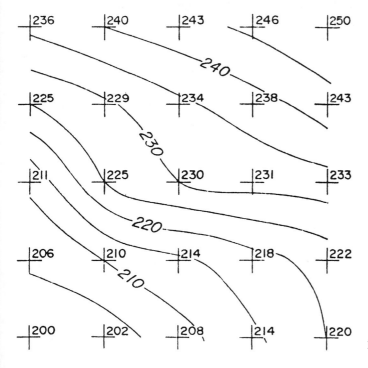

Figure 7-15. Contour lines.

fail to note that the reading is in the 1-ft interval of the rod. The Philadelphia leveling rod, when used with target, will catch this type of blunder.

Leveling instruments that contain stadia cross wires can also result in the wrong wire being read. Even setting the target will not eliminate this error unless the reader turns his or her head aside and conscientiously resights the reticle with this error in mind. Manufacturers have shortened the stadia wires in an attempt to help distinguish them from the horizontal cross wire, which runs over the entire field of view.

7-24. SYSTEMATIC ERRORS AND ADJUSTMENTS

All surveying instruments have systematic errors that result from improperly assembling components to meet certain geometric conditions, such as the line of sight being parallel to the axis of the level bubble vial. Manufacturers have built adjustable components into the

equipment, so field or laboratory adjustments can be performed to place the equipment in near-perfect alignment. Systematic errors are reduced or eliminated by instrument adjustments and proper field procedures.

Modern leveling equipment has two geometric conditions that must be corrected or serious systematic errors will accumulate on a long level line. The first requirement is to make the axis of the bull's-eye level on an automatic level, or the axis of the spirit level on a dumpy level perpendicular to the vertical axis (see Figure 7-16). As shown, the vertical axis of the level is tilted and the bubble centered in the tube. The level bubble axis *AB* is now rotated 180° so end *A* goes to *A'*, and *B* to *B'*. The bubble will not stay in the tube center and the distance that the bubble is off center is twice the error present. The bubble should be brought halfway back with an adjusting pin and then placed in the center of the vial using the leveling screws. This adjustment will make the vertical axis truly vertical.

Only after the first adjustment is completed may the second condition be adjusted—i.e., to

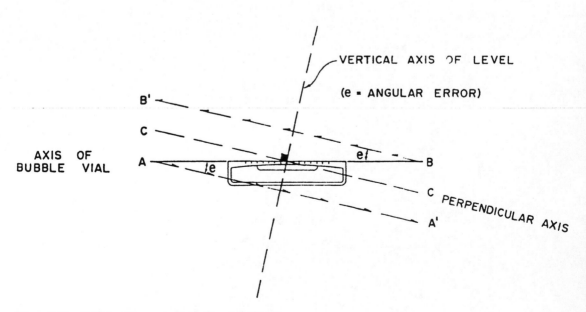

Figure 7-16.   Bubble axis perpendicular to vertical axis.

make the telescope's line of sight parallel to the axis of the level bubble vial (see Figure 7-17).

If the line of sight is not parallel to the bubble-vial axis, it will sight up or down and give an erroneous rod reading. In differential leveling, this error can be canceled out by keeping the backsight and foresight distances equal. In profile leveling, the error increases with distance from the instrument. *Different-leveling* procedures should therefore be used and the bubble always placed in the tube center for each sighting. This means the leveling screws on older four-screw dumpy levels can be readjusted for each sight, and the sight distances balanced by some technique—e.g., pacing, stadia, counting rail lengths or concrete highway expansion joints.

Before and during a run of profile levels, it is best to level the instrument and perform the level-bubble-axis test by rotating the telescope 180° and bringing the bubble *one-half* way back to the center with the leveling screws. Then, take the profile readings and the values should be satisfactory except for atmospheric and refraction corrections.

A check on this second geometric condition is called the *peg test* and can be performed in a variety of ways. One precise variation of the technique is to set up the level midway between two stakes 200 ft apart and determine the true difference in elevation between the stakes. The instrument is then placed at some convenient distance, say, 20 ft, behind one of the stakes and a reading taken on the first one. If we assume that the instrument is in adjustment, the rod reading on the second stake is calculated, and for a precise determination, a correction is made for curvature and refraction. The rod is then placed on the distant stake and read. It will equal the calculated value for an instrument in adjustment or be greater or smaller if the line of sight is up or down. The telescope reticle can be adjusted to a correct position based on the measured value. For more details on instrument adjustment, see Chapter 8.

7-25. SPECIFICATIONS

In order to keep systematic errors at a minimum, agencies have developed specifications for various orders of work. Table 7-6 has been used by the NGS, NOAA, and generally accepted by persons wishing to fulfill first- or second-order vertical-control surveys. These specifications show three orders of work: first-, second-, and third-order, with classes I and II listed under the first two. First-order, first-class pertains to the basic framework of the national network and is the most precise work performed. Third-order work is used on small engineering projects.

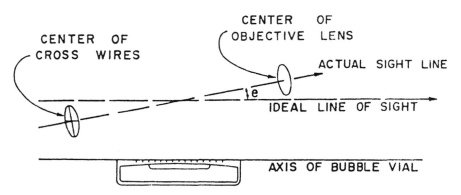

Figure 7-17. Line-of-sight condition.

Table 7-6. Classification, standards of accuracy, and general specifications for vertical control (1974)

Order Class	First I	First II	Second I	Second II	Third
Classification	Class I, Class II	Class I	Class II		Center wire
Minimal observation method	Micrometer	Micrometer	Micrometer or 3-wire	3-wire	Center wire
Section running	SRDS* or DR or SP	SRDS or DR or SP	SRDS or DR† or SP	SRDS or DR‡	SRDS or DR§
Difference of forward and backward sight lengths never to exceed					
Per setup (m)	2	5	5	10	10
Per section (m)	4	10	10	10	10
Maximum sight length (m)	50	60	60	70	90
Minimum ground clearance of line of sight (m)	0.5	0.5	0.5	0.5	0.5
Even number of setups when not using leveling rods with detailed calibration	Yes	Yes	Yes	Yes	—
Determine temperature gradient for the vertical range of the line of sight at each setup	Yes	Yes	Yes	—	—
Maximum section misclosure (mm)	$3\sqrt{D}$	$4\sqrt{D}$	$6\sqrt{D}$	$8\sqrt{D}$	$12\sqrt{D}$
Maximum loop misclosure (mm)	$4\sqrt{E}$	$5\sqrt{E}$	$6\sqrt{E}$	$8\sqrt{E}$	$12\sqrt{E}$

Single-run methods
Reverse direction of single runs every half-day

Nonreversible compensator leveling instruments
Offlevel/relevel instrument between observing the high and low rod scales

3-wire method
Reading check (difference between top and bottom intervals) for one setup not to exceed (tenths of rod units)
Read rod 1 first in alternate setup method

Double scale rods
Low-high scale elevation difference for one setup not to exceed (mm)
With reversible compensator
Other instrument types
Half-centimeter rods
Full-centimeter rods

Reverse direction of single runs every half-day	Yes	Yes	Yes	—	—
Offlevel/relevel instrument between observing the high and low rod scales	Yes	Yes	Yes	—	—
Reading check... for one setup not to exceed (tenths of rod units)	—	—	2	2	3
Read rod 1 first in alternate setup method	—	—	Yes	Yes	Yes
With reversible compensator	0.40	1.00	1.00	2.00	2.00
Half-centimeter rods	0.25	0.30	0.60	0.70	1.30
Full-centimeter rods	0.30	0.30	0.60	0.70	1.30

* SRDS, single-run, double simultaneous procedure; DR, double-run; SP, spur, less than 25 km, double-run; D, shortest length of section (one way) in km; E, perimeter of loop in km.
† Must double-run when using 3-wire method.
‡ May single-run if line length between network control points is less than 25 km.
§ May single-run if line length between network control points is less than 10 km.

Instruments acceptable include automatic, tilting, or geodetic types. The rods comprise those containing Invar scales and steel-face rods. The lines should be double-run forward and backward, and monuments spaced at a maximum of 1- to 2-km sections in larger loops. The limiting length of sight is 50 to 90 m.

To minimize possible tilt in the line of sight, the maximum difference in length between a forward and backward sight is restricted, as well as the cumulative difference in lengths. A factor based on inclination of the line of sight is applied to the final unbalanced length to correct fieldwork for any tilted sight error.

When a survey starts at a point and loops back to the same point, the accidental errors in reading, sighting, and atmospheric conditions are proportional to the number of setups and/or distances between bench marks. The combined error can be estimated based on the maximum-length-of-sight restriction, a closure calculated using a coefficient for the order of work, and the square root of the distance around the loop or length of a section.

In large networks, the area loops intersect at certain common station points. Work done in each loop will contain the separate loop errors, and elevations at the common points must be adjusted to provide consistent data. These adjustments are usually performed by the method of least squares suited to surveying data that follow normal error-theory distribution (see Chapters 3 and 6).

7-26. PRECISE THEORIES

When performing precise first-order surveys, special note should be made of the fact that the earth is not spherical but lumpy and shaped like an oblate spheroid. Because the earth is spinning, gravity values vary in the north-south direction from 978 gal at the equator to 983 gal at the poles. Named for Galileo, a *gal* is a gravity unit equal to 1 cm/sec^2. The earth has a molten core and hard crust that bulges at the equator, making distance from the center of the earth to the equator longer than that from the center to the pole (see Figure 7-18 on p. xxx). A review of the figure shows that gravity values decrease as you go farther from the center of the earth, and the gravity value at the pole is higher than at the equator. Because of the difference in the *a* semimajor and *b* semiminor lengths, the

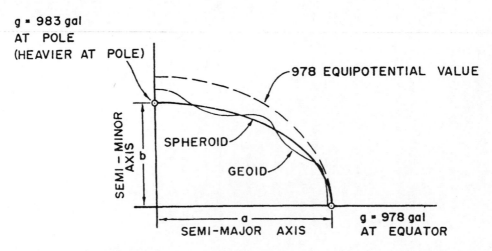

Figure 7-18.   Earth's bulge.

rate of change of the gravity field streamlines in the north-south direction of the pole. A problem results because leveling instruments are set up on the earth's surface, and the gravity field changes at a different rate, with elevation in going from the equator to the pole.

7-27. GEOIDAL HEIGHT

Leveling data should be referred to the *geoid*. The difference between the topographic surface and geoid is known as a *geoidal height* (see Figure 7-19). For horizontal surveying computations, positions are referred to the pure mathematical spheroidal surface. A geoidal height and spheroidal height differ by undulations in the geoid. A geoidal height expressed in linear units is called an *orthometric* height. For first-order work run in the north-south direction, a special *orthometric correction* is required. It can be taken from nomograms, which require using a section's mean elevation and mean latitude to get orthometric correction per minute of change in latitude.

A second and more precise method of determining the size of correction is to measure the gravity values at the bench marks on the level line. This can lead to a new system of measurements that better explains the physical earth situation (see Figure 7-20). It assumes a north-south direction and shows a change in latitude on the earth. No corrections are applied to measurements taken in the east-west direction.

Points *a* and *b* on the topographic surface are vertically above points *A* and *B* on the geoid. The *equipotential gravity surfaces* converge going northerly. This means that plumb lines *curve* because they are perpendicular to the gravity field and hence lines *aA* and *bB* are shown slightly curved in the figure. It also means that length *aA* does not equal *bB* because the gravity surfaces streamline.

Leveling from *B* to *A* on the geoid will show no change in elevation. However, leveling from *B* to *b*, then along the level surface *ba*, which is in the north-south direction, and finally leveling from *a* to *A* will result in an elevation change equal to the distance (*bB* − *aA*). The second result will be different from direct leveling of *BA*. Thus, by using normal recording techniques, two different route-depended elevations will be obtained by leveling from *B* to *A*.

To overcome this recording problem, a different measuring system has been derived that accounts for the linear change in height and change in the gravity field. The result is a

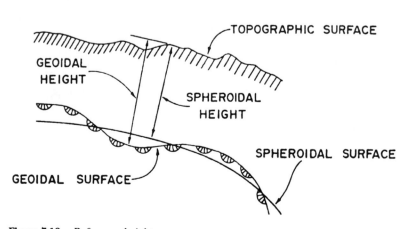

Figure 7-19. Reference heights.

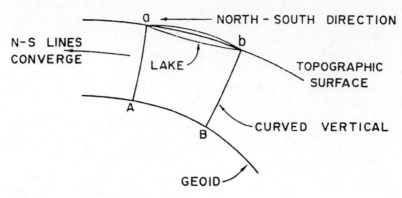

Figure 7-20. Orthometric problem.

Table 7-7. Adjusted surface gravity, leveling elevations, geopotential numbers

Station	Surface Gravity (milligal)	Leveling Elevations (m)	Geopotential Number (geopotential unit)
Corvallis, OSU-PC	980 573.14*	77.142	75.643
$\tau\beta\pi$	980 575.49	71.337*	69.951
U54	980 573.21	76.787†	75.295
Corvallis, OSU-KL	980 573.81	73.336	71.911
College	980 573.31	72.219	70.816
RM2 College	980 573.28	71.880	70.484
RM1 College	980 573.28	72.757	71.344

* Fixed from published data before adjustment.
† Fixed through leveling observations.

geopotential number

$$\text{GPU} = \int_{A}^{a} g\,dh$$

where g is a variable acceleration due to gravity and dh the linear elevation change. This record system requires knowledge of the gravity field by direct measurement or interpolation.

An example GPU calculation follows (see Table 7-7). The leveling elevation of station $\tau\beta\pi$ is given as 71.337 m and the adjusted surface gravity value, in millegal, is 980,575.49. The product of these numbers yields a GPU value for station $\tau\beta\pi$ of 69.951. This number better expresses the leveling condition at $\tau\beta\pi$ than the given elevation, but a great deal of work must be expended to determine an ad-

justed surface gravity value. Hence, this type of recording is generally used only for first-order leveling.

REFERENCES

ANDERSON, E. G. 1979 Are primary leveling networks useless? 39th Annual ACSM Meeting. Washington, D.C.

BALAZS, E. I. 1981 The 1978 Houston-Galveston and Texas Gulf Coast vertical control surveys. *Surveying and Mapping 41*(4), 401.

BALAZS, E. I., and C. T. WHALEN. 1977 Test results of first-order class III leveling. *Surveying and Mapping 37*(1), 45.

BERRY, R. M. 1976 History of geodetic leveling in the United States. *Surveying and Mapping 36*(2), 137.

_____ 1977 Observational techniques for use with compensator leveling instruments for first-order levels. *Surveying and Mapping 37*(1), 17.

BOAL, J. D., F. W. YOUNG, and R. MAZAACHI. 1984 Geometric aspects of vertical datums. 44th Annual ACSM Meeting. Washington, D.C.

CADDESS, H. N., and G. M. COLE. 1983 Precision leveling river crossing technique using reciprocal vertical angles. 43rd Annual ACSM Meeting. Washington, D.C.

COOK, K. L., and R. J. SCHULTZ. 1971 Understanding geopotential numbers. 37th Annual ACSM Meeting. Washington, D.C.

HOLDAHL, S. R. 1983 Correction for leveling refraction and its impact on definition of the North American datum. *Surveying and Mapping 43*(2), 23.

———— 1984 Aspects of a new height system for North America. 44th Annual ACSM Meeting. Washington, D.C.

HUETHER, G. 1983 Employment and experiences with NI 002 from Jenoptik Jena for the motorized leveling. 43rd Annual ACSM Meeting. Washington, D.C.

HUSSAIN, M., and R. D. HEMMAN. 1985 Accuracy evaluation of laser levels. 45th Annual ACSM Meeting. Washington, D.C.

KIVIOJA, L. A. 1985 Hydrostatic leveling. 45th Annual ACSM Meeting. Washington, D.C.

LIPPOLD, H. R., JR. 1980 Readjustment of the national geodetic vertical datum. *Surveying and Mapping 40*(2), 155.

QUINN, F. H. 1976 Pressure effects on Great Lakes vertical control. *ASCE Journal of the Surveying and Mapping Division 102*(SU1), 31–38. New York.

REMONDI, B. W. 1986 Performing centimeter-level surveys in seconds with GPS carrier phase: Initial results. 46th Annual ACSM Meeting. Washington, D.C.

8

Instrument Adjustments

Gerald W. Mahun

8-1. INTRODUCTION

Surveying instruments are very durable, but delicate and precise pieces of equipment. No matter how well an instrument has been adjusted, rough handling, temperature variations, humidity, and a host of other factors can quickly affect its precision. The safest rule a surveyor can follow is to keep an instrument adjusted, but then use it as if it is not adjusted.

A surveyor should not adjust equipment unless it is needed, since minor corrections may not be necessary or possible. The rules in this chapter outline procedures whereby surveyors first test equipment to determine if an adjustment is necessary, and the corrections are then carried out only if needed.

Nearly all adjustments of levels, transits, and theodolites are based on the *principle of reversion*. Reversing the instrument position by rotation in a horizontal or vertical plane doubles any error present, enabling a surveyor to directly determine how much correction is needed. If the instrument is badly out of adjustment, it may be necessary to repeat all steps to reduce the size of error each time.

As a rule, older levels and transits are easy to work with because of their simple, open construction. Automatic levels and theodolites

are more difficult, owing to their use of *compensators*, prisms, and glass circles. Many of their adjustments must only be done by qualified specialists; however, this chapter will outline those procedures surveyors can, with practice, perform.

A log book should be maintained for each instrument, stored in the instrument case or office, and an entry made noting the date and type of adjustment each time one is performed. This serves two purposes: (1) It reminds surveyors to periodically check an instrument's adjustment and (2) If one particular adjustment is consistently required, it indicates repair is necessary. A good instrument, properly adjusted and handled, should last a lifetime and spend a minimum number of hours in a repair facility.

8-2. CONDITIONS FOR ADJUSTMENT

Before making an adjustment, it is wise to ensure that any instrumental error tested for and found is a result of the equipment's condition and not the test's deficiency. To properly test and adjust equipment, the following rules should be followed:

1. Perform adjustments on a cloudy windless day, free of heat, if possible. Avoid situations where the sight line passes alternately through sun and shadow. Allow up to 30 min for the instrument temperature to stabilize, if there is a significant difference between the temperature at the storage and adjustment locations. The instrument should be shaded from any direct sun rays.

2. Make sure that all tripod hardware is snug, so the tripod will not shift under the instrument weight. Spread the tripod feet well apart and press the shoes firmly into the ground. Do not set up on a hard surface, as there is a chance a leg could either slide or get kicked out.

3. Choose a relatively flat area that provides flat sights for at least 200 ft in opposite directions.

4. Locate all adjusting nuts and screws and clean any threads that might be dirty. Most tools needed for older instruments consist of adjusting pins of various sizes. Test-fit the pins to see which adjusting nuts they are for. Do not use undersize pins, as they will ream out the holes in the adjusting nuts. Screwdrivers and wrenches, if needed, should be test-fit also. In any case, do not use a pair of pliers to grip a nut or screw. Adjusting pins can be readily fashioned from flush-cut nails that are carefully filed down to size. If any adjusting nut hole has been reamed, it can be carefully drilled out with a twist drill bit. The opposite end of the bit can then be used as an adjusting pin. When adjusting, do not overtighten the screws or nuts.

5. Perform adjustments in the proper sequence, as most are dependent on previous ones.

6. During and after the adjustments, handle the instrument carefully. Rough handling may negate any adjustments performed.

7. Refer to the instrument manual for any special adjustments. This is especially true for theodolites and automatic levels.

8-3. BREAKDOWN OF ADJUSTMENTS

Two types of adjustments are made on most surveying instruments: (1) preliminary and (2) principal. Preliminary adjustments are those performed each time an instrument is used and should habitually be checked each time the instrument is set up.

Principal adjustments are more detailed and are made only when a test indicates a need for them. They should be checked periodically to determine any possible instrument errors.

Sections of this chapter explain the different types of surveying equipment and their preliminary and principal adjustments. Surveyors should be capable of successfully adjusting most equipment by following the procedures explained.

8-4. GENERAL DEFINITIONS

Most instrument adjustments are partially dependent on the position of an air bubble or the intersection of a set of cross hairs under a given condition. To interpret these positions and relate them to an adjustment procedure, it is important to understand some of the mechanical aspects and common terms associated with the various surveying instruments.

The bubble tube or level vial is a sealed glass tube nearly filled with a nonfreezing, fast-moving, quite stable liquid—commonly, purified synthetic alcohol. The upper inner surface of the bubble tube is circular in the tube length's direction. The tube top is etched with graduations used to center the bubble and determine how far the bubble moves off center when the bubble tube is reversed. Generally, one end of the tube is fixed in position when mounted on the instrument; the other end can be raised or lowered by adjusting nuts or screws. The bubble-tube axis is an imaginary longitudinal line tangent to the midpoint of the upper inner curved surface of the bubble tube (Figure 8-1).

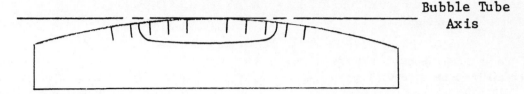

Figure 8-1. Bubble tube.

The cross hairs consist of very fine filaments of etched lines on a flat glass plate. They are placed on a reticle ring inside the telescope, forward of the eyepiece. The reticle ring is held in position by four capstan-headed screws (Figure 8-2) that pass through elongated holes in the telescope tube, so if one or more screws is loosened, the reticle ring can be rotated through a small angle. The reticle ring is moved vertically by loosening the top (or bottom) screw and then tightening the bottom (or top) screw, and horizontally by loosening the left (or right) screw and tightening the right (or left) screw.

An instrument's line of sight is defined as a line passing through the cross hairs' intersection and the optical center of the telescope's objective lens. The optical center is a fixed point, but the line of sight can be moved by shifting or rotating the cross hairs.

The vertical axis of an instrument is defined as the line about which the instrument rotates in a horizontal plane. It coincides with the spindle axis and a freely suspended plumb line attached to the instrument.

Transits and theodolites also have a horizontal axis. It is a line about which the telescope rotates in a vertical plane. This axis coincides with that of the horizontal cross arm supporting the telescope. The correct axes relationships for a properly adjusted level, transit, and theodolite are shown in Figures 8-3 and 8-4.

8-5. ADJUSTMENT OF LEVELS

The two preliminary adjustments required of all levels are to (1) eliminate parallax and (2) properly position the cross hairs.

Parallax

When working with any instrument telescope, an observer simultaneously views two images. One is the object focused on by the

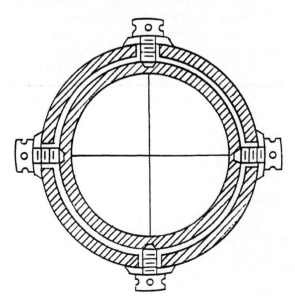

Figure 8-2. Reticle ring.

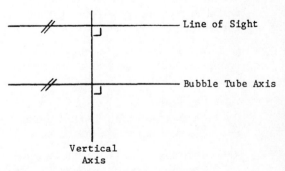

Figure 8-3. Axes relationship of a level.

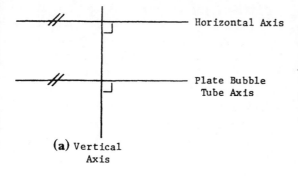

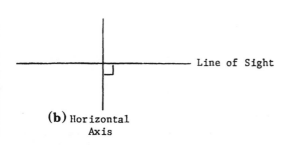

Figure 8-4. Axes relationship for a transit/theodolite. (a) Front view. (b) top view.

telescope, and the second is the cross hairs' image. Both must come to focus on a single plane—i.e., at the back of the observer's eye in order to be seen clearly. If this condition is not met, parallax exists.

To test for parallax, the telescope should be focused on some distant well-defined object. While viewing the object through the telescope, the observer's eye is shifted slightly horizontally and vertically to check for any movement of the cross hairs relative to the object. If the cross hairs do not so move over the object, an adjustment is not necessary. If the cross hairs do appear to move on the object, parallax exists and must be corrected.

To eliminate parallax, the telescope's focusing knob is rotated so everything is out of focus except the cross hairs. Then, using the eyepiece focusing ring located at the telescope's rear, adjust the focus of the cross hairs until they are sharp and well defined. Refocus the telescope on a distinct object and again

check for parallax. Repeat the procedure as necessary until parallax is eliminated, after which the instrument will not need to be adjusted again. However, if another surveyor uses the equipment, he or she should also test for and clear any parallax, since vision varies from person to person.

Cross Hairs

Equipment manufacturers attempt to make the vertical and horizontal cross hairs truly perpendicular to each other, but for older instruments this condition is less likely to have been met. It is important that the horizontal cross hair of a level be truly horizontal when the instrument is leveled so, if necessary, any part of this cross hair can be used to obtain true rod readings. If under these conditions the vertical cross hair is not exactly vertical, surveyors should understand that this situation will not affect the performance of the level in determining elevations.

To test the horizontal cross hair, first level the instrument, then check for and eliminate any parallax. Using one end of the horizontal cross hair, take a reading on a level rod. With the horizontal slow motion of the instrument, rotate the level so the horizontal cross hair is sighted to its other end on the level rod. Check to see if there has been any change in the vertical position of the cross hair with respect to the initial rod reading. If there is no change, the cross hair is truly horizontal. If the cross hair has moved above or below the initial rod reading, then it is not truly horizontal and must be adjusted.

To adjust, note the distance the cross hair has moved above or below the initial rod reading. Then, slightly loosen two adjacent reticle adjusting screws. While sighting through the telescope, rotate the reticle until the end of the cross hair is moved back half the length it was off the initial rod reading. Tighten the two reticle adjusting screws.

Check the adjustment by repeating the test. It is important to note that the horizontal cross hair was rotated about its center so a new initial rod reading must be taken. Repeat as necessary until the rod reading at both ends of the horizontal cross hair is the same.

If the vertical and horizontal cross hairs are truly perpendicular, the vertical cross hair must now be truly vertical. To check this, after the horizontal cross hair has been adjusted, a sight is made on a freely suspended plumb line with the vertical cross hair. If the vertical cross hair does not coincide exactly with the plumb line, a note should be made in the instrument's log book.

8-5-1. Dumpy Level: General Information

The dumpy level consists of a telescope, bubble tube, and leveling head containing the spindle (Figure 8-5).

8-5-2. Dumpy Level: Principal Adjustments

Due to their simple construction, dumpy levels have only two principal adjustments: (1) bubble tube and (2) line of sight.

Bubble Tube

The purpose of this adjustment is to make the bubble-tube axis perpendicular to the instrument's vertical axis. To test this condition, set up the instrument so the bubble tube is directly over two opposite leveling screws and carefully center the bubble. Rotate the instrument 90° to place the bubble tube over the remaining pair of leveling screws and again center the bubble. Rotate the instrument 180° to reverse the tube's position. If the bubble runs off center, an adjustment is necessary.

The distance the bubble moves represents twice the error present. To correct, bring the bubble back halfway by turning the adjusting nuts at one end of the bubble tube. Recenter the bubble using the two leveling screws in line with the tube. Rotate the instrument 90° and center the bubble using the other pair of leveling screws. Provided the adjustment was done correctly, the bubble will remain centered as the instrument is rotated. If the bubble runs again, repeat the adjustment until it stays centered.

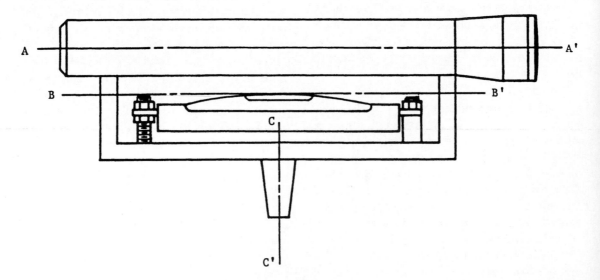

A–A'	Line of Sight
B–B'	Bubble Tube Axis
C–C'	Vertical Axis

Figure 8-5. Dumpy level.

Line of Sight

The purpose of this adjustment is to make the line of sight perpendicular to the instrument's vertical axis. The method used to test this condition is called the two-peg test or, simply, the *peg test*.

Level the instrument at a point C midway between two stakes A and B, which should be at least 200 ft apart (Figure 8-6). Assume an elevation for A and take a backsight (BS) on a level rod held there. Rotate the level and read a foresight (FS) on a rod held on B. Because the instrument is halfway between A and B, any error caused by an inclined or depressed line of sight is the same in both rod readings. The true elevation of B with respect to the assumed elevation at A is obtained by adding the BS to A's elevation, to get the height of the instrument (HI), then subtracting the FS. The error is both added and subtracted, thereby canceling itself out. The only elevation in error is the HI.

The instrument is then moved to a point D on the opposite side of B from A with the eyepiece end of the telescope within a few inches of a rod held on B. After leveling the instrument, a short BS is taken on B, looking backward through the telescope objective lens. The cross hairs will not be visible, but a pencil point held against the rod can be centered in the field of view to get a reading. After rotating the instrument, a normal FS is taken on A. The HI is the elevation of B plus the BS and since the distance is very short, the HI is essentially without error. The computed FS for A is then simply the HI minus A's elevation. If the computed and observed FS are not the same, an adjustment is necessary.

To correct the error, loosen the top (or bottom) and tighten the bottom (or top) reticle screws to move the cross hairs vertically until the observed FS matches the computed one. Always loosen one screw first and then tighten the second to prevent reticle ring warping. After the cross hairs have been moved to the correct position, test to make sure the horizontal cross hair is still horizontal. To check the adjustment, repeat the test, setting up the instrument behind A.

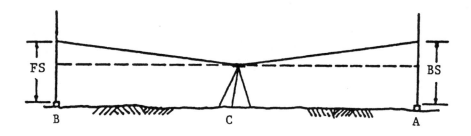

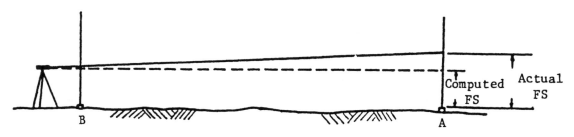

Figure 8-6. Peg test.

8-7-3. Wye Level: General Information

The wye level differs from the dumpy in that the telescope, with the bubble tube, is removable from the two wyes in which it is mounted. The wyes are circular clamps mounted on a support bar that is attached to the vertical axis of the instrument (Figure 8-7). One or both wyes has two adjusting nuts, which allow it to be moved up or down with respect to the support bar. When the clamps are opened, the telescope is free to roll in the wyes. An imaginary line connecting the center of each wye defines their axis, i.e., the line about which the telescope rolls.

8-5-4. Wye Level: Principal Adjustments

Since the wye level has more components than the dumpy, it has additional sources of error. The conditions to test are the line of sight, lateral adjustment, bubble tube, and wyes.

Line of Sight

The purpose of this adjustment is to make the line of sight coincide with the axis of the wyes. To test this condition, open the wyes, level the instrument, and sight a distinct point with the cross hairs. While sighting, roll the

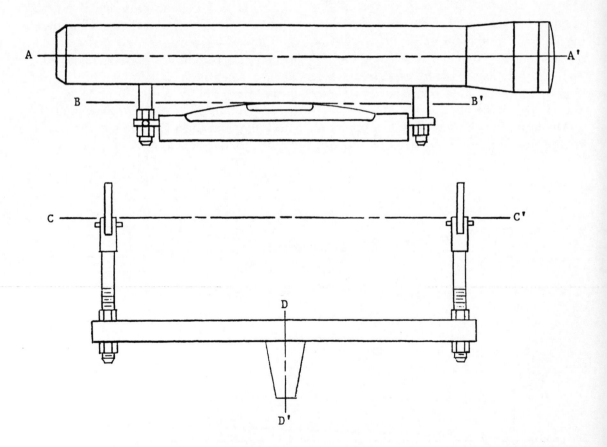

A–A'   Line of Sight
B–B'   Bubble Tube Axis
C–C'   Axis of the Wyes
D–D'   Vertical Axis

Figure 8-7. Wye level.

telescope 180° in the wyes so the bubble tube is above the telescope. An adjustment is necessary if the cross hairs have moved off the sighted point.

The length the cross hairs move above or below the point represents twice the error. Bring the cross hairs back half the distance by loosening the top (or bottom) and tightening the bottom (or top) reticle screws. Check by resighting a distinct point and repeating the test. Repeat as necessary until the adjustment is complete. Roll the telescope back to its correct position, clamp the wyes, and check to ensure the horizontal cross hair is still horizontal.

Lateral Adjustment

This adjustment makes the axis of the wyes, line of sight, and bubble-tube axis lie in the same vertical plane when the instrument is leveled.

At one end of the bubble tube is a set of adjusting nuts (as on a dumpy) and a set of capstan-headed screws perpendicular to them. These screws shift one end of the tube horizontally when the telescope is in its normal position. To check this adjustment, open the wyes and level the instrument. Roll the telescope approximately 30° in the wyes so the bubble tube, viewed from the rear, is in the five o'clock position. Any length the bubble runs off center represents the full error. To adjust, loosen one and tighten the other capstan-headed screw until the bubble is brought back to center. Check by rolling the telescope so the bubble tube is in the seven o'clock position. Repeat the adjustment if the bubble runs.

Bubble Tube

The purpose of this adjustment is to make the bubble-tube axis parallel to both the line of sight and axis of the wyes. To test, open the wyes, rotate the instrument placing the bubble tube directly over two opposite leveling screws, and center the bubble. Rotate the instrument 90°, placing the tube over the remaining pair of leveling screws, and again center the bub-

ble. Carefully remove the telescope from the wyes, turn it end for end, and replace it. An adjustment is necessary if the bubble runs off center.

The length the bubble runs represents twice the error. To correct, bring it back halfway using the bubble-tube adjusting nuts. Check by releveling the instrument and repeating the test. If the bubble runs, repeat the adjustment until it remains centered.

Wyes

This adjustment makes the axis of the wyes perpendicular to the instrument's vertical axis. If the preceding adjustments have been carried out correctly, this will also make the line of sight and bubble-tube axis perpendicular to the vertical axis.

Center the bubble first over one pair of opposite leveling screws and then over the remaining pair. Rotate the instrument 180° and check the bubble run. The length of movement represents twice the error present. Correct by bringing the bubble back halfway with the wye adjusting nuts. Readjust as necessary until the bubble stays centered in all positions.

8-5-5. Automatic Level: General Information

Automatic levels differ from dumpy and wye levels in having a compensating device that maintains a horizontal line of sight when the instrument is approximately leveled. Automatic levels also have three leveling screws, instead of four, and a circular bubble whose upper inner surface is spherical and has etched a bull's-eye on it. This bull's-eye generally defines the limits within which the compensator will maintain a horizontal line of sight.

At first glance, automatic levels appear to be complicated devices that a surveyor should not attempt to adjust. However, except for the compensator, the instrument is relatively simple in design and a few adjustments can be easily performed with satisfactory results.

8-5-6. Automatic Level: Preliminary Adjustments

Preliminary adjustments for automatic levels are the same as those for the dumpy and wye: (1) parallax and (2) cross hair. On dumpy or wye levels, the reticle adjusting screws are easy to find; on automatic levels, they tend to be elusive, but generally are located under a cover just forward of the telescope eyepiece. Some automatic levels will have only one or two reticle screws. If there is only one, the horizontal cross hair has been preset at the factory and should not be rotated. If there are two, a surveyor may be able to rotate the reticle. Generally speaking, since these screws are well-shielded, the cross hairs will stay in adjustment but should be periodically checked. If this condition is not met, then the cross-hair intersection should be used in taking rod readings or the instrument should be sent to a repair facility.

An additional preliminary check for automatic levels concerns the compensator. If dust or humidity enters the compensator or the instrument is excessively jarred, the compensating mechanism may stick and give erroneous rod readings. To test for this, carefully level the instrument and take a rod reading on a solid point. While sighting, tap a leg of the tripod. This will cause the compensator to swing, moving the cross hairs off the reading and then back to it. If the cross hairs return to the original reading, the compensator is working properly. If they do not, the compensator is sticking. Tap again to check.

In the event the compensator sticks, the surveyor should not attempt to fix it. After a cover is removed, the problem will worsen more as dust or moisture find their way into the compensator, and the level must be sent to a repair facility for proper adjustment.

8-5-7. Automatic Level: Principal Adjustments

Due to their simple design, automatic levels have only two principal adjustments: (1) circular bubble and (2) line of sight.

Circular Bubble

Unlike the ordinary bubble-tube axis, a circular bubble has a plane tangent to the midpoint of its upper inner surface. For proper adjustment, this plane must be perpendicular to the instrument's vertical axis. To test this condition, center the bubble in the bull's-eye using the leveling screws, then rotate the instrument 180°. It requires adjustment if the bubble moves out of the bull's-eye.

In order to correct this error, the bubble must be brought back half the distance it ran. The circular bubble housing should have a set of three or four adjusting screws located on its top or bottom. By turning one or more of these screws, bring the bubble back halfway. Relevel the instrument and repeat the test.

Line of Sight

The purpose of this adjustment is to make the line of sight perpendicular to the instrument's vertical axis. The test and adjustment procedure are the same as those used for the dumpy level line-of-sight adjustment (peg test). To move the cross hairs vertically on instruments having only one or two reticle adjusting screws, the screw at the six or twelve o'clock position is turned. The reticle is spring-loaded at the opposite side, so it is forced to move when the screw is turned. This adjustment can only be performed correctly if it has been determined that the compensator is functioning properly.

8-5-8. Tilting Level: General Information

Tilting levels are three-screw instruments consisting of a telescope, circular bubble, sensitive bubble tube, and leveling head. The telescope is mounted so that it can be tilted by rotating a drum located beneath the eyepiece. This feature allows the instrument to be precisely leveled each time a reading is taken.

The preliminary adjustments are the same as those for the dumpy level, except that there may be an auxiliary telescope for observing the bubble tube—in which case, it too must be checked for parallax.

8-5-9. Tilting Level: Principal Adjustments

Because of its simple design, there are only two principal adjustments for the tilting level: (1) circular bubble and (2) precise bubble tube.

Circular Bubble

The purpose of this adjustment is to make the plane of the circular bubble perpendicular to the instrument's vertical axis.

The test and adjustment of the circular bubble are dependent on how it is mounted on the instrument. If it rotates with the telescope about the vertical axis, use the same procedure as that for an automatic level. If it does not rotate with the telescope, an adjustment is not really necessary, since the precise bubble tube is used to obtain a horizontal line of sight.

Precise Bubble Tube

This adjustment makes the precise bubble-tube axis parallel to the instrument's line of sight. To test for this condition, use the dumpy level peg-test procedure. When the correct FS to give a horizontal line of sight is computed, the cross hairs are brought to that reading by rotating the telescope's tilting drum. The precise bubble is then centered using the bubble-tube adjusting nuts. If the bubble is the coincident type, the adjustment makes the bubble's two ends coincide.

8-6. ADJUSTMENTS OF TRANSITS AND THEODOLITES

The two preliminary adjustments for transits and theodolites are to (1) eliminate parallax and (2) properly position the cross hairs.

Parallax

This adjustment is the same as for levels.

Cross Hairs

In leveling, it is important to have the horizontal cross hair truly horizontal. Since transits and theodolites are used primarily for angle measurement, it is more important to have the vertical cross hair truly vertical in order to use any part of it for sighting.

To test the vertical cross hair, first level the instrument, then check for and eliminate any parallax. Sight on a freely suspended non-swinging plumb line. If the vertical cross hair does not coincide with the plumb line, an adjustment must be made by rotating the reticle. Loosen two adjacent reticle screws and rotate the cross hairs until the vertical one coincides with the plumb line. Retighten the screws and check the adjustment by reversing the position of the telescope and repeating the test.

After the adjustment has been made, perpendicularity of the horizontal and vertical cross hairs should be checked by using the horizontal cross-hair test described for levels. The result should be recorded in the instrument's log book.

8-6-1. Transit: General Information

The primary function of transits is to measure horizontal and vertical angles. Figure 8-8 shows the instrument axes to be adjusted.

Traditional transits are of an open design with all adjusting screws and nuts exposed. Basic transit design has been modified on newer instruments to incorporate additional features. Some use an optical plummet in lieu of a plumb line, whereas others have been modified to the point where they resemble a theodolite more than a transit. These instruments have some advantages of both, and since they are a cross between the two designs, their adjustments are a cross between those for transits and theodolites.

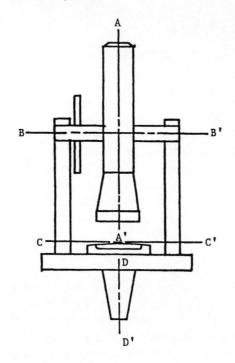

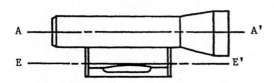

```
A-A'   Light of Sight
B-B'   Horizontal Axis
C-C'   Plate Bubble Tube Axis
D-D'   Vertical Axis
E-E'   Telescope Bubble Tube Axis
```

Figure 8-8.   Transit/theodolite.

8-6-2. Transit: Principal Adjustments

Transits have more axes than do levels; therefore, there are more adjustments to be made. The principal adjustments for transits are (1) plate bubble tubes, (2) line of sight, (3) horizontal axis, (4) telescope bubble tube, (5) vertical vernier, and (6) horizontal vernier test.

Plate Bubble Tubes

Most traditional transits have two-plate bubble tubes mounted at right angles to each

other. Each tube has its own axis, which must be adjusted to make it perpendicular to the instrument's vertical axis. Newer instruments may have only a single-plate bubble tube.

To test the plate bubble tubes, set up the instrument so each bubble tube is in line with two diagonally opposite leveling screws. Center each bubble separately using the corresponding pair of leveling screws, then rotate the transit 180°, and check the bubbles' runs. If one or both bubbles run, an adjustment is necessary—the amount of movement representing twice the error present.

To correct, bring each bubble back halfway, using the bubble-tube adjusting nuts or screws. Relevel the instrument and repeat until the bubbles remain in place as the transit is rotated.

Line of Sight

The purpose of this adjustment is to make the line of sight perpendicular to the horizontal axis. This will allow true straight-line extension when transiting the telescope.

The method used to test this adjustment is to extend a straight line on relatively flat terrain by *double centering* (Figure 8-9). Set up the instrument and select or set a distinct point A at a distance of at least 100 ft. Point A and the instrument point define a straight line that is to be extended. Backsight on A locking both horizontal motions; then reverse the telescope and set a point B at a distance of at least 200 ft. If the transit is in adjustment, point B will be on the extension of the straight line. Rotate the transit about its vertical axis and backsight on A with the telescope now reversed. Lock both horizontal motions; then reverse the telescope and set a point C at a distance $AC = AB$. If the transit is in adjustment, points B and C will coincide exactly.

If B and C do not coincide, the true extension of the straight line will pass through a point D halfway between B and C. Distance CB represents four times the error present.

To adjust the transit, while still sighting on C, move the cross hairs horizontally one-fourth

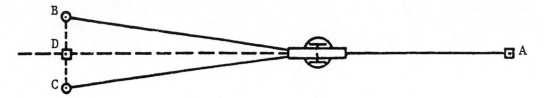

Figure 8-9. Double centering.

distance *CB* in the direction of *B*. To move the cross hairs, slightly loosen the top reticle adjusting screw, then alternately loosen and tighten the side reticle screws; check by repeating the test. After the adjustment is completed, recheck the vertical cross hair and adjust if necessary.

Horizontal Axis

This adjustment makes the horizontal axis perpendicular to the vertical axis of the transit. To test this condition, set up and level the instrument approximately 20 ft from a tall vertical wall. Raise the telescope to a vertical angle of approximately 30° and sight some distinct point *A* on the wall. Plunge the telescope to horizontal—i.e., to a vertical angle of approximately 0°—and mark a point *B* on the wall. Rotate the instrument 180°, reverse the telescope, and resight *A*. Plunge the telescope to horizontal and mark a point *C* on the wall (Figure 8-10). If *B* and *C* do not coincide, the horizontal axis needs to be adjusted.

One end of the horizontal axis must be raised or lowered by means of an adjusting screw at the end of the horizontal cross arm. This moves a saddle in which the cross arm is seated and held in place by a clamp. Loosening the clamp and turning the adjusting screw shifts the horizontal axis.

To determine the required length of movement, mark a point *D* halfway between *B* and *C*. This places *D* vertically beneath *A*. Sight *D*, then raise the telescope to *A* where, because of the instrument error, the cross hairs' intersection will miss point *A*. Using the horizontal-axis adjusting screw, raise or lower the end of the cross arm until the cross hairs are brought to *A*. Tighten the clamp and

check the adjustment by repeating the test. After completion, recheck the vertical cross hair and adjust as necessary.

A word of caution: Do not overtighten the clamp, since this can apply too much pressure on the cross arm, preventing the telescope from rotating freely.

Telescope Bubble Tube

The purpose of this adjustment is to make the axis of the telescope bubble tube parallel to the line of sight.

To perform this adjustment, follow the peg-test procedure described for dumpy levels. The telescope bubble must be centered, using

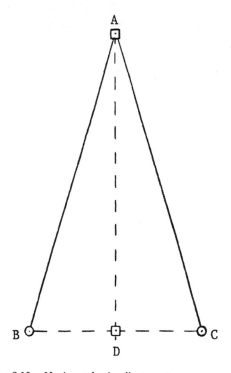

Figure 8-10. Horizontal-axis adjustment.

the vertical slow motion, for each backsight and foresight reading. When the correct FS to give a horizontal line of sight is computed, the cross hairs are tilted up or down to that reading using the vertical slow motion. The telescope bubble is then brought to center using the bubble-tube adjusting nuts or screws. Check by repeating the peg test.

Vertical Vernier

The purpose of this adjustment is to ensure that the vertical vernier is reading exactly 0°, when the line of sight is horizontal. This adjustment is performed in one of two ways, depending on the type of vertical vernier arrangement used on the transit.

METHOD 1. This is used if there is no vertical vernier bubble tube. Set up and level the instrument; then using the vertical lock and slow motion, center the telescope bubble. If the vertical vernier does not read exactly 0°, carefully loosen the vernier mounting screws, move it to a reading of exactly 0°, and retighten.

This adjustment is much more difficult than it seems because when the mounting screws are loosened, the vernier "falls away" from the vertical circle. When retightening, it is important that the vernier not rub against the circle and a gap is not left between the two.

To avoid shifting the vernier, reading a vertical angle direct and reverse and using the average will cancel out any errors. In cases where it is not possible or practical to read angles direct and reverse, as in stadia work, an *index error* should be applied. An index error is the angle on the vertical vernier, when the telescope bubble is centered. It should be recorded, with its correct mathematical sign and telescope orientation, in the instrument's log book and case. On newer transits where the vertical vernier is not readily accessible, an index error should be used or the instrument sent to a repair facility.

METHOD 2. This is used if the vertical vernier has a bubble tube. Set up and level

the instrument; then using the vertical lock and slow motion, center the telescope bubble. Set the vernier to a reading of exactly 0° using the vernier slow-motion screw and center the vernier bubble using the bubble-tube adjusting screws.

Horizontal Verniers Test

The purpose of this test is to determine if the two horizontal verniers on a transit are truly 180° apart. The transit is designed to provide the option of reading one or the other of the A and B horizontal verniers when measuring angles. An error in horizontal-angle measurement is introduced if the two verniers are read alternately, and they are not 180° apart. To test for this condition, lock the A vernier at exactly 0° and read the B vernier. Record any error. Repeat the procedure for at least three more readings spread evenly around the horizontal circle.

If the error is consistent, within reading ability, one of the verniers is off. An adjustment is not easily made, so the transit should be sent to a repair facility, or only one vernier should be used consistently when measuring angles.

If the error is not consistent, this may indicate that the spindles are worn or the plates are warped. If that is the case, the transit must be sent to a repair facility.

8-6-3. Theodolite: General Information

A theodolite's main function is the same as that of a transit: measuring horizontal and vertical angles. The instrument differs from the transit in having an optical plummet, optical-reading system, a circular bubble, only a one-plate bubble tube, and three leveling screws. Most adjusting screws and nuts are located under protective covers.

Figure 8-8 shows the basic axes to be adjusted. Some of these adjustments also apply to the newer-style transits, which physically resemble theodolites. In addition to the preliminary transit adjustments for parallax and cross hairs' position, a theodolite must also have the parallax cleared in its optical plummet and angle-reading telescope.

8-6-4. Theodolite: Principal Adjustments

The principal adjustments of theodolites are similar to those of transits: (1) plate bubble tube, (2) circular bubble, (3) line of sight, (4) horizontal axis, (5) telescope bubble tube, (6) vertical circle, and (7) optical plummet.

Plate Bubble Tube

The purpose of this adjustment is to make the axis of the plate bubble tube perpendicular to the vertical axis of the theodolite. To test, set up the theodolite and roughly level it using the circular bubble. Rotate the instrument so the plate bubble-tube axis is parallel to a line through two leveling screws, and carefully center the bubble using them. Rotate the instrument 90° and center the bubble using only the remaining leveling screw. Rotate the theodolite 180° and check for bubble run. An adjustment must be made if the bubble runs for a distance representing twice the error present.

To correct, bring the bubble back halfway using the bubble-tube adjusting nuts or screws. Repeat until the bubble remains stationary as the theodolite is rotated.

Circular Bubble

The circular bubble is used to roughly level a theodolite and allow the use of the optical plummet. This adjustment makes the plane of the circular bubble perpendicular to the instrument's vertical axis.

To test the circular bubble, level the theodolite using the plate bubble tube. If the circular bubble is not centered in the bull's-eye, it needs to be adjusted. Carefully center the circular bubble in the bull's-eye using the circular bubble adjusting nuts or screws.

Line of Sight

The purpose of this adjustment is to make the line of sight perpendicular to the horizon-

tal axis. The procedure used is the same as that described for transit line-of-sight adjustment.

Horizontal Axis

This adjustment makes the horizontal axis perpendicular to the instrument's vertical axis. The procedure used is the same as that described for a transit horizontal-axis adjustment.

Telescope Bubble Tube

This adjustment makes the axis of the telescope bubble tube parallel to the line of sight. Generally, theodolites do not have telescope bubble tubes. For those theodolites and newer-style transits that do, use the procedure described for a transit telescope bubble-tube adjustment.

Vertical Circle

The purpose of this adjustment is to ensure that the vertical circle of a theodolite is correctly oriented, with respect to gravity, when vertical angles are read. This is accomplished either by (1) an automatic compensator or (2) a vertical-circle bubble tube.

To test a theodolite, level it and if it has a vertical-circle bubble tube, carefully center it using the bubble centering screw. Read a direct and reverse vertical angle to a selected point A. A vertical angle, for the purposes of this test, is defined as measured with respect to the horizon in a vertical plane. Angles of inclination are considered positive, angles of depression negative. The instrument is in need of adjustment if the direct and reverse vertical angles are not equal. Averaging the two readings gives the correct vertical angle.

If the theodolite has a compensator for circle orientation, it should be properly adjusted at a repair facility. An index error can be computed, recorded in the log book, and applied to each single vertical angle.

If the theodolite has a vertical-circle bubble tube, resight on point A and, using the bubble centering screw, set the correct vertical angle on the reading system. The effect of this is to slightly rotate the vertical circle while leaving the cross hairs set on *A*. This will also cause the vertical-circle bubble to run. Recenter the bubble using the bubble-tube adjusting nuts or screws. Check by reading direct and vertical angles to *A* again, repeating the adjustment as necessary.

Optical Plummet

The optical-plummet sight will only be truly vertical if the instrument is level, and the line of sight of the plummet is coincident with the instrument's vertical axis. An optical plummet is either built into the tribrach or upper instrument assembly. In the first case, the plummet remains fixed in position as the instrument is rotated; in the second, it rotates with the theodolite.

Both types can be tested using a plumb bob. Level the instrument and hang a plumb bob below it. Carefully mark a point on the ground directly beneath the plumb bob and remove it. Sight through the optical plummet and check the plummet reference mark with respect to the ground mark. An adjustment is necessary if the marks do not coincide. Four plummet adjusting screws are located just forward of the eyepiece and may be under some sort of cover. Turn the appropriate adjusting screws—first loosening, then tightening—to bring the plummet reference mark to the ground mark.

If an optical plummet is mounted in the upper assembly, it can also be tested by leveling the theodolite over a ground point using the optical plummet, then rotating 180°. If the plummet mark moves off the ground point, use the adjusting screws to bring it back halfway. Check by repeating the test.

8-7. OTHER SURVEYING EQUIPMENT

8-7-1. Tribrach

Tribrachs are the most versatile of surveying instruments and should be periodically tested and adjusted. Tribrachs use a circular bubble

for leveling and may or may not have a built-in optical plummet.

The tribrach should be attached to a compatible theodolite, if possible. To test and adjust the circular bubble and optical plummet, follow the procedures explained under principal adjustments of theodolites.

If a compatible theodolite is not available, a tribrach can still be tested and adjusted if an extra circular bubble or striding level is available. The tribrach is leveled using one of these, and its circular bubble is brought to center using the adjusting screws. The optical plummet can be adjusted using a plumb line, as previously explained under principal adjustments of theodolites.

8-7-2. Rod Level

To test a *rod level*, hang a plumb bob from a firm overhead support and mark a point on the ground directly beneath it. Raise the plumb bob high enough to just clear a short section of range pole. Attach the rod level by screwing or taping to the pole and then place the range pole tip on the ground point, centering its top beneath the plumb bob. Use the bubble adjusting screws to center the bubble if necessary.

8-7-3. Striding Level

Striding levels should be tested and adjusted on the transits or theodolites with which they were designed to be used. The transit or theodolite should first have its horizontal axis adjusted. Carefully level the instrument and place the striding level on its cross arm. If the bubble runs toward one end or the other, center it using the adjusting nuts at one end of the bubble tube.

8-7-4. Tripod

The tripod is an often-overlooked piece of surveying equipment. It serves as a platform for various instruments. For example, a theodolite, no matter how well-adjusted, cannot be expected to give good results if the tripod supporting it is unstable.

Shoes must be rigidly attached to the tripod's legs to prevent shifting under the weight of an instrument. Secure fastening clamps are necessary to avoid leg slippage on extension-leg tripods. Bolts connecting legs to the tripod head should be tightened firmly but not tight enough to disallow easy folding of the legs. Metal tripods should be checked for dents that could affect sliding of the extension legs. Wooden tripods must be inspected for cracks and flat spots under the clamps, worn areas on the wood's protective coating refinished, and to prevent swelling, any moisture wiped off immediately.

8-8. CLEANING EQUIPMENT

Surveying equipment is frequently used in relatively hostile environments; dirt and water are its worst enemies. Proper maintenance includes not only periodic adjustments, but also regular cleaning. Instruments can be sent to a repair facility for a thorough cleaning and lubrication, but a surveyor *can* do a few things to keep equipment in good condition.

As soon as possible, dirt and water must be removed from external instrument parts with a mild general household cleaner, cotton swabs, and pipe cleaners. Pay particular attention to clamp screws, leveling screws, and exposed metal joints. If water gets inside a telescope, resist the temptation to go in after it; opening a telescope allows dust to get in, and on older instruments the cross hairs are fragile and easily destroyed.

A soft camel's hair brush works best for cleaning lenses. If lenses are streaked, a lint-free cloth and some optical-quality glass cleaner are necessary. Newer optics are coated and can be damaged by excessive rubbing or using a household glass cleaner.

Surveyors should avoid the temptation to oil or grease equipment. Lubricants attract dust like a magnet, accelerating wear. Thorough lubrication and internal cleaning should be done by a repair facility.

8-9. SHIPPING EQUIPMENT

When shipping equipment to a repair facility, it is important to pack it properly. Instrument cases alone are not designed for shipping purposes and, therefore, should be put in a sturdy container with a generous amount of packing material. If the container is dropped, the packing material rather than the instrument will absorb the shock.

The shipping and return addresses should appear in at least two different places on the exterior of the container and be included inside in the event the external addresses are destroyed or obliterated. Labels identifying the contents as *fragile precision equipment* should also appear in multiple locations on the exterior. A letter explaining in detail the problems with the equipment should be inside the container.

Equipment being shipped must be insured. Surveying equipment represents a large investment, and all possible measures should be taken to protect it.

REFERENCES

BOUCHARD, H., and F. H. MOFFITT. 1992. *Surveying*, 9th ed. New York: Harper Collins.

LAPCZYNSKI, D. 1980. Keep it clean. *P.O.B. Magazine* 6(1), 34.

LAPCZYNSKI, D. 1981. Sending instruments for service. *P.O.B. Magazine* 6(2), 25.

LOMMEL, G. E., H. RUBEY, and M. W. TODD. 1958. *Engineering Surveys: Elementary and Applied*, 2nd ed. New York: Macmillan.

SMITH, F. R. 1982. How to check the adjustment of a rod level. *P.O.B. Magazine* 7(5), 30.

WOLF, P. R., and R. C. BRINKER. 1994. *Elementary Surveying*, 8th ed. New York: Harper Collins.

9

Traversing

Jack B. Evett

9-1. INTRODUCTION

A traverse is a series of consecutive straight lines along the path of a survey, the lengths and directions of which are or have been determined by field measurements. The surveying performed to evaluate such field measurements is known as *traversing*. Although often used in land and route surveying, it is also employed in other types of surveying.

The end points of traverse lines, known as traverse stations or "hubs," are commonly marked in the field by wooden stakes with tacks in the top, steel rods, or pipes driven into the ground. On blacktop or concrete pavement, traverse stations can be located by driving a nail into the blacktop or by chiseling or painting an "X" or other mark on the concrete. On a map or plat, traverse stations may be marked with a small circle. A small triangle denotes a control station.

There are two basic types of traverses: (1) *open* and (2) *closed*. Both originate at a point of known location. An open traverse terminates at a point of unknown position; a closed traverse finishes at a point of fixed location. Figure 9-1 illustrates an open traverse that might represent a proposed highway or pipeline location. Figure 9-2 shows two closed

traverses. In Figure 9-2a, *ABCDE* represents a proposed highway route, but the actual traverse begins at known location 1 and ends at fixed location 2. This type of closed traverse is known as "geometrically open, mathematically closed." In Figure 9-2b, *ABCDEA* represents a parcel of land for which the actual traverse begins and ends at known point *A*. This type of closed traverse is "geometrically and mathematically closed." (Subsequent citations to closed traverses in this chapter refer to geometrically and mathematically closed ones.)

Although open traverses sometimes are used on route surveys, such as highway or pipeline locations, they should be avoided because an independent check for errors and mistakes is not available. The only means of verifying an open traverse is to repeat all measurements and computations (not an independent check). For closed traverses, independent mathematical means of checking both measured angles and distances are available (see Sections 9-3 and 9-4) and should be utilized to verify survey accuracy. Whenever open traverses of the type shown in Figure 9-1 are encountered, if possible they should be transformed to either (1) geometrically open, mathematically closed ones (Figure 9-2a) by extending the traverse to beginning and ending points of known loca-

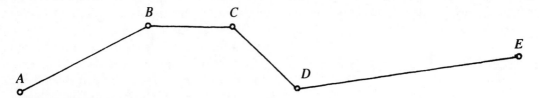

Figure 9-1. Open traverse.

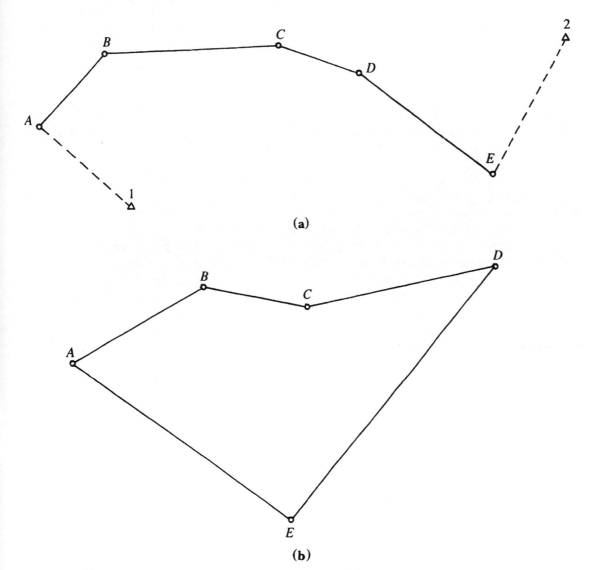

(a)

(b)

Figure 9-2. (a) Geometrically open, mathematically closed traverse. (b) Geometrically and mathematically closed traverse.

tion, or (2) geometrically and mathematically closed ones (Figure 9-2b), by continuing the traverse so it ends on its beginning point.

9-2.   FIELD MEASUREMENTS IN TRAVERSING

As stated in the previous section, traversing involves measuring both the lengths and directions of lines. Details regarding these procedures in the field have been presented in earlier chapters; however, some considerations applicable to traversing are presented in this section.

Lengths of traverse lines can be determined by any convenient method, but most measurements are made with electronic devices or by taping. In closed traverses, the lengths of lines are measured, recorded, and shown on a map or plat. On open traverses, it is common practice to locate stations by their total distances from the starting point. Distances are then noted in "stations" and "pluses" (a full station is 100 ft). In Figure 9-1, if *A* is station 0 + 00 and the distance from *A* to *B* is 569.8 ft, station *B* becomes 5 + 69.8. For length *BC* equal to 744.5 ft, station *C* is 13 + 14.3.

Directions of traverse lines can be determined relative to a reference direction (such as north) by reading bearings or azimuths, or measuring interior angles, deflection angles, or angles to the right (preferred) or to the left. Bearings and azimuths are obtained by sighting the transit's telescope along a line and noting the compass reading. A deflection angle is formed at a traverse station by an extension of the previous line and the succeeding one. The numerical value of a deflection angle must always be followed by R or L to indicate whether it was turned right or left from the previous traverse line extended. An angle to the right is read at a traverse station by backsighting along the previous line and measuring the clockwise angle to the next point. In Figure 9-3, the deflection angle is 33°33′ R; the angle to the right is 213°33′.

Closed traverse—e.g., land boundary surveys—are usually run by measuring and recording interior angles, such as *ABC*, *BCD*, etc., in Figure 9-2b. Open traverses—e.g., route surveys—are more commonly run using either deflection angles or angles to the right.

Since bearings read in the field are not highly accurate, traversing is generally done by measuring and recording interior angles, deflection angles, or angles to the right, all of which are determinable to the nearest minute or smaller relatively quickly with an ordinary transit or theodolite. However, bearings are generally used in computing latitudes and departures as well as closure (see Section 9-4)

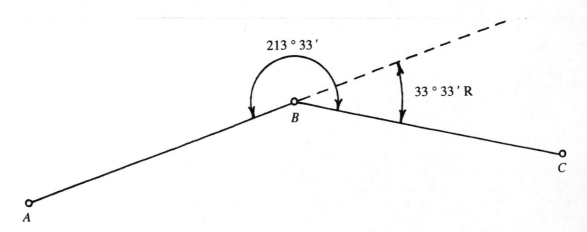

Figure 9-3.   Deflection angle and an angle to the right.

and directions are frequently indicated on maps and plats by bearings. Therefore, although bearing (compass) readings may be made and recorded in the field for checking purposes only, actual bearings used in making closure calculations are often computed from appropriate angle readings—e.g., interior angles; those given on maps and plats are generally secured from the closure calculations.

9-3. ANGLE MISCLOSURE AND BALANCING

For closed traverses, an excellent verification of angular measurements is available, as the sum of the interior angles of a closed polygon Σ is

$$\Sigma = (n - 2)180° \qquad (9\text{-}1)$$

where n is the number of sides (or angles) in the polygon. Hence, if the sum of the measured interior angles of a closed traverse is equal to Σ as computed using Equation (9-1), the accuracy of each measured angle is assured with reasonable certainty. (It is always possible that compensating errors or mistakes were made.)

Because of imperfections in equipment and errors made by surveyors, it is not unusual for the sum of the measured angles to differ from Σ. The numerical difference between the computed sum and Σ is known as the *angle misclosure*. An angle misclosure of 1 or 2 min might ordinarily be considered tolerable, but larger values are not. Permissible misclosure c can be computed using the formula

$$c = K\sqrt{n} \qquad (9\text{-}2)$$

where n is the number of sides (or angles) and K a fraction of the least division of a transit vernier or the smallest graduation on a theodolite scale. A commonly used value of K is 1 min. If this value is reasonable, permissible misclosure for a nonagon is $1'\sqrt{9}$, or $3'$, and

an angle sum that falls within the range 1259°57′ to 1260°03′ would be acceptable.

If the angle misclosure for a closed traverse is greater than the permissible figure as determined from Equation (9-2), a surveyor should remeasure each angle in order to achieve acceptable misclosure. If the angle misclosure is within the permissible range, the angles should be balanced or adjusted so their sum is equal to the correct geometric total—i.e., the number determined by Equation (9-1). Angle balancing can be done utilizing arbitrary adjustments, average adjustments, or adjustments based on measuring conditions. Details of these methods follow.

An *arbitrary* adjustment of traverse angles is commonly used for most ordinary traverses. Thus, if the misclosure is 1′, that figure is put in a suspect angle (if there is one), otherwise in any angle. For a 2′ misclosure, the entire correction might be inserted in one angle, or 1′ each in two angles.

The *average* adjustment method divides misclosure by the number of angles and applies the result to all angles. When following this system, care must be taken not to give a false impression of angle precisions. For example, if the misclosure of a nonagon is 3′, the average adjustment would be 3′/9 = 20″. For original measurements made to the nearest minute, it is inappropriate to change each angle by 20″. Instead, a correction of 1′ can be applied to every third angle, thereby avoiding more serious distortion of the traverse. If, however, original measurements were made to the nearest 20″ —by "repetition" or using a better instrument—it is reasonable to apply corrections of 20″ to every angle.

When warranted, adjustments can be made based on known measuring conditions. If the sight line along one traverse side is partially obstructed, thereby making accurate sighting difficult, the angle misclosure can be divided into two equal parts and applied to each angle having this line as a common side. Or, if two lines forming an angle are both much shorter than all other traverse sides, a larger error is more likely to occur there. Hence, that angle deserves the total adjustment.

An important factor in most surveys is to maintain comparable precision in angle and distance measurements (see Table 6-2). Note that errors in angle measurements are not related to their size, whereas errors in distance measurements increase as lines lengthen.

9-4. TRAVERSE MISCLOSURE AND BALANCING

Once a closed traverse survey has been completed, its accuracy must be checked. If required, the survey should be balanced or adjusted to effect perfect closure—i.e., geometric consistency among angles and lengths. The first step in this process is to determine angle misclosure and balance the angles (see Section 9-3). This step ensures the correct total for angular measurements, but additional computations are needed to assess the effects on traverse accuracy by including distance measurements and probably balancing the survey for them. This step is normally done by computing "latitudes" and "departures" for use in various computations.

9-4-1. Latitudes and Departures

The *latitude* of a line is its orthographic projection on the north-south axis of the survey. In terms of an ordinary rectangular coordinate system, latitude is the y-coordinate of a line secured by multiplying its length by the bearing angle cosine. North latitudes are considered positive, south ones negative.

The *departure* of a line is its orthographic projection on the east-west axis of the survey —i.e., the x-coordinate on an ordinary rectangular system found by multiplying its length by the bearing angle sine. East departures are considered positive, west ones negative.

The basis for using latitudes and departures to check and adjust a traverse survey is that the algebraic sums of both latitudes and departures must equal zero for a closed traverse.

If both algebraic sums are zero, the survey is balanced, and its overall accuracy accepted. As with angle misclosure, it is not unusual for latitude and departure algebraic sums to differ from zero, so the discrepancy is the *latitude misclosure* and, for departures, the *departure misclosure*. Their combined result, known as *linear misclosure*, is determined by computing the square root of the sum of the squares of latitude and departure misclosure. A final parameter used in analyzing traverse surveys is their *precision*, determined by dividing linear misclosure by traverse perimeter and expressing the quotient in reciprocal form. Typically, the denominator of the precision is rounded to the nearest 100, or the nearest 10 if the denominator is relatively small.

Precision is used to judge whether or not the linear misclosure of a traverse is permissible. For a given survey, permissible precision may be prescribed by state or local law. For example, North Carolina requires a minimum of 1/10,000 for "urban land surveys" and 1/5000 for "rural and farmland surveys." In some cases, permissible precision may be specified in the contract under which a surveying project is being performed.

Example 9-1. Table 9-1 gives the lengths and bearings, as computed from measured interior angles, of a five-sided closed traverse survey. Determine the latitudes and departures, linear misclosure, and precision of the survey.

The computations for solving this problem are shown in Table 9-1. The latitude of AB was determined by multiplying its length by the cosine of its bearing angle—i.e., $647.25 \times \cos 56°25' = 358.03$. Its sign is negative because the latitude is south. The departure of AB was determined similarly using the sine function. Computations for the other four lines were made in the same manner. Note from Table 9-1 that the linear misclosure and precision were determined to be 0.85 ft and 1/2400, respectively.

Table 9-1. Computation of latitudes and departures

Station	Bearing	Length (ft)	Latitude (ft)	Departure (ft)
A				
	S 56°25′ W	647.25	− 358.03	− 539.21
B				
	N 32°00′ E	300.95	255.22	159.48
C				
	N 28°52′ W	318.18	278.64	− 153.61
D				
	N 82°02′ E	555.02	76.92	549.66
E				
	S 3°49′ W	252.61	− 252.05	− 16.81
A				
		2074.01	+ 0.70	− 0.49

Linear misclosure $= \sqrt{(0.70)^2 + (-0.49)^2} = 0.85$ ft.
Precision $= 0.85/2074.01$ or $1/2400$.

9-4-2. Traverse Balancing

If linear misclosure for a closed traverse survey is greater than the permissible limit prescribed by law, contract, or the like, lengths and, if necessary, angles of the traverse must be remeasured in order to get more accurate information and a permissible misclosure. If linear misclosure is within the permissible amount, the survey should be balanced, or adjusted, by distributing linear misclosure throughout the traverse to close the figure. Methods for balancing traverses include (1) arbitrary method, (2) Crandall method, (3) least-squares method, (4) transit rule, and (5) compass rule.

When an arbitrary method is used, latitudes and departures are adjusted based on a surveyor's judgment. If there is justification to believe the measurement of one traverse line is less reliable than all others, it would be reasonable to adjust only the latitude and departure of that line, forcing the latitude and departure algebraic sums to zero. The Crandall and least-squares methods follow prescribed computations based on probability theory. Both the transit and compass rules apply proportional adjustments.

With the transit rule, adjustments are applied to respective latitudes in proportion to their lengths; thus the longer a latitude, the greater is its adjustment, and vice versa. Similarly, adjustments are applied to respective departures in proportion to their lengths. Adjustments can be computed using the following formulas:

$$\frac{\text{Adjustment in latitude } AB}{\text{Latitude misclosure}}$$
$$= \frac{\text{latitude of } AB}{\text{absolute sum of latitudes}} \quad (9\text{-}3)$$

$$\frac{\text{Adjustment in departure } AB}{\text{Departure misclosure}}$$
$$= \frac{\text{Departure of } AB}{\text{Absolute sum of departures}} \quad (9\text{-}4)$$

For simplicity in computations, the formulas can be rearranged to the form

Adjustment in latitude AB
$$= \text{latitude of } AB \times \frac{\text{latitude misclosure}}{\text{absolute sum of latitudes}}$$

since the misclosure/absolute sum ratio is a constant for all latitudes in a particular traverse.

Similarly, adjustments by the compass rule are applied to both latitudes and departures in proportion to the lengths of the lines. In other words, the longer a line, the greater are its latitude and departure adjustments, and vice versa, as shown in the following formulas:

$$\frac{\text{Adjustment in latitude } AB}{\text{Latitude misclosure}}$$

$$= \frac{\text{length of } AB}{\text{perimeter of traverse}} \quad (9\text{-}5)$$

$$\frac{\text{Adjustment in departure } AB}{\text{Departure misclosure}}$$

$$= \frac{\text{length of } AB}{\text{perimeter of traverse}} \quad (9\text{-}6)$$

The compass rule, relatively simple to apply, is the most often employed method for balancing traverses.

Example 9-2.   Balance the traverse of Example 9-1 by the compass rule.

The computations for solving this problem are shown in Table 9-2. The adjustment for latitude *AB* was determined according to Equation (9-5) by multiplying the latitude misclosure by the length of *AB* and dividing by the perimeter of the traverse—i.e., 0.70 × 647.25/2074.01 = 0.22. Its sign, as well as those for adjustments of all other latitudes, is negative because the latitude misclosure is

positive; therefore, each individual latitude must be made algebraically smaller. The adjustment for departure *AB* was determined similarly; its sign is positive because the departure misclosure is negative.

After adjusted latitudes and departures have been determined, revised lengths and bearings for the various traverse lines can be computed trigonometrically. The adjusted length of a line may be determined by finding the square root of the sum of the squares of the adjusted latitude and departure of that line. The adjusted bearing angle may be computed as the arctangent of the quotient of departure divided by latitude. The quadrant in which the bearing falls can be determined by observing the signs of the latitude and departure.

Example 9-3.   Determine the adjusted lengths and bearings of the traverse lines for which adjusted latitudes and departures were computed in Example 9-2.

The computations for solving this problem are shown in Table 9-3. The adjusted length of *AB* was computed by taking the square root of the sum of the squares of its latitude and departure, i.e.,

$$\sqrt{(-358.25)^2 + (-539.06)^2} = 647.25 \text{ ft}$$

Table 9-2.   Adjusted latitudes and departures by the compass rule

Station	Computed		Adjustment		Adjusted	
	Latitude	Departure	Latitude	Departure	Latitude	Departure
A						
	−358.03	−539.21	−0.22	+0.15	−358.25	−539.06
B						
	255.22	159.48	−0.10	+0.07	255.12	159.55
C						
	278.64	−153.61	−0.11	+0.08	278.53	−153.53
D						
	76.92	549.66	−0.19	+0.13	76.73	549.79
E						
	−252.05	−16.81	−0.08	+0.06	−252.13	−16.75
A						
	+0.70	−0.49			0.00	0.00

All values are in ft.

Table 9-3. Adjusted lengths and bearings

Station	Adjusted Latitude (ft)	Adjusted Departure (ft)	Adjusted Length (ft)	Adjusted Bearing
A				
	−358.25	−539.06	647.25	S 56°24′ W
B				
	255.12	159.55	300.90	N 32°01′ E
C				
	278.53	−153.53	318.04	N 28°52′ W
D				
	76.73	549.79	555.12	N 82°03′ E
E				
	−252.13	−16.75	252.69	S 3°48′ W

Its bearing angle was determined by finding the arctangent of the quotient of departure divided by latitude—i.e.,

$$\arctan (539.06/358.25) = 56°24'.$$

Since both latitude and departure are negative, this bearing falls in the southwest quadrant, S 56°24′ W. Computations for the remaining lines were made in the same manner.

9-5. RECTANGULAR COORDINATES

In map plotting, area computing, as well as in other applications, it is sometimes convenient to locate a line by giving rectangular coordinates for its end points with respect to a reference coordinate system. Rectangular coordinates for a point with respect to a common *x-y* coordinate system have two numbers separated by a comma and enclosed in parentheses. The first number indicates distance to the point measured from the *y*-axis parallel to the *x*-axis; the second gives distance to the point from the *x*-axis parallel to the *y*-axis. In surveying, coordinates may be referred to north-south and east-west axes (meridians) with the north coordinate given first.

Coordinates of each corner of a traverse can be determined readily if (adjusted) latitudes and departures are known. In comput-

ing coordinates, it is necessary to have a starting point—i.e., one corner having known coordinates. The starting point may be referenced to a known coordinate system (such as the state plane coordinate system), or assumed coordinates may be used.

It should be clear from Figure 9-4 that, given the coordinates of one point, say, *A*, the *x*-coordinate of *B* is equal to the *x*-coordinate of *A* plus (or minus) the departure of *AB*. Similarly, the *y*-coordinate of *B* can be found by adding (or subtracting) the latitude of *AB* to the *y*-coordinate of *A*. Coordinates of all corners of a closed traverse can be calculated in the manner just described by beginning at a point *A* with known (or assumed) coordinates and proceeding around the traverse to point *A*. If latitudes and departures were "balanced," the original and calculated coordinates of *A* should be the same, thereby affording a good, but not perfect, check on the computations.

Example 9-4. If coordinates of traverse point *A* in Example 9-3 are N 2000.00, E 1200.00 ft, determine the coordinates of the other traverse corners.

The computations for solving this problem are shown in Table 9-4. The N-coordinate of *B* was determined by adding the latitude of *AB* to the N-coordinate of *A*—i.e., −358.25 + 2000.00 = 1641.75. The E-coordinate of *B* was

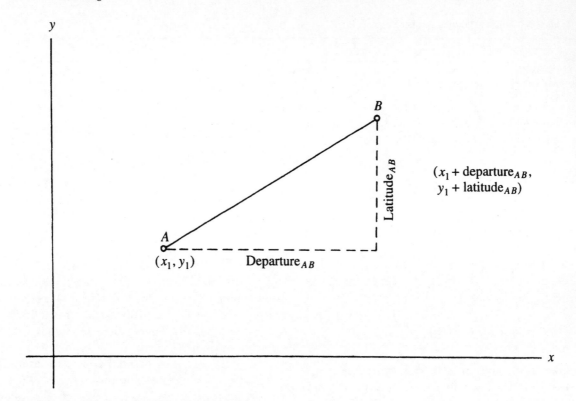

Figure 9-4. Illustration of computation of coordinates.

computed by adding the departure of AB to the E-coordinate of A—i.e., $-539.06 + 1200.00 = 660.94$. Coordinates of remaining points were found in the same manner. Starting and calculated coordinates of point A are the same, indicating that the computed coordinates are probably correct.

9-6. MISSING DATA

Preceding sections demonstrated how traverse misclosure can be determined if lengths and bearings of all lines have been measured. The premise for computing misclosure is that the algebraic sums of both latitudes and depar-

Table 9-4. Computation of station coordinates

Station	Latitude	Departure	N-coordinate	E-coordinate
A			2000.00	1200.00
	−358.25	−539.06		
B			1641.75	660.94
	255.12	159.55		
C			1896.87	820.49
	278.53	−153.53		
D			2175.40	666.96
	76.73	549.79		
E			2252.13	1216.75
	−252.13	−16.75		
A			2000.00	1200.00

All values are in ft.

tures are zero for perfect closure. This premise can be used to calculate a maximum of two "missing data" for a closed traverse—lengths of two lines, bearings of two lines, length and bearing of the same line, or length of one line and bearing of another—if *all other* bearings and lengths are known. The algebraic sums of both latitudes and departures, some of which will be unknown or in terms of unknowns, must equal zero. Then two simultaneous equations with two unknowns can be solved to find the unknown values.

Probably the most common application of this procedure is calculating the bearing and length of a single traverse line, when all others have been measured. This problem is easy to solve since the line's latitude and departure must force the traverse total latitudes and departures to equal zero. After the latitude and departure of the missing line have been computed, its length and bearing are readily obtained by the methods described in Section

9-4-2 and illustrated in Example 9-3. Example 9-5 illustrates this type of procedure.

If lengths of two different lines of a traverse are unknown, the method described yields two simultaneous equations with two unknowns, which can generally be solved directly. In some cases, however, two equations with two unknowns do not provide a unique solution. An example occurs when the bearings of two adjacent traverse lines are unknown. As illustrated in Figure 9-5, two sets of bearings for *BC* and *CD* will close the figure for the same values of all line lengths and other bearings. Both sets of bearings can be obtained by solving the simultaneous equations.

The procedure described here to solve for missing data should not ordinarily be used for surveying traverses because it negates any check on their accuracy. Surveyors might measure the lengths and bearings of all traverse lines except one, then compute them for the remaining line. But any errors in the field

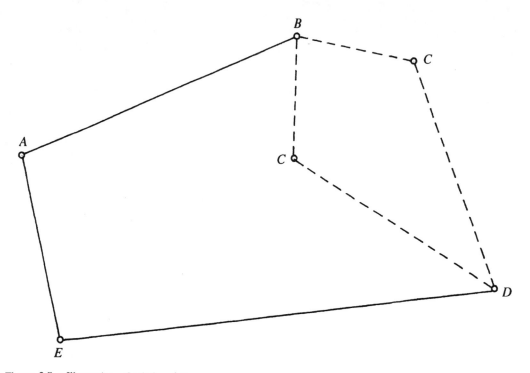

Figure 9-5. Illustration of missing data.

work are thrown into the computed bearing and length of the "unknown" line, so surveyors might be unaware of large mistakes made in the measurements. Unfortunately, this practice is sometimes followed with catastrophic results, if large measurement mistakes were made.

Solving for missing data may be warranted in some cases, however. For example, suppose it is necessary to establish a field line between two points A and B through heavily wooded terrain, as in Figure 9-6, and neither the length nor direction from A to B is known. One possible solution would be to run a random line from A toward B. If it does not pass precisely through B (a likely prospect), compute a corrected direction and rerun the line. This could be difficult if obstructed by trees and/or brush. An alternative might be to run a random traverse from A to B along a rela-

tively clear path, then compute a length and direction for line AB to lay it out in the field. Such a random traverse is shown in Figure 9-6 (A to D to C to B).

Another example of effective use when computing missing data is partitioning land into separate tracts. For example, suppose *ABCDEFGHA* in Figure 9-7 has been surveyed, and balanced latitudes and departures are known for each line. It is desired to divide this tract into three smaller tracts by cutoff lines H to C and G to D. Lengths and directions of these cutoff lines—to stake them in the field —can be ascertained by the methods presented in this section.

Example 9-5. Suppose the scenario described previously to define a direction and length for line AB in the field is followed by running a random traverse along the path BCDA shown in Figure 9-6. Bearings computed from deflection angles and the measured lengths of lines BC, CD, and DA are given in Table 9-5. Find the length and bearing of line AB. All lengths are in feet.

Computed latitudes and departures are shown in Table 9-5. Equate latitude and departure sums to zero.

$$x + 522.75 + 735.18 + 232.21 = 0$$
$$y - 352.82 - 60.87 + 190.29 = 0$$
$$x = \text{latitude of } AB = -1520.14$$
$$y = \text{departure of } AB = 233.40$$

$$\text{Length of } AB = \sqrt{(-1520.14)^2 + (233.40)^2}$$
$$= 1536.47 \text{ ft}$$

$$\text{Bearing angle of } AB = \arctan(223.40/1520.14)$$
$$= 8°22'$$

Since latitude is negative and departure is positive, the bearing of AB is S 8°22′ E.

9-7. AREA COMPUTATIONS

One of the reasons for running and computing closed traverses is to define areas. Land is

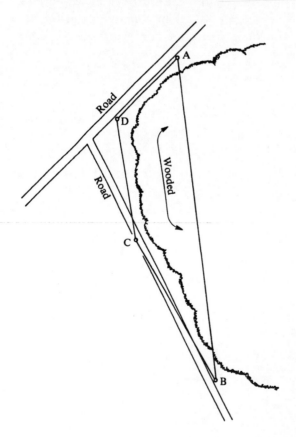

Figure 9-6.   Illustration of missing data.

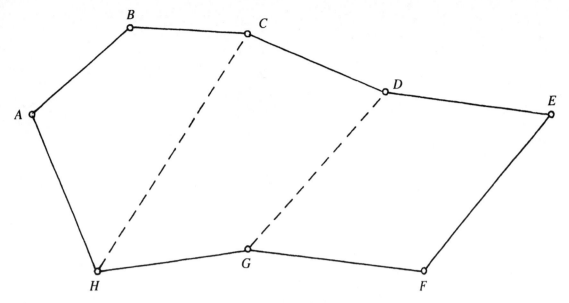

Figure 9-7. Illustration of missing data.

ordinarily bought and sold on a basis of cost per unit area. For this reason as well as many others, an acurate determination of a tract area is often necessary.

If distances are measured in feet, area is generally computed in square feet. For large areas, particularly those related to land in the United States, area is commonly expressed in *acres:* 1 acre = 43,560 ft². When distance is measured in meters, area is computed in square meters or *hectares:* 1 hectare = 10,000 m². The relationship between acres and hectares is 1 hectare = 2.471 acres or 1 acre = 0.4047 hectare.

Three means of computing traverse area are presented in this section: (1) the *double meridian distance* (DMD) method, (2) coordinate method, and (3) the use of a *planimeter*.

9-7-1. DMD Method

The DMD method requires that latitudes and departures of traverse boundary lines be known, as they are after a traverse has been checked for misclosure and balanced.

The meridian distance of a line is the perpendicular distance from the line's midpoint to a reference meridian (north-south line). In Figure 9-8, *FD* is

Table 9-5. Data for finding length and bearing of line *AB*

Station	Bearing	Length (ft)	Latitude (ft)	Departure (ft)
A				
			x	*y*
B				
	N 32°33′ W	655.75	552.75	−352.82
C				
	N 4°44′ W	737.70	735.18	−60.87
D				
	N 39°20′ E	300.22	232.21	190.29
A				

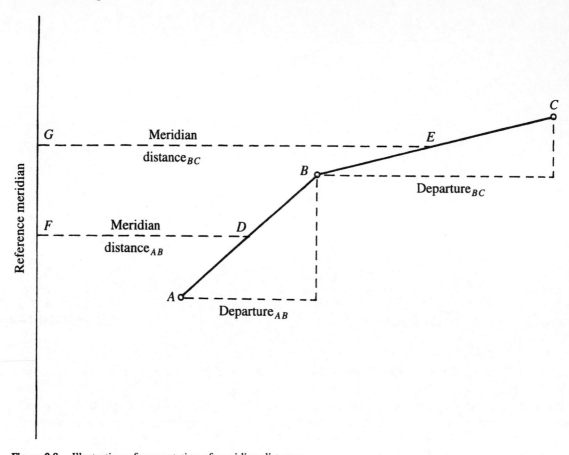

Figure 9-8. Illustration of computation of meridian distance.

the meridian distance of line *AB*, and *GE* the meridian distance of line *BC*. Mathematically, the meridian distance of *BC* is equal to the meridian distance of *AB*, plus half the departure of *AB* plus half the departure of *BC*.

In order to avoid working with half-departures, surveyors use the double meridian distance—i.e., twice the meridian distance—in making computations. Obviously, the DMD of *BC* is equal to the DMD of *AB* plus the departure of *AB* plus the departure of *BC*. This can be generalized to say that *the DMD of a traverse line is equal to the DMD of the previous line plus the departure of the previous line plus the departure of the line itself*. If the reference meridian is moved to pass through *A* in Figure 9-8, the DMD of *AB* is equal to its departure.

Summarizing the preceding discussion gives the following rules for computing DMDs for a closed traverse:

1. The DMD of the first line is equal to the departure of the first line. (If the "first line" is chosen as the one that begins at the westernmost corner, negative DMDs can be avoided.)

2. The DMD of each succeeding line is equal to the DMD of the previous line plus the departure of the previous line plus the departure of the line itself.

As a means of providing a check on DMD computations, if departures have been balanced, the last line's DMD should be equal in magnitude but opposite in sign to its departure.

Once DMDs have been determined, traverse area can be computed by multiplying the DMD of each line by its latitude, summing the products, and taking half the absolute value of the sum. This computation gives traverse area, but the proof is not demonstrated here.

Example 9-6. Find the area of the closed traverse of Example 9-3 by the DMD method.

Computed DMDs and DMD × latitude products are shown in Table 9-6. The DMD of *AB* was set equal to the departure of *AB* (-539.06), and the DMD of *BC* determined by adding the DMD of *AB*, departure of *AB*, and departure of *BC*,

$$-539.06 - 539.06 + 159.55 = -918.57$$

DMDs of remaining lines were calculated in the same manner. Note that the DMD of the last line *EA* (16.75) is equal in magnitude but opposite in sign to the departure of *EA* (-16.75). Values in the last column were obtained by multiplying the latitude of each line by its DMD—i.e., for line *AB* (-358.25) × (-539.06) = 193,118.

The traverse area is $339,239/2 = 169,620$ ft^2, or 3.894 acres. The minus sign indicates only that the DMDs were calculated by a sequence around the traverse in a clockwise direction (instead of counterclockwise). Carrying out acreage beyond three decimal places is probably the limit, since 0.0001 acre represents 4.36 ft^2. Note that in many deeds, the acreage stated is qualified by "more or less" to cover *small* errors only.

A check on the calculated area can be made by employing *double parallel distances* (DPDs). The DPD for any traverse course is equal to *the DPD of the previous line plus the latitude of the previous line plus the latitude of the line itself.* (The DPD of the first course may be set equal to the latitude of the first course.) The traverse area can be computed by multiplying the DPD of each line by its departure, summing the products, and taking half the absolute value of the total.

Standard tabular forms are available for computing latitudes and departures, adjusted latitudes and departures, DMDs and areas on a single sheet.

9-7-2. Coordinate Method

The area of a closed traverse can be determined by this method if the coordinates of each corner are known.

The computational procedure in applying the coordinate method is to multiply the *x*-coordinate of each corner by the difference between adjacent *y*-coordinates, add the resulting products, and take half the absolute value of the sum; *y*-coordinates must be taken in the same order when obtaining the differ-

Table 9-6. Computation of area by DMDs

Station	Latitude (ft)	Departure (ft)	DMD (ft)	DMD × Latitude (ft^2)
A				
	-358.25	-539.06	-539.06	193,118
B				
	255.12	159.55	-918.57	$-234,346$
C				
	278.53	-153.53	-912.55	$-254,173$
D				
	76.73	549.79	-516.29	$-39,615$
E				
	-252.13	-16.75	16.75	$-4,223$
A				
				$-339,239$

ence between adjacent y-coordinates. This process can be expressed in equation form as

$$A = [x_1(y_2 - y_n) + x_2(y_3 - y_1)$$
$$+ \cdots + x_n(y_1 - y_{n-1})]/2 \qquad (9\text{-}7)$$

The coordinate method may be applied by substituting coordinates into Equation (9-7), but the method is expedited by listing them in the following form and securing sums of the products of all adjacent diagonal terms taken: (1) down to the right, i.e., $x_1 y_2$, $x_2 y_3$, etc., and (2) up to the right, i.e., $y_1 x_2$, $y_2 x_3$, etc.

$$\frac{x_1 \; x_2 \; x_3}{y_1 \; y_2 \; y_3} \cdots \frac{x_n \; x_1}{y_n \; y_1}$$

The traverse area is equal to half the absolute value of the difference between these two sums. In applying this procedure, note that the first coordinate listed must be repeated at the end of the list.

Example 9-7.   Find the area of the closed traverse of Example 9-3 (and Example 9-4) by the coordinate method.

First, list the coordinates determined in Example 9-4 in the format indicated previously.

$$\frac{2000.00 \; 1641.75 \; 1896.87}{1200.00 \; \; 660.94 \; \; 820.49}$$

$$\times \frac{2175.40 \; 2252.13 \; 2000.00}{\; 666.96 \; \; 1216.75 \; \; 1200.00}$$

The sum of the products of adjacent diagonal terms taken down to the right

[i.e., $(2000.00)(660.94)$

$$+ (1641.75)(820.49) + \dots]$$

is 9,283,530, and that of products taken up to the right

[i.e., $(1200.00)(1641.75)$

$$+ (660.94)(1896.87) + \dots] = 8,944,292$$

The difference between sums is $9,283,530 - 8,944,292$, or $339,238$. Thus, the traverse area is $339,238/2 = 169,619$ ft^2, or 3.894 acres.

9-7-3. Area by Planimeter

Applying the DMD method requires knowing latitudes and departures, where the coordinate method needs coordinates for each corner. Both methods are limited to use with areas bounded by straight lines, but many parcels have some curved boundaries. For example, one or more boundaries of a land tract may follow a meandering roadway or creek. Such curved boundaries can be converted to a number of small straight-line segments suitable for the DMD or coordinate method.

An alternative means of finding land area utilizes a *planimeter*. Its operation does not require latitudes and departures, coordinates, or straight-line boundaries. Unlike the other methods, however, a scale drawing of the tract for which area is to be determined must be available.

A planimeter (see Figure 9-9) is a mechanical device that integrates area and records the answer as an operator traces the boundary of a figure with the pointer. An ordinary planimeter consists of two arms. One, the anchor arm, has a weight with a sharp point at its free end. The other, a scale bar, has a pointer at its free end. Near where the two arms join are a graduated drum, disk, and vernier.

To measure the area of a tract on a scale drawing, the anchor point is secured at some convenient location on the drawing, preferably outside the area to eliminate applying a polar constant, and the pointer set over a specific traverse boundary corner. An initial four-digit reading is taken; the first digit is read from the disk, the next two from the drum, and the last one on the vernier. The operator then carefully moves the pointer around the traverse boundary until the starting point is reached. A straightedge may be used to guide the pointer around the traverse, but ordinarily it is moved meticulously around the boundary freehand. At this time, another

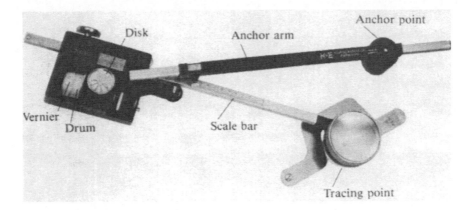

Figure 9-9. Mechanical planimeter. (Courtesy of Cubic Precision, K & E Electro-Optical Products.)

reading is taken. The difference between initial and final readings, scaled if necessary, gives the traverse area. The boundary is then traced in the opposite direction back to the starting point, where the reading should be within a few digits in the fourth place as a check.

Although the procedure for finding area by planimetering sounds simple, caution must be exercised if accurate results are to be obtained. Since the area obtained by a planimeter is not necessarily an exact value (the same area measured twice will often yield slightly different results), it is good practice to trace a figure several times and take an average of the results thus obtained. It is also desirable to trace the figure one or more times in the opposite direction and average these values also. Unless absolutely sure of the planimeter

scale constant, its value should be verified prior to determining a desired area. This can be accomplished easily by tracing a figure of known area, such as a 5-in. square drawn to scale by the user and the diagonals measured to assure exactly a 25-in.^{2} area. One final admonition: As noted previously, the anchor point is preferably positioned outside the traverse. If positioned inside, a polar constant (usually provided by the manufacturer) must be added.

A *mechanical planimeter* is shown in Figure 9-9. *Electronic planimeters*, similar in operation to mechanical planimeters, present results in digital form on a display console with the ability to give answers directly in units of acres or hectares. Figure 9-10 displays an electronic planimeter.

Figure 9-10. Planix electronic planimeter. (Courtesy of the Lietz Co.)

```
C    THIS PROGRAM BALANCES A CLOSED-TRAVERSE SURVEY BY THE COMPASS RULE.
C    IT CAN BE USED FOR DISTANCES MEASURED IN EITHER FEET OR METERS.
C
C    INPUT DATA MUST BE SET AS FOLLOWS.
C
C    DATA LINE 1    COLUMN 1       ENTER 0 (ZERO) OR BLANK IF DISTANCES
C                                  ARE IN FEET. ENTER 1 (ONE) IF DISTANCES
C                                  ARE IN METERS.
C                   COLUMNS 2-9    ENTER DATE, IF DESIRED.
C                   COLUMNS 10-80  ENTER TITLE, IF DESIRED.
C
C    DATA LINE 2    COLUMNS 1-2    ENTER DESIGNATOR FOR BEGINNING POINT OF
C                                  FIRST TRAVERSE LINE.
C                   COLUMNS 4-5    ENTER DESIGNATOR FOR ENDING POINT OF
C                                  FIRST TRAVERSE LINE.
C                   COLUMNS 6-20   ENTER NUMBER INCLUDING DECIMAL GIVING
C                                  LENGTH OF FIRST TRAVERSE LINE.
C                   COLUMN 25      ENTER N IF BEARING OF FIRST TRAVERSE
C                                  LINE IS NORTH; ENTER S IF IT IS SOUTH.
C                   COLUMNS 27-28  ENTER NUMBER (RIGHT-ADJUSTED WITHOUT
C                                  DECIMAL) GIVING NUMBER OF DEGREES IN
C                                  BEARING ANGLE OF FIRST TRAVERSE LINE.
C                   COLUMNS 30-31  ENTER NUMBER (RIGHT-ADJUSTED WITHOUT
C                                  DECIMAL) GIVING NUMBER OF MINUTES IN
C                                  BEARING ANGLE OF FIRST TRAVERSE LINE.
C                   COLUMN 33      ENTER E IF BEARING OF FIRST TRAVERSE
C                                  LINE IS EAST; ENTER W IF IT IS WEST.
C                   COLUMNS 35-44  ENTER NUMBER INCLUDING DECIMAL GIVING
C                                  X-COORDINATE OF BEGINNING POINT.
C                   COLUMNS 45-54  ENTER NUMBER INCLUDING DECIMAL GIVING
C                                  Y-COORDINATE OF BEGINNING POINT.
C
C    DATA LINE 3    ENTER DATA FOR SECOND TRAVERSE LINE (OMIT COORDINATES
C                                  OF BEGINNING POINT) IN SAME FORMAT AS
C                                  THAT FOR FIRST TRAVERSE LINE (SEE DATA
C                                  LINE 2 ABOVE).
C
C    SIMILARLY, ENTER DATA FOR SUCCEEDING TRAVERSE LINES IN SAME FORMAT.
C
      INTEGER UNITS
      DIMENSION DATE(2),TITLE(12),NL(100),NR(100),DS(100),BL(100),
     SNDEG(100),NMIN(100),BR(100),ALAT(100),ADEP(100),XCOOR(100),
     SYCOOR(100)
99    FORMAT(I1,2A4,11A6,A5)
98    FORMAT(A2,1X,A2,F15,2,4X,A1,1X,I2,1X,I2,1X,A1,1X,2F10.2)
97    FORMAT(1,1X,A2,5,2A4,///,11X,'GIVEN DATA',1X,11X,'=======',///
     S  LINE', FT     BEARING',/,'  ----- ------ -------  -------')
```

Figure 9-11. Program for analyzing traverses. (Adapted from J. B. Evett, 1979, *Surveying*, New York: John Wiley & Sons, p. 147.)

```
90    FORMAT(...)
95    FORMAT(//,29X,'SURVEY ADJUSTED BY COMPASS RULE',/,29X,'==============',
     $'=============',//,' POINT LINE    LENGTH    BEARING',
     $' LATITUDE   DEPARTURE   X-COORDINATE   Y-COORDINATE',/,
     $'-------------------------------')
94    FORMAT(2X,A3,58X,F10.2,2,5X,F10.2,/,6X,A2,'-',A2,2X,F9.2,4X,A1,
93    $2I3,1X,A1,2X,F10.2,2X,F10.2)
      $FORMAT(//,' AREA = LINEAR MISCLOSURE =',F6.2,' FT',//,' PRECISION = 1/'
87    $IS,//, AREA IN SQ FT OR',F11.3,' ACRES)
      $FORMAT('1',11A6,A5,2A4,//,11X,'GIVEN DATA',/,/,1X,'==============',///
      $'-----')
      $'   LINE    LENGTH   M   BEARING',/,/,
83    FORMAT(//,' LINEAR MISCLOSURE =',F6.2' M',//,' PRECISION = 1/',
      $IS,//,' AREA = ',F11.D,' SQ M    OR',F11.3,' HECTARES')
      DSUM=0.
      SUMD=0.
      SUML=0.
      AREA=0.
      PI=3.14159265
      READ(5,99)UNITS,DATE,TITLE
      DO 100 J=1,100
      N=J-1
      IF(J.EQ.1)READ(5,98)NL(J),NR(J),D(J),BL(J),NDEG(J),NMIN(J),BR(J),
     $XBEG,YBEG
      IF(J.NE.1)READ(5,98,END=101)NL(J),NR(J),D(J),BL(J),NDEG(J),NMIN(J)
     $,BR(J)
      DEG=NDEG(J)
      AMIN=NMIN(J)
      BRAD=(DEG+AMIN/60.)*PI/180.
      ALAT(J)=D(J)*COS(BRAD)
      IF(BL(J).EQ.'S')ALAT(J)=-ALAT(J)
      SUML=SUML+ALAT(J)
      ADEP(J)=D(J)*SIN(BRAD)
      IF(BR(J).EQ.'W')ADEP(J)=-ADEP(J)
      SUMD=SUMD+ADEP(J)
      DSUM=DSUM+D(J)
100   CONTINUE
101   IF(UNITS.EQ.0)WRITE(6,92)TITLE,DATE
      IF(UNITS.EQ.1)WRITE(6,87)TITLE,DATE
      WRITE(6,96)(NL(J),NR(J),D(J),BL(J),NDEG(J),NMIN(J),BR(J),J=1,N)
      WRITE(6,95)
      EC=SQRT(SUML*SUML+SUMD*SUMD)
      PC=DSUM/EC
      IPC=PC+.5
      DO 102 J=1,N
      ALAT(J)=ALAT(J)-SUML*D(J)/DSUM
      ADEP(J)=ADEP(J)-SUMD*D(J)/DSUM
      D(J)=SQRT(ALAT(J)**2+ADEP(J)**2)
```

Figure 9-11. *(Continued)*

173

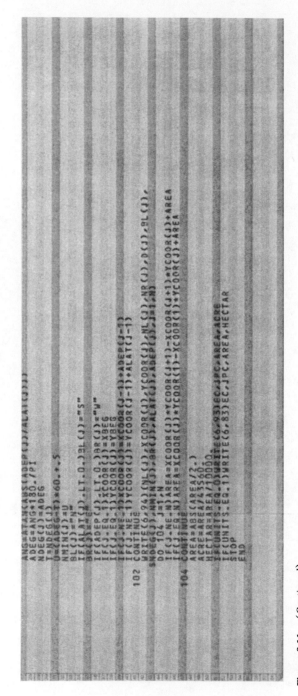

```
        ANG=ATAN(ABS(ADEP(J)/ALAT(J)))
        ADEG=ANG*180./PI
        NDEG(J)=ADEG
        T=NDEG(J)
        U=(ADEG-T)*60.+.5
        NMIN(J)=U
        BL(J)=N
        IF(ALAT(J).LT.0.)BL(J)="S"
        BR(J)=EC
        IF(ADEP(J).LT.0.)BR(J)="W"
        IF(J.EQ.1)XCOOR(J)=XBEG
        IF(J.EQ.1)YCOOR(J)=YBEG
        IF(J.NE.1)XCOOR(J)=XCOOR(J-1)+ADEP(J-1)
        IF(J.NE.1)YCOOR(J)=YCOOR(J-1)+ALAT(J-1)
102     CONTINUE
        WRITE(6,94)(N,(J),XCOOR(J),YCOOR(J),NL(J),NR(J),D(J),BL(J),
       SNDEG(J),NMIN(J),BR(J),ALAT(J),ADEP(J),J=1,N)
        DO 104 J=1,N
        IF(J.NE.N)AREA=(XCOOR(J+1)-XCOOR(J))*YCOOR(J+1)*YCOOR(J)+AREA
        IF(J.EQ.N)AREA=(XCOOR(1)-XCOOR(J))*YCOOR(1)*YCOOR(J)+AREA
104     CONTINUE
        AREA=ABS(AREA/2.)
        ACRE=AREA/43560.
        HECTAR=AREA/10000.
        IF(UNITS.EQ.0)WRITE(6,93)EC,JPC,AREA,ACRE
        IF(UNITS.EQ.1)WRITE(6,83)EC,JPC,AREA,HECTAR
        STOP
        END
```

Figure 9-11. (*Continued*)

174

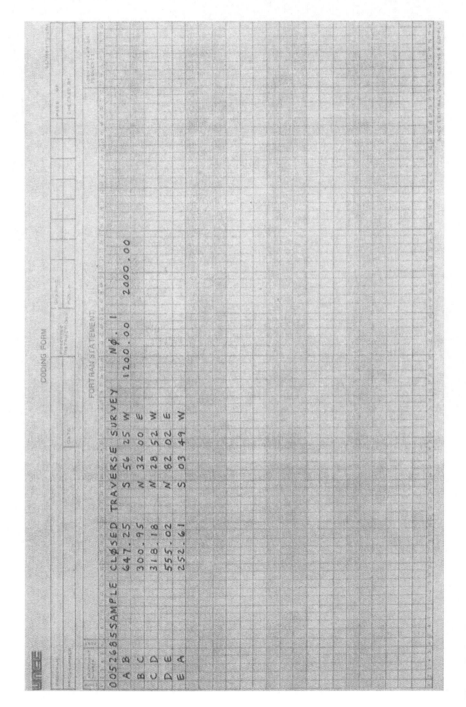

Figure 9-12. Input data for sample closed traverse survey no. 1.

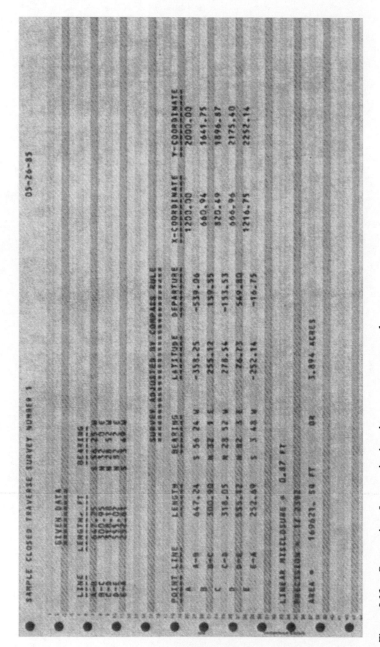

Figure 9-13. Output data for sample closed traverse survey no. 1.

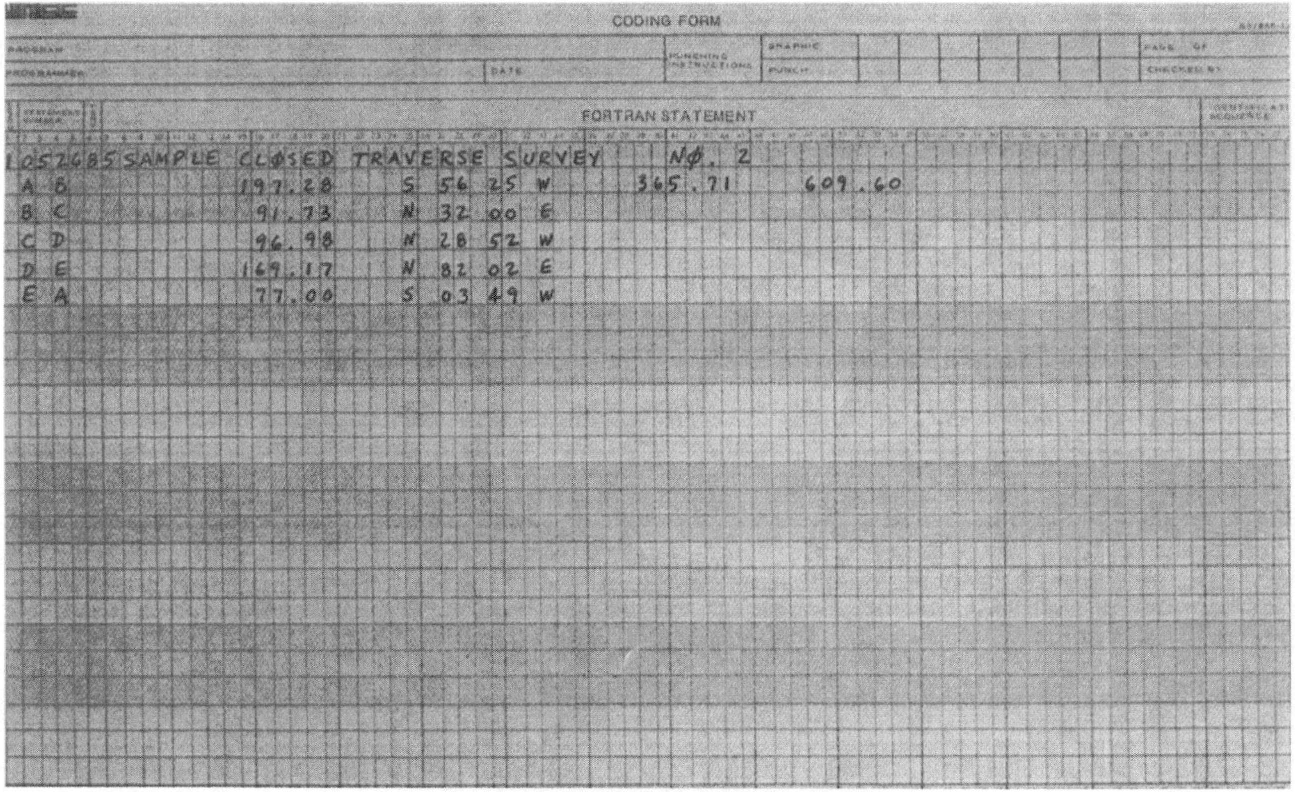

Figure 9-14. Input data for sample closed traverse survey no. 2.

SAMPLE CLOSED TRAVERSE SURVEY NUMBER 2   05-26-85

        GIVEN DATA
        ==========

LINE   LENGTH, M   BEARING
----   ---------   -------
A-B   197.28   S 56 25 W
B-C   91.73   N 32 00 E
C-D   96.93   N 28 52 W
D-E   169.17   N 32 40 E
E-A   77.00   S 3 40 N

                        SURVEY ADJUSTED BY COMPASS RULE
                        ===============================

POINT LINE   LENGTH   BEARING   LATITUDE   DEPARTURE   X-COORDINATE   Y-COORDINATE
----- ----   ------   -------   --------   ---------   ------------   ------------
  A   365.76   609.60
       A-B   197.28   S 56 24 W   -109.19   -164.30
  B   201.46   500.41
       B-C   91.72   N 32 1 E   77.76   48.63
  C   250.09   578.17
       C-D   96.94   N 28 52 W   84.90   -46.80
  D   203.29   663.07
       D-E   169.20   N 32 3 E   23.39   167.58
  E   370.87   686.46
       E-A   77.02   S 3 48 N   -76.86   -5.11

LINEAR MISCLOSURE =   0.26 M

PRECISION = 1/ 2434

AREA =   15759. SQ M   OR   1.576 HECTARES

9-8. PROGRAMMED TRAVERSE COMPUTATIONS

The various traverse computations presented in this chapter are among the most common, extensive, and important ones made in surveying. Practicing surveyors prepare these calculations on a daily basis.

Many years ago, surveyors had to make such computations manually, using slide rules or trigonometric tables and logarithms. Subsequently, large mechanical calculators capable of performing addition, subtraction, multiplication, and division became available, but trigonometric tables were still required. In the 1960s and 1970s, high-speed digital computers and hand-held calculators, some programmable, greatly increased the surveyor's computational capability. In addition to making rapid computations, both computers and calculators had built-in trigonometric functions, so trigonometric tables were no longer needed. In terms of routine computations, subsequent advances in computer hardware and software continue to make life easier for surveyors.

There are numerous computer programs available for surveyors to use in analyzing traverses. For example, the program in Figure 9-11 is written in FORTRAN and designed for card input, but it could easily be modified to another language for other kinds of input or

for use on microcomputers. For input, the program receives the length and bearing of each line and gives as output adjusted lengths, bearings, latitudes, departures, and coordinates, as well as linear misclosure, precision, and traverse area. The program can be used for distances measured in either feet or meters; comments at the program's beginning tell how input data must be arranged to utilize the program.

For demonstration purposes, the computer program was run using input data from Example 9-7 (lengths and bearings of a closed traverse survey). Input data prepared for use in the program are shown in Figure 9-12. Output from the program, which includes answers found previously in Examples 9-1, 9-2, 9-3, 9-4, and 9-7 (the program computes the area by the coordinate method), is given in Figure 9-13. Examination of this output reveals that the answers closely verify those obtained by calculation in the examples, with some slight variations resulting from round-off errors in the manual computations.

As a final demonstration, lengths of the traverse lines in Example 9-1 (used as input to the computer program) were converted from feet to meters and the program run again for the same traverse with lengths in meters, along with original bearings. Input data for this application of the program are shown in Figure 9-14 and output in Figure 9-15.

10

Survey Drafting

Edward G. Zimmerman

10-1. INTRODUCTION

A sketch, map, or graphic display, is often the only visible product of a surveyor's work. Therefore, the importance of presenting the client with a nice-appearing, professionally done graphic product cannot be overemphasized. Attractiveness, accuracy of plot, legibility, and clearly imparted information are vital in creating a survey drafting product worthy of professional respect.

Survey drafters differ from their engineering and architectural counterparts in being not only familiar with applying principles of drafting and graphics, but also able to comprehend survey instrumentation and methods of measurement. Field notes must be reduced and interpreted and information sources researched so a drafter can construct a comprehensive graphic representation of actual facts and conditions.

A survey drafter is a multitalented technician. He or she should have training in the basics of mechanical drawing and mathematical ability extending through trigonometry. Since much of the work prepared will originate from either a set of field notes or rough sketch with some penciled-in instructions, the drafter needs to be familiar with fundamental principles of survey operations. Actual field experience is important in developing an awareness of the methods of recognizing, collecting, and recording field data and will assist the draftsperson in translating the data into a finished drawing.

Drafters do not spend all their time drawing or tracing maps. Depending on the type and priority of workloads, survey drafters may also calculate and check traverse surveys, reduce and plot cross-section and topography field notes, calculate earthwork quantities, and prepare material estimates for construction projects. Often, with a smaller firm, the draftsperson may be assigned to field duties as workloads respond to seasonal fluctuations.

10-2. SURVEY DRAWINGS

Survey drawings fall into three general categories: (1) property and control maps reflecting surveys made to establish or reestablish ownership lines or survey control networks; (2) topographic maps showing elevations, natural and artificial features, and form of the earth's surface; and (3) construction maps made to provide and control horizontal and vertical location, alignment, and configuration of construction work.

An accurate plot of survey information plus legibility and attractiveness determine a map's usefulness. Most maps show few dimensions, and a person using them must rely on a scale, protractor, or drafting machine to determine intermediate dimensions. Unlike mechanical or civil-engineering drawings, survey maps are irregular and not readily drawn by traditional "T square and triangle" methods.

All survey drawings are made to be copied or reproduced. Therefore, drafters should be aware of the infinite variety of reproduction possibilities and applications available through graphic processes, photographic enlargement and reduction, and exhibit preparation. The quality of a finished drawing, type of reproduction equipment, quality of the printed product, and economic requirements are all factors influencing the choice of reproduction method.

Computer-assisted drafting (CAD) is changing mapmaking much as electronic instrumentation is revolutionizing field surveying. Interactive drafting systems, including a cathode-ray display tube and an automatic plotter, are combined with a computer for processing data recorded in the field on an interfaceable data collector. In larger agencies and companies, automated drafting units are fast becoming standard equipment. Operation of a CAD system, such as the one shown in Figure 10-1, requires yet another level of training and ability in a survey drafter but will not be addressed in this chapter.

10-3. MAP SCALE

Map scale is the term used to define the ratio of distances represented on a map to actual ground distances. When a drawing is made to a chosen scale, all dimensions—distance, direction, and difference in elevation—will be in correct relationship and accurately represent the actual figure.

The scale of a map should be indicated by both numerical and graphic means. Numerical scales may be either representative, in which one unit on the map represents a certain number of the same units on the ground —e.g., 1/400, 1 : 400—or equivalent, in which a statement indicates that 1 in. on a map equals a whole number of feet on the ground —e.g., 1 in. = 4000 ft.

Figure 10-2 shows an example of graphic scales. Since drawing paper may change dimensions over time or be distorted by reproduction processes, a graphic scale should be placed on maps to provide a constant check of the exact scale.

The ranges of scales are defined as (1) large scale, 1 in. = 100 ft or larger; (2) medium scale, 1 in. = 100 to 1000 ft; and (3) small scale, 1 in. = 1000 ft or smaller. Table 10-1 provides a guide in choosing the scale for a particular map.

10-4. MAP DRAFTING

Most maps fall into two general classifications: (1) those maps that show land ownership and become part of the public record and (2)

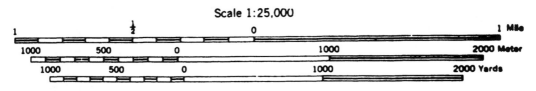

Scale 1:25,000

Figure 10-1. Typical computer-aided drafting system. (Courtesy of Hewlett-Packard Inc.)

Figure 10-2. Typical graphic or bar scale.

those that show land form and are used as the basis for design and construction of structural facilities, both private and public.

Preparation of a preliminary map or manuscript is the first two phases of map drafting. First, the manuscript is carefully laid out, plotting all control lines and points with the utmost accuracy. This map is drawn with a hard-grade pencil on a high-quality drafting film that produces sharp and precisely located line work. Features comprising the final map should be plotted on the manuscript in the following sequence: (1) Lay out control points and lines; (2) plot details; (3) compile topography and other detail work; and (4) finish the map, complete with lettering and all notes.

Lettering and symbolization need not be of finished quality, just accurately located.

Second, following checking and revisions, the manuscript can be placed on a "light table" and the final map traced in ink on a stable-base mylar drafting film. The preferred sequence is to (1) complete all the lettering, notes, and title and (2) finish the line work or topography.

Scribing, another method of producing a final map, is gaining in popularity. In this process, lines from the manuscript are photographically transferred onto a sheet of drafting film coated with an opaque surface. Using specially designed scribing tools, the drafter scrapes and cuts the coating to reproduce all

Table 10-1. Selection of map scales

Type of Map and Use	Equivalent (ft per in.)	Representative (ratio, 1/ ...)
Design		
Civil improvements	10–50	120–600
General construction	40–200	480–2000
Property/boundary (dependent on figure size)	50–500	600–6000
Topographic/planimetric		
Small site	10–50	120–600
Large site	40–200	480–2400
Urban	200–1000	2400–12,000
Regional	500–2000	6000–24,000

the original manuscript's features and line work.

10-5. DATUMS FOR MAPPING

All measurements made by surveyors to determine and depict elevations and horizontal positions should relate to a datum of reference. In the 48 contiguous United States and Alaska, the *American Datum of 1983* is used. These reference figures are made available through the state plane coordinate system of a particular state. Most states have a plane coordinate system based on either a Lambert conformal-conic projection, or transverse Mercator projector system. Adopted in the 1930s, the systems use the U.S. survey foot (1 ft = 1200/3937 m) as the standard of measurement unit.

Elevations in the United States are referred to a vertical datum or reference surface based on mean sea level—i.e., the *North American Vertical Datum of 1988* (NAVD 88). This datum is determined from the average elevations of 26 sea-level tidal stations in the United States and Canada. Coordinates of horizontal control stations and elevations of benchmarks throughout the United States are available in published form from the National Ocean Survey (NOS). Instructions and specifications for the use of both datums are also available from the NOS.

10-6. TOPOGRAPHIC MAPS

A topographic map is a graphic representation of a portion of the earth's surface as it existed on a certain date. It is drawn from field survey data or aerial photographs and shows, by notation or symbol, all natural or artificial land features, including boundaries, cities, roads, railroads, pipelines, electric lines, buildings, and vegetation. Land forms are depicted by contour lines. A topographic map without contour lines is defined as a planimetric map. See Figure 10-3 for a typical map.

10-7. TOPOGRAPHIC MAP CONSTRUCTION

A topographic map should be drawn in three phases: (1) Develop horizontal control, producing a framework for plotting details; (2) plot all points of known elevation and locations of artificial or natural features; and (3) construct contour lines from plotted points of elevation, drawing all features and symbols.

Discussions in the following sections address the work involved in conventional line drawings. The availability of photogrammetrically based plotting systems makes production of larger topographic maps by automated equipment more economical. Most large surveying and engineering firms and agencies use this method. Automated drafting will be covered in later sections.

10-7-1. Plotting Control

Control points and lines for topographic as well as other survey maps can be plotted by one of several methods. The selection of which method to use is guided by the field survey format and form in which field information is forwarded to the drafter.

A traverse or control survey can be plotted by laying out a series of angles and scaled lines. Angles are plotted by a drafting machine, protractor, or coordinates, or constructed by methods described in Section 10-7-5.

10-7-2. Drafting Machine

A drafting machine combines all the functions of a straightedge, triangle, scale, and protractor in one convenient unit (Figure 10-4). It is constructed so that any movement of the machine head is in a parallel motion, but the horizontal and vertical scales retain an initial base-line orientation, ready for use at every position on a drawing. The protractor and vernier scale allow the ruling scales to be rotated to any desired angular value, usually

184   *Survey Drafting*

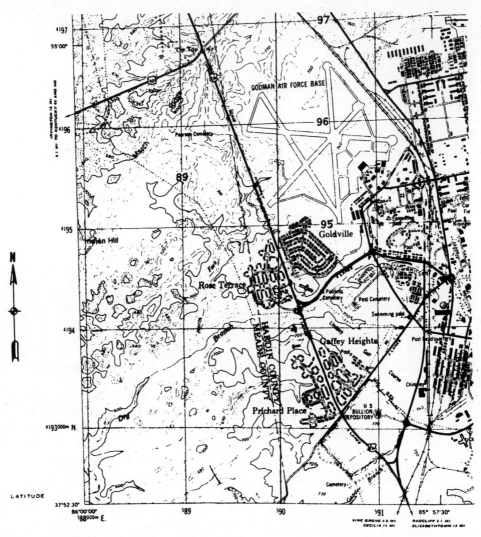

Figure 10-3. Portion of typical topography map.

within ±1 min for either angle construction or instrument orientation.

10-7-3. Protractor

Protractors are a tool vital to mapping and topographical plotting work, and are available in several configurations (Figure 10-5). Most are graduated in half-degree increments and vary in size from 4- to 10-in. diameters. To use, simply align the protractor base and center mark with a base line and angle point, then scale the required angle on the protractor's periphery. Although quick and convenient to use, protractors are not accurate enough for plotting precise control networks.

10-7-4. Coordinates

Plotting by coordinates is a simple and accurate method, although it requires that all information be coordinated and an accurately gridded base map used for plotting. An accurate grid pattern is laid out at an appropriate scale with squares having an even dimension, say, 100, 500, or 1000 ft. Label each x and y line with its grid value to avoid plotting blunders.

Figure 10-4. Drafting machine and table. (Courtesy of Alvin & Co., Inc.)

By scaling from the appropriate x and y line, each point can be plotted to an accuracy of 0.01 to 0.02 in. Each point is marked by a small circle and connected with straight lines to outline the figure. After plotting, measure the bearing and distance between plotted points to check plotting accuracy and detect any blunders. Every point is plotted independently of the others so compounding plotting errors cannot occur.

10-7-5.   Alternate Methods of Plotting Angles

Tangent Method

This method consists of extending the back tangent a scaled number of units beyond a point. The length of extension times the natural tangent of the angle to be constructed yields an offset distance, which is then scaled along a perpendicular erected at the end of the extension (Figure 10-6).

Chord Method

This method is similar to the tangent procedure except that a tangent arc is struck, rather than a perpendicular. The chord distance for the angle to be plotted is calculated and scaled along the arc (Figure 10-7).

10-7-6.   Plotting Details

Although details need not be plotted with the same accuracy as control points, objects must be placed on the map within allowable error standards. The choice of plotting methods for details depends on field procedures for gathering the information. For example, if topography was obtained by a radiation survey, then details should be plotted with a drafting machine or protractor and scale. If a survey was made by the *checkerboard* method, a grid constructed on the map sheet is used to plot details. Surveys from a total station can be plotted by radiation or coordinates.

10-7-7.   Characteristics of Contours

For proper delineation and interpretation of a topographic map, it is important for a drafter to be familiar with the nature of contour lines, as shown in Figure 10-8. Their

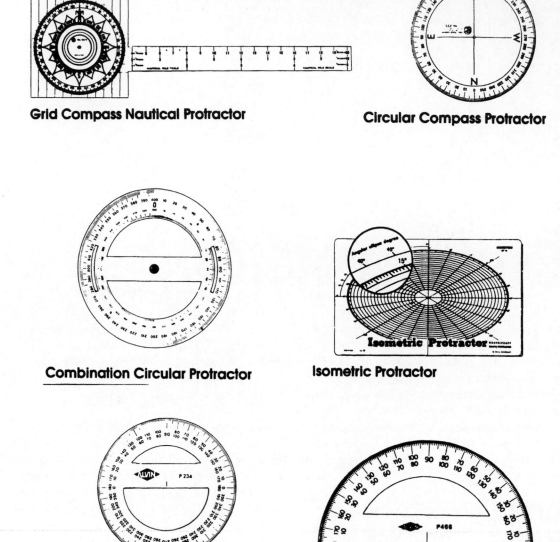

Grid Compass Nautical Protractor

Circular Compass Protractor

Combination Circular Protractor

Isometric Protractor

Academic Circular Protractor

Academic Protractors

Figure 10-5.   Surveying engineering protractors. (Courtesy of Alvin & Co.. Inc.)

principal characteristics are as follows:

1. Contour lines spaced closed together represent steeper slopes; lines farther apart indicate a gentler slope.
2. Uniform spacing of lines represents a uniform, even slope.
3. Mounds or depressions are portrayed by closed contour lines. Depression contours

will have inward-facing, radial tick marks to avoid confusion.

4. Contour lines always close.
5. Contour lines never cross or merge into each other, except, e.g., in the case of overhanging rock ledges.
6. Contour lines are perpendicular to the direction of the represented slope.

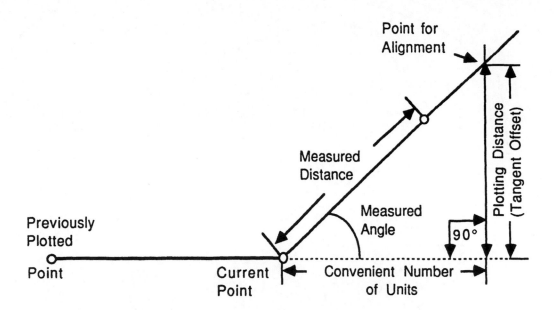

TAN Measured Angle X Number of Units = Plotting Offset.

Figure 10-6.   Construct angle: Tangent method.

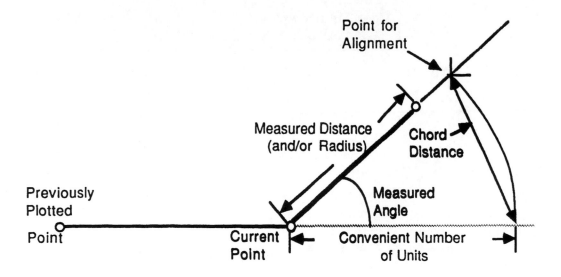

Measured Angle = Central Angle of Arc.
Chord = 2R X SIN 1/2 Central Angle.

Figure 10-7.   Construct angle: Chord method.

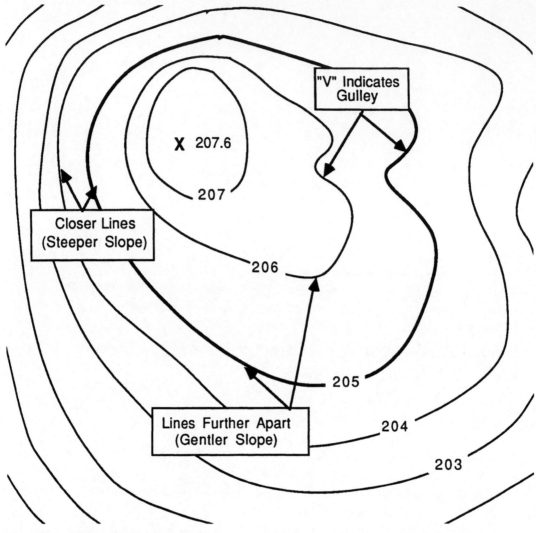

Figure 10-8. Characteristics of contours.

10-7-8. Plotting Contours

When detail plotting is completed, the map sheet will show planimetric detail and elevations of pertinent points. The drafter's next operation is to draw contour lines, guided by previously plotted elevations. A contour line represents an even unit of elevation, generally in multiples of 1, 2, 5, 10, 20, 50, or 100 ft. Selection of a contour interval depends on map scale, details to be presented, and severity of the terrain being mapped. Table 10-2 shows suggested contour intervals. Contour lines are drawn only for those elevations divisible by the contour interval.

Each line is sketched freehand, using smooth flowing curves at direction changes to more nearly represent a natural formation. Every fifth line should be drawn heavier and numbered with the elevation. When a fairly level area is depicted, it is advisable to number all contour lines.

10-7-9. Drawing Contours by Interpolation

Interpolation is a procedure to locate contour lines in their correct proportional position between adjacent points of elevation. This is done by assuming the terrain's slope be-

Table 10-2.   Selection of contour interval

Use of Topographic Map	Equivalent Scale (ft per in.)	Contour Interval (in ft)
Smaller sites	10–50	1–5
Larger sites	40–200	1–10
Urban	200–1000	2–20
Regional	500–2000	5–100

tween any two elevations is constant, and therefore, intervals in elevation will translate to horizontal distances. As illustrated in Figure 10-9, a drafter can accomplish this in one of several ways:

1. *Estimation.* An experienced topographer can make mental calculations as well as estimate positions of the lines.

2. *Direct calculation.* Measure the intervening horizontal distance and proportion the correct position for each line.

3. *Mechanical interpolation.* A rubberband marked with a uniform series of marks can be stretched to find the correct interval for each line. Also, spacing dividers can be used to proportion contour lines. This drafting instrument is constructed with 11 legs arranged to subdivide a distance spanned into equal parts.

4. *Graphic.* A transparent piece of drafting film with converging lines, as shown in Figure 10-10, can be pivoted between points of elevation to find correctly proportioned positions.

10-7-10.   Topographic License

Engineering is tempered with art to create a contour map convening the natural appearance of land forms to the map viewer. Mapping art is defined as using discretion and judgment in expressive placement of lines between control points, which reflect locations determined by engineering methodology. Topographic license is a drafting talent that can be attained by only a proper combination of field experience, drawing practice, and training.

10-8.   CONSTRUCTION MAPS

Maps of this type can vary from a simple plot plan for a residence to a major engineering project, such as a dam or freeway, and maps assisting civil engineers in project design. There are as many different layouts for construction maps as there are for other types of maps; however, all types have a title, typical scale, lettering and symbols, or other recognized guidelines.

10-8-1.   Earthwork Cross Sections

Cross sections are generally used for design purposes and to prepare estimates for earthwork projects. They are easily plotted on preprinted paper having 1-in. grids divided into 0.1-in. increments (see Figure 10-11). Cross sections can be plotted from field books or scaled from contour maps.

When working with field notes, plot sections in a vertical column, using the same stationing sequence shown in the notes. Also, follow the indicated right/left placement. Vertical plotting scale should be at least four times as large as horizontal scale to exaggerate and achieve useful vertical separation within each section. When cross sections are scaled from a topographic map, draw the center or base line, mark full stations on it, and lay out cross-section lines at each station. Scale along every cross-section line, noting distances where these lines cross a contour, then plot the distance out and elevation as though they came from a field book.

Cross sections with lines plotted on them representing finished grade become closed

MATHEMATICAL RELATIONSHIP:

$$\frac{\text{Horiz. Dist.}}{\text{Vert. Diff.}} = \frac{42\text{-ft}}{6.5\text{-ft}} = \frac{6.5\text{-ft}}{1\text{-ft}} = \frac{0.65\text{-ft}}{0.1\text{-ft}}$$

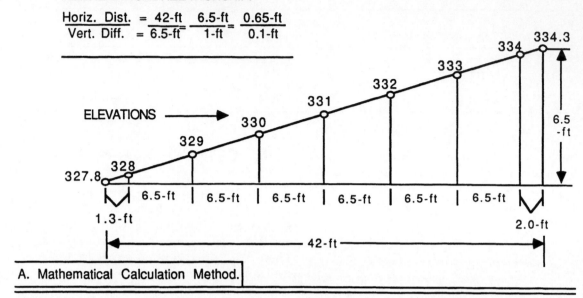

A. Mathematical Calculation Method.

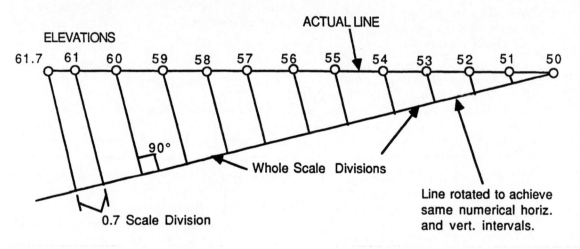

B. Proportionate Triangles Method.

Figure 10-9. Interpolation of contour intervals.

figures, so their areas can be determined by computation, planimetering, and other methods. Additional computations by the *average-end-area* method produce reasonably accurate volumes of a series of adjoining sections. A plot of this type should also include field-book references, referrals to the surveyor's vertical and horizontal control data, and scale.

10-8-2. Plan and Profile Maps

This type of map—drawn to depict details of linear-type construction projects, such as streets and highways, railroads, and pipelines —is used by builders to guide them in construction. As shown in Figure 10-12, the *plan* portion of the drawing is the horizontal layout

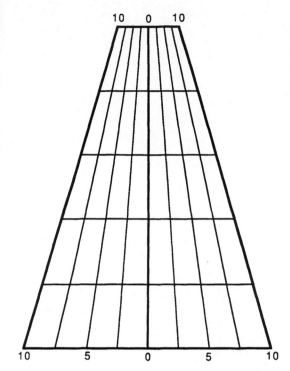

Figure 10-10. Clear plastic interpolator (may be rotated to fit the distance between two controlling points of elevation).

of the project, the *profile* view shows the vertical plane or *grade* information. Either preprinted paper or mylar sheets can be used. Generally, the sheet's top half is blank for a plan view, and the bottom half ruled for a profile view. Preruled lines facilitate rapid and accurate plotting of vertical features, such as flow lines of pipes, grade lines of highways, and natural ground profiles. The following items should be included on a plan and profile map:

1. Basis of stations (usually on centerline)
2. Vertical datum and available bench marks
3. Centerline alignment and control information
4. Lengths and widths of all features
5. Underground structural details
6. Construction notes pertaining to required materials and methods

The plan view portion is drawn by tracing previously plotted control maps or plotting survey data from field books or maps. Features of the project are clearly outlined, along with survey control and dimensional information sufficient for construction layout. Topographic features and contour lines are sometimes included to give further information for estimating and layout purposes.

Profile views include a natural ground line drawn along the project centerline and also grade lines with slope percentages for project features. Grades for aboveground construction are shown for the tops of finished surfaces, but underground grades control the structure's invert.

10-8-3. Site or Grading Plans

In contrast to plan and profile maps depicting long narrow strips of development, a site plan represents multisided figures showing parking lots, buildings, shopping centers, etc. This variety of map is drawn as a plan view but must have sufficient elevation information to enable the designer, estimator, and builder to complete their work. Site plans are developed from boundary and topographic surveys and must reflect proper design criteria conforming to local agency standards covering drainage, ingress and egress, etc. (see Figure 10-13).

Generally, a site or grading plan is prepared in two stages: (1) a preliminary map showing topography and boundary information for use by the engineer to create and finalize his or her design and (2) the final map reflecting completed design information and employed to actually construct the project.

10-8-4. As-Built Maps

At the conclusion of construction projects, *as-built* field surveys are made to locate all features of the project. Rarely is a project completed without making "field adjustments" or modifications to the original design. The survey map can usually be made by highlight-

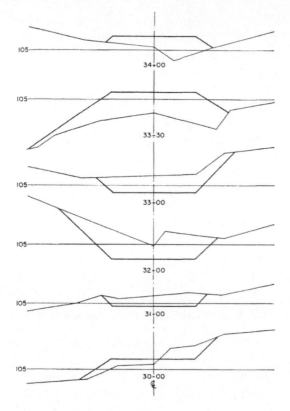

Figure 10-11. Cross-section plot. Grid lines have been omitted for clarity.

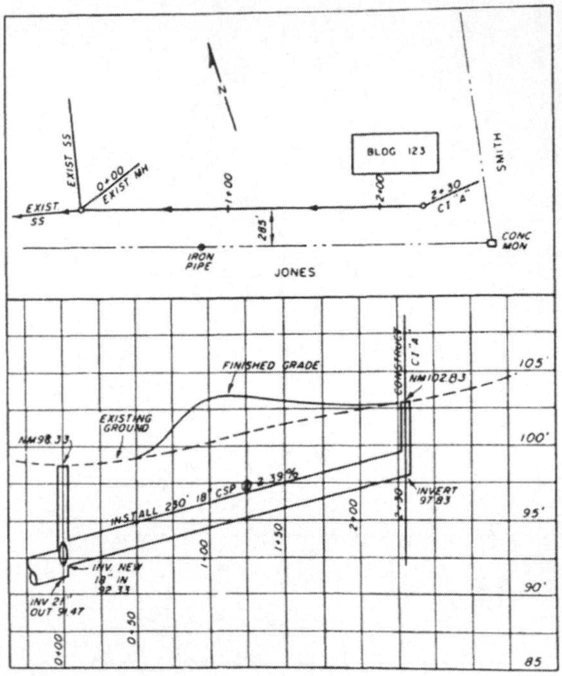

Figure 10-12. Portion of a plan and profile map.

ing or superimposing the changes on a copy of the original construction plans.

The map should show any changes or modifications to the original project design, including notations, locations, and dimensions, and referencing to the primary survey control system. To avoid confusion, as-built maps should be clearly designated as such. In future engineering relative to the completed project, they will become a permanent and useful information source.

10-9. BOUNDARY MAPS

A boundary map is drawn to depict ownership facts disclosed by research from public records and/or field surveys. They usually delineate currently owned parcels and parcels being created for future sales. When relatively large portions of the earth are shown, boundary maps are sometimes called cadastral maps.

10-9-1. Subdivision Maps

Subdivision maps portray a tract of land that has been divided into several smaller parcels. The effect of the map, when made part of the public record, creates a basis of ownership for each individual parcel of lot as it is sold. A typical subdivision map is shown in Figure 10-14; however, maps of this type may vary in conformance to local agency requirements.

Subdivision maps are best traced in ink on mylar from a preliminary map that has been plotted by coordinates. Information shown on these maps should include (1) the scale, (2) survey data references, (3) ownership and title

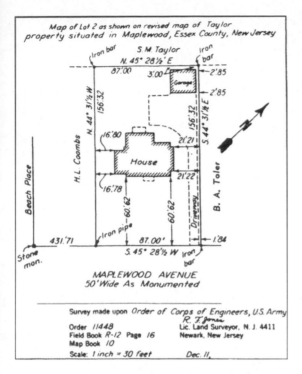

Figure 10-13. Typical site plan.

data, and (4) zoning and other regulatory statements.

10-9-2. Parcel Maps

This map is similar in format and purpose to a subdivision map, but is intended for either a combined or single parcel of land. It must also conform to local regulations and is drawn in the same format as a subdivision map (see Figure 10-15).

10-9-3. Record of Survey

A record of survey map generally does not show a transfer of property. It is intended to show a survey performed and then made part of the public record. This map, although similar to those described in the preceding sections, does not have to conform to the same regulatory requirements (see Figure 10-16).

10-10. INFORMATION SHOWN ON MAPS

Information provided on survey maps obviously varies, depending on their purpose. However, some important items are required on all surveying and engineering maps and are discussed in this section.

10-10-1. Scales

Map scale should be indicated both as a statement—e.g., 1 in. = 400 ft—and shown graphically. It should be placed on the map for quick location by the viewer, preferably close to the title block. See the example in Figure 10-17.

10-10-2. Meridian Arrows

If at all possible, the top edge of a map should represent north, but configuration or size may require a different orientation for the drawing. Regardless of the figure's position, the basis of the map's reference meridian is indicated by a stylized arrow or arrows such as shown in Figure 10-18. The true meridian is defined by a full arrow and other meridians —grid, magnetic, etc.—can be identified with a half arrow and letter designation. It is advisable to show angular difference (rotation) between two or more meridians.

10-10-3. Lettering

Survey drawings can be lettered in two different styles, the choice generally determined by their prospective use. For professional appearance, the lettering style and methods used on a map should be consistent throughout the entire drawing. Single-stroke Reinhardt-style machine-guided lettering is used for drawings intended to remain in the office or be used strictly by other surveyors or engineers. Maps produced for or utilized by the general public can be lettered freehand, using either Rein-

PLAT OF

CARNATION VILLA

LOTS 13 & 14, CAMELLIA ACRES (15 BM 21)
CITY OF SACRAMENTO, CALIFORNIA
JANUARY, 1986   SCALE: 1"=40'
TIMOTHY S. TRAIN, LAND SURVEYOR
SHEET 2 OF 2

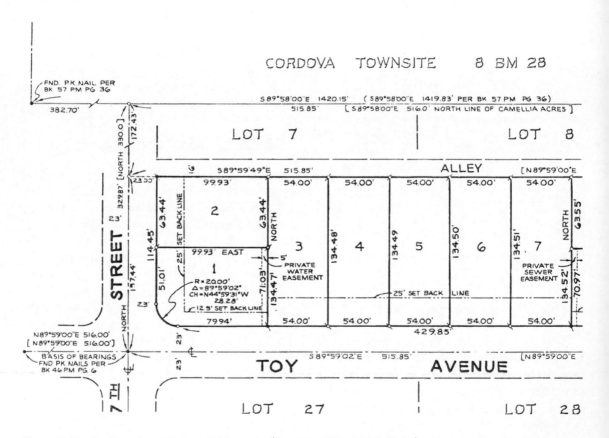

Figure 10-14.   Portion of a typical subdivision map. (Courtesy of Timothy S. Train.)

hardt letters or any attractive style that can be produced speedily and consistently (see Figure 10-19 for examples).

Adhesive-backed film transfer media can be typed on, cut out, and pasted on a drawing. This process is particularly time-saving for tables or lengthy statements on a map. A machine that prints a wide variety of fonts and

sizes on opaque or transparent adhesive-backed tape is shown in Figure 10-20.

10-10-4. Titles

The purpose of a title is to identify the drawing and thus provide information for indexing. It may be neatly lettered freehand,

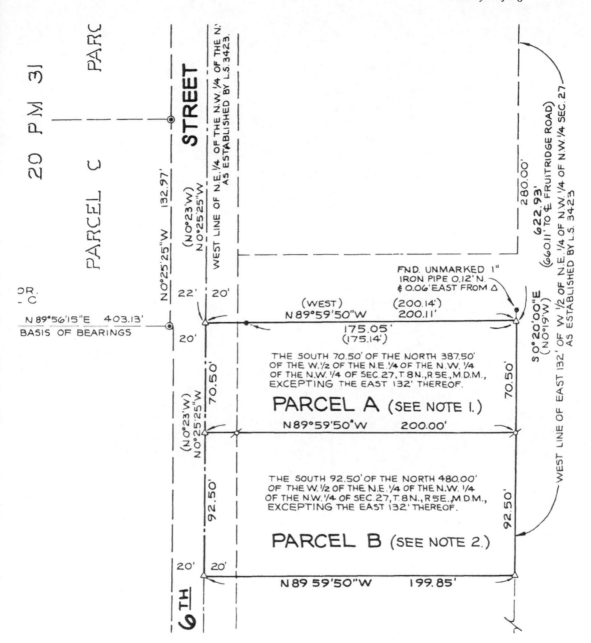

Figure 10-15. Portion of a typical parcel map. (Courtesy of Timothy S. Train.)

mechanically lettered, or preprinted commercially on standard size drawing sheets.

To be esthetically pleasing, a title block may be placed anywhere on the map sheet, but is ordinarily located in the lower right-hand corner. Care should be taken to keep the title

block size symmetrical and in proper proportion to the map size. Use conventional lettering with variations in weight and size to emphasize important parts of the title.

Elements of a title block include (1) name of the firm or agency, (2) type of map, (3)

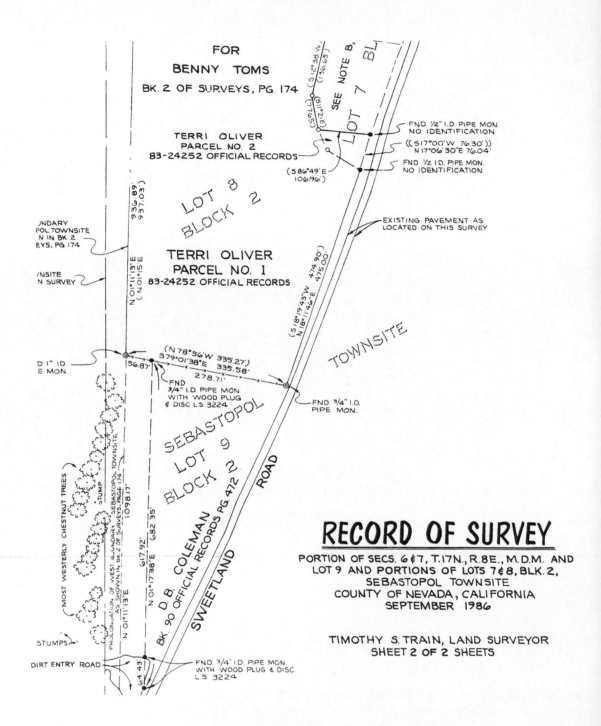

Figure 10-16. Portion of a record of survey map. (Courtesy of Timothy S. Train.)

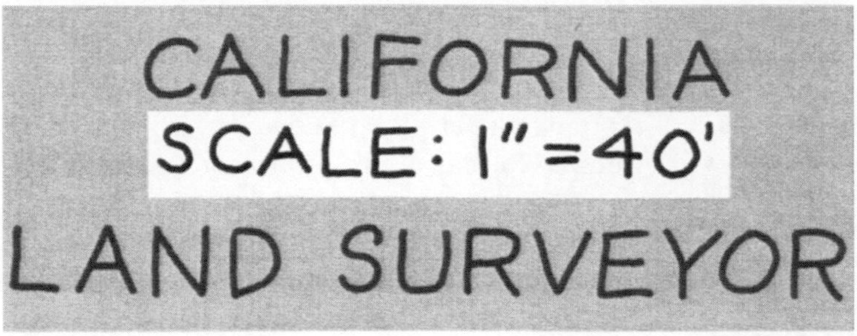

CALIFORNIA
SCALE: 1"=40'
LAND SURVEYOR

Figure 10-17. Statement of scale placed on map. (Courtesy of Timothy S. Train.)

indexing information, (4) name and location of the project, (5) scale and contour interval, (6) names of key personnel responsible for the map, and (7) map completion date (see Figure 10-21). It is common practice for most firms and agencies to use preprinted map sheets with frames and title blocks in a standard format and location. All the drafter need do is fill in the blank spaces.

10-10-5. Notes and Legends

Notes are placed on a map to provide explanations for special features unique to it. They should be brief yet impart adequate information to preclude misinterpretation. Information contained in notes should refer to items such as (1) project basis of bearings, (2) datums of both horizontal and vertical con-

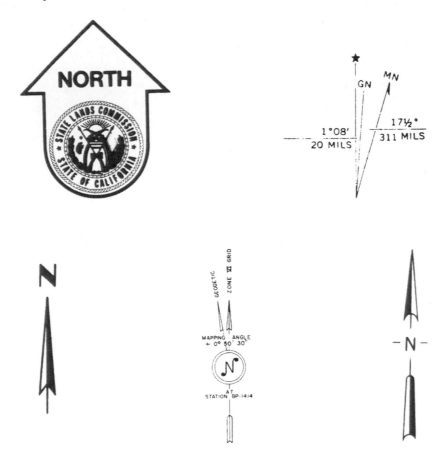

Figure 10-18. Several examples of meridian arrows.

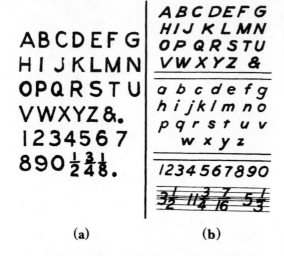

(a)   **(b)**

Figure 10-19. General lettering examples. (a) Vertical single-stroke Gothic capitals, and numerals. (b) Inclined single-stroke Gothic.

trol, and (3) references to old maps or data referred to in compiling the map.

A legend provides the key to symbols representing topographic or other features that otherwise would require lengthy or repetitive explanatory notes. Notes and legends should be located in close proximity to the title to aid the viewer in quickly locating them (see Figure 10-22 for an example).

10-10-6. Symbols

Because it is not possible to show all features on a small map, many can be represented by symbols. Several hundred standard symbols have been developed to represent topographic and other features unique to surveying. A few of these are shown in Figure 10-23 and inside the back cover. Employing symbols permits plotting many features in their correct locations without hopelessly crowding the map. Accurate positions of the features are plotted and a representative symbol drawn or conveniently cut from a preprinted sheet of adhesive-backed material and simply pasted on the map.

10-11. MAP PLACEMENT

A map's pleasing appearance helps to inspire confidence and dependability in the user's mind. Nothing detracts more from an otherwise acceptable product than omission of a heavy line border, and/or an out-of-balance placement of a drawing. Generally, a 1-in. margin is used on all four sides, although a 2-in. space may be necessary on the left edge if a number of sheets are to be bound together or hung on a rack. The title block is usually integrated with the lower right-hand corner border lines.

To properly place and fit a traverse and topographic details on a map sheet, first compute the maximum total of N-S latitudes and E-W departures. For a 24 × 30-in. sheet with 1-in. margins on three sides, and 2 in. on the left edge, the available mapping space is 22 × 27 in.

Assume that the total of the largest N-S latitudes is 1280 ft and for the E-W departures is 1810 ft. The largest scale possible is 1280/22 or 1 in. = 58 ft in the N-S direction and 1810/27 or 1 in. = 67 ft in the E-W space. Rounding off to a usable multiple of 10, 1 in. = 80 ft fits a standard scale.

Having the most westerly, northern, and southern stations (or topographic details) at the same distance from the borders produces an orderly appearance. The title box in the lower right-hand corner and any special notes placed directly over it will counterbalance the "weight" of the traverse and details. Special symbols, if any, are located above the notes. This arrangement enables a map user to quickly select the desired map by title and then check for special information before moving on to the drawing.

Placing the north arrow near the top right-hand side of the sheet also helps to balance it. Maps are read from the bottom or right-hand side, so the meridian is generally parallel with the right-hand border. On maps that are not required to match others on one or more sides, the meridian can be rotated to accom-

Helvetica Medium
ABCDEFGHIJKLMNOPQRSTUV
WXYZ abcdefghijklmnopqrstuv
wxyz 0123456789 .,?!&“ ”()’-:/#
%$£+−×÷=±□;@

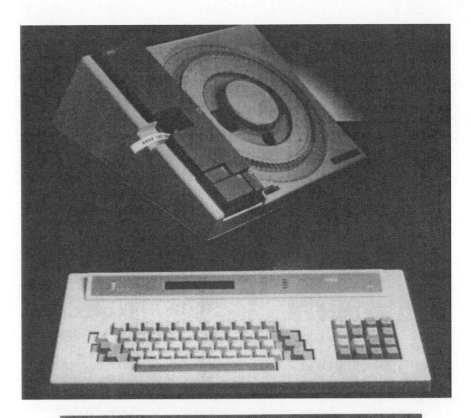

Figure 10-20. Kroy 190 ™ Lettering System and typical text. (Courtesy of KROY Inc.)

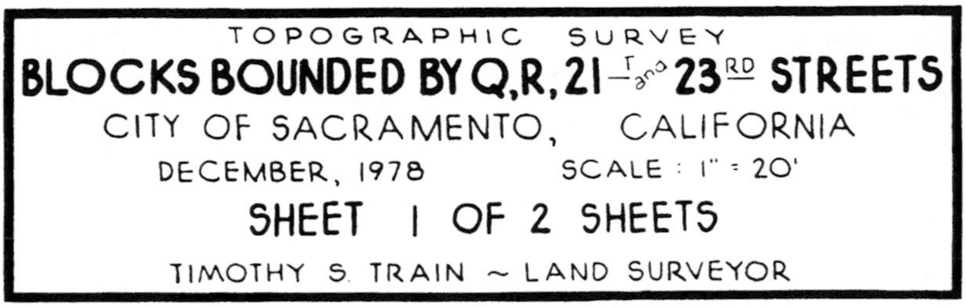

TOPOGRAPHIC SURVEY
BLOCKS BOUNDED BY Q,R, 21ST and 23RD STREETS
CITY OF SACRAMENTO, CALIFORNIA
DECEMBER, 1978
SCALE : 1" = 20'
SHEET 1 OF 2 SHEETS
TIMOTHY S. TRAIN ~ LAND SURVEYOR

Figure 10-21. Typical title block for map. (Courtesy of Timothy S. Train.)

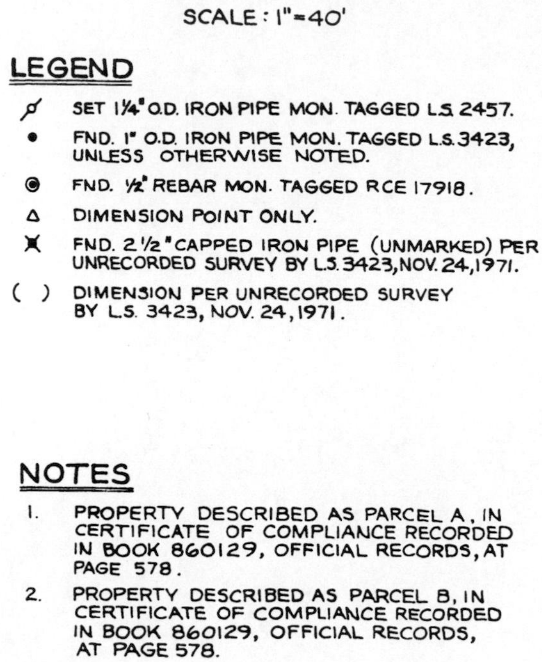

SCALE : 1"=40'

LEGEND

⌀   SET 1¼" O.D. IRON PIPE MON. TAGGED L.S. 2457.

●   FND. 1" O.D. IRON PIPE MON. TAGGED L.S. 3423, UNLESS OTHERWISE NOTED.

◉   FND. ½" REBAR MON. TAGGED RCE 17918.

△   DIMENSION POINT ONLY.

✖   FND. 2 ½" CAPPED IRON PIPE (UNMARKED) PER UNRECORDED SURVEY BY L.S. 3423, NOV. 24, 1971.

( )   DIMENSION PER UNRECORDED SURVEY BY L.S. 3423, NOV. 24, 1971.

NOTES

1.   PROPERTY DESCRIBED AS PARCEL A, IN CERTIFICATE OF COMPLIANCE RECORDED IN BOOK 860129, OFFICIAL RECORDS, AT PAGE 578.

2.   PROPERTY DESCRIBED AS PARCEL B, IN CERTIFICATE OF COMPLIANCE RECORDED IN BOOK 860129, OFFICIAL RECORDS, AT PAGE 578.

Figure 10-22. Typical map notes and legend. (Courtesy of Timothy S. Train.)

modate an odd-shaped figure and/or to have property, street, or traverse lines parallel with the border.

Preparing a scaled sketch map on tracing paper to outline the most distant features is often desirable. The sketch can then be shifted and rotated on the final map sheet, assuring good positioning.

10-12. DRAFTING MATERIALS

Three choices of drafting media are available for today's survey drafter: (1) paper, (2) cloth, and (3) film. Sufficient differences exist between and within each medium to allow the draftsperson a wide choice of material to fit the requirements of any drafting application.

10-12-1. Paper

Drafting papers are manufactured in a wide selection. They are classified by strength, longevity, and erasability, and are categorized as opaque (drawing) or transparent (tracing) paper. For sketching single-use exhibits, or for a high-quality map, opaque paper is virtually

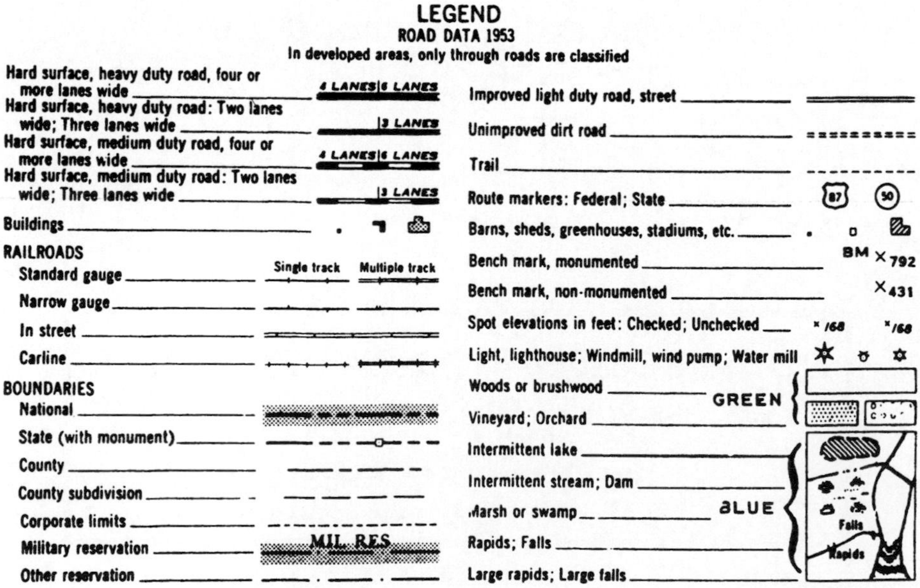

Figure 10-23. Sampler of symbols.

ignored. Transparent papers have wide applications as tracing paper on which a master drawing is to be produced; after checking and revision, it is used in reproduction processes to provide any number of copies. In addition to good actinic transparency, a high-quality tracing paper must withstand repeated handling and have a high degree of permanence.

10-12-2. Cloth

Cloth or "linen" combines features of transparency, surface quality, strength, and permanence. This medium resists repeated erasures, with little or no loss in surface quality. Linen provides a highly receptive surface for pencil, ink, and typing. Ink erasures can be made with a vinyl eraser or a gentle abrasive, such as Bon-Ami, and then wiped clean with a moist cloth. Although the advantages of cloth outweigh those of paper, it is generally not used for economic reasons except to meet a particular project specification or application.

10-12-3. Film

The advantages of a polyester drafting medium include high transparency, dimensional stability, and resistance to tearing, heat, and aging. In addition, it is waterproof, highly receptive to pencil, ink, typewriting, and paste-up processes, easily erased, and can be coated with an opaque material for use in scribing.

Recognizing the time and expenses accumulated in any survey project, we see that it is advisable to invest in the slightly more costly best-quality medium to guarantee excellent appearance and the highest permanence of survey drawings.

10-13. REPRODUCTION OF MAPS AND DRAWINGS

Two distinct processes are used for reproduction purposes: (1) copying or (2) duplicating. Copy machines are suitable for either line or pictorial work, but they operate at relatively slow speeds; hence, cost per copy is accordingly high.

By comparison, duplicating machines are designed to produce large numbers of copies at high speed, resulting in a lower cost per copy. Offset presses and stencil machines are examples of duplicating processes. Although in some instances these machines can be employed in combination with a copy process, reproduction requirements dictated by survey drawings are best met by one of the various copying procedures. Therefore, duplication processes will not be addressed in this text.

10-13-1. Copying Processes

Diazo, photographic, and electrostatic are the most popular types of copy processes. At one time, "blueprinting" was the most popular method, but it has now been largely replaced by the diazo process.

A complete line of reproduction and copy services is available from companies that specialize in this work. If volume warrants, a survey firm might consider the convenience afforded by using an in-house reproduction unit. Many small diazo units are available, as are limited-format copy cameras that do not require darkroom development processes.

10-13-2. Diazo

In this process, paper coated with diazonium salt is exposed to light that has passed through a transparent original. The exposed sheet is then developed by exposure to an alkaline-based solution such as ammonia. Where light has passed through the tracing, the diazonium salt breaks down, leaving a blank area on the print; on remaining areas not exposed to light, the salts and ammonia combine to produce an opaque dye, leaving an image of the original tracing lines. This image is not absolutely permanent, will fade when exposed to sunlight, and the exposed paper does not have very good drafting qualities. A positive original produces a positive copy and likewise a negative tracing will yield a negative copy. A diazo copy is a high-contrast image with fine detail and ideal for document reproduction.

10-13-3. Photographic

This process consists of using a precise camera to photograph an original document or drawing. The exposed copy film is developed into a negative from which a variety of prints can be made. A copy camera is usually mounted on a track along with a vacuum-frame copy board to hold the document being copied. The camera has a large format and is equipped with a high-quality lens. The copy board is movable relative to the camera, allowing a wide latitude of reduction of enlargement possibilities.

Film and paper prints, both positive or negative, clear or matte film positives, and a photographic negative are products available through this process. Film prints are provided on a stable-based material that is also a high-quality drafting surface. This process yields the most accurate and versatile product, but it is also the most expensive; a large, complex camera and darkroom facilities are required.

10-13-4. Electrostatic

This process, also known as *xerography*, depends on an electrostatically charged aluminum drum to deposit powder onto copy paper. The drum is given a positive charge that is partially dissipated by light reflected from the document to be copied. The charge portion remaining attracts negatively charged powder, creating an intermediate image on the drum. Copy paper with a negative charge is brought into contact with a drum that transfers the image onto the paper, which is then fixed or permanently fused into the paper by heat.

Electrostatic copy machines require no chemical processing, so a dry finished copy is obtained in very few seconds. The more expensive machines have lens systems capable of reducing and enlarging originals. Copies can be made on virtually any type of paper, including plain paper, offset masters, and transparent paper or film. Recently developed machines are capable of producing full-color copies.

Proper selection of a copy procedure should be based on intended use, availability, and permanence. Consideration should also be given to the cost and time required to produce a copy.

10-14. AUTOMATED DRAFTING

CAD is transforming mapmaking much as electronic instrumentation is revolutionizing field surveying. Interactive drafting systems include a cathode-ray display tube and automatic plotter, and are driven by a computer to process field data. Digital information is recorded in the field on an electronic data collector that interfaces with the CAD system. This system is fast becoming standard equipment in larger agencies and companies.

Advantages offered by various automated machines now available are many: Human error is all but eliminated, production is greatly increased by the obvious speed advantages, and a consistent and accurate map product results. Once a map has been complied from information input on type or magnetic disk, some or all of the data may be used on future projects. For example, initial data from a topographic/boundary survey can be stored and later retrieved to make earthwork estimates, develop a site/grading plan, produce a boundary map, and compile an as-built map.

State-of-the-art survey instrumentation collects huge quantities of information on tape or other data-storage devices. This information in the x, y, and z dimensions can be channeled into a computer to produce a digitized terrain model. The model, in turn, is fed to an automated drafter, which automatically turns out a map in preselected format and specifications by the computer operator.

10-15. SOURCES OF MAPPING ERRORS

Mapping errors can generally be traced to the following sources:

1. Inaccurate linework from use of a blunt or too-soft pencil
2. Inaccurate angular plotting with a protractor
3. Inaccurate linear plotting with the scale
4. Selection of mapping scale or contour interval unsuitable for map requirements
5. Drafting media affected by moisture or climatological change

10-16. MISTAKES AND BLUNDERS

Blunders differ from errors and result from carelessness and poor judgment. The following are a few examples of mistakes:

1. Poor linework or lettering, creating ambiguities
2. Using the wrong scale
3. Setting incorrect angles on a drafting machine or protractor
4. Misinterpretation of field notes
5. Misorientation in all or portion of a plot
6. Inappropriate choice of drafting media

10-17. MAPPING STANDARDS AND SPECIFICATIONS

Federal mapping agencies have adopted standards to control expected map accuracy by specifying the maximum error permitted in horizontal positions and elevations shown on maps. A map conforming to these specifications can use the statement, "This map complies with national map accuracy standards."

The mapping standards state that the following maximum errors are permitted on maps:

Horizontal Accuracy

For maps at scales larger than 1 : 20,000 (1 in. = 1667 ft), not more than 10% of the well-defined points tested shall have a plotting error in excess of 1/30 in. For maps of smaller scales, the error factor is 1/50 in. Well-defined points are characterized as easily located in the field and capable of being plotted to within 0.01 in.

Vertical Accuracy

No more than 10% of elevations tested shall be in error more than one-half the contour interval.

REFERENCES

ASCE. 1972 Selection of maps for engineering and planning. Task committee for preparation of a manual on selection of map types, scales, and accuracies for engineering and planning. *Journal of the Surveying and Mapping Division* (SUI).

———. 1983 Map uses: scales and accuracies for engineering and associated purposes. Report of the ASCE Surveying and Mapping Division Committee on Cartographic Surveying, New York.

MOFFITT, F. H., and H. BOUCHARD. 1992 *Surveying*, 9th ed. New York: Harper Collins.

DAVIS, R. E., F. S. FOOTE, J. M. ANDERSON, and E. M. MIKHAIL. 1981 *Surveying—Theory and Practice*, 6th ed. New York: McGraw-Hill.

SLOANE, R. C., and J. M. MONTZ. 1943 *Elements of Topographic Drawing*. New York: McGraw-Hill.

WATTLES, G. W. 1981 *Survey Drafting*. Orange, CA. Wattles Publications.

WOLF, P. R., and R. C. BRINKER. 1994 *Elementary Surveying*, 9th ed. New York: Harper Collins.

11

Triangulation

M. Louis Shafer

11-1. INTRODUCTION

Triangulation is the surveying technique in which unknown distances between stations may be determined by trigonometric applications of a triangle or triangles. In triangulation, one side called the baseline and at least two interior angles of the triangle must be measured. When all three interior angles are measured, accuracy of the calculated distances is increased and a check provided against any measurement error.

The most basic use of triangulation can be found in surveys of the public domain. Although the use of electronic measuring instruments has eliminated most requirements for this type of triangulation, the 1973 *Manual of Surveying Instructions* made the following statement:

> Triangulation may be used in measuring distances across water or over precipitous slopes. The measured base should be laid out so as to adopt the best possible geometric proportions of the sides and angles of the triangle. If it is necessary to determine the value of an angle with a precision of less than the least reading of the vernier, the method of repetition should be employed.

A complete record of the measurement of the base, the determination of the angles, the location and direction of the sides, and other essential details is entered in the field tables, together with a small diagram to represent the triangulation. In the longer and more important triangulations, all of the stations should be occupied, if possible, and the angles should be repeated and checked to a satisfactory closure; the latter may be kept within 0'20" by careful use of the one-minute transit.

In line practice the chainmen are frequently sent through for taped measurement over extremely difficult terrain, but with the length of the interval verified by triangulation. This is done to ensure the most exact determination of the length of the line while also noting the intervening topographic data.[1]

The use of triangulation or *trigonometry* has been addressed by various public-land survey instructions since "Instructions for Deputy Surveyors, E. Tiffin, Surveyor General, United States, 1815" for ascertaining distances across "insuperable obstacles" such as rivers and canyons. If it is necessary to retrace an original survey across such obstacles, the original field notes are essential to determine how the distance was measured. The following table is an example of public-land survey field notes for triangulation across a lake:

Field Record	Chains	Final Field Notes
At A $\dfrac{54°29'}{3}$ = 18°09'40″(−02″) At B $\dfrac{245°13'}{3}$ = 81°44'20″(−09″) At C $\dfrac{240°19'}{3}$ = $\dfrac{80°06'20''(-09'')}{180°00'20''(-20'')}$	27.80	To the south shore of Grand Lake, bears N 62° E and S 48° W. Set an iron post, 3 ft long, 1 in. in diam., 28 in. in the ground, for meander cor. of frac. secs. 13 and 18, with brass cap marked. To make a triangulation across the lake, I designate the above meander cor. point A and set a flag B at point for meander cor. on north shore of lake, also a flag C on the north shore that from point A bears N 18°09'38″ E; the base BC bears S 81°44'11″ E, 16.427 chs. dist., the mean by two sets of chainmen, by 1st set = 16.425 ch, by 2nd set = 16.429 ch, longer base impracticable; the angle subtended at point C = 80°06'11″; all angles by three repetitions with error of 0'20″ balanced to 180°. Distance across lake = 51.92 ch.
Dist. = 16.427 $\dfrac{\sin 80°06'11''}{\sin 18°09'38''}$ log 16.427 = 1.215558 ″ sin 80°06'11″ = 9.993488 1.209046 ″ sin 18°09'38″ = 9.493710 ″ 51.92 = 1.715336 +27.80 79.72	79.72	The north shore of lake, bears S 82° E and N 75° W.

11-2. GEODETIC TRIANGULATION

A wider spread and intricate application of triangulation are used for the horizontal control required over a vast area, when traversing does not provide the high uniform accuracy desired. A geodetic triangulation survey, in which stations are miles apart, must consider the earth's size and shape. It is performed primarily by the National Geodetic Survey (NGS) (formerly the U.S. Coast and Geodetic Survey, a branch of the U.S. Department of Commerce). Over the past two centuries, a net of triangulation stations, related to each other, has been developed over the entire continen-

tal United States. The system can be used as starting control for any triangulation survey that may be undertaken by a private surveyor, engineer, or public entity.

Triangulation is also employed for control in large metropolitan areas and on major construction projects. This chapter describes the procedures and instructions necessary to develop a triangulation network and retrieve information on stations already established. All procedures and instructions shown in this chapter follow NGS standards, and any triangulation conforming to these procedures and standards can be indexed into its system.

This chapter is written with the knowledge that modern technology is on the verge of

PERMISSIBLE FIGURES

(A) Simple quadrilateral.—The simple quadrilateral is the best figure, and it should be employed wherever possible. It combines maximum strength and progress with a minimum of essential geometrical conditions when approximately equilateral or square and therefore the square quadrilateral is the perfect figure. It has a strength factor,

$$\frac{D - C}{D} \quad \text{of} \quad 0.6.$$

(B) Four-sided central-point figure with one diagonal.—When one diagonal of the quadrilateral is obstructed, a central point, which is visible from the four corners can be inserted. This figure requires the solution of two side equations and five angle equations, and hence adds to the labor of adjusting. Its strength factor is 0.56.

(C) Four-sided central-point figure without diagonal.—At times, neither diagonal can be made visible and the figure becomes a simple four-sided central-point quadrilateral with a strength factor of 0.64. The central point in this case should be carefully located to maintain the strength of the R_1 chain of triangles. An excellent location is near one side line and about midway along it. If too near the side line, however, refraction errors may be almost the same for the closely adjacent lines, and furthermore, the R_2 value will be so large as to be of little value as a check on lengths computed through the R_1 triangles.

(D) Three-sided central-point figure.—This is a simple and usually very strong figure. It is often used to compensate for a great variation in length of the side lines of adjacent quadrilaterals, and to quickly change the direction of the scheme. Its strength factor is 0.60 and the equations required for its adjustment are the same as for a regular quadrilateral.

(E) Five-sided figure with four diagonals.—This figure may be considered as a four-sided central-point figure with one diagonal, in which the central point falls outside the figure. It is used to afford a check when either a diagonal or a side line is obstructed. It has the same strength factor, 0.56, as the above four-sided central-point figure with one diagonal, (*B*), and requires the same adjustment equations and precautions against making any of the angles too small. This figure can often be used by the observing party when a side line of a quadrilateral is found to be obstructed.

(F) Five-sided figure with three diagonals.—This figure is similar to the four-sided central-point figure, (C), except that the central point falls outside the figure. The strength factor is 0.64.

(G) Five-sided central-point figure with two diagonals.—This figure is an overlap of a central-point quadrilateral and a simple quadrilateral, and is the most complicated figure employed. It has been used to carry the scheme over difficult or convex areas. This figure can generally be made very strong. Its strength factor is 0.55.

(H) Five- and six-sided central-point figures without diagonals.—Any polygon with a central point, having separate chains of triangles on either side of the central point, will give a double determination of length, since it is permissible to carry the two lengths through the same triangle provided different combinations of distance angles are employed. However, the five- and six-sided central-point polygons are the only ones that should receive consideration, and they are inferior to the simpler quadrilaterals. The factors of strength are 0.67 for five sides and 0.68 for six.

Figure 11-1. Types of figures used in triangulation and strength factors. (*Manual of Reconnaissance for Triangulation.* SP225, Government Printing Office: Washington, D.C., 1938.)

making conventional triangulation obsolete in most instances. The use of the global positioning system (GPS) is becoming commonplace in the establishment of control networks. GPS involves the use of satellites.

GPS not only develops higher accuracy, but also does not require lines of sight between stations. It does, however, require a clear overhead horizontal view. The use of GPS is discussed in Chapter 15.

Almost all geodetic triangulation involves a series of triangles called a triangulation system or triangulation network to complete the control of a selected area. Control is carried from one known base line through several triangles before another base line must be established or checked into. The number depends on the "strength of figures," which will be discussed later. Tighter control is obtained by using a series of quadrilaterals, requiring three other stations to be visible from each station instead of the two necessary when using triangles.

Figure 11-1 is an excerpt from the *Manual of Reconnaissance for Triangulation*, special Pub-

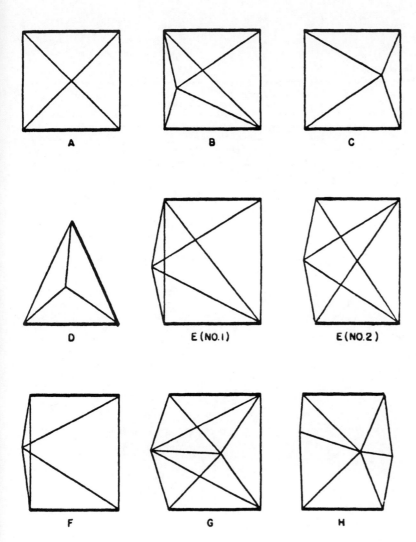

A   B   C

D   E (NO.1)   E (NO.2)

F   G   H

Figure 11-1. (*Continued*).

lication No. 225.[3] It shows various permissible figures and their strength factors for determining the strength of figures and how often base lines or check-in distances are needed. This is covered later in the strength-of-figures section.

11-3. CLASSIFICATION AND SPECIFICATIONS OF TRIANGULATION

In 1984, the Federal Geodetic Control Committee (FGCC), with approval of the Office of Management and Budget (OMB), published *Classification, Standards of Accuracy, and General Specifications of Geodetic Control Surveys*.[2] This publication outlines permissible tolerances for triangulation surveys acceptable by order and class. Three orders of accuracy are listed, with the second and third orders subdivided into class I and class II. The purpose and accuracy required dictate the tolerances allowed.

First-order classification (Table 11-1) demands the highest accuracy and is recommended for primary national networks, special surveys to study movements in the earth's crust, and metropolitan area surveys. Minimum rela-

Table 11-1. Classification of triangulations

Classification	First-Order	Second-Order		Third-Order	
		Class I	Class II	Class I	Class II
Recommended spacing of principal stations	Network stations seldom less than 15 km. Metropolitan surveys 3 to 8 km and others as required.	Principal stations seldom less than 10 km. Other surveys 1 to 3 km or as required.	Principal stations seldom less than 5 km or as required.	As required.	As required.
Strength of figure					
R_1 between bases					
Desirable limit	20	60	80	100	125
Maximum limit	25	80	120	130	175
Single figure					
Desirable limit					
R_1	5	10	15	25	25
R_2	10	30	70	80	120
Maximum limit					
R_1	4 10	25	25	40	50
R_2	15	60	100	120	170
Base measurement					
Standard error	1 part in 1,000,000	1 part in 900,000	1 part in 800,000	1 part in 500,000	1 part in 250,000
Horizontal direction					
Instrument	0″.2	0″.2	0″.2} { 1″.0	1″.0	1″.0
Number of positions	16	16	8 } or {12	4	2
Rejection limit from mean	4″	4″	5″ } { 5″	5″	5″
Triangle closure					
Average not to exceed	1″.0	1″.2	2″.0	3″	5″.0
Maximum seldom to exceed	3″.0	3″.0	5″.0	5″.0	10″.0
Side checks					
In side equation test, average correction to direction not to exceed	0″.3	0″.4	0″.6	0″.8	2″
Astro azimuths					
Spacing-figures	6–8	6–10	8–10	10–12	12–15
No. of obs./night	16	16	16	8	4
No. of nights	2	2	1	1	1
Standard error	0″.45	0″.45	0″.6	0″.8	3″.0
Vertical angle observations					
Number of and spread between observations	3 D/R–10″	3 D/R–10″	2 D/R–10″	2 D/R–10″	2 D/R–20″
Number of figures between known elevations	4–6	6–8	8–10	10–15	15–20
Closure in length					
(also position when applicable) after angle and side conditions have been satisfied, should not exceed	1 part in 100,000	1 part in 50,000	1 part in 20,000	1 part in 10,000	1 part in 5000

Source: FGCC. 1974. "Classifications, Standards of Accuracy, and General Specifications of Geodetic Control Surveys."

tive accuracy between directly connected points for first-order is 1 part in 1000,000. Spacing between first-order stations should be more than 15 km for primary network stations, 3 to 8 km in a metropolitan survey.

Second-order, class I standards are recommended for control surveys established between tracts bounded by the primary national control network. They strengthen the network and provide more stations for local applications. These standards should also be used in metropolitan areas where a closer net is required than allowable for first-order. The minimum relative accuracy between adjacent stations for second-order, class I is 1 part in 50,000. Stations strengthening the primary network should be not closer than 10 km apart, whereas those in metropolitan areas should be 1 to 3 km apart.

Second-order, class II standards are used to establish control along coasts and inland waterways, interstate highway systems, and large land subdivisions and construction projects. Second-order, class II standards can also be used for the further breakdown of second-order, class I nets in metropolitan areas. Second-order, class II triangulation contributes to, but is supplemental to, the primary national network. The minimum relative accuracy between adjacent points is 1 part in 20,000. Supplemental stations to the primary net should be at least 5 km apart while other second-order, class II stations are set up as required.

Third-order, class I and II standards establish control for local area projects, such as small engineering jobs on local improvements and developments and small-state topographic mapping. Third-order triangulation extends higher-order control and can be adjusted. National network stations set for third-order work may be spaced as required to satisfy project needs. The minimum relative accuracy between adjacent stations for third-order, class I is 1 part in 10,000; for third-order class II, 1 part in 5000.

In order to obtain a certain classification, definite specifications must also be met. For triangulation, the following four specifications are required:

1. Specifications for Instruments

Classification	Type of Instrument
First-order	Optical-reading theodolite micrometer readings, smaller than 1 sec, Examples: Wild T-3 and Kern DKM-3.
Second-order class I and II	Optical-reading theodolite micrometer reading of 1 sec. Examples: Wild T-2, Kern DKM-2, and Zeiss Th-2.
Third-order	Good-quality transit or repeating theodolite. Same quality theodolite required for second-order is recommended to eliminate the extra effort needed to obtain the specified accuracies for repeating instruments.

2. Specifications for Number of Observations Required (One observation includes direct and reverse reading.)

Classification	Number of Observations
First-order	16*
Second-order, class I	16*
Second-order, class II	8 with 0'.2" instrument* 12 with 1".0 instrument*
Third-order, class I [†]	4 with 1".0 instrument*
Third-order, class II [†]	2 with 1".0 instrument*

*To minimize collimation circle errors and micrometer irregularities, initial settings utilizing the entire circle are necessary. (See Table 11-2 for initial settings.)
†When we use a transit or repeating theodolite, one to eight sets of six observations are required, depending on instrument type (Table 11-3).

3. Rejection Limit Specifications

Classification	Rejection Limit from Mean
First-order	±4"
Second-order, class I	±4"
Second-order, class II	±5"
Third-order, class I	±5"
Third-order, class II	±5"

(Continued)

Table 11-2. Circle settings: directional theodolites

Two positions of circle

	10-min Micrometer Drum		
1	0°	00'	10"
2	90	05	40

Four positions of circle

	5-min Micrometer Drum			10-min Micrometer Drum			Wild T-3 Circle	Micrometer Readings	(units)
1	0°	00'	40"	0°	00'	10"	0°	00'	15
2	45	01	50	45	02	40	45	00	45
3	90	03	10	90	05	10	90	02	15
4	135	04	20	135	07	40	135	20	45

Six positions of circle

	5-min Micrometer Drum			10-min Micrometer Drum			Wild T-3 Circle	Micrometer Readings	(units)
1	0	00	10	0	00	10	0	00	15
2	30	01	50	30	01	50	30	00	35
3	60	03	30	60	03	30	60	00	50
4	90	00	10	90	05	10	90	00	15
5	120	01	50	120	06	50	120	00	35
6	150	03	30	150	08	30	150	00	50

Eight positions of circle

	5-min Micrometer Drum			10-min Micrometer Drum			Wild T-3 Circle	Micrometer Readings	(units)
1	0	00	40	0	00	10	0	00	10
2	22	01	50	22	01	25	22	00	25
3	45	03	10	45	02	40	45	00	35
4	67	04	20	67	03	55	67	00	50
5	90	00	40	90	05	10	90	00	10
6	112	01	50	112	06	25	112	00	25
7	135	03	10	135	07	40	135	00	35
8	157	04	20	157	08	55	157	00	50
1	0	00	40	0	00	10	0	00	10
2	15	01	50	15	01	50	15	00	25
3	30	03	10	30	03	30	30	00	35
4	45	04	20	45	05	10	45	00	50
5	60	00	40	60	06	50	60	00	10
6	75	01	50	75	08	30	75	00	25
7	90	03	10	90	00	10	90	00	35
8	105	04	20	105	01	50	105	00	50
9	120	00	40	120	03	30	120	00	10
10	135	01	50	135	05	10	135	00	25
11	150	03	10	150	06	50	150	00	35
12	165	04	20	165	08	30	165	00	50

Sixteen positions of circle

	5-min Micrometer Drum			10-min Micrometer Drum			Wild T-3 Circle	Micrometer Readings	(units)
1	0	00	40	0	00	10	0	00	10
2	11	01	50	11	01	25	11	00	25
3	22	03	10	22	02	40	22	00	35
4	33	04	20	33	03	55	33	00	50
5	45	00	40	45	05	10	45	00	10
6	56	01	50	56	06	25	56	00	25
7	67	03	10	67	07	40	67	00	35
8	78	04	20	78	08	55	78	00	50
9	90	00	40	90	00	10	90	00	10
10	101	01	50	101	01	25	101	00	25
11	112	03	10	112	02	40	112	00	35
12	123	04	20	123	03	55	123	00	50
13	135	00	40	135	05	10	135	00	10
14	146	01	50	146	06	25	146	00	25
15	157	03	10	157	07	40	157	00	35
16	168	04	20	168	08	55	168	00	50

Table 11-3. Number of observations using a transit and circle settings

Accuracy Class	Transit	Number of Observations	Number of Sets	Spread between D & R and Sets Not to Exceed
Third-order, class I triangulation	10"	6 D & R	2–3	4"
	20"	6 D & R	4–5	5"
	30"	6 D & R	6–8	6"
Third-order, class II triangulation	10"	6 D & R	1–2	5"
	20"	6 D & R	2–3	6"
	30"	6 D & R	3–4	7"

Transit and repeating type instruments

The circle settings

Sets	Instrument 10" Setting			Instrument 20" Setting			Instrument 30" Setting		
1	0°	00'	00"	0°	00'	00"	0°	00'	00"
2	90	05	30	90	10	20	90	10	30
1	0	00	00	0	00	00	0	00	00
2	60	03	30	60	06	20	60	06	30
3	120	07	00	120	13	00	120	13	00
1				0	00	00	0	00	00
2				45	05	20	45	05	30
3				90	10	00	90	10	00
4				135	15	20	135	15	30
1				0	00	00	0	00	00
2				36	04	20	36	04	30
3				72	08	00	72	08	00
4				108	12	20	108	12	30
5				144	16	00	144	16	00
1							0	00	00
2							30	03	30
3							60	07	00
4							90	10	30
5							120	14	00
6							150	17	30
1							0	00	00
2							25	02	30
3							51	05	30
4							76	08	00
5							102	10	30
6							128	14	30
7							153	17	00
1							0	00	00
2							22	02	30
3							45	05	00
4							67	07	30
5							90	10	00
6							112	12	30
7							135	15	00
8							157	17	30

4. Triangle Closure Specification for Triangulation Net*

Classification	Average Not to Exceed	Maximum Seldom to Exceed
First-order	1″.0	3″.0
Second-order, class I	1″.2	3″.0
Second-order, class II	2″.0	5″.0
Third-order, class I	3″.0	5″.0
Third-order, class II	5″.0	10″.0

*This is a simple field check to indicate the accuracies of triangulation observations.

11-4. PLANNING

After the triangulation project limits are determined, it is necessary to select the station sites. The first step is to collect all pertinent data, including various scaled maps and information on any existing triangulation stations in or near the area.

One of the handiest tools for planning the project and field reconnaissance is a set of topographic maps. U.S. Geological Survey Quadrangle (quad) maps are excellent and easily obtained from local surveying or map supply firms. They show contours, roads, trails, improvements, and some primary triangulation stations in the area. Quad sheets also have geodetic and rectangular coordinate control and are available for the entire United States.

In most cases, any triangulation project considered today has enough existing primary triangulation stations in the general vicinity for beginning control and data for necessary checks. These data can be obtained from the Director, National Geodetic Information Center, NOS, NOAA, Rockville, MD. Included diagrams will show locations of geodetic control stations, lines of sight between stations, and horizontal control data sheets giving directions to monuments, other visible stations, and grid and geodetic stations' coordinates. With these data, existing triangulation stations not given on a topographic map can be plotted to show their location and availability for a proposed project. If existing control is not available, base lines must be established, as discussed later in the chapter.

After all existing horizontal control on topographic maps is plotted, the general location of all triangulation stations needed for the project will be known before setting foot in the field. Contours determine the probable visibility between proposed stations, so profiles can be drawn to assure that the required clearance is available, and exact heights of towers required to provide acceptable clearance considering obstructions and the earth's curvature. Obstructions can become problems in all types of terrain, but curvature of the earth is noted mainly in flat lands and over long distances. The earth's curvature and refraction are discussed in detail in Chapter 7. For planning a triangulation project, the effect of curvature and refraction, which have an approximate relation to each other, can be determined by the formula

$$h \ \text{(ft)} = 0.574M^2 \ \text{(mi)} \tag{11-1}$$

where h is the height in feet that a line, horizontal at the point of observation, will be above a level surface at a distance of M statute miles. Table 11-4 lists the corrections of curvature and refraction for 1 to 60 mi.

If the line of sight between two stations extends across flat terrain and towers of equal heights can be constructed at the stations, the heights necessary to compensate for curvature and refraction can be determined by the formula

$$h = 0.574\left(\frac{M}{2}\right)^2 \tag{11-2}$$

in which h is the height above ground at both stations (in ft), and M the distance between stations (in mi).

The formula for working in the metric system is

$$h \ \text{(m)} = 0.0675\left(\frac{k}{2}\right)^2 \ \text{(kilometers)} \tag{11-3}$$

By constructing towers at both stations, the height needed is only one-fourth that required if a tower is constructed at only one station.

Table 11-4. Correction for earth's curvature and refraction

Distance	Correction	Distance	Correction	Distance	Correction	Distance	Correction
Miles	Feet	Miles	Feet	Miles	Feet	Miles	Feet
1	0.6	16	146.9	31	551.4	46	1214.2
2	2.3	17	165.8	32	587.6	47	1267.7
3	5.2	18	185.9	33	624.9	48	1322.1
4	9.2	19	207.2	34	663.3	49	1377.7
5	14.4	20	229.5	35	703.0	50	1434.6
6	20.6	21	253.1	36	743.7	51	1492.5
7	28.1	22	277.7	37	785.6	52	1551.6
8	36.7	23	303.6	38	828.6	53	1611.9
9	46.4	24	330.5	39	872.8	54	1673.3
10	57.4	25	358.6	40	918.1	55	1735.8
11	69.4	26	388.0	41	964.7	56	1799.6
12	82.7	27	418.3	42	1012.2	57	1864.4
13	97.0	28	449.9	43	1061.0	58	1930.4
14	112.5	29	482.6	44	1111.0	59	1997.5
15	129.1	30	516.4	45	1162.0	60	2065.8

The height calculated by this formula corrects only for curvature and refraction. The total heights necessary to have a satisfactory line of sight need to be increased enough to reduce the horizontal refraction caused by unequal air currents along the terrain.

Whether a line of sight will clear an obstruction can be determined by the formula

$$e = e_1 + (e_2 - e_1)\frac{d_1}{d_1 + d_2} - 0.574d_1d_2 \quad (11\text{-}4)$$

where

e = elevation of the line at obstruction (in ft)

e_1 = elevation of the lower station (in ft)

e_2 = elevation of the higher station (in ft)

d_1 = distance from the lower station to obstruction (in mi)

d_2 = distance from obstruction to the higher station (in mi)

$0.574d_1d_2$ = correction for curvature and refraction

Example 11-1. From a topographic map, it is planned to locate stations A and B 10 mi apart. The elevation of A is 20 ft and the elevation of B 90 ft. On line, 3 mi from A, is a ridge with an elevation of 40 ft. To determine the elevation of the line of ob-

struction, use the following formula:

$$e = 20 + (90 - 20)\frac{3}{(3 + 7)}$$
$$- 0.574(3)(7)$$

where e = 28.1 ft for the line of obstruction and the elevation of obstruction is 40 ft; therefore, stations A and B are not intervisible from the ground.

To determine the height above stations A and B necessary to see over the obstruction, subtract the elevation of the line of obstruction from the obstruction's elevation. In this case, $40 - 28.1 = 11.9$ ft. Thus, a tower of 11.9 ft, plus a required clearance necessary over the obstruction to reduce horizontal refraction, should be constructed over each station. If it is feasible to construct towers over these stations, the locations are satisfactory. Otherwise, new sites must be selected.

11-5. STRENGTH OF FIGURES

The accuracy of a triangulation net depends on not only the methods and precision used in making observations, but also the shapes of figures in the net. The system to measure the accuracy of shapes is known as *strength of figures*. Distance angles are those opposite the known and required sides. The accuracy or relative strength of a triangle is expressed as a number —i.e., the smaller the number, the greater the relative strength. To qualify as a certain classification, the sum of all numerical values of relative strength through a series of triangles between base lines cannot exceed the set standard. These standards are listed in Table 11-1 under strength of figures.

R is the standard symbol for strength of figures. When figures of a net are other than triangles (Figure 11-1), more than one scheme can be used to calculate the required side. R_1 and R_2 indicate that the strength of figure through the best-shaped triangles and second-best-shaped triangles, respectively. The summation $(\Sigma)^1$ of the R_1 and R_2 values determines when a base line is necessary to comply with the classification specifications.

The formula for strength of each figure (known side to required side) is

$$R = \frac{D - C}{D}(A^2 + AB + B^2) \qquad (11\text{-}5)$$

where

  D = number of directions observed in each figure
  C = number of conditions to be satisfied in each figure
  A and B = logarithmic differences for 1 sec of distance angles A and B in units of the sixth decimal place

In determining D, directions along the known side of the figure are not included. C can be determined by the following formula:

$$C = (n' - s' + 1) + (n - 2s + 3) \qquad (11\text{-}6)$$

where

  n = number of lines observed in both directions (including known side)
  s' = number of stations occupied
  n = total number of lines (including known side)
  s = total number of stations

In triangle *ABC* (Figure 11-2) where *AB* is known, all stations are occupied, and $D = 4$

$$C = (3 - 3 + 1) + (3 - 2(3) + 3) = 1$$
$$\frac{D - C}{C} = \frac{4 - 1}{4} = 0.75$$

In the strength-of-figure formula, $D - C/D$ is referred to as the strength factor and given for the various figures listed in Table 11-1. The values for $\delta_A^2 + \delta_A \delta_B + \delta_B^2$ have been tabulated and listed in Table 11-5 To use this table, find the factor by locating the smaller distance angle across the top of the table and large distance angle down the left side.

Example 11-2. Determination of strength of figures (Figure 11-3). If side *MN* and all interior angles are known, find R_1, R_2 to required side *OP*.

Possible schemes are (1) (*MNP, NOP*), (2) (*MNO, MOP*), (3) (*MNP, MOP*), and (4) (*MNO, NOP*).

$$D = 10 \text{ and } C = (6 - 4 + 1)$$
$$+(6 - 2(4) + 3) = 4$$
$$\text{Strength factor} = D - C/D = 10 - 4/10 = 0.60$$

For scheme *MNP, NOP*: *MNP* distance angles are 75°, 64° and *NOP* distance angles are 83°, 53°.

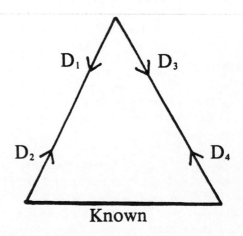

Figure 11-2. Determination of strength factors.

Table 11-5. The strength of figure

°	10°	12°	14°	16°	18°	20°	22°	24°	26°	28°	30°	35°	40°	45°	50°	55°	60°	65°	70°	75°	80°	85°	90°
10	428																						
12	359	359																					
14	315	295	253																				
16	284	253	214	187																			
18	262	225	187	162	143																		
20	245	204	168	143	126	113																	
22	232	189	153	130	113	100	91																
24	221	177	142	119	103	91	81	74															
26	213	167	134	111	95	83	74	67	61														
28	206	160	126	104	89	77	68	61	56	51													
30	199	153	120	99	83	72	63	57	51	47	43												
35	188	148	115	94	79	68	59	53	48	43	40	33											
40	179	137	106	85	71	60	52	46	41	37	33	27	23										
45	172	129	99	79	65	54	47	41	36	32	29	23	19	16									
50	167	124	93	74	60	50	43	37	32	28	25	20	16	13	11								
55	162	119	89	70	57	47	39	34	29	26	23	18	14	11	9	8							
60	159	115	86	67	54	44	37	32	26	24	21	16	12	10	8	7	5						
65	155	112	83	64	51	42	35	30	24	22	19	14	11	9	7	5	4	4					
70	152	109	80	62	49	40	33	28	23	21	18	13	10	7	6	5	4	3	2				
75	150	106	78	60	48	38	32	27	21	19	17	12	9	7	5	4	3	2	1	1			
80	147	104	76	58	46	37	30	25	20	18	16	11	8	6	4	3	2	1	1	1	1		
85	145	102	74	57	45	36	29	24	19	17	15	10	7	5	4	3	2	1	1	1	0	0	
90	143	100	73	55	43	34	28	23	19	16	14	9	7	5	3	2	2	1	1	0	0	0	0
95	140	98	71	54	42	33	27	22	18	16	13	9	6	4	3	2	1	1	0	0	0	0	
100	138	96	70	53	41	32	26	22	17	15	13	8	6	4	3	2	1	1	0	0	0		
105	136	95	68	51	40	31	25	21	17	14	12	8	6	4	3	2	1	1	0	0			
110	134	93	67	50	39	30	25	20	16	14	12	7	5	4	2	2	1	1	1				
115	132	91	65	49	38	30	24	19	16	13	11	7	5	3	2	2	1	1					
120	129	89	64	48	37	29	23	19	15	13	11	7	5	3	2	2	1						
125	127	88	62	46	36	28	22	18	15	12	10	7	5	3	3	2							
130	125	86	61	45	35	27	22	18	14	12	10	7	5	4	3								
135	122	84	59	44	34	26	21	17	14	12	10	7	6	4									
140	119	82	58	43	33	26	21	17	14	12	10	8	6										
145	116	80	56	42	32	25	21	17	14	13	11	9											
150	112	77	55	41	32	25	21	18	15	13	13												
152	111	75	54	40	32	26	22	19	16	16													
154	110	75	53	40	33	27	23	21	17														
156	108	74	54	41	34	28	25	22															
158	107	74	54	42	35	30	27																
160	107	74	56	45	38	33																	
162	107	76	59	48	42																		
164	109	79	63	54																			
166	113	86	71																				
168	122	98																					
170	143																						

215

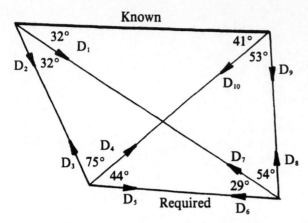
Known

Figure 11-3. Determination of strength of figures.

In Table 11-5, locate distance angles $75°$ on the left side, $64°$ across the top, and by intersection of these two angles and interpolation, the factor is found to be 2. Next, locate $83°$ on the left side, $53°$ across the top; the factor is listed as 3. Add $2 + 3 = \delta_A^2 + \delta_A \delta_B + \delta_B^2 = 5$.

$$R(MNP, NOP) = D - C/D(\delta_A^2 + \delta_A \delta_B + \delta_B^2)$$

$$= 0.60 \times 5 = 3.0$$

If we complete the same process for the remaining three schemes, the following tabulation can be made:

Scheme	Common Side	Distance Angles	$\delta_A^2 + \delta_A \delta_B + \delta_B^2$	$\dfrac{D - C}{D}$	R
MNP	NP	$75°, 64°$	2		
NOP		$83°, 53°$	3		
			$\Sigma = 5$	$\times 0.60$	$3 = R_1$
MNO	MO	$54°, 94°$	2		
MOP		$119°, 32°$	9		
			$\Sigma = 11$	$\times 0.60$	$7 = R_2$
MNP	MP	$75°, 41°$	8		
MOP		$29°, 32°$	39		
			$\Sigma = 47$	$\times 0.60$	28
MNO	NO	$54°, 32°$	19		
NOP		$44°, 53°$	11		
			$\Sigma = 30$	$\times 0.60$	18

Since the more accurate scheme has a smaller R value, scheme MNP, NOP becomes R_1 and scheme MNO, MOP is R_2. Referring to the standards of accuracy shown in Table 11-1, we see that this figure qualifies for first-order triangulation. If a network was developed with all figures having these same R_1 and R_2 values, then a base line is necessary after two figures to qualify as first-order, or a base line is required after eight figures to meet the standards of second-order, class I.

11-6. BASE LINES

Measured lines that control a triangulation net are referred to as base lines. The frequency with which base lines are needed depends on strength of the geometric figures, and since lengths of all required lines are derived from the base lines, great care must be taken to ensure their accuracy. A base line may be between two existing control stations or independently measured within the triangulation net.

Plotting triangulation schemes on the reconnaissance maps can determine when base lines are necessary, and whether existing stations are adequate or new ones must be established.

With the acceptance of electronic distance-measuring instruments (EDMIs) for measuring base lines, the procedure has been greatly simplified. Pre-EDMI base-line measurements required a fairly level leg between stations and use of special Invar tapes and supports. The favorite location was along a railroad tangent, thereby restricting their selection. With EDMIs, a base line can be placed wherever there is a clear line of sight. Care must be taken that the EDMI used meets all requirements and specifications for the class order of triangulation desired (see Chapter 5 on EDMIs).

11-7. FIELD RECONNAISSANCE

Field reconnaissance crews are responsible for contacting property owners, finding all existing control, and selecting the location of the proposed stations.

It is important to use proper methods in contacting owners of lands to be entered. Generally, an owner will consent to the use of his or her property if asked beforehand but will resent the intrusion without prior consent. If visited prior to any field reconnaissance, an owner will probably have no objection to future use as long as all conditions agreed on are met.

A property owner should be fully informed on the procedure to be followed in completing the triangulation survey, including (1) monuments to be set, (2) type of stands and towers required, (3) probability of night observations, (4) number of times the station will be used, and (5) settlement for any damage that may occur to his or her property as a result of monumenting the station. A good motivation for allowing placement of a station on private

land is to name it after the owner. Having a set of monuments with one's name gives a sense of importance and pride.

The owner should be contacted each time before land is entered unless a blanket consent has been given, especially if there are livestock, buildings, or equipment on the land. It is also advisable to talk with the local law enforcement agency, as this could prevent later embarrassment.

After descriptions of existing control stations have been acquired, field recovery can begin. All stations within the National Geodetic Survey control networks have a *station description* that includes distances and directions of reference azimuth marks from a main station mark, geodetic and state coordinate position of the mark, and a "to-reach" description tied to a nearby permanent landmark such as the local post office (Figure 11-4). Since these descriptions may have been written and not updated for more than 50 years, finding stations can require a diligent search, or even include looking for accessory marks as well as the main station.

When a station is found, a "Report on Condition of Survey Mark" form should be filled out and sent to the NOS (Figure 11-5). If the station sought is not found, the thoroughness of the search should be reported; if recovered, the marks condition should be reported along with any differences discovered in the to-reach description. When approved by the NOS, changes are incorporated into the station description. For a station recovered exactly as previously described, with the monuments and marks in good condition, and measurements between marks in agreement, a statement to this effect will suffice. Any changes concerning station marks or to-reach descriptions must be addressed on the "Report of Condition" note.

Proposed triangulation station sites can be researched after existing stations have been recovered and proposed schemes laid out on the map. Initially, these sites are visited to see if they are acceptable as station locations. Fac-

DEPARTMENT OF COMMERCE
U.S. COAST AND GEODETIC SURVEY
Form 825
Rev. Aug. 1964
NAME OF STATION: BIG BAR
CHIEF OF PARTY: Walter R. Helm

DESCRIPTION OF TRIANGULATION STATION

STATE: California   COUNTY: Butte
YEAR: 1949   Described by: L. A. Critchlow

| Note | Height of Telescope above Station Mark 15.67 Meters | | Height of Light above Station Mark 18.56 Meters | | |

| 4 | Surface-station mark, Underground-station mark | DISTANCES AND DIRECTIONS TO AZIMUTH MARK, REFERENCE MARKS AND PROMINENT OBJECTS WHICH CAN BE SEEN FROM THE GROUND AT THE STATION |

	OBJECT	BEARING	feet (DISTANCE)	meters	DIRECTION (° ′ ″)
	BUCK (USGS)				00 00 00.0
12c	Azimuth Mark	E	about 1.0 mile		40 52 19.2
12c	R.M.# 2	W	57.093	17.402	222 47 01
	Big Bar Lookout tower	W	128.001	(39.015)	231 59 09.0
12c	R.M.# 1	NE	65.493	19.962	355 32 44

Detailed description:

The station is located on a high, timbered ridge on the east side of the Feather River Canyon, about 20 miles airline north-northeast of Oroville, about 10 miles airline east of Paradise, about 2 miles airline southeast of Pulga and on the same high point on which the *Big Bar Lookout* is located.

To reach the station from the post office in Oroville, go northwest on Oak Street for 2 blocks to a stop sign. Turn right on State Highway # 24 and go 28.0 miles to a bridge and a sign "NORTH FORK FEATHER RIVER BRIDGE 12-38." Keep straight ahead on State Highway # 24, crossing the bridge, and go 0.4 mile to a side road on the right. Turn right as per sign "BUCKS LAKE 23" and go 6.5 miles to a 5-way intersection of roads. Turn sharp right as per sign "BIG BAR L.O." and go 4.1 miles to a triangle blazed tree and the azimuth mark on the right. Keep straight ahead and go 1.2 miles to a fork. Turn right as per sign "BIG BAR LOOKOUT 1" and go 0.8 mile to the summit and the station.

The station mark is a standard disk stamped *"big bar 1949"* set in a drill hole in a boulder that projects about 4 inches above the ground.

Reference mark number 1 is a standard disk stamped "BIG BAR NO 1 1949" set in a drill hole in a boulder that projects about 3 inches above the ground. It is about 2 feet lower than the station.

Reference mark number 2 is a standard disk stamped "BIG BAR NO 2 1949" set in a drill hole in a boulder that projects about 6 inches above the ground. It is 58.6 feet east of the southeast leg of the lookout tower and about 2 feet higher than the station.

The azimuth mark is a standard disk stamped "BIG BAR 1949" set in a drill hole in a boulder that projects about 16 inches above the ground. It is about 21 feet west of the center of the road and 23.2 feet north of a 24-inch pine tree with a triangle blaze.

Big Bar Lookout Tower is located on the highest point of the hill and is built on ground that is about 2 feet higher than the station. The tower is in the early stages of construction. The point measured to and cut in is a punch mark in the center and top of the middle base I-beam.

Figure 11-4. Station description (horizontal control data from USC & GS).

tors to be checked include (1) visibility to the other station sites, (2) accessibility, (3) probable permanence of the station marks, and (4) acceptance and future plans of the property owners.

The most important item to verify is visibility. Numerous factors affecting visibility may not be evident on the reconnaissance maps, but do show up on a field check of the site. If we consider the height of towers available for

REPORT ON CONDITION OF SURVEY MARK

Form Approved: OMB No. 41—R1923
Approval Expires: April 1978

Name or Designation: _____ Year Established: _____

State: _____ County: _____ Organization Established by: _____

Distance and direction from nearest town: _____

Description published in: *(Line, book, or quadrangle number)* _____

Mark searched for or recovered by: Name - _____

Organization - _____

Date of report _____ Address - _____

Condition of marks: List letters and numbers found stamped in (not cast in) each mark.

Mark stamped:	Condition:		

Marks accessible? ☐ Yes ☐ No   Property owner contacted? ☐ Yes ☐ No

Please report on the thoroughness of the search in case a mark was not recovered, suggested changes in description, need for repairing or moving the mark, or other pertinent facts:

Witness Post? Yes _____ No _____
Witness Post set _____ feet _____ of _____ mark.
Witness Post set _____ feet _____ of _____ mark.

If additional forms are needed, indicate number required. _____

NQAA FORM 76—91   U.S. DEPARTMENT OF COMMERCE · NATIONAL OCEANIC AND ATMOSPHERIC ADMINISTRATION · NATIONAL OCEAN SURVEY
(3—77)

Figure 11-5.   Report on condition of survey mark.

use, new construction, tall trees, or other intervening obstructions can make sight lines to proposed or existing stations impossible. Visibility must be checked thoroughly, since any error can cause time-consuming delays when observations begin, possibly disrupting the entire triangulation scheme.

The ability to get materials and equipment to a station location for setting monuments and occupying stations is another important factor. Many triangulation stations require some walking to reach the mark, but this should be kept to a minimum. A station requiring excessive hiking or climbing not only adds an element of danger for surveyors, but also makes the station useless to all but those few willing to assume the extra effort and risk. It is not always possible to avoid inaccessible stations, but the main idea is to make them usable.

The property owner is a principal factor in determining accessibility. If contacted beforehand, the owner will probably permit the vehicular entry of his or her property and explain or show the easiest way to a specific location, in addition to providing keys or combinations to any locked gates.

Permanence of marks is never assured, but many factors should be considered to make their location more reliable. Things to look for include (1) soil condition, (2) possibility of construction or development, (3) farming activities, and (4) changing ownership of private or public land.

Ground condition where a station is placed should be stable. Sand hills, large gravel deposits, and unstable rock outcroppings should be avoided, as well as areas subject to continuous frost action or erosion. Reliability of monuments set in these conditions is greatly diminished owing to the high probability of movement.

Any area subject to proposed development or construction is an unfavorable site for a station, and although it is impossible to guarantee this will not happen, the odds can be greatly increased with a little research. By checking master plans and zoning from the planning agency with jurisdiction over the area, the likelihood of future development can be determined. The topography of an area will also indicate possible future development. In addition, always contact property owners to find out their future plans for the land.

In farming areas, station locations are favorable along a fence line between property owners, near farm buildings, and along groves maintained for shade or wind breaks. Cultivated fields are dangerous since crops may be destroyed every time the station is used, and a surface monument cannot be maintained because of repeated cultivation.

Lands owned by public entities are generally ideal station locations. Locating a monument along the right-of-way fence of established roadways or railroads and within a public park usually ensures its permanence. It is always necessary for property owners to be satisfied with station sites; otherwise, their permanence and availability may be in jeopardy. This policy holds true with a public entity for stations on public land, as well as for private owners.

11-8.   SETTING STATION MARKS

After the triangulation net has been established and approved, station monuments are set. Each station should consist of a station mark, at least two reference marks, and one azimuth mark. The azimuth mark may be eliminated if another station is readily visible from the ground.

Station marks, the main monuments from which observations are made, have disks placed both underground and on the surface whenever possible. Both of the bronze disks have the station name and year established stamped on them before being set. The station mark can be set in a concrete monument, boulder, or rock outcropping, or brazed to the end of a pipe positioned in the ground.

The concrete monument is poured in place and consists of a subsurface and surface monument. Figure 11-6 shows the dimensions and configuration of a concrete monument set in the ground. The surface monument is positioned over the subsurface one by use of a plumbing bench. Care must be taken while pouring the concrete not to disturb this bench. When using a rock outcropping to set a station mark, it should be a hard, solid part of the main ledge, not a detached fragment. A hole is drilled deep enough to accommodate the disk stem, and the rock surface chipped in a diameter large enough to countersink the disk surface. The disk is attached with cement or epoxy.

At least two surface reference marks should be set for each station: bronze disks marked with an arrow showing direction to the station mark and labeled REFERENCE MARK. The disks are stamped with the station name and date and numbered serially clockwise from north—i.e., RM No. 1 would be the first reference mark looking angularly clockwise from

the north direction. Lines from reference marks should intersect the station mark as close to 90° as possible, ensuring a good intersecting angle to recover or relocate the station mark. The lines from reference marks to a station monument should be clear and kept shorter than 30 m to make taping easy, but long enough to ensure direct visibility from an observing tower. Since the function of reference marks is to check or relocate the station mark, they are placed in locations least susceptible to being disturbed, e.g., set in rock outcrops, solid permanent concrete structures such as retaining walls, or concrete monuments at least 30 in. long and 12 in. in diameter.

An azimuth mark is set for each station, primarily for usage by local surveyors. It should be in a location visible from the station mark, using an ordinary ground tripod setup and approximately $\frac{1}{4}$ to $\frac{1}{2}$ mi from the station mark. Azimuth marks are usually set along a right-of-way leading to the station mark. Construction of the monument is the same as for reference marks, except the disk is labeled AZIMUTH MARK. It has an arrow, like the reference-mark disks, set pointing to the station mark. Although an accurate distance may be measured between the azimuth mark and station mark, it is not mandatory and its line is basically to establish direction.

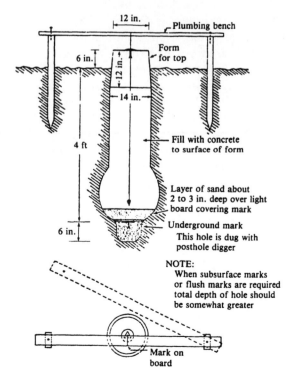

Figure 11-6. Diagram of concrete triangulation station monument.

11-9. SIGNAL BUILDING

Before actual observations can begin, observing stands or towers (signals) may have to be constructed over the station mark. Usually, these signals are only constructed at stations involved with immediate observations and removed as the observing crew moves on. Types of signals vary from a basic surveying tripod to a 100-ft-plus steel or Bilby tower.

Steel or Bilby towers are specially designed for triangulation. They consist of an inner and a structural steel tripod that are independent of each other. The inner tripod supports the theodolite; the outer tripod is for the observer's platform and signal lamp, which is placed 10 ft above the base plate of the inner tripod. An observer's tent is also placed around the outer tower. Both towers are built simultaneously from the ground up. These towers are classified by the heights of the inner tower; they come in heights of 37, 50, 64, 77, 90, 103, and 116 ft.

Bilby towers were a necessity to the NGS for extending triangulation over heavily wooded parts of the country, but for most localized projects wooden towers should be sufficient. The NGS has separate building crews that construct towers ahead of the observing crews. Smaller public surveying departments and private surveying companies will probably rely on the same personnel to both observe and build towers. Depending on the size and purpose of a project, stands and towers may be constructed over every station before any observations are made or constructed in conjunction with observations. A time schedule should be worked out in either case to ensure the even flow of the project and determine the personnel necessary. Care should be taken when constructing a tower to be certain that a vertical leg does not obstruct the lines of sight to other stations.

11-10. OBSERVATIONS

The NOS has a standard format for notekeeping on a triangulation survey, including *Observations of Horizontal Directions* (NOAA 76-52), *Observations of Double Zenith Distances* (NOAA 76-156), *Abstracts of Directions* (NOAA 76-86), *Abstracts of Zenith Distances* (NOAA 76-135), and *List of Directions* (NOAA 76-72).[4-8]

11-10-1. Description of Triangulation Station

Figure 11-4 describes a station mark and how to find it, and lists bearings and distances to the azimuth mark and reference monuments.

11-10-2. Observations of Horizontal Directions

Notes are entered in this book as stations are occupied. All data are completed on top of each page along with other pertinent remarks as notes are taken. A complete set of notes is entered for each station before notes on another one are begun. A complete set of notes in recorded order consists of (1) to-reach description and schematic of the station monuments (Figure 11-7), (2) description of and measurements between monuments (Figure 11-8), (3) observation of the marks and intersecting stations (Figure 11-9), and (4) observations of the main and supplemental stations (Figure 11-10).

Figure 11-8 describes each mark at the station, including measurements to topographic features and between the station mark and reference marks. The identification stamped on a monument must be stated exactly—i.e., do not note "stamped WYNECOOP RM NO 2, 1974." if the disk shows "WYNECOOP NO 2, 1974." The notes "1a, 7a" referred to are standard numbered notes used by the NGS and its predecessor, the U.S. Coast and Geodetic Survey, in its publication *Manual of Geodetic Triangulation*[9] (Table 11-6).

Figure 11-7 lists observations on marks and intersection stations taken at station Wynecoop. Pertinent remarks to be recorded include (1) names and duties of the observing party members, (2) weather conditions, and (3) height and type of signal. The remarks column is used to record conditions affecting observations. Positions that need to be reobserved are noted at the end of regular observations. The position number is recorded in the first column and refers to the initial circle settings listed in Table 11-2.

Complete names and descriptions of station marks and intersecting stations observed are entered in the second column for the first position; abbreviated names may be used on subsequent positions, but numbers or letters cannot be used in place of abbreviations. Intersecting stations—e.g., Williams Water Tank

(*Text continues on p. 229*)

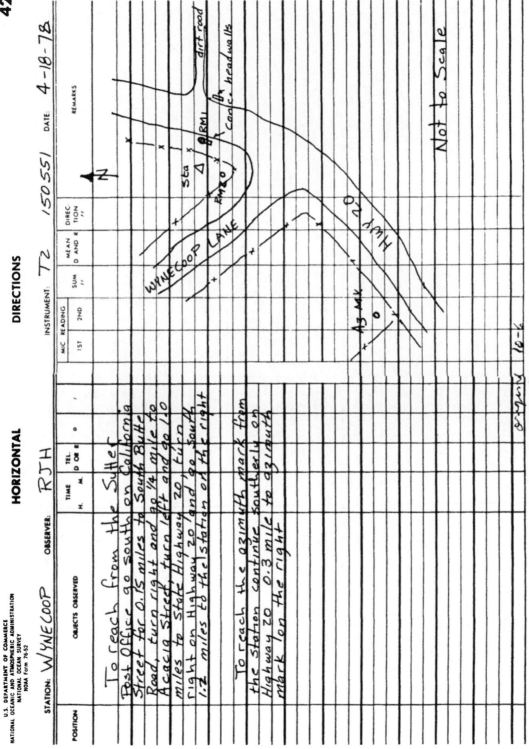

Figure 11-7. To-reach description and sketch.

U.S. DEPARTMENT OF COMMERCE
NATIONAL OCEANIC AND ATMOSPHERIC ADMINISTRATION
NATIONAL OCEAN SURVEY
NOAA Form 76-52

HORIZONTAL   DIRECTIONS

STATION: WYNECOOP   OBSERVER: RJH   INSTRUMENT: TZ 150551   DATE: 4/18/78

POSITION	OBJECTS OBSERVED	TIME H. M.	TEL. D OR R	°	'	MIC. READING 1ST	MIC. READING 2ND	SUM ''	MEAN D AND R	DIREC-TION ''	REMARKS
	STATION										
	Stamped: WYNECOOP	1972									Stamped: WYNECOOP NO 1 1972
	Note: 1b, 7a										Set flush in New end Concrete Headwall
	Projects: 0.2 feet										Projects: 0.2 ft. above Edge pavement
	48.5 ft W of center Hwy 20										15.2 ft E of center HWY 20
	18.3 ft W of wire Fence										8.5 ft W wire fence
	20 ft S center dirt road on E side Hwy 20										
	89.8 ft NE Power Pole						Held	8.200 m			26.82 ft
								.025			X.3048
	AZIMUTH MARK							8.175 m			2.1456
	Stamped: WYNECOOP	1972									1.0728
	Note: 17a										8.0460
	Projects: 0.8 ft above 0.G										8.174736 m
	34.2 ft NW Center Hwy 20										
	4.0 ft NW wire fence										
	4.0 ft NE wire fence						RM 2					
							Stamped: WYNECOOP NO 2 1972					
							Note: 1b					
	RM 1 to RM 2						Projects: 0.2 ft.					
							60.2 ft W of Center HWY 20					
			91.41 ft				3.0 ft NE of wire fence					
	Held 27.900 m	X.3048					5.6 ft E of power pole					
	Cut .039	7312.8					Held	26.600 m			82.27 ft
	27.86	3656.4					Cut	.001			X.3048
		274.230						26.599 m			6.9816
		27.80/76.8 m									3.4908
											2.61800
											26.599896 m

Figure 11-8. Mark descriptions.

36

U.S. DEPARTMENT OF COMMERCE
NATIONAL OCEANIC AND ATMOSPHERIC ADMINISTRATION
NATIONAL OCEAN SURVEY
NOAA Form 76-52

HORIZONTAL   **DIRECTIONS**

STATION: WYNECOOP   OBSERVER: RJH   INSTRUMENT: T 2 150551   DATE: 4/18/78

POSITION	OBJECTS OBSERVED	TIME H M	TEL. D OR R	°	'	MIC. READING 1ST	2ND	SUM "	MEAN D AND R "	DIREC. TION "	REMARKS
1	BALDY (1939)	1640	D	00	00	09	10	19			Rec. C Nelson
			R	180	00	13	13	26	11.2		
	Williams Water Tank (gray) 3.0 mi NE		D	92	12	46	46	92			
			R	272	12	49	49	98	45.5	34.3	92° 12'
	WYNECOOP RM No 1 (E) (1972)		D	119	56	21	22	43			
			R	299	56	23	22	45	22.0	10.8	119° 56'
	WYNECOOP RM No 2 (S) (1972)		D	210	10	55	56	111			
			R	30	10	56	56	112	55.8	44.6	210° 10'
	WYNECOOP AZ MK (SW) (1972)		D	250	12	03	03	06			
			R	70	12	04	05	09	03.8	52.6	250° 11'
2	BALDY		D	45	02	36	37	73			
			R	225	02	39	39	78	37.8		
	Williams Water Tank		D	137	15	16	16	32			
			R	317	15	16	16	32	16.0	38.2	92° 12'
	RM No 1		D	164	58	44	45	89			
			R	344	58	46	46	92	45.2	07.4	119° 56'
	RM No 2		D	255	13	20	20	40			
			R	75	13	21	21	42	205	42.7	210° 10'
	AZ MK		D	295	14	33	32	65			
			R	115	14	35	34	69	33.5	55.7	250° 11'

RJH

Figure 11-9. Observation of marks and intersection stations.

37

U.S. DEPARTMENT OF COMMERCE
NATIONAL OCEANIC AND ATMOSPHERIC ADMINISTRATION
NATIONAL OCEAN SURVEY
NOAA Form 76-52

HORIZONTAL   **DIRECTIONS**

STATION: WYNECOOP   OBSERVER: RJH   INSTRUMENT: T-2 150551   DATE: 4/18/78

POSITION	OBJECTS OBSERVED	TIME H. M.	TEL. D OR R	°	'	MIC. READING 1ST	2ND	SUM ''	MEAN D AND R	DIREC-TION ' ''	REMARKS
1	BALDY 1939	1810	D	00	00	10	12	22			Weather: Clear & Cool
		1817	R	180	00	14	14	28	12.5	12.5	Obs: RJ Hand
											Rec: C Nelson
	GAS 1972		D	21	14	46	47	93			Ht Stand 3.61 m
			R	201	14	47	48	95	47.0	34.5	Ht Inst 3.85 m
											21° 14'
	SALT 1972		D	68	22	17	17	34			
			R	238	22	20	21	41	188	063	58° 22'
	CACHE 1939		D	247	57	03	03	06			
			R	67	57	04	06	10	064.0	51.5	247° 56'
2	BALDY	1829	D	22	02	39	39	78			
		1821	R	202	02	41	41	82	40.0		
	GAS		D	43	17	12	12	24			
			R	223	17	13	15	28	13.0	33.0	21° 14'
	SALT		D	80	24	45	45	90			
			R	260	24	48	49	97	46.8	06.8	58° 22'
	CACHE		D	269	59	32	33	65			
			R	89	59	33	33	66	328 52.8		247° 56'

JRH

Figure 11-10. Observation of main scheme stations.

Table 11-6.   Standard numbered notes for description of marks.

The following notes have been used for many years in published descriptions and other publications of the CGS and later by NGS.

Surface marks

Note 1. A standard triangulation-station disk set in the top of (a) a square block or post of concrete, (b) a concrete cylinder, (c) an irregular mass of concrete.

Note 2. A standard triangulation-station disk cemented in a drill hole in outcropping bedrock, (a) and surrounded by a triangle chiseled in the rock, (b) and surrounded by a circle chiseled in the rock, (c) at the intersection of two lines chiseled in the rock.

Note 3. A standard triangulation-station disk set in concrete in a depression in overcropping bedrock.

Note 4. A standard triangulation-station disk cemented in a drill hole in a boulder.

Note 5. A standard triangulation-station disk set in concrete in a depression in a boulder.

Note 6. A standard triangulation-station disk set in concrete at the center of the top of a tile (a) that is embedded in the ground, (b) that is surrounded by a mass of concrete, (c) that is fastened by means of concrete to the upper end of a long wooden pile driven into the marsh, (d) that is set in a block of concrete and projects from 12 to 20 in. above the block.

Underground marks

Note 7. A block of concrete about 3 ft below the ground containing at the center of its upper surface (a) a standard triangulation-station disk, (b) a copper bolt projecting slightly above the concrete, (c) an iron nail with the point projecting above the concrete, (d) a glass bottle with the neck projecting a little above the concrete, (e) an earthenware jug with the mouth projecting a little above the concrete.

Note 8. In bedrock, (a) a standard triangulation-station disk cemented in a drill hole, (b) a standard triangulation-station disk set in concrete in a depression, (c) a copper bolt set in cement in a drill hole or depression, (d) an iron spike set point up in cement in a drill hole or depression.

Note 9. In a boulder about 3 ft below the ground (a) a standard triangulation-station disk cemented in a drill hole, (b) a standard triangulation-station disk set in concrete in a depression, (c) a copper bolt set with cement in a drill hole or depression, (d) an iron spike set with cement in a drill hole or depression.

Note 10. Embedded in earth about 3 ft below the surface of the ground (a) a bottle in an upright position, (b) an earthenware jug in an upright position, (c) a brick in a horizontal position with a drill hole in its upper surface.

Reference marks

Note 11. A standard reference-mark disk, with the arrow pointing toward the station, set at the center of the top of (a) a square block or post of concrete, (b) a concrete cylinder, (c) an irregular mass of concrete.

Note 12. A standard reference-mark disk, with the arrow pointing toward the station, (a) cemented in a drill hole in outcropping bedrock, (b) set in concrete in a depression in outcropping bedrock, (c) cemented in a drill hole in a boulder, (d) set in concrete in a depression in a boulder.

Note 13. A standard reference-mark disk, with the arrow pointing toward the station, set in concrete at the center of the top of a tile (a) that is embedded in the ground, (b) that is surrounded by a mass of concrete, (c) that is fastened by means of concrete to the upper end of a long wooden pile driven into the marsh, (d) that is set in a block of concrete and projects from 12 to 20 in. above the block.

Previously used notes 14 and 15 referred to seldom used types of witness marks and are purposely omitted.

Azimuth marks

Azimuth-mark notes are almost identical to reference-mark notes 11 through 13, which have been previously used for azimuth marks. The following numbers 16, 17, and 18 refer specifically to azimuth-mark disks.

Note 16. A standard azimuth-mark disk, with the arrow pointing toward the station, set at the center of the top of (a) a square block or post of concrete, (b) a concrete cylinder, (c) an irregular mass of concrete.

Note 17. A standard azimuth-mark disk, with the arrow pointing toward the station, (a) cemented in a drill hole in outcropping bedrock, (b) set in concrete in a depression in outcropping bedrock, (c) cemented in a drill hole in a boulder, (d) set in concrete in a depression in a boulder.

Note 18. A standard azimuth-mark disk, with the arrow pointing toward the station, set in concrete at the center of the top of a tile (a) that is embedded in the ground, (b) that is surrounded by a mass of concrete, (c) that is fastened by means of concrete to the upper end of a long wooden pile driven into the marsh, (d) that is set in a block of concrete and projects from 12 to 20 in. above the block.

—should be described completely, including the structure's local name and description of the part observed, approximate distance from the occupied station, and bearing (E, NE, etc.) from the occupied station. Beginning and ending times for each position are entered under the time head in the third column, using military time from 0000 (midnight) to 2359.

The fourth column describes the telescope position—D for direct or R for reverse—followed by the direction observed with two micrometer readings taken for each position (see Section 11-11-3). The sum of the micrometer readings is entered under the sum heading while the mathematical mean, to the nearest 0.1 sec, is noted in the following column, under mean. When using the Wild T-3 theodolite, the sum of the two direct micrometer readings and that of the two reverse micrometer readings are meaned. On the Wild T-2 and other comparable directional theodolites, the entry is the mean of all four micrometer readings. Seconds of direction observed from the initial station to each object are entered under "direction" (subtract the mean of initial from the mean of observed).

All pertinent information discussed earlier is entered under remarks. Degrees and minutes of direction observed from the initial station are also listed under this column for each direction observed.

After observations of marks and intersecting stations are completed, observations of the main and supplemental scheme begin (Figure 11-9). Recording main and supplementary observations is done in the same manner as for mark and intersecting station observations. Reobservations are entered after the original set. When the theodolite is checked for level, it should be recorded under remarks and any adjustment noted.

After each page or set of observations is recorded, it must be completely reviewed, all items checkmarked, and each page initialed, and corrections also checked and initialed. Entries must be in ink, and no erasures are permitted. Errors are deleted by a single line through the incorrect entry and the correction placed above the deletion. When filled, an alphabetized index is recorded in the front of the book.

11-10-3. Observations of Double Zenith Distances

Typical recordings for vertical angles or double zenith distances observed as given in Figure 11-11 are self-explanatory.

11-10-4. Abstracts

After each set of horizontal distances and zenith distances has been observed and the field book checked, an abstract is recorded to make certain all observations are within required tolerances. Abstracts should be completed in ink, without erasures.

The "abstract of directions" (Figure 11-12) lists all observations taken directly from the "horizontal directions" book, except those taken on a wrong object or any resulting from a kicked tripod, which are rejected immediately in the book. Any observation so far from the apparent mean as to be considered a blunder may be rejected before calculating the initial mean. Once all observations have been entered on the abstract, the columns are summed and a mean angle is derived. Any observation outside the required tolerance is rejected. That position is then reobserved and the new observation entered on the abstract. The mean of the observations is recalculated using the new observation. If, when we use the new mean, the rejected observation and new observation both fall within the required tolerance, they are meaned and used as the observed seconds for that position. A rejected observation still outside required tolerance is enclosed in parentheses and an R written behind, meaning that it is rejected and not used to calculate the mean direction.

Station WYNECOOP State California Instrument T2 150551

Observer R JH County Sutter Date 4/18/78

Object Observed	Time	Level O.	Level E.	Circle Right or Left	Circle Reading		Verniers A	Verniers B	Verniers Mean	Zenith Distance °	Zenith Distance '	Zenith Distance "	Remarks
BALDY (1939) (light) 1.33m	1205			L	89° 58'		58	58	58.0				Rec. C Nelson Cool
				R	270 00		56	56	56.0				Ht of Stand 3.61m
				DZD 179 58					02.0	89	59	01.0	Ht of Inst 3.85m
				L	89 59		03	03	03.0				
				R	270 00		51	51	51.0				
				DZD 179 58					12.0	89	59	06.0	
GAS (light) 1.27m	1220			L	89 07		19	19	19.0				
				R	270 52		36	36	36.0				
				DZD 178 14					43.0	89	07	21.5	
				L	89 07		05	05	05.0				
				R	270 52		45	45	45.0				
				DZD 178 14					20.0	89	07	10.0 (R)	
				L	89 07		21	21	21.0				
				R	270 52		29	29	29.0				
				DZD 178 14					52.0	89	07	26.0	

Do not write in this margin

JDRG

Figure 11-11. Recording double zenith distances.

POSITION NO.	BALDY (1939)	GAS	SALT	CACHE (1939)	Williams Water Tank	WYNECOOP RM No 1	WYNECOOP RM No 2	WYNECOOP AZ MK
	WYNECOOP							
	STATE California	COMPUTED BY CN	DATE 4/18/78					
	(INITIAL) 0° 00'	21 14	58 22	247 56	92 12	119 56	210 10	250 11
1	0.00	34.5	06.3	51.5	34.3	10.8	44.6	52.6
2	0.00	33.0	06.8	52.8	38.2	07.4	42.7	55.7
3	0.00	28.8	10.2	47.2	39.2	12.2	48.6	51.0
4	0.00	31.0	07.5	50.8	37.0			56.6
5	0.00	30.2	11.2	52.0				
6	0.00	33.5	09.5	48.0				
7	0.00	29.1 (25.2)R	08.0	49.8				
8	0.00	32.8	08.2	53.2				
9	0.00							
10	0.00							
11	0.00							
12	0.00							
13	0.00							
14	0.00							
15	0.00							
16	0.0							
SUM		12.9	67.7	5.3	28.7	30.4	15.9	15.9
MEAN		31.61	08.46	50.66	37.2	10	45	54.0
COR. FOR ECC.								
DIRECTION								

OBSERVER RJH   CHECKED BY DRG   INSTRUMENT NO. T2 150551   SHEET___ OF___   VOLUME NO.

ABSTRACT OF DIRECTIONS

Figure 11-12. Abstract of directions.

The following is a summary of the rejection limits:

1. First-order: 4 sec
2. Second-order, third-order, and azimuth marks: 5 sec
3. Reference marks: 20 sec
4. Intersection station: 5 to 10 sec, depending on the sharpness and nearness of an object

The "abstract of zenith distances" (Figure 11-13) lists all the vertical angles observed from a station. Besides the headings, the data in columns 1, 2, 3, 5, and 8 are entered by the observing party from information recorded in the "observations of double zenith distances" field book. Heights above the station mark, dates, and stations the lights are shown to are entered at the bottom of the abstract, as the information becomes available. The rest of the information is entered or calculated later.

11-11. FIELD OBSERVATIONS

An observing party should plan to arrive at the station early enough to (1) write the station description; (2) measure the distances between the reference and station marks; (3) check for stability and collimate the stand or tower; (4) set up the instrument; (5) observe vertical angles; and (6) observe horizontal directions to the azimuth mark, reference marks, and intersection stations before dark.

11-11-1. Station Setup

While driving to the station, a final check should be made of the to-reach description, including the approximate distance from azimuth mark to a station mark. On reaching the station mark, the to-reach description and schematic are recorded in the "observation of horizontal directions" book.

Next, distances between reference marks and the station mark are measured independently, in hundredths of a foot and thousandths of a meter. A 30-m tape, marked in meters on one side and feet on the other, is used to make the measurements. They are made horizontally and recorded immediately in observation of horizontal directions. Measurements are repeated until a check of 0.003 m is obtained between the meter and feet readings (Figure 11-12).

11-11-2. Vertical Angles

Normally, the first set of observations is for the vertical angles, measured from the zenith and referred to and recorded as double zenith distances in "observations of double zenith distances." Reciprocal observations should be made at all occupied stations, simultaneously if possible. If reciprocal observations are made on different days, it is best to try to make them under the same climactic condition and during the same hour of the day. This will lessen the effect caused by varying refraction. The best time to observe vertical angles and the hours of least refraction are between noon and 4:00 PM.

First-order and second-order, class I triangulation require three individual sets of observations of the double zenith distance. All other classifications require two. Observations are made with the middle of the horizontal cross hair sighted on the object. A full direct and reverse set of observations must be completed before another set is started to the same station. Two or more consecutive direct or reverse pointings to the same station are not made. This is to ensure that each pointing is separate and distinct. Sets may be completed on one object at a time, or completed with the telescope in one position on several objects in succession and then reversed.

11-11-3. Horizontal Directions to Reference Marks, Azimuth Marks, and Intersection Stations

Part of the daylight activities at a station is observing horizontal directions to the reference and azimuth marks and any intersection stations. Whenever possible, the initial station used for these marks should be the same as that used on the main scheme observations.

ABSTRACT OF ZENITH DISTANCES

Station WYNECOOP   State CALIFORNIA

Observer RJH   Instr. T-2 150551   Height of Stand 3.61 M

Date	Hour	Object Observed	Object Above Station =o	Telescope Above Station =t	Diff. of Heights t−o	Reduction to Line Joining Stations	Observed Zenith Distance	Corrected Zenith Distance
			Meters	Meters	Meters	"	° ′ ″	° ′ ″
4/18/78	1205	BALDY (1939)	1.33	3.85	+2.52		89 59 01.0	
							06.0	
							89 59 03.5	
4/18/78	1220	GAS	1.27	3.85	+2.58		89 07 21.5	
							10.0 (R)	
							26.0	
							89 07 23.8	
4/18/78	1242	SALT	3.77	3.85	+0.08		90 01 15.0	
							22.0	
							90 01 18.5	
4/18/78	1255	CACHE	7.63	3.85	−3.78		89 56 47.0	
							42.0	
							89 56 44.5	

Date	Light Shown to Station	Height of Light Above Station	Date	Light Shown to Station	Height of Light Above Station
		Meters			Meters
4/16/78	BALDY (1939)	3.78	4/23/72	CACHE	3.78
4/19/78	GAS	3.78			
4/18/78	SALT	2.95			✓ DRG

DO NOT WRITE IN THIS MARGIN.

Figure 11-13. Abstract of zenith distances.

Three positions are observed on reference marks; four positions are observed on azimuth marks and intersection stations. Any positions on the reference marks in excess of 20 sec from the mean are rejected and reobserved, whereas any positions of the azimuth mark or intersection stations in excess of 5 sec from the mean are rejected and reobserved.

Any necessary reobservations are made after the primary four positions have been completed and checked. A circle setting for the reobserved position will be the same as for the rejected position.

11-11-4. Horizontal Directions to Main and Supplemental Stations

The best time to start observing on main and supplemental stations is at dusk, as soon as all sighting lights become visible. Care must be taken to make sure that lights are pointed as accurately as possible. By this time, all other duties of the observing party should be completed. If vertical angles could not be observed earlier, they should be measured after the main scheme directions are taken. Stations used for the initial observation should have an easily discernible and reliable light for uninterrupted use. The distance to it should be long enough to ensure no effect due to wind, local refraction, or a slightly mispointed light, but not so far away as to cause an interruption because the light is faint or hazy. If possible, pick an initial that is convenient for observing directions in a clockwise rotation.

Observation is completed in the same manner as that used for reference marks and intersection stations, using the required number of positions for the classification sought. Pointing should be quick but not hurried. Deliberate pointing should be avoided as this tends to reduce accuracy and greatly increases the time required to complete a position.

Lightkeepers should keep their own set of notes, listing what sights were used on which stations. They should also list the sight height above the station mark. They are responsible

for making sure lights are constant and pointed as accurately as possible, and they should be in constant touch with the observing crew.

11-12. CONCLUSION

Many factors of triangulation have not been considered in this chapter, such as eccentricity of stations, reduction to center, and the use of striding levels for observation with a vertical angle over 2°. When we consider a triangulation project, the *Manual of Geodetic Triangulation*, Publication No. 247, is available from the U.S. Government Printing Office, or any successor to this publication will be helpful. Publication No. 248 is the guide for NGS triangulation crews and covers the complete procedures required in the field and not reduction process.

NOTES

1. U.S. Department of Interior Bureau of Land Management. 1973. *Manual of Surveying Instructions*. *U.S.* Government Printing Office, Washington, D.C.

2. FGCC. 1984. Classification, standards of accuracy, and general specifications of geodetic control surveys. Rockville, MD.

3. *Manual of Reconnaissance for Triangulation.* SP225, U.S. Government Printing Office: Washington, D.C., 1938.

4. NOS. *Observations of horizontal directions*. 1976. NOAA 76-52, Rockville, MD.

5. NOS. *Observations of double zenith distances*. 1976. NOAA 76-52, Rockville, MD.

6. NOS. *Abstracts of directions*. 1976. NOAA 76-86, Rockville, MD.

7. NOS. *Abstracts of zenith distances*. 1976. NOAA 76-135, Rockville, MD.

8. NOS. *List of directions*. 1976. NOAA 76-72, Rockville, MD.

9. *Manual of Geodetic Triangulation*. Publication No. 247, U.S. Government Printing Office, 1971. Washington, D.C.

10. *Manual of Geodetic Triangulation*. 1971. Publication No. 247, U.S. Government Printing Office, Washington, D.C.

12

Trilateration

Bryant N. Sturgess and Frank T. Carey

12-1. INTRODUCTION

Trilateration is a method of control extension, control breakdown, and control densification that employs electronic distance-measuring instruments (EDMIs) to measure the lengths of triangles sides rather than horizontal angles, as in triangulation. The triangle angles are then calculated based upon measured distances by the familiar law of cosines. Trilateration consists of a system of joined and/or overlapping triangles usually forming quadrilaterals or polygons, with supplemental horizontal angle observations to provide azimuth control or check angles. Zenith angles are required when elevations have not been established or differential leveling is not contemplated, in order to reduce slope distances to a common reference datum.

With the development of EDMIs, trilateration has become a very practical highly accurate, and precise means of establishing and/or expanding horizontal control.

12-2. USE OF TRILATERATION

Trilateration is commonly employed to study gradual and secular movements in the earth's crust in areas subject to seismic or tectonic activity, to test and construct defense and sci-

entific facilities, and on high-precision engineering projects. It is also used in control expansion or densification for future metropolitan growth; coastline control; inland waterways; control extension; densification for land subdivisions and construction; and deformation surveys of dams, geothermal areas, structures, regional/local tectonics, and landslides. Trilateration can be used for a simple low-order topographic survey covering a small area, or on large projects for the design and/or construction of highways, bridges, dams, or even to extend topographic mapping control from small local tracts to regional areas. It can be a simple process with single-line measurements using ordinary off-the-shelf EDMIs and support equipment. Or, it can be a complex process employing highly refined EDMIs, with special measures for determining the refractive index correction and eccentric measurements to an eccentric or offset bar, at either the reflector or instrument station with instrument occupations at both ends of the line.

12-3. ADVANTAGES OF TRILATERATION

Trilateration is a practical and highly accurate means of rapid control extension. When properly executed, it is superior to both triangulation and traverse for special-purpose precise surveys and often is the preferred method, because of its advantageous cost-benefit ratio and potential.

Basic trilateration is less expensive than classical triangulation and, under most conditions, more accurate. Trilateration permits controlling large and small geographical areas with a minimum number of personnel. It is not required to measure lines with all sights simultaneously in position, as with triangulation, unless the procedure of line pairs is being employed. Trilateration also provides necessary scale control lacking in triangulation.

12-4. DISADVANTAGES OF TRILATERATION

Trilateration has a smaller number of internal checks compared with classical triangulation, in which each quadrilateral contains two diagonals called *braces* (hence, braced quadrilateral) and has four triangle closures, three of which are independent. Additionally, there are other checks consisting of agreements between common sides of the triangles (*side-equation* tests). The number of checks or redundancies in a braced quadrilateral in triangulation is four. In trilateration, there is only one.

Trilateration can be reinforced by modified observational techniques to provide the same number of redundancies as triangulation by employing the group or ratio method of length measurement. In order to employ this technique, the length of a line must be measured from both ends. Cost then becomes the limiting factor. Geometric restrictions, which limit the selection of locations where stations can be established, may cause difficulty in fully utilizing trilateration networks to place control at specially needed sites. Pure trilateration cannot be performed when precise angle measurements must be taken to intersect reference objects, azimuth marks, and reference monuments. In such cases, a combination of trilateration and triangulation has to be employed.

Higher-order trilateration (first- and second-order, class I) requires sampling of meteorological conditions to be commensurate in precision with the distance measurements. It is paramount that appropriate techniques be employed to guarantee reasonable sampling. First-order and super first-order trilateration necessitate implementing meteorological sampling techniques, perhaps by flying the sight line with a light aircraft to sample temperature, humidity, and barometric pressure, or utilizing an EDMI with a two-color laser system to determine the refractive index correction.

Trilateration is inherently more expensive than traverse. In addition, on large geodetic control projects the inventory of reflectors for measuring long lines must be sufficient in numbers to guarantee maximum signal strength to the EDMI. Reflector arrays have to be manned and are significantly more expensive than *show lights* required for triangulation.

12-5. COMPARISON OF TRIANGULATION AND TRILATERATION

Triangulation and trilateration are both forms of control extension, control expansion, or control breakdown (densification). Triangulation is a method of surveying in which the stations are points on the ground, forming vertices of triangles comprising chains of quadrilaterals or polygons. Within these triangles the angles are observed by theodolite, and the lengths of sides determined by successive computations through the chain of triangles. Scale is provided by at least two stations having known positions in the first quadrilateral of the chain, or by a base line connecting two stations in the first quad. The positions of all remaining stations in the first quad, and any in a chain of quads, are computed in terms of measured angles and known positions at the chain's beginning and end.

In triangulation, distances are computed from angle observations; in trilateration, angles are calculated from distance observations. Triangulation gives impressive redundancy when compared with trilateration. For example, in a simple triangle with all angles measured, one redundant measurement yields one condition equation or one degree of freedom. Given the same triangle in trilateration, unless three distances are measured from each end, there are no redundant measurements.

Similarly, a braced quadrilateral with all angles measured provides four redundancies; the same figure in pure trilateration yields only one redundancy. Table 12-1 tabulates the redundancies possible with various figures. All points given in the example are presumed to be intervisible.

It can be seen that the number of redundancies in triangulation is significantly higher than in trilateration. This does not necessarily mean triangulation provides a more accurate or precise positional solution than trilateration. On the contrary, when trilateration networks are given an adequate design configuration by avoiding angles smaller than the specified minimum, by using proper field techniques and properly matched calibrated equipment, and by adding check angles when appropriate, the final results can be significantly superior to triangulation, with lower cost.

12-6. CONSTRUCTION OF BRACED QUADRILATERALS

Well-shaped geometrical figures are required for both arc and area networks. For arcs, quadrilaterals must approximate a square with both diagonals measured. When only a single diagonal can be observed, a center point must be visible from the four vertices of the quad. In area systems, well-shaped triangles containing angles seldom smaller than 15° are mandatory for first- and second-order surveys. If these conditions cannot be met, then one or more of the large angles in the quad must be observed by theodolite. On occasion, geographical constraints will not allow the "ideal" quadrilateral configuration having angles larger than 25°.

Figures can contain angles of 5 to 7° smaller than specified while maintaining line accuracies. However, this is permissible only on second-order, class II or third-order, class I and II surveys. It is desirable to include horizontal-angle readings to increase redundancy and improve accuracy in cases where conditions of deficient angle sizes are present. First-order and second-order, class I surveys should not have deficient angle sizes (within the same range of 18 to 20°) unless compensated for by angle measurements.

Since few engineers and surveyors are engaged in projects that establish new primary positions on arc or transcontinental networks,

Table 12-1. Geometrical redundancies

	Quadrilateral	Pentagon	Hexagon
Number of lines	6	10	15
Number of triangles	4	10	20
Number of triangles used in the computations	3	6	10
Number of check triangles	1	4	10
Number of geometrical conditions in trilateration	1	3	6
Number of geometrical conditions in triangulation	4	9	16

Source: FGCC 1979. "Classification, Standards of Accuracy, and General Specifications of Geodetic Control Survey." Silver Spring MD.

only area networks and small-area-type projects will be addressed,

Figures 12-1, 12-2, and 12-3 show typical polygons that can be used in trilateration. The center-point polygon and center-point quadrilateral are more expensive to design, but may be required owing to topographical constraints. In a high-order control network, the familiar braced quadrilateral, reinforced with angular measurements at selected stations, provides an economical means of increasing redundancies in trilateration.

A trilatered center-point polygon composed of single triangles with no diagonals observed contains only a single condition or degree of freedom, regardless of the number of triangles contained in the figure. For each additional line measured, one more condition is added.

The following are methods of increasing redundancy by one condition or degree of freedom, thus improving reliability by adding:

1. An azimuth measurement other than the one required to orient the system.

2. An independent distance measurement accomplished through triangulation, e.g.

3. Two degrees of freedom by including a station of known position, other than the station required to index and originate the computations.

4. An angle observation; for each one, an additional condition or degree of freedom is introduced.

The more conditions designed into a figure network, the greater reliability obtained for

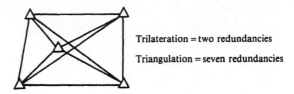

Figure 12-2.   Center-point quadrilateral.

the final position of each point within the system.

A good policy to follow is a semimarriage of trilateration and triangulation by measuring all distances and incorporating angle observations generally at the station with the largest angle. Sufficient redundancy is gained by one such occupation to make the expenditure cost-effective. For instance, one occupation of this type on a conventional braced quadrilateral increases the redundancy from one to two. However, it must be remembered that when certain EDMIs are properly tuned and calibrated, they will surpass the performance of any theodolite (short distances excluded) in determining angles through distance observation. Consequently, when check angles are specified, the proper angle observation weight must be determined accurately to avoid biasing the figure. Therefore, unless the check angle is placed in the proper perspective, it will yield an extra redundancy but can degrade the mathematical fit of the figure unless properly weighted.

Another means of increasing redundancies in a conventional braced quadrilateral for higher-order trilateration surveys, such as first

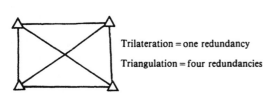

Figure 12-1.   Braced quadrilateral.

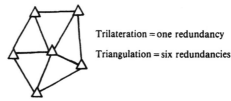

Figure 12-3.   Center-point polygon.

and second, class I, is to incorporate distance measurements taken alternately from line ends with measurements to EDMI reflectors mounted on an eccentric bar. Observing distances from each end provides a field check on a prior line measurement, without lengthy computations. Ideally, lengths should be measured from each end on different days. However, if production or scheduling will not permit this, measurements can be taken on the same day, but at a spaced interval such as morning and afternoon, or day and night. The advantage of this method is that different meteorological conditions exist at various times, so when meaned, they should yield a result closer to the true distance.

A full marriage of trilateration and triangulation, in which all distances are measured from each end and every station is occupied by theodolite, will provide the strongest and theoretically most precise means of establishing horizontal control. This method predictably is the most costly; however, for those projects demanding the ultimate in accuracy and precision, there is no better method with the possible exception of the global positioning system (GPS).

12-7.   NETWORK DESIGN

The geometric figures used in trilateration are usually braced quadrilaterals and center-point polygons. However, because of a lack of internal checks, established specifications and standards must be adhered to rigorously. The minimum specified angle sizes should not be reduced unless check angles are observed at those locations.

Theoretically, the basic quadrilateral should be square in configuration, and a central polygon should have equal angles. Field conditions may make achieving this desirable arrangement difficult or impossible. The sample trilateration networks shown in Figures 12-4 and 12-5 are typical of new networks.

12-8.   RECONNAISSANCE

Field reconnaissance for trilateration is similar to traversing and triangulation reconnaissance. It is equally essential in trilateration and triangulation that all existing and proposed new stations be plotted on a base map, so the network configuration can be verified and checked for conformance with specifications. In areas where the triangles fall below specifications, owing to geographical constraints, occupation for angle observations can be planned to add reinforcement to the figures and redundancies in the mathematical solution. For a detailed description of reconnaissance, refer to Section 11-7, "Field Reconnaissance."

General guidelines to follow in layout of a trilateration net are:

1. Avoid situations where the sight line passes in close proximity to a ridge, saddle, tree top, or other obstruction between the EDMI and reflector station. Unless adequate clearance is maintained, the light beam will be reflected and weakened by heat waves, thus reducing the range and diminishing the reliability of distance measurements.

2. Line selections that go over large bodies of water must clear them by at least 50 ft; otherwise, range and accuracy suffer.

3. Attempting EDMI measurements near high-voltage transmission lines or microwave relays (even two-way radio communication with the EDMI station) can cause erroneous measurements. Avoid EDMI setups close to such energy sources. In most cases, a distance of 200 ft is a reasonable minimum.

In station-to-station communication between survey party members, avoid any transmissions from the EDMI station during line measurements. The precautions mentioned in item 3 do not necessarily apply to all EDMIs. It is prudent to first determine if the equipment selected for the project is susceptible to these electromagnetic and static forces, before choosing a questionable site.

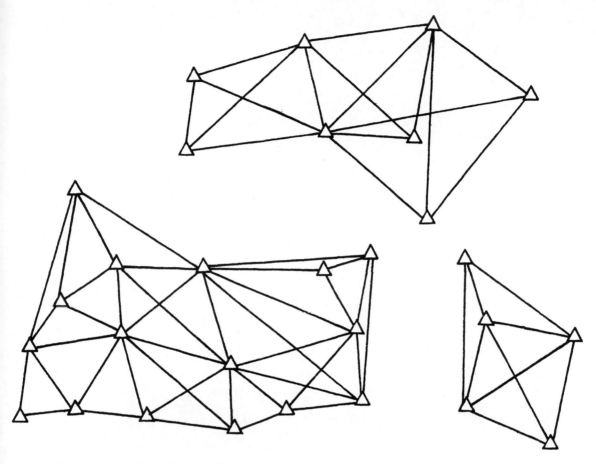

Figure 12-4.   Existing trilateration networks.

12-9. SPECIFICATIONS AND TRILATERATION

In the United States, standards of accuracy for geodetic control surveys are prepared by the Federal Geodetic Control Committee (FGCC) with subsequent review by the American Congress on Surveying and Mapping (ACSM), the American Society of Civil Engineers (ASCE), and the American Geophysical Union (AGU).

Table 12-1, which is taken from the FGCC publication entitled *Classification, Standards of Accuracy, and General Specifications of Geodetic Control Surveys*, provides a summary of various standards of accuracy for triangulation; Table 12-2 shows standards for trilateration control. They are available from the National Ocean

Survey (NOS) in Rockville, MD Table 12-3 lists trilateration standards for field operations covering various orders of surveys.

It is paramount that reference be made to another publication, *Specifications to Support Classification, Standards of Accuracy, and General Specifications of Geodetic Control Surveys*, by the FGCC, also published by the NOS.[2] Both publications should be consulted before designing a trilateration control network. The standards provided in these publications are for three orders of accuracy: first, second, and third. Second- and third-order are subdivided into two classes: I and II.

First-order trilateration is used mainly in (1) arc networks, (2) transcontinental primary control, (3) network densification in large metropolitan areas, (4) highly accurate surveys

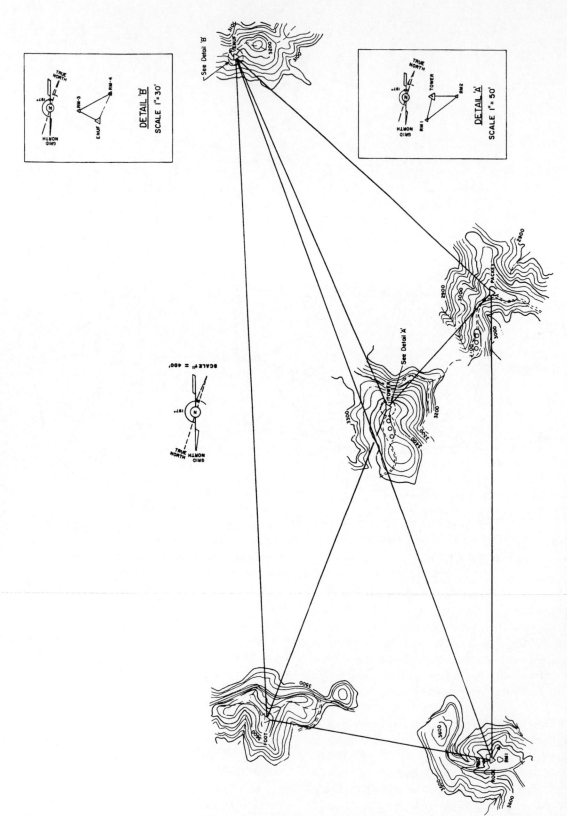

Figure 12-5. Plat of geysers geothermal field (descriptions of stations have been omitted).

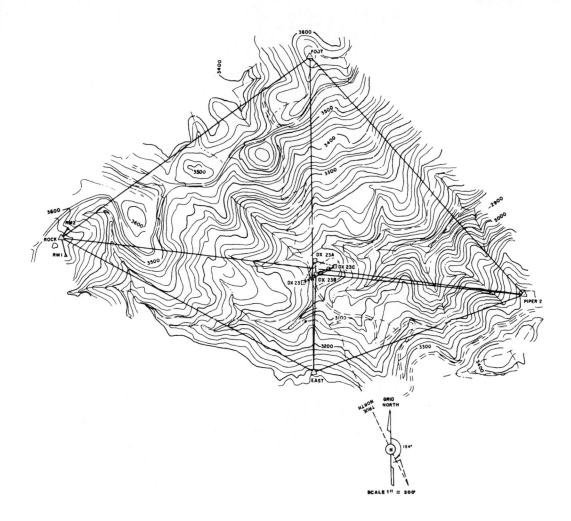

Figure 12-5. (*Continued*).

NOTE: The source of contours is drawing number 3250 entitled "Unit 17 Project Area, Upper Squaw Creek" by Union Oil Co. of California, Geothermal Division.

for defense, (5) sophisticated engineering control projects, and (6) surveys for monitoring earth deformation. First-order is the *primary* horizontal control and provides the basic framework for a national control net.

Second-order, class I horizontal control connects first-order arcs and is the principal framework for control densification. This order is also employed for large engineering projects as well as control expansion for metropolitan areas and, in some cases, deformation surveys.

Second-order, class II surveys are used as the basic framework for big photogrammetric projects, large-scale subdivisions, and spacious construction projects. This order is occasionally utilized to control interstate highway systems.

Third-order, class I and II trilateration surveys are a supplemental-type control. This lowest order is employed primarily to control small engineering projects, construction, and photogrammetric, hydrographic, and topologic projects.

Table 12-2. Classification, standards of accuracy, and general specifications for horizontal control

| | | TRILATERATION | | | |
| | | Second-Order | | Third-Order | |
Classification	First-Order	Class I	Class II	Class I	Class II
Recommended spacing of principal stations	Network stations seldom less than 10 km. Other surveys seldom less than 3 km.	Principal stations seldom less than 10 km. Other surveys seldom less than 1 km.	Principal stations seldom less than 5 km. For some surveys, a spacing of 0.5 km between stations may be satisfactory.	Principal stations seldom less than 0.5 km.	Principal stations seldom less than 0.25 km.
Geometric configuration§ Minimum angle contained within, not less than	25°	25°	20°	20°	15°
Length measurement Standard error*	1 part in 1,000,000	1 part in 750,000	1 part in 450,000	1 part in 250,000	1 part in 150,000
Vertical angle observations† Number of and spread between observations	3 D/R—5"	3 D/R—5"	2 D/R—5"	2 D/R—5"	2 D/R—5"
Number of figures between known elevations	4–6	6–8	8–10	10–15	15–20

Astro azimuths**	6–8	6–10	8–10	10–12	12–15
Spacing-figures					
No. of obs./night	16	16	16	8	4
No. of nights	2	2	1	1	1
Standard error	0″.45	0″.45	0″.6	0″.8	3″.0
Closure in position‡					
After geometric conditions have been satisfied should not exceed	1 part in 100,000	1 part in 50,000	1 part in 20,000	1 part in 10,000	1 part in 5000

*The standard error is to be estimated by

$$\sigma_m = \sqrt{\frac{\Sigma y^2}{n(n-1)}}$$

where σ_m is the standard error of the mean, y a residual (i.e., the difference between a measured length and the mean of all measured lengths of a line), and n the number of measurements. The term "standard error" used here is computed under the assumption that all errors are strictly random in nature. The true or actual error is a quantity that cannot be obtained exactly. It is the difference between the true value and measured value. By correcting each measurement for every known source of systematic error, however, one may approach the true error. It is mandatory for any practitioner using these tables to reduce to a minimum the effect of all systematic and constant errors so that real accuracy may be obtained. (See page 267 of Coast and Geodetic Survey, Special Publication No. 247, *Manual of Geodetic Triangulation* Revised edition, 1959, for definition of "actual error.")

**The standard error for astronomic azimuths is computed with all observations considered equal in weight (with 75% of the total number of observations required on a single night) after application of a 5-sec rejection limit from the mean for first- and second-order observations.

†See FGCC *Detailed Specifications on Elevation of Horizontal Control Points* for further details. These elevations are intended to suffice for computations, adjustments, and broad mapping and control projects, not necessarily for vertical network elevations.

‡Unless the survey is in the form of a loop closing on itself, the position closures would depend largely on the constraints or established control in the adjustment. The extent of constraints and actual relationship of the surveys can be obtained through either a review of the computations, or a minimally constrained adjustment of all work involved. The proportional accuracy or closure (i.e., 1/100,000) can be obtained by computing the difference between the computed value and fixed value, and dividing this quantity by the length of the loop connecting the two points.

§See FGCC *Detailed Specifications on Trilateration* for further details. *Source:* FGCC *Specifications to Support Classification, Standards of Accuracy, and General Specifications of Geodetic Control Surveys*, Silver Spring, MD.

Table 12-3. Trilateration standards for field operations

	First-Order	Second-Order Class I	Second-Order Class II	Third-Order Class I	Third-Order Class II
Length measurement standard error	1 part in 1,000,000	1 part in 750,000	1 part in 450,000	1 part in 250,000	1 part in 150,000
PPM (total error budget)	1 ppm	1.3 ppm	2.2 ppm	4 ppm	6.7 ppm
"Atmospheric sampling accuracy requirements"					
Temperature (T)	±0.2°F/±0.1°C (+0.1 ppm)	±0.9°F ±0.5°C (±0.5 ppm)	±1.8°F ±1.0°C (±1.0 ppm)	±1.8°F ±1.0°C (±1.0 ppm)	±3.6°F ±2.0°C (±2.0 ppm)
Barometric pressure (P)	±0.01" HG,10' alt. ±0.25 mm HG (±0.1 ppm)	±0.05" HG, 50' alt. ±1.3 mm HG (±0.5 ppm)	±0.10" HG 100' alt. ±2.5 mm HG (±1 ppm)	±0.10" HG,100' alt. ±2.5 mm HG (±1 ppm)	±0.20" HG, 200' alt. ±5.0 mm HG (±2 ppm)
Humidity (E)	Suf. sampling to yield an accuracy of 0.1 ppm	Suf. sampling to yield an accuracy of 0.1 ppm	Sampling is not nec. Use avg. +0.4 ppm	Sampling is not nec. Use avg. +0.4 ppm	Sampling is not nec. Use avg. +0.4 ppm
EDMI Centering	±0.5 ppm ±1/2 mm	±1 ppm ±1/2 mm	±2 ppm ±1 mm	±3 ppm ±1 mm	±5 ppm ±1 mm
Recommended atmospheric sampling procedure	5	5	5	6	6
Distance measurement procedure					
Eccentric bar 4 positions measurements (0, +, −, 0) 10 REPS each position	Yes	Yes	Yes	Yes	No (10 repetitions on center)
Measure from both ends of the line	Yes	Yes	Yes	No	No
Measure on different days	Yes	Yes	No	NA	NA
Different time of day (morning/afternoon) or (day/night)	Yes	Yes	Yes	NA	NA
Use of conventional tripods	No	No	Yes preferable	Yes	Yes
Use of 4-ft stands	Yes	Yes	Yes	No	No
Minimum distance **	10,000 m†	5000 m†	5000 m	2500 m	1250 m

*Represents the lowest acceptable method.
**EDMI specification of ±(S mm + 1 ppm).
†With frequency monitoring.*Source:* FGCC *Specifications to Support Classification, Standards of Accuracy, and General Specifications of Geodetic Control Surveys,* 1980) Silver Spring, MD.

12-10. CHECK ANGLES

It is good practice to include economically justified angle observation in a trilateration network to increase the number of redundancies within the scheme. For first- and second-order, class I networks, only first-order universal theodolites such as the Wild T-3, Wild T-2000S, or Kern DKM-3 should be employed. The level of precision and accuracy required for this order of work precludes the use of any lower-order instruments. First-order instruments on second-order, class II surveys can be useful and productive if suitable equipment and thoroughly trained personnel are available. The Wild 2000, Wild T-2E, Kern DKM-2AE, Aus Jena 010A, or Lietz TM-1A are representative second-order universal theodolites.

Table 12-3 recommends that 4-ft wooden stands (see Figure 12-4b) be utilized for all first- and second-order trilaterations. Table 12-1 lists the number of positions and least count required in various classes of triangulation to be compatible with the trilateration specifications shown in Table 12-2. Targets used in conjunction with check-angle observations should be good quality, with no phase. Ideally, directional lights or 360° lights should be used exclusively (see Figure 12-13).

On lines where the inclination (vertical angle) of the observed line is over 5° from the horizontal, striding-level readings should be made and corrections applied to the observations for this error. Because of their course divisions, most newer second-order theodolites do not have a sufficiently sensitive plate level to allow meaningful corrections.

Older second-order instruments permit a striding-level vial to sit atop (astride) the theodolite vertical axis. For the most part, modern second-order instruments with automatic verticle-circle indexing do not provide a striding level, but can be precisely leveled using special procedures noted in their respective manuals. These methods employ an automatic vertical-circle indexing provision of the theodolite. First-order geodetic theodolites, such as the Wild T-3 and Kern DKM-3, have plate levels sufficiently sensitive to determine corrections commonly called the "correction for inclination of the standing axis." It is a function of the following:

1. Value of one division on the plate (or striding level) is usually in seconds of arc.
2. Number of graduations the standing axis is out of plumb.
3. Tangent of sight-line inclination.

The correction formula for C is

$$C = d(W - E)\tan h \qquad (12\text{-}1)$$

where

C = correction in arc seconds

d = bubble value in seconds of arc per level-vial graduation

W = arithmetic difference of the direct and reverse readings of the west (or left) end of the bubble

E = arithmetic difference of the direct and reverse readings of the east (or right) end of the bubble

h = vertical angle of the sight line to the target, positive if the station is above the horizon, negative if below

Example 12-1. Given: Arc seconds per vial division $d = 6.4''$, inclination of the sight line $h = 10°20'$, and the following bubble readings, compute W, E, and C.

	Circle Left (Direct)	Circle Right (Reverse)
	+	−
Direct	3.5	30.0
Reverse	28.0	1.0

$W: 28.0 - 3.5 = 24.5$

$E: 30.0 - 1.0 = 29.0$

$W - E = 24.5 - 29.0 = -4.5$

$C = d/4[(W) - (E)]\tan h$

$C = 6.4''/4(-4.5)(0.182)$

$C = -1.31''$

The correction applied to the line (direction) on which the inclination of the standing axis is read. If the correction sign is minus, the correction is subtracted; if plus, it is added.

When bubble readings are taken for inclination correction, it is important that the instrument be kept as nearly level as possible, caused by possible inconsistencies in uniformity of the level vial. For the example given, the dislevelment angle is extreme, so the instrument should be releveled before the next round begins. As noted, divisions should be estimated to 0.1 of its smallest graduation.

The recorder can perform a helpful check to guard against errors in reading the plate and striding levels by noting the difference between the left- and right-hand bubble readings. This indicates the relative *length* of the bubble, which should be consistent for both direct and reverse orientations and not disagree by more than 0.3 divisions. If it does not check, then either an observing or recording error has occurred, and the source should be identified by reobserving the bubble if the error is detected during the occupation.

It is possible for the bubble length to vary during a night's work because of changes in temperature and barometric pressures; however, these changes will be slow and practically unnoticeable. Bubble lengths should remain relatively static, enabling an observer to get a feel for the general trend, so errors in readings are spotted instinctively.

12-11. ZENITH ANGLE OBSERVATIONS

Determining elevation differences by employing zenith angle measurements and slope distance is accurate and rapid, provided that (1) the timing of observations is carefully considered, (2) atmospheric conditions are favorable, (3) good sights and observing techniques are employed, and (4) the second-order (or better) universal theodolite is in good repair and ad-

justment. The times of day when refraction is at its worst should be avoided.

Minimum refraction occurs from noon to 3 PM and the maximum from 9 PM to midnight. During the period from noon to 3 PM, refraction tends to be relatively constant. The poorest times for observing are between 8 and 9 AM, and between 6 and 7 PM, because during these periods refraction is changing most rapidly. When simultaneous reciprocal observations are used, the paired observations at each end should be completed within a total elapsed time of 15 min or less to avoid changing refraction conditions. This technique practically eliminates the effects of curvature and refraction.

The following list should be followed in selecting techniques and parameters for observations. It gives the order of preferred methods with the best listed first:

1. Simultaneous reciprocal observations, noon to 3 PM. (Two instruments, combined ET for all observations is $Z = 15$ min.)

2. Reciprocal observations, noon to 3 PM, different days.

3. Simultaneous reciprocal observations, 9 PM to midnight. (Two instruments, combined ET for all observations is $Z = 15$ min.)

4. Reciprocal observations, 9 PM to midnight, different days.

5. Simultaneous reciprocal observations at any other time. (Two instruments, combined ET for all observations is $Z = 15$ min.)

6. Reciprocal observations, with no coordination of time.

7. Nonreciprocal observations.

- (a) Noon to 3 PM.
- (b) 9 PM to midnight.
- (c) Any other time.

Table 12-2 lists the recommended number of repetitions and rejection limits for each class of trilateration. However, users are advised to use 3 D/R or more and a rejection limit of ± 5 arc seconds or less in all classes.

12-12. TRIGONOMETRIC LEVELING

In order to reduce observed slope distances to a common datum, elevations must be known at a line's terminal points. Since trilateration establishes the position of new stations, elevations are seldom known. The most common method of determining elevations is, of course, differential leveling. It is the most accurate but unfortunately the most expensive. An acceptable alternative to differential leveling is trig leveling. Both methods are very successful when sound specifications and operational procedures are followed (see Section 7 for a discussion of trig leveling and the computation of differences in elevations based on zenith angle and slope distance observations).

Table 12-3 lists the number of observations and allowable spread between zenith angle or vertical angle sets, together with the allowable variation from known elevations.

12-13. ELECTRONIC DISTANCE MEASURING INSTRUMENTS

An EDMI is the heart of any trilateration network. It is the workhorse of any modern surveying project and ultimately controls the raw data secured. EDMIs employed in the United States are generally the light-wave and microwave types. Visible-light-wave EDMIs include the Rangemaster and Ranger series marketed by Keuffel & Esser Company and the geodimeter produced and marketed by the AGA Corporation of Stockholm, Sweden. Nonvisible infrared light-wave types are produced by nearly all EDMI manufacturers.

Microwave is a second type of carrier wave. However, the effects of meteorological conditions on this type of electromagnetic wave are very extensive, compared with visible and non-

visible light waves, so microwave does not have widespread usage in geodetic control surveying. For example, humidity contributes, as a worst possible case, about $\frac{1}{2}$-ppm error to light-wave-type EDMIs, but a comparable level of humidity can cause 70 ppm or more in microwave instruments.

Most short-range (< 3 km) EDMIs on the U.S. market today employ the gallium-arsenide electroluminescent diode (GA-AS diode) as an infrared light source for the projected beam. No separate light modulation is necessary because the desired intensity or pattern of modulation is obtained directly from the diode by RF (reference frequency) controlled voltage. Some medium-range instruments employ a modified GA-AS diode as a semiconductor lasing infrared light source.

12-14. EDMI INTERNAL ERRORS

EDMI manufacturers list an error statement typically in the form $\pm(A + B)$, where A is expressed in millimeters or decimal feet and B in representative parts per million (ppm). These specified errors are usually ± 1 standard deviation (the standard error or one-sigma, σ) that is the error to be normally expected for a single observation when the EDMI is in good working order and properly calibrated. The expression is an informative message from manufacturer to consumer, indicating the relative reliability to be expected from a single observation. The true value of A as determined by base-line testing should generally be equal to or smaller than the stated standard error.

The range of all errors should never exceed 3σ. If this limit is surpassed, it could indicate that the EDMI is in need of repair or maintenance. The unique error statement for an individual EDMI is not static and will change with time due to the aging of electronic components, faulty optical alignment of the instrument caused by an external shock or blow, or a frequency change resulting from mechanical and/or electronic sources. It is imperative that sufficient base-line calibration checks be made and regular shop maintenance scheduled at a factory-authorized repair facility. Normally, a base-line check should be done after any shop work is completed and at the completion of a project.

In considering the two groupings of internal errors inherent in EDMIs, the first one *A* consists of a family of errors present for all measurements within the operating range of the instrument that is not a function of distance. Thus, the error is the same whether the EDMI is measuring 1 m or 6 km. These internal errors are listed as follows:

1. Instrument offset.
2. Cyclic errors (nonlinear).
3. Instrument resolution.
4. Instrument repeatability.
5. Pointing error.
6. Reflector, instrument calibration.

The second group of internal errors *B* results from a drift of the crystal frequency standard used to control the modulation frequency. Several of the more common causes of this shift are due to (1) loosening of an adjustment screw, (2) electronic aging, (3) extreme temperature conditions in the case of a crystal not provided with an oven or a temperature compensating system, and most commonly (4) inadequate warm-up time to allow the crystal and other electronic components to stabilize. For example, laboratory frequency monitoring has shown that inadequate warm-up time for various infrared and lasing instruments can produce frequency errors of up to 4 ppm or more. Usually, a warm-up period of 20 mm is sufficient for frequency stabilization.

Naturally, good judgement must prevail when determining which lines need a full warm-up period and which do not. A line of 10,000 m, for instance, should get full warm-up since an arbitrary error of 4 ppm would represent $10,000 \times 4 = 0.040$ m. A line of 100 m, on the other hand, would produce only 0.0004 m, which is considerably outside the sensitivity range of most EDMIs.

Consider first the *A* group of internal errors. Items 1 and 6 are systematic in nature and should be treated together. Comparison measurements on an NGS-established base line will give an indication of the sign and magnitude of this error, provided that items 2 through 5 are not excessive. Cyclic error is systematic and normally measured before an EDMI leaves the factory. Data are available from some manufacturers on request. Cyclic error is not constant throughout an EDMI's lifespan and will change with age. A repair facility should be available that can accomplish this calibration, as well as determine instrument resolution, repeatability, and pointing errors by applying appropriate tests. The reduction or elimination of these error sources is a project for a factory or repair facility.

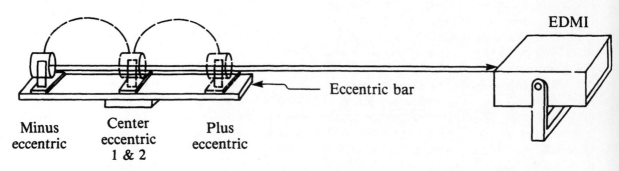

Minus eccentric   **Center eccentric 1 & 2**   **Plus eccentric**   Eccentric bar   EDMI

Figure 12-6.   Eccentric measurements.

Instrument resolution can be tested by using an eccentric bar (Figure 12-6). The EDMI and reflector array are set up at a convenient distance of approximately 300 m and multiple measurements taken to the prism on center, repointing for each shot, while very carefully noting maximum signal response. The prism is moved to a minus eccentric position and a similar series of shots taken, then turned to a plus eccentric position, and finally back to center. An average of the minus eccentric measurements reduced by the offset distance should equal the centers' average. The plus eccentric measurements' average added to the offset distance must also equal the average of the centers. Any difference between the measured center—those centers determined by applying the offset correction to the minus and plus eccentrics—should agree to within a millimeter or smaller. (See Figure 11-14a for a detail of the eccentric bar.)

The measurement procedure and check of conditions in using the eccentric bar are as follows:

1. Order of measurement is CTR1, −, +, CTR2.

2. Each eccentric position is read 10 times.

3. Repoint the EDMI after each set of five shots.

4. Center ECC_1 must equal center ECC_2 within $\pm A$ group error (see Section 12-15).

5. Minus eccentric − the offset distance (0.150 m) must equal center ECC_1 within $\pm A$ group error.

6. Plus eccentric + the offset distance must equal center ECC_1 within $\pm A$ group error.

7. When all three checks are made, all four observations are averaged.

8. If A group errors are shown to be ± 5 mm, then the corrected plus and minus eccentric shots should be equal to or smaller than center ECC_1 ± 5 mm.

Instrument repeatability and pointing error can be checked during the same setup for instrument resolution. The repeatability test is performed by pointing the instrument for maximum signal response. Then activate the EDMI and record a series of multiple measurements without disturbing the instrument pointing. There would be no change of distance on the display greater than $\pm$ one least-count unit. Statistically, 95% of the measurements should remain unchanged, with 5% having no greater difference than $\pm$ one least-count number on the display.

Pointing error is an apparent change in the distance display caused by the bundle of light rays returning to the EDMI not focusing at the proper location on the diode. The technique for determining the magnitude of pointing error is shown in Figure 12-7.

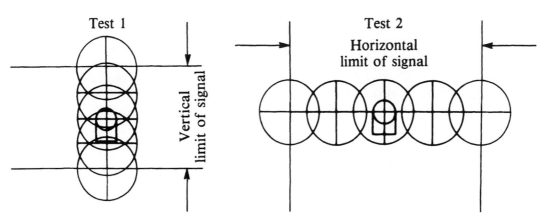

Figure 12-7. Magnitude of pointing error.

Test 1 begins by maximizing the signal horizontally with the horizontal cross hair below the prism. Using the vertical motion, drop the cross hair until the signal fades out. Then raise the vertical wire until the signal is strong enough to cycle the EDMI and record the distance and vertical angle (or vote the vertical position of the cross hair with respect to the prism). Continue to test with approximately five distance measurements until the signal is lost on the upper side of the prism. Test 2 is identical to test 1, except the signal is maximized vertically, and the EDMI turned in azimuth left to right with the horizontal motion. Distance measurements are recorded, and the vertical cross-hair position is noted with respect to the prism.

If there is a change in the distance readout within ± one prism width or height, special pointing techniques should be standardized for the EDMI to normalize all distance measurements. It is very important that the pointing technique used in a repair facility for calibration be identical to the procedure employed on the base line and in the trilateration net. The technique chosen should be mutually decided on by the repair technician and operator in order to capitalize on the unique design characteristics of an individual EDMI and the way the instrument will be utilized in the field.

A technique that proved very successful on the HP 3800A and HP 3810B, and should work on all EDMIs of similar design for systematic pointing, is as follows: (1) Using the vertical wire, slit the prism (or center the vertical cross hair on the maximum signal, using the horizontal motion) and drop the cross hair vertically until the signal stops. (2) Elevate the cross hair very slowly with the vertical motion only; watch the signal return and stop at the point of the *first* maximum signal return.

If test 1 and 2 do not show a change in distance with ± one prism width or height, the standard pointing technique recommended by the manufacturer should be followed.

It is not necessarily possible to totally eliminate the first group of internal errors (group A), but they can be minimized. Items 1, 2, and 6 are systematic and may approach zero, but 3, 4, and 5 can be reduced by a savvy repair technician. If item 5 is a problem, it can be eliminated with better pointing techniques.

The second element to consider is the *B* group of internal errors—those caused by a drift in the frequency standard of the crystal used to control the modulation frequency. If the instrument is thoroughly warmed-up and the crystal frequency is part of a temperature compensating system, the frequency error is minimal; however, all frequency errors are systematic. The resulting distance error can be corrected if the magnitude of frequency error is known. On higher-order projects, it might be advisable to monitor the frequency periodically throughout the job, with an appropriately sensitive frequency counter. On other projects, checking the frequency at its beginning and end is sufficient.

A good policy to follow immediately prior to initiating a trilateration project is to send the EDMI to a reputable shop for maintenance and calibration, with special emphasis given to the items just discussed. After shop work, the next step in the calibration sequence is a trip to an NGS-established base line for further testing and calibration. It is imperative that the same techniques of distance measurement are utilized on the base line as on the trilateration network, preferably with the same prism or prism type and the same observer.

A series of measurements is taken (usually 10) with repointing after each group of five in all four eccentric positions of the eccentric bar (center, minus, plus, then center). After the first line is measured, the prism is advanced to the next baseline monument. When every line has been measured and all refractive index corrections and systematic errors are applied, each measured distance is compared with that established by the NGS and the difference computed for every line. The resulting error is tabulated and a standard deviation computed.

If a systematic error remains, it will show up by analyzing the variances. The scatter should be smaller than ±5 mm, and roughly an equal number of lines will measure short as long. If the variances are predominately plus or predominately minus, the error plot should be adjusted up or down to establish a line centered on zero error. The distance a graph moved up or down is then treated as another total offset correction for measured distances.

The error plot should not have a sloping shape—i.e., an error with respect to distance showing a rate of change, since this suggests the presence of a frequency error.

Selecting an EDMI for trilateration is not a simple task: There are many variables to consider, such as the following:

1. The range of distances within which the EDMI will be utilized.
2. The manufacturer's stated error for the EDMI.
3. Reliability and reputation of the manufacturer.
4. Durability and accuracy of the EDMI, as reported by users and repair facilities.
5. Temperature range within which the EDMI will maintain its design error specification — i.e., is its crystal temperature compensated to maintain frequency standard through a wide range of temperatures?
6. Price.

A common error specification claimed by various manufacturers for medium-range infrared EDMIs is ±(5 mm + 5 ppm). The EDMI specifications are not adequate for third-order, class II trilateration as demonstrated in Example 12-2.

Example 12-2. Can an EDMI with a specification of ±(5 mm + 5 ppm) be utilized on a third-order, class I trilateration project in which the minimum distance is 3000 m?

The answer is no! Referring to Table 12-3 under ppm (total error budget), we can see

that third-order, class I has a total error budget of ±4 ppm. The *B* group error statement of ±5 ppm as claimed by the manufacturer already exceeds the allowable error of ±4 ppm for third-order, class I, without even considering the *A* group errors, centering errors, or refractive index correction (RIC) errors.

Example 12-3. Can the same EDMI be used on a third-order, class II project with the minimum distance being 3000 m?

If we assume that the maximum EDMI error does not exceed the manufacturer's claimed error statement, and further, that the ppm's listed for T, P, e^1, and centering, etc., per Table 12-3 are held constant, then

STEP 1. Convert the *A* group error to ppm and combine with *B* group error.

$$\text{ppm}_{\text{EDMI}_E} = \pm(5 \text{ mm} + 5 \text{ ppm})$$

$$A \text{ group error}_{\text{ppm}} = \frac{0.005 \text{ m} \times 10^6}{3000 \text{ m}}$$

$$= 1.67 \text{ ppm}$$

$A \text{ group error}_{\text{ppm}} + B \text{ group error}_{\text{ppm}}$

$$= \pm(1.67 \text{ ppm} + 5 \text{ ppm})$$

$$= \pm 6.67 \text{ ppm}$$

Therefore, $\text{ppm}_{\text{EDMI}_E}$ at 3000 m = ±6.67 ppm.

STEP 2.

Σppm_E

$$= \sqrt{\text{ppm}^2_{\text{EDMI}} + \text{ppm}^2_T + \text{ppm}^2_P + \text{ppm}^2_e 1 + \text{ppm}^2_{CE}}$$

where Σppm_E is the total error budget in ppm, ppm_{EDMI} the error statement for the

Table 12-4. Parts per million changes in length for pressure and temperature

1 ppm change in length	Pressure error or change of $\pm 0.10''$ HG $\pm 100'$ ± 2.5 mm HG Temperature error or change of $\pm 1.8°F$ $\pm 1.0°C$

EDMI as determined by baseline calibration and laboratory measurement of frequency, ppm_T the estimate of temperature measurement uncertainities in ppm (see Table 12-4), ppm_P the estimate of barometric pressure uncertainities in ppm (see Table 12-4), $ppm_e 1$ the estimate of vapor pressure uncertainities (humidity) in ppm (see Table 12-4), and ppm_{CE} the centering error in ppm (see Tables 12-3 and 12-5).

$$\left(\frac{10^6 \sqrt{0.001^2 \text{m} + 0.001^2 \text{ m}}}{3000 \text{ m}} \right.$$

$$\left. = \text{ppm for } \pm 1 \text{ mm centering error} \right)$$

$$\Sigma ppm_E = \sqrt{6.67^2 + 2^2 + 2^2 + 0.4^2 + 0.5^2}$$

$$\Sigma ppm_E = \pm 7.3 \text{ ppm}$$

The EDMI specifications are not adequate for third-order, class II trilateration. An inspection of the data shows that the element contributing most to the high error of ± 7.3 ppm is the *B* group specification concerning frequency. If it is possible to lower this source of error by reading the actual frequency and calculating corrections, or adjusting the frequency to near perfection, this EDMI would be capable of both third-order, class II and class I measurements at a minimum distance of 3000 m.

If the error specification can be adjusted to $\pm(5 \text{ mm} + 1 \text{ ppm})$, the following improvement could be realized:

$$ppm = 2.67^2 + 2^2 + 2^2 + 0.4^2 + 0.5^2$$

$$pp = \pm 3.94 \text{ ppm}$$

The new ppm resulting from improving the error specification of the EDMI now has shown dramatic results, improved its predicted per-

Table 12-5. Distance versus centering error

Distance (m)	Centering Error of ± 1 mm		Centering of $\pm 1/2$ mm	
	$ppm_{CE}^{\pm}$	Error Ratio $1_\pm$ Part in ...	$ppm_{CE}^{\pm}$	Error ratio $1_\pm$ Part in ...
10,000	0.1	7,071,068	0.1	14,142,136
5000	0.3	3,535,534	0.1	7,071,068
2500	0.6	1,767,767	0.3	3,535,534
1000	1.4	707,107	0.7	1,414,214
500	2.8	353,553	1.4	707,106
300	4.7	212,132	2.4	424,264
200	7.1	141,421	3.5	282,842
100	14.1	70,711	7.1	141,422

formance in line measurement, and conforms to third-order, class I specifications.

12-15. EDMI EXTERNAL ERRORS

A number of factors causing errors in EDMI measurement will be discussed.

12-15-1. Refractive Index Correction

Table 12-4 lists the magnitude of errors in pressure and temperature sampling that cause a ±1-ppm change in the distance. Chapters 4 and 6 deal with the equations for determining the refraction index correction resulting from changes of temperature, atmospheric pressure, and water vapor pressure, from the standard. A discussion of these equations will not be repeated. However, a discussion on sampling techniques will be addressed.

The accuracy and precision of an EDMI measurement are a direct function of how carefully the atmospheric conditions are determined along the beam path. It is typical procedure to simply measure temperature at the EDMI height and pressure with an inexpensive barometer, probably uncalibrated, and ignore the effect of water-vapor pressure. This method will not be effective for even the lowest order of trilateration unless distances are very short.

12-15-2. Temperature Measurement

Measuring temperatures with a thermometer gives figures only for the instrument and not necessarily the air temperature. Unless shielded from sunlight, a thermometer is heated by radiation and shows an erroneous temperature. A thermometer carelessly hung on a low bush can pick up thermal radiation from the ground, as will dangling it from the survey truck's door handle or exterior mirror.

Nearby heated objects affect temperature readings. If we assume that the thermometer is shielded from radiation, standardized, and the sight line from the EDMI to the reflector is equally distant above the ground throughout the beam length, sampling tripod-height temperatures at both ends of a line should be adequate. Shorter distances will ensure smaller temperature errors. Sight lines are seldom a constant height above the ground. Within the first 30 ft above the ground, temperatures can vary as much as 9°C (16.2°F) or more. This corresponds to a 9-ppm error if a ground temperature is erroneously used.

Isotherms are imaginary lines connecting points of equal temperature. A characteristic of an isotherm is that it generally tends to follow the ground profile. Typically, an isotherm for a given temperature is likely to be at a lower height above the ground on a hilltop than on the valley floor. Figure 12-8 is a graphic representation of this phenomenon.

In most situations, the temperatures will be more accurate if taken on each end of the line at a height of 25 to 30 ft above the EDMI and reflector. The most representative temperature samplings are obtained on overcast days, with light to moderate winds mixing the atmosphere.

The primary source of error in determining the *index of refraction correction* is the difference that exists between the observed temperature and actual average temperature over the line's length. A typical sight line passes from one high point to another, or from a high point to a low point at variable heights above the earth's surface. Since these variations in height above the ground occasionally have large differences and least since in atmospheric temperature varies with altitude, temperature measurements at the end points may not create an accurately representative model. Hence, scheduling field operations during periods when meteorological conditions are more conducive to accurate temperature sampling will provide improved representative temperature models.

Figure 12-9 charts temperature versus height above the ground surface and time of day in

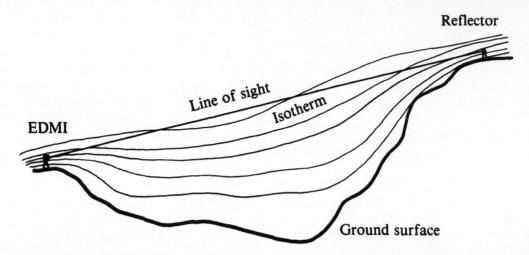

Figure 12-8.   Isotherms.

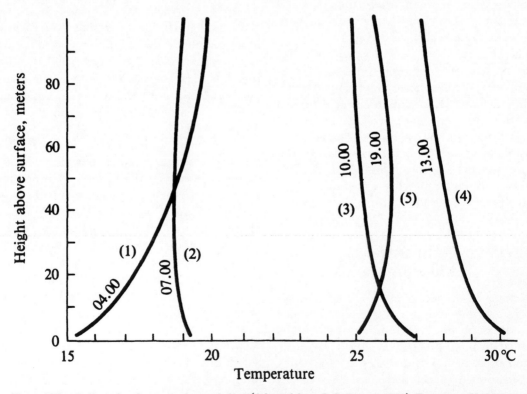

Figure 12-9.   Daily cycle of temperation variation. (Adapted from P. S. Carnes, 1961). *Temperature Variations in the First 200 Feet of the Atmosphere in an Arid Region*, Missile Meteorology Division, U.S. Army Signal Missile Support Agency, New Mexico.

lowest 100 m at White Sands Missile Range, New Mexico. At an air sampling height of 10 m—e.g., an average line height above the ground of 40 m—the largest error in temperature during the daytime is 1°C (1 ppm) around 1300 hr, and the smallest error occurring in the daytime is $\frac{1}{4}$°C ($\frac{1}{4}$ ppm) around 0700 hr.

Given overcast conditions with sufficient wind to stir and mix the air, the graph approaches 0700, as shown in Figure 12-9, where near-ground temperatures in the 10 m range would closely approximate those temperatures much higher up (to 80 m ±). If elevation differences are extreme and weather conditions not conducive to good sampling at 10 m, other tactics must be employed to determine the refractive index corrections, such as a two-color laser EDMI, multiwavelength distance-measuring instrument (MWDMI); the ratio method; or aircraft monitoring, etc. Using temperature-sampling procedures at the 10-m height should suffice for second-order, class II and lower.

The following conditions should be considered before writing specifications for field operations:

1. Night temperatures will tend to be cooler near the ground than above it. Consequently, if the average height of the sight line above ground is higher than the sampling of height, temperature readings taken at night under clear skies will be cooler than the true average sight-line temperature, resulting in EDMI distances that are shorter than the true lengths.

2. Daytime temperatures tend to be warmer near the ground than higher up. If the average height of the sight line above ground is higher than the sampling height, temperature readings taken during the daytime under sunny skies will be warmer than the true average temperature along the sight line. Therefore, the resulting EDMI distance is longer than the true distance.

3. Overcast days or nights, with moderate winds, provide the best conditions for accurate temperature sampling.

4. Temperature and pressure samplings are most representative when taken along the line of sight.

The following list notes the preferred methods of sampling the atmosphere to determine the refractive index correction. Items tallied are shown in the order of accuracy from best to worst. The methods correlate with Table 12-3.

1. MWDMI.

2. Ratio method.

3. Aircraft monitoring of temperatures and pressure along the line of sight.

4. Balloon-suspended temperature thermistors.

5. Thermistors and aspirators erected on a mast at the EDMI and reflector.

6. Temperature and pressure samplings at the height of the EDMI and reflector.

7. Temperature and pressure sampling at the EDMI only.

Temperature readings can be taken with thermometers, but such use is not recommended except in measuring relative humidity. Portable digital-reading battery-operated temperature indicators manufactured by Weathertronics Qualimetrics, Inc. are relatively inexpensive. Such temperature sensors are accurate to ±0.1°F or 0.16°C and have ventilated sensors and cables that can reach short distances, up to 50 ft. Also available are similar units that measure both temperature and relative humidity.

Some fabrication is necessary to properly utilize the temperature-sensing equipment. First, an appropriate aspirator must be designed that has the following features: (1) circulates air past the temperature sensor, (2) shields the sensor from solar radiation, (3) is lightweight, and (4) is mountable on an extending mast. Second, an appropriate extending mast must have the following features: (1)

is lightweight, (2) is extendable to 30 ft or higher, and (3) is mountable on a truck bumper or pickup bed.

12-15-3. Aspirator

A design was developed incorporating the use of a 5-in. PVC elbow with a reducer to 4 in. on one end and 12-in. section of 4-in. PVC glued into the reducer. A 12-V DC motor with fan was mounted inside the elbow and a piece of $1\frac{1}{2}$-in. PVC was glued to the 12-in section of the 4-in. PVC to act as a receiver for the mast. Then, all were covered with insulating foam to guard against solar radiation. A hole was bored to receive the temperature-indicator sensor and placed to position it in the aspirator's airstream. The motor was powered via a long extension attaching to the truck battery with alligator clips; it is designed to exhaust from the elbow when the sensor was placed upwind from the motor (see Figure 12-11).

Figure 12-10.   12-V aspirator; extended mast; truck-mounting assembly; meteorological box.

12-15-4. Mast

The mast is an aluminum extension pole originally designed as a handle for a tree pole saw. It consists of five or more sections of aluminum tubing, 5 to 6 ft in length, with different diameters to permit the unit to "telescope." Each section is held aloft by a friction collar. (See Figure 12-10.) Use caution when extending the mast. *Do not extend the mast when in the vicinity of power lines.*

12-15-5. Truck Mounting

A piece of $1\frac{1}{2}$-in. pipe, 36 in. long with a $1\frac{1}{2}$-in. floor flange bolted to a truck bumper, front or rear, or bolted to the bed of a pickup, serves as a holder for the mast and aspirator. The aspirator, mast, and truck mount can be assembled using locally available materials. The temperature assembly described has been used, under ideal conditions, with good results in performing first-order trilateration; however,

the project was a minigeodetic network consisting of lines no longer than 4 km and non-critical changes in elevation. The described system is adequate for second-order, class II and lower-order surveys. Under favorable atmospheric conditions—overcast, light-to-moderate winds—the system can exceed second-order, class II specifications. (See Figure 12-11.)

12-15-6. Barometric Pressure Measurement

Barometric pressure must be measured at each end of the line with a sensitive, high-quality temperature-compensated barometer. Also, the barometer should reliably differentiate the least division shown in Table 12-3 under atmospheric sampling accuracy requirements for barometric pressure. The barometers should be periodically calibrated against a mercury column, preferably at the beginning and end of the project—more often if the

Figure 12-11. Another example of instruments shown in Figure 12-10.

project lasts longer than several weeks or the barometers have been subjected to hard usage. (See Figure 12-12.)

During the project, a daily check should be made on all barometers used for line measurement operations to determine if they are reading the same ($\pm$ one least count). For example, if one of the instruments does not agree with the others, it can be safely reindexed in the field to match them. If more than one barometer reads differently, all should be recalibrated before proceeding with the project. Surveying barometers are delicate precision instruments and should be shaded from the sun and treated with the same care due any scientific equipment.

12-15-7. Humidity

The effect of humidity on visible light and infrared EDMIs is admittedly small and does not exceed 1×10^6 under the worst conditions. However, second-order, class I and first-order trilateration must have atmospheric observations made at each end of the line to determine the size of correction for water-vapor pressure e^1. The equipment is inexpensive and consists basically of two thermometers, one a regular unmodified thermometer and the other adapted to include an apparatus, usually a sleeve with an absorbent material that is moistened with water.

Both thermometers are usually aspirated by a small battery-powered fan to evaporate the water and cool the thermometers to an appar-

Figure 12-12. Meteorological box containing surveying barometer, thermometer, and electronic temperature-sensing equipment with 40-ft cable.

ent lower temperature when a difference between the thermometers is noted. A correction is then computed by the equation given in Section 12-4. A choice is available between a conventional sling psychrometer or newer state-of-the-art portable battery-powered types with digital readout and electronic sensor.

12-15-8. Electronic Interference

Care must be taken to avoid potential EDMI errors when operating in the proximity of transmission lines, microwave, or radio communication. Radio transmission also can severely affect electronic atmospheric-sensing equipment, so all such transmission should be shunned as well during temperature measurements.

12-15-9. Reflector Correction

Reflector correction for individual differences between units due to slight differences in their offsets has been discussed previously in Section 12-14. The reflector correction is both a systematic and an external error, relative to the EDMI, with a sign that can be either plus or minus.

12-16. REFLECTORS

Any reflecting surface can be used from a common shaving mirror to precision-ground planoparallel optical surfaces, highway reflectors, reflective tape, mercury surfaces, and, of course, the conventional retroreflecting prism, provided that the following conditions are met: (1) The exact reflector constant/offset is known, (2) the reflected light source is sufficiently strong to enable an EDMI to function within the expected accuracy and precision standards, and (3) the reflected beam is free of ambiguous stray light from a surface other than the intended interior ones of the reflector.

Typically, surfaces such as reflective tape, plane mirrors, and highway reflectors are offset no more than 1 mm. On the other hand, retroreflective prisms vary from -40 mm to -70 mm to -30 mm, depending on the manufacturer. It must be emphasized that these offsets are *nominal*. The true offset can be in disagreement by as much as ± 1 mm to ± 3 mm by actual measurement. If the project standards—i.e., error budget—can tolerate this uncertainty, then by all means disregard the true offset and employ the nominal. However, it is prudent surveying practice to know the exact magnitude of an individual reflector offset. This can be measured in combination with calibration tests given an EDMI while on the test base line, but only if the EDMI has sufficient resolution to measure a least count of ± 0.001 ft, and can reliably detect relative displacements of the same size.

The HP 3800 has been utilized for this purpose, with good results. This test is performed at relatively short distances of about 300 ft.

The same reflector submitted with an EDMI for laboratory calibration, zeroing (making the total offset correction equal zero), and later taken to a base line for validation measurements is selected for standardizing all other reflectors which need calibration. After approximately 20 measurements are made to the calibrated reflector, all others are inserted successively and the measurements repeated. The difference between the mean of the first measurements using a calibrated reflector is subtracted from those taken with each individual unit. The result is a reflector correction *RC* added to or subtracted from the slope distance to correct for the offset difference from the standardized prism.

The next step is to (1) run the trilateration EDMI and calibrated reflector through an NGS base line to validate the total offset correction *TOC* established during laboratory calibration earlier or (2) determine the magnitude and sign of the *TOC* for the EDMI and calibrated reflector as a unit in the event there has been a change.

Multiple shots are taken, with the EDMI being repointed after each distance measurement with a sufficient number of repetitions read to arrive at a realistic mean value. Then the difference is computed between the mean of the measured distance minus the NGS-established distance published for the base line. This results in one *TOC* determination for each base-line distance. Several different distances on the base line must be measured to ensure that the best value of the EDMI/reflector combination *TOC* has been determined. Once the *TOC* and reflector correction for the calibrated reflector are known, the corrected slope distance for a line is computed by the following relationship:

Corrected slope distance = measured slope distance
$$\pm\ TOC \pm RC$$

Each reflector should be individually numbered for identification. The corresponding value of *RC* established for the specific reflector is then recorded for future use. Usually, it will be observed that all reflectors from a manufacturer tend to have *RC*s approximately the same size. Differences of as much as 7 mm have been found between reflectors of different manufacturers, although each claims to have the same offset. Caution, therefore, should be exercised when mixing reflectors, unless the *RC* value is known.

If maximum-rated distances are to be measured, the reflectors must be premium quality, clean, reasonably accurately pointed, and in good repair. For lines inclined more than 15°, the use of tilting prisms is recommended.

Centering an EDMI and reflectors over the marks is extremely important and requires the tribrachs to be in perfect adjustment; otherwise, precision calibration will be meaningless. Centering the reflector over a sufficient number of base-line monuments ensures getting the best possible value for the instrument/reflector combination *TOC*.

12-17. CENTERING

The Wild-type tribrachs with attached optical plummet, manufactured by Wild, Geotech, Lietz, Topcorn, and others, are very strong,

well-designed, and durable. However, unless reasonable care is taken in transporting and day-to-day handling during field usage, the bull's-eye bubble and mechanical adjustment system of the optical plummet can be jarred out of calibration. Extreme variations in temperature contribute to shifts in the tribrach zero, owing to extremely unbalanced tension on the adjusting screws. Read the manufacturer's adjustment instructions carefully and then follow them precisely. The tolerable centering error ± 1 mm. (0.003′) more or less, depending on the class of survey and line length (see Table 12-5 on page 252).

Since centering errors are only a part of the total error allowed in distance measurements, it naturally follows that the influence of centering errors on the total must be held to a minimum. Table 12-5 serves as a guideline for determining when distances become critical in considering the effect of centering and establishes centering standards for project control specifications. For instance, Tables 12-2 and 12-4 show that trilateration standards for a second-order, class II distance of 500 to 1000 m *might* be attained if the centering is done to $\pm 1/2$ mm or lower. As a practical note, it is very difficult to adjust a Wild-type tribrach finer than 1/2 mm because the cross hairs on some optical plummets appear nearly 1/2 mm wide at nominal tripod heights.

If extremely high precision is required on short lines, necessitating a centering error of 0.5 mm or smaller, operators should incorporate a centering-rod-type system instead of the familiar tribrach style with a fixed optical plummet. Centering-rod types are currently manufactured by both Wild and Kern. Other tribrach designs allow the optical plummet to be rotated in azimuth while viewing the cross hair, and they can be centered to within 1/2 mm without much difficulty.

The following is a list of general centering guidelines:

1. Check the bull's-eye level bubble optical plummet twice each week or more often as conditions dictate.

2. Tribrachs should be stored and transported —even from station to station—in a suitably protective container that safeguards the tribrach from moisture, physical shocks, and vibration.

3. Do not press the surface of the bull's-eye bubble or touch the vial unless the bubble is being adjusted.

4. When leveling for final centering, look squarely down on the bull's-eye to avoid any parallax and make absolutely sure the bubble is concentric with the reference circle on the glass.

5. If a tribrach is dropped or given any form of physical shock, it must be checked before being used for a distance or angle measurement.

6. Follow the manufacturer's instructions for centering adjustment. Pay particular attention to the tightening and loosening sequence for capstan screws and locks.

7. In some cases where high precision is important, the use of a level vial, such as one in a tribrach carrier or theodolite, can be utilized for leveling and centering the tribrach. This is always superior to using the bull's-eye on the tribrach.

Example 12-4. Given a centering error of ± 1 mm and line of 5000 m in a length, what is the probable effect of centering errors, expressed in parts per million (ppm) and error ratio (ER)?

To find a solution, two setups are required for each line: one at the EDMI station and another at the reflector station. If we employ the standard statistical method of error propagation, which states that the final error is equal to the square root of sums of the squares of the individual errors, the following expressions apply:

$$\text{ppm}_{CE} = \frac{10^6 \sqrt{CE_1^2 + CE_2^2}}{D} \tag{12-2}$$

$$ER = \frac{D}{\sqrt{CE_1^2 + CE_2^2}} \tag{12-3}$$

where ER is the error ratio expressed in the familiar form " 1 part in..." resulting from the assigned centering error, ppm_{CE} the parts per million resulting from the assigned centering error, D the measured distance, CE_1 the assigned centering error of the EDMI, and CE_2 the assigned centering error of the reflector. Then for $D = 5000$ m,

$$\text{ppm}_{CE} = \frac{10^6 \sqrt{0.001^2 \text{ m} + 0.001^2 \text{ m}_c}}{5000 \text{ m}}$$

$$= \pm 0.28 \text{ ppm}$$

$$ER = \frac{5000 \text{ m}}{\sqrt{0.001^2 \text{ m} + 0.001^2 \text{ m}}}$$

$$= 3,535,534 \text{ or } 1 \text{ part in } 3,500,000 \text{ parts}$$

Table 12-5 tabulates a centering error of ± 1 mm and $\pm 1/2$ mm, respectively, utilizing Equations (12-2) and (12-3) with distance as the argument.

12-18. INSTRUMENT SUPPORTS

NGS-type instrument supports should be employed for first-order and second-order, class I control surveys. These so-called 4-ft stands are economical and easy to fabricate and provide a superbly stable instrument and target base for measuring both distances and angles. (See Figure 12-14b for construction details.) According to Table 12-3, standard tripods are advisable for only third-order, class I, class II and perhaps second-order, class II projects. If a theodolite setup is required for check-angle measurements on second-order, class II, or third-order, class I and II, it is advisable to construct a 4-ft stand for theodolite occupations.

It is possible to forego the use of stands in pure trilateration, if available conventional tripods are (1) in good repair; (2) in good adjustment; and (3) substantial and sturdy enough to support the EDMI, reflectors, etc., without displacement from weight and/or wind.

If any angle observations are contemplated, 4-ft stands should be built. The increased accuracy and precision, plus ease of setting up, will offset the time and cost of their construction. An additional advantage of employing the stand is that its height above every mark remains constant throughout the project. Thus, an instrument or reflector height always equals the height of the stand above the mark plus the incremental height of the instrument or reflector above the stand. After the stand height has been measured and checked, it need not be remeasured throughout the project, as would be required each time a new setup is made with a conventional tripod.

It is always good practice to measure all heights in feet and meters, independently, to provide a necessary cross-check. It is not difficult to obtain figures that agree within ± 1 mm (0.003 ft). High-order, small-scale precise trilateration networks (minigeodetic networks) of 2 mi or shorter require precision in measurement of the instrument and reflector heights above the marks. In trig leveling, the accuracy of HIs and HSs are especially critical on short lines. The heights of instruments and sights should be taken to ± 1 mm. On lines longer than 2 mi, the requirement can be relaxed to ± 1 cm or more, depending on precision needs.

12-19. ACCESSORIES

If 4-ft stands are used for instrument supports, several items are needed to attach the angle- and distance-measuring equipment to the stand. The suggested designs shown on Figures 12-13 and 12-14a have evolved through many years of trial and error and are patterned from equipment manufactured by the NGS.

Tribrach Plate

The tribrach plate is attached by wood screws directly to the wooden cap of the stand. Before fastening it, the plate is centered exactly over the monument with a vertical colli-

mator or the optical plummet of a tribrach, and then screwed down and checked. The plate is constructed of $\frac{3}{8}$-in. aluminum stock and cut triangular in shape, 14 in. on each side, with a $\frac{5}{8}$-in. hole drilled slightly oversized in the center of the plate. Additionally, six holes are drilled around the perimeter and countersunk to allow loose passage of no. 8 wood screws, while permitting firm attachment of the tribrach plate to the stand's wooden cap.

Attachment Bolt

The attachment bolt is a $\frac{5}{8} \times 11 \times 1$-in. hex-head bolt, which serves as a means of attaching all prisms, lights, and tribrach of the eccentric bar to a tribrach plate. The only machining necessary is a $\frac{5}{16}$-in. hole bored through the bolt from end to end (Figure 12-9). This hole allows the optical plummet sight line on the tribrach to pass through the bolt and permit viewing the mark below.

Eccentric Bar

The eccentric bar is constructed of $\frac{5}{8} \times 2\frac{3}{4} \times 15$-in. aluminum stock (Figures 12-10 and 12-11). Three $\frac{5}{8}$-in.-diameter holes are bored on the centerline at a spacing of 0.492 ft. (0.150 m). The spacing of 0.150 m was chosen to keep the bar a manageable size and weight. The purpose of the eccentric bar is to (1) provide redundant measurements, (2) provide an internal check on the EDMI, (3) detect blunders, and (4) check EDMI resolution.

Stud Bolt

The purpose of a stud bolt is to secure the prism in the eccentric bar. It is a $\frac{5}{8} \times 11$-in. threaded bolt $1\frac{1}{16}$-in. long, turned to a proper diameter to allow a slip-fit into each hole in the eccentric bar. The slip-fit must not permit any wobble or lateral movement of the prism. The bolt is designed with one threaded end to screw into a prism case, and one unthreaded end to allow the prism to be lifted vertically and changed from one eccentric position to another.

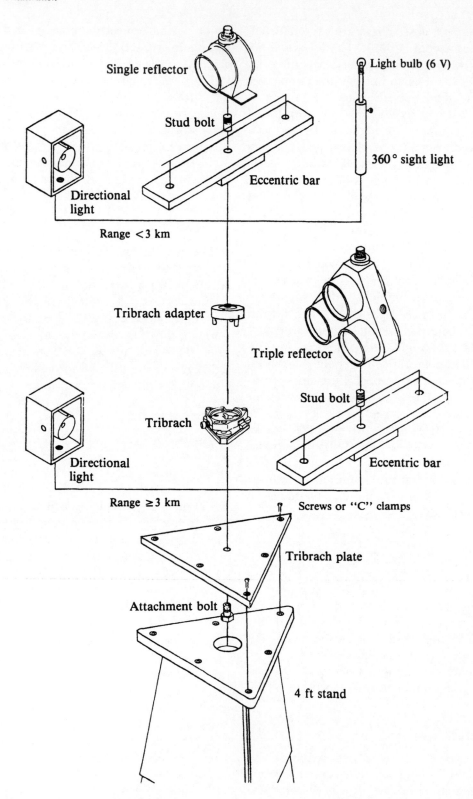

Single reflector

Light bulb (6 V)

Stud bolt

360° sight light

Directional
light

Eccentric bar

Range <3 km

Tribrach adapter

Triple reflector

Stud bolt

Tribrach

Directional
light

Eccentric bar

Range ≥3 km

Screws or "C" clamps

Tribrach plate

Attachment bolt

4 ft stand

Figure 12-13.   Survey target system.

TARGET SYSTEM DETAIL

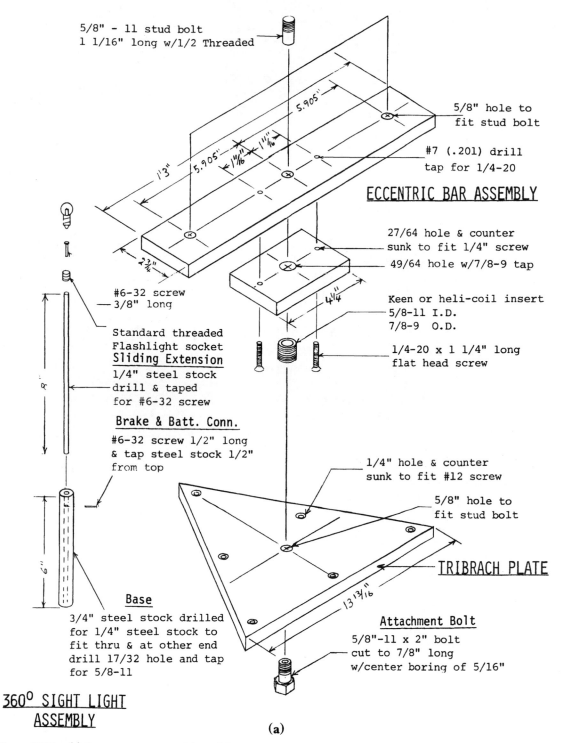

5/8" - 11 stud bolt
1 1/16" long w/1/2 Threaded

5.905"

5.905"

1'3"

2 3/4"

5/8" hole to
fit stud bolt

#7 (.201) drill
tap for 1/4-20

ECCENTRIC BAR ASSEMBLY

27/64 hole & counter
sunk to fit 1/4" screw

49/64 hole w/7/8-9 tap

4 1/4"

Keen or heli-coil insert
5/8-11 I.D.
7/8-9 O.D.

1/4-20 x 1 1/4" long
flat head screw

#6-32 screw
3/8" long

Standard threaded
Flashlight socket
Sliding Extension
1/4" steel stock
drill & taped
for #6-32 screw

Brake & Batt. Conn.
#6-32 screw 1/2" long
& tap steel stock 1/2"
from top

1/4" hole & counter
sunk to fit #12 screw

5/8" hole to
fit stud bolt

TRIBRACH PLATE

13 13/16"

Base
3/4" steel stock drilled
for 1/4" steel stock to
fit thru & at other end
drill 17/32 hole and tap
for 5/8-11

Attachment Bolt
5/8"-11 x 2" bolt
cut to 7/8" long
w/center boring of 5/16"

360° SIGHT LIGHT ASSEMBLY

(a)

Figure 12-14.   (a) Survey target system details. (b) Stand assembly.

4 FOOT STAND

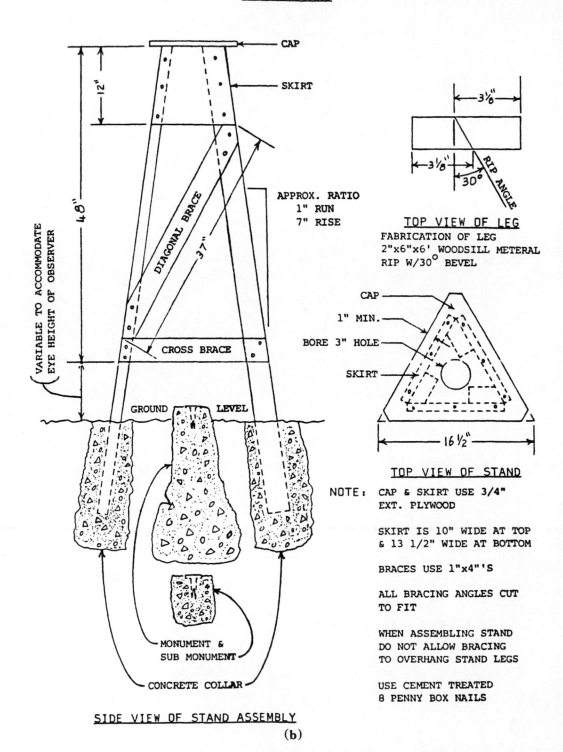

VARIABLE TO ACCOMMODATE
EYE HEIGHT OF OBSERVER

12"

48"

CAP

SKIRT

DIAGONAL BRACE

37"

APPROX. RATIO
1" RUN
7" RISE

CROSS BRACE

GROUND   LEVEL

MONUMENT &
SUB MONUMENT

CONCRETE COLLAR

SIDE VIEW OF STAND ASSEMBLY

3⅛"

3⅛"

30°

RIP ANGLE

TOP VIEW OF LEG

FABRICATION OF LEG
2"x6"x6' WOODSILL METERAL
RIP W/30° BEVEL

CAP

1" MIN.

BORE 3" HOLE

SKIRT

16½"

TOP VIEW OF STAND

NOTE: CAP & SKIRT USE 3/4"
EXT. PLYWOOD

SKIRT IS 10" WIDE AT TOP
& 13 1/2" WIDE AT BOTTOM

BRACES USE 1"x4"'S

ALL BRACING ANGLES CUT
TO FIT

WHEN ASSEMBLING STAND
DO NOT ALLOW BRACING
TO OVERHANG STAND LEGS

USE CEMENT TREATED
8 PENNY BOX NAILS

(b)

Figure 12-14. (*Continued*).

360° Sight Light

The 360° sight light is designed to be used as a triangulation target. The light is powered by a 6-V battery source with a rheostat-controlled power lead. It provides a very superior target for short- to medium-length lines. The maximum distance for daytime use is approximately 1 km; for nighttime work, approximately 6 km or more with full power.

The sight light consists of two basic parts: (1) the base and (2) a sliding extension. The base, a $\frac{3}{4}$-in.-diameter steel stock, bored $\frac{1}{4}$-in.-diameter longitudinally and on one end bored and threaded to $\frac{5}{8} \times 11$ in. for attaching to a standard tribrach adaptor, or for connecting directly to the tribrach plate by the attachment bolt. Additionally, the base is drilled and tapped for a no. 6×32 screw, which is used for a brake on the extension, and as an electrical pole for one side of the battery. The second principal part is a sliding extension, turned for a smooth slip-fit into the base, with one end drilled and tapped to receive the 6×32 screw attaching the threaded flashlight socket to the extension.

On long lines, where centering can be more lax, the show lights or reflectors may be bolted directly to the tribrach plate if the plate is sufficiently level to prevent eccentricities detrimental to the desired classification standard being followed. Typically, the 4-ft stand cap should be set level when the stand is installed, regardless of the project accuracy requirement.

These accessories are inexpensive to fabricate and require only simple machining any home craftsman can do. The aluminum stock is available in sheet form and can be sheared or cut to the proper size for a nominal charge at most sheet-metal facilities.

12-20. DISTANCE REDUCTION AND TRILATERATION ADJUSTMENTS

The two adjustments commonly used in trilateration are (1) condition equations involving differences in angles or areas and (2) the variation of coordinates method. For low-order networks 1 : 15,000 or lower, adjustments similar to the compass and transit rules can be employed but are not recommended.

Condition equations for adjusting trilateration were developed by the late Earl S. Belote, a geodesist with the U.S. Coast and Geodetic Survey. Condition equations use the differences of the angles calculated from the sides and evolved by differentiating the basic equations to compute those angles. Changes in the lengths can be expressed as follows:

$$dA''$$
$$= \frac{(ada - a \cos Cdb - a \cos BDc)}{2}(\text{area}) \sin 1''$$

$$db''$$
$$= \frac{(-b \cos Cda + bdb - b \cos Adc)}{2}(\text{area}) \sin 1''$$

$$dc''$$
$$= \frac{(-c \cos Bda - c \cos Adb - Cda)}{2}(\text{area}) \sin 1''$$

In a trilateration network, the number of equations is equal to $N - 2S$, N being the number of lines measured and S the number of new stations. If ties to additional stations are contemplated or required, additional equations will be necessary. Therefore, it is suggested that if connecting ties are to be utilized, the variation of coordinates method should be used. In Chapter 16 the theory and application of the variation of coordinates method of adjusting a quadrilateral are described in detail.

12-21. FIELD NOTES

Figure 12-15 shows the field notes for a meteorological data observation, Figure 12-16 covers zenith angle field data, and Figure 12-17 is a set of EDMI field measurements for multireadings/zenith angles.

DISTANCE MEASUREMENT DATA
METEOROLOGICAL DATA OBSERVATION   DATE: _4 APRIL 1984_

STA. _ROCK_   TO   STA. _ENUF_
       (INSTRUMENT)   (REFLECTOR)

INITIAL DATA   FINAL DATA

1735	TIME	1750
45.8 °F	TEMP	45.1 °F
26.65	PRESSURE INCHES / HG	SAME
3240 FT	PRESSURE ALTITUDE	SAME
4.960 FEET	HS	1.512 METERS

REFLECTORS

TYPE _WILD TILTING SINGLE_

NO. OF REFLECTORS _1_

REFLECTOR NO. _WILD #2_

EXTENSION USED: YES   (NO)

STAND HEIGHT   4312   1314

RFL. HEIGHT 0.648   198
            HS = 4.960   1.512

COMMENTS: _GROUND TEMP 46° F_
          ASP. HEIGHT = 30 FT.

STATE LANDS DIVISION

Figure 12-15.   Distance-measurement data-meteorological data.

266

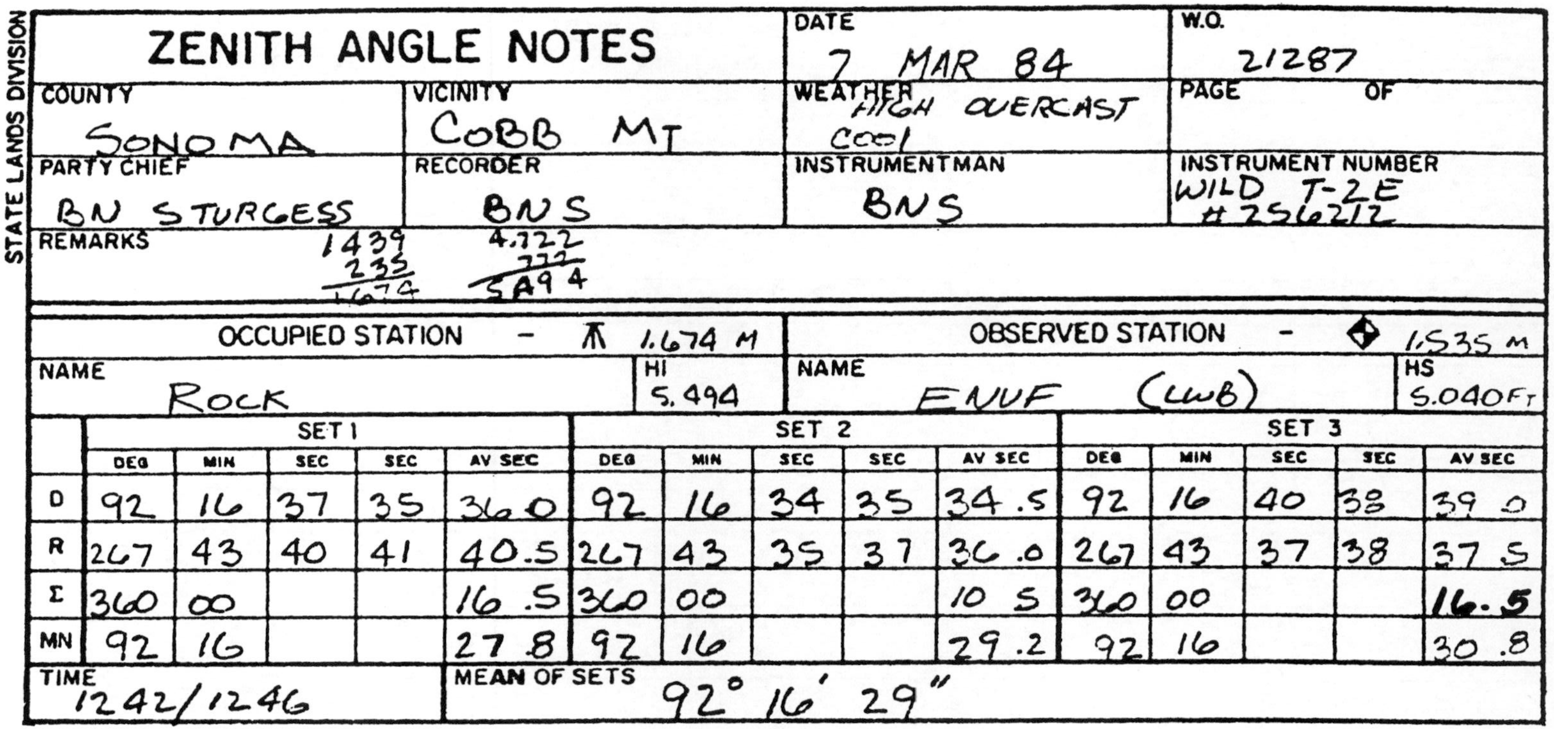

ZENITH ANGLE NOTES

COUNTY	VICINITY	DATE	W.O.
SONOMA	COBB MT	7 MAR 84	21287

	WEATHER	PAGE OF
	HIGH OVERCAST cool	

PARTY CHIEF	RECORDER	INSTRUMENTMAN	INSTRUMENT NUMBER
BN STURGESS	BNS	BNS	WILD T-2E #256272

REMARKS:

1439
235
1674

4.722
772
5.494

OCCUPIED STATION — 🔭 1.674 M   HI 5.494

NAME: ROCK

OBSERVED STATION — ◈ 1535 m   HS 5.040 FT

NAME: ENUF (wb)

	SET 1					SET 2					SET 3				
	DEG	MIN	SEC	SEC	AV SEC	DEG	MIN	SEC	SEC	AV SEC	DEG	MIN	SEC	SEC	AV SEC
D	92	16	37	35	36.0	92	16	34	35	34.5	92	16	40	38	39.0
R	267	43	40	41	40.5	267	43	35	37	36.0	267	43	37	38	37.5
Σ	360	00			16.5	360	00			10.5	360	00			16.5
MN	92	16			27.8	92	16			29.2	92	16			30.8

MEAN OF SETS   92° 16' 29"

TIME   1242/1246

Figure 12-16. Zenith angle notes.

EDM FIELD MEASUREMENT
FOR MULTI READINGS/ZENITH ANGLES

STATE LANDS COMMISSION

DATE 20 MAR 84	W.O. 21287	
CHECKED	DATE	
		PAGE OF

COUNTY	VICINITY	WEATHER
LAKE	COBB MTN.	CLEAR COOL W'LY O/5

CHIEF OF PARTY	RECORDER	OBSERVER	SIGHT SETTER
B.N. STURGESS	BNS	BNS	C. WISHMAN

OCCUPIED STATION	HI 5.157 FT / 1.571 M	ELEV. 4000.000	EDM MODEL & NO.	THEO. MODEL & NO.
ENCJF CSLC 1976			HP 3810 B	

OBSERVED STATION	HS 5.382 FT / 1.640 M	ELEV. 4423.405	PRISM TYPE & NO. WILD SINGLE #1 NO EXTENSION	PRISM CONSTANT
Rock CSLC 1980				

ECC. POSI.	CENTER	σ	− ECC	σ	+ ECC	σ	CENTER
1	10,740.657		10,741.145		10,740.171		10,740.653
2	.653		.145		.168		.653
3	.650		.139		.155		.650
4	.650		.152		.148		.650
5	10,740.653		10,741.139		10,740.145		10,740.650
6	10,740.653		10,741.145		10,740.158		10,740.647
7	.650		.155		.151		.647
8	.647		.145		.145		.653
9	.657		.159		.148		.653
10	10,740.650		10,741.155		10,740.151		10,740.653
MEAN	10,740.652		10,741.148		10,740.154		10,740.651
ECC. COR.	0		−.492		+.492		0
DISTANCE	10,740.652		10,740.656		10,740.646		10,740.651

EDM READINGS

Figure 12-17. EDM field measurement.

STATE OF CALIFORNIA

	START		FINISH		MEAN (4 SETS)	
	EDM	PRISM	EDM	PRISM		10,740.651
Time	0827		0842		PPM CORR.	-.001
Temp. Air	60.8°F	59.7°F	61.5°F	59.5°F	PRISM CORR.	-.011
Temp. Grd.	62°	59°	61°	62°	FINAL SLOPE DIST.	10,740.639
Press.	26.78"HG	26.36HG	NC	NC		
Press. Alt.	3100	3530	NC	NC		
PPM Calc.	+31.88		+31.04			
PPM Mean			+31.04			
PPM Dialed	+32					

NOTES: SIGNAL =65 NO ATTEN.
TEMP @ +30'

HEIGHT OF STAND   4.312 FT.   1.314 M
HEIGHT OF EDM ABOVE TOP OF STAND   .845 FT.   .257 M
HEIGHT OF EDM ABOVE MARK (HI)   5.157 FT.   1.571 M

ZENITH ANGLE	START	FINISH	MEAN
HI	TIME		
HS	MEAN OF SET		

		°	'	"

	SET 1						SET 2						SET 3				
	DEG.	MIN.	SEC₁	SEC₂	AV. SEC.		DEG.	MIN.	SEC₁	SEC₂	AV. SEC.		DEG.	MIN.	SEC₁	SEC₂	AV. SEC.
D																	
R																	
Σ																	
MN																	

Figure 12-17. (Continued).

269

NOTES

1. FGCC 1979. "Classification Standards of Accuracy, and General Specifications of Geodetic Control Survey." Silver Spring MD.
2. FGCC 1975. *Specifications to Support Classification, Standards of Accuracy, and General Specifications of Geodetic Control Surveys.* [1975] 1980 (revised). Federal Geodetic Control Committee Washington DC.

REFERENCES

BOMFORD, G. [1971] 1977 *Geodesy*, 3rd ed. New York: Oxford University Press.

BRYAN, D. G. 1980 Scale variation in trilateration adjustment applied to deformation surveys. M. S. Thesis, Virginia Polytechnic Institute and State University, Blacksburg, VA.

BURKE, K. F. 1971 Why compare triangulation and trilateration. 31st Annual ACSM Meeting, March 7–12. Washington D.C.

CARNES, P. S. 1961 Temperature variations in the first two hundred feet of the atmosphere in an arid region. Missle Meteorology Division, U.S. Army, Signal Missile Support Agency, N.Y. Alamogordo.

CARTER, W. E., and J. E. PETTEY 1981 Report of survey for McDonald Observatory, Harvard Radin Astronomy Station and vicinity. NOAA Technical Memorandum NOS NGS 32. Rockville MD.

DAVIS, R. E., F. S. FOOTE, J. M. ANDERSON, and E. M. MIKHAIL. 1981 *Surveying Theory and Practice.* New York: McGraw-Hill.

DRACUP, J. F. [1969] 1976 Suggested specifications for local horizontal control surveys (revised).

_____ 1976 Tests for evaluating trilateration surveys. Proceedings of the ASCM Fall Convention, Seattle, WA, Sept. 28–Oct. 1.

_____ 1980 Horizontal control. NOAA Technical Report NOS 88 NCS 19. Rockville MD.

DRACUP, J. F., and C. F. KELLEY. [1973] 1981 Horizontal control as applied to local surveying needs. ASCM Publication (reprint).

DRACUP, J. F., C. F. KELLEY, G. B. LESLEY, and R. W. TOMLINSON. 1979 *Surveying Instrumentation and Coordinate Computation Workshop Lecture Notes*, 3rd ed. ACSM. Falls Church VA.

FRONCZEK, C. J. 1977 Use of calibration base lines. NOAA Technical Memorandum NGS-10. Rockville MD.

GOSSETT, CAPTAIN F. R. 1959 Manual of geodetic triangulation. Special Publication No. 247. USGPO Wash. D.C.

GREENE, J. R. 1977 Accuracy evaluation in electro-optic distance measuring instruments. *Surveying and Mapping*. Qtrly, September Falls Church VA. Vol. 37, No. 3 P. 247.

INGHAM, A. E. 1975 *Sea Surveying*. New York: John Wiley & Sons.

KELLY, M. L. 1979 Field calibration of electronic distance measuring devices. Proceedings of the ACSM, 39th Annual Meeting, March 18–March 24. Washington D.C.

LAURILA, S. H. 1976 *Electronic Surveying and Navigation*, New York: John Wiley & Sons.

MEADE, B. K. 1969 Corrections for refractive index as applied to electro-optical distance measurements. U.S. Department of Commerce, Environmental Science Services Administration, Coast and Geodetic Survey. Rockville MD.

_____ 1972 Precision in electronic distance measuring. *Surveying and Mapping*. Vol. 32, No. 1 p. 69.

MOFFITT, F. H., and H. BOUCHARD. 1975 *Surveying*, 6th ed. New York: Harper Collins.

ROBERTSON, K. D. 1975 A method for reducing the index of refraction errors in length measurement. *Surveying and Mapping Journal*.

_____ 1979 *The Use and Calibration of Distance Measuring Equipment for Precise Measurement of dams*. Fort Belvoir, VA: U.S. Army Corps of Engineers, E.T.L. (revised).

13

Geodesy

Earl F. Burkholder

13-1. INTRODUCTION

Literally, the word geodesy means "dividing the earth"; however, by usage its meaning now includes both science and art. The science of geodesy is devoted to determining the earth's size, shape, and gravity field. The art of geodesy utilizes scientific data in a practical way to (1) obtain latitude, longitude, and elevation of points; (2) compute lengths and directions of lines on the earth's surface; and (3) describe the trajectory of missiles, satellites, or other spacecraft. It is not intended here to designate a given activity as being either science or art, but to recognize a legitimate difference in emphasis that may exist in various areas of geodesy.

13-2. DEFINITIONS

1. *Geometrical geodesy*. Concerned with the size and shape of the earth's mean-sea-level surface.
2. *Physical geodesy*. Relates the earth's geophysical internal constitution to its corresponding external gravity field.
3. *Satellite geodesy*. Deals with satellite orbits, tracking existing satellites, and predicting the trajectory of a given missile, satellite, or spacecraft.

4. *Geodetic astronomy*. Chronicles the changing position of the stars and other celestial objects. Although listed separately, it overlaps geometrical and satellite geodesy and is not discussed further here. Additional information can be found in Chapter 17 in the section on field astronomy and in other texts on geodetic astronomy.

13-3. GOALS OF GEODESY

It is not practical to mathematically describe the earth's entire topographical surface. However, one goal of geodesy is to obtain a mathematical model that best approximates the earth's mean-sea-level surface. The model most commonly used is an *ellipsoid*, formed by rotating an ellipse about its minor axis. (In the past, such a figure has been referred to as a *spheroid*. For the purposes of this handbook, the two terms can be used interchangeably.) The ellipse major axis is in the equatorial plane; the minor axis coincides with the earth's spin axis. Often an ellipsoid is defined by the length of its semimajor axis a and the semiminor axis b, Figure 13-1a. However, in Section 13-6, the ellipsoid is also defined in other ways.

The earth's mean-sea-level surface is called the *geoid*, shown in Figure 13-1b. The geoid extends under the land masses and is the

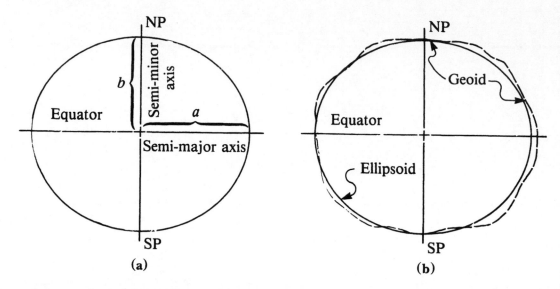

Figure 13-1. (a) Mathematical ellipsoid. (b) Physical geoid.

mean equilibrium level to which water would rise in a transcontinental canal. The geoid does not follow the ellipsoid exactly, but undulates from it by as many as 100 m. For this reason, the earth's mean-sea-level shape has been referred to as a lumpy potato. Additionally, there is an identifiable bulge in the geoid of 10 to 15 m in the southern hemisphere, giving rise to the earth being described as pear-shaped. On the other hand, despite mountains and ocean trenches, the earth is nearly spherical and by comparison is smoother than an orange. If its diameter at the equator was 10 m, the distance South Pole to North Pole would be shorter by only 0.034 m.

A second goal of geodesy is to describe the location of points on the earth's surface relative to the equator (latitude), an arbitrary meridian (longitude), and mean sea level (elevation). Thus, the rotational ellipsoid is indispensable in providing a framework for geodetic control networks. These networks, such as that in Figure 13-2, define the latitude and longitude of control points throughout the world. Previously, geodetic surveying operations were confined to continental land masses, and precise intercontinental ties were

impossible. Now with space-age technology, geodetic surveying activities are conducted on a global scale, so any two points on the earth can be tied together.

A third goal of geodesy is to determine the earth's external gravity field. Isaac Newton, Christian Huygens, and others in the middle 1600s recognized that the earth's shape is influenced by gravity. Since then, much scientific research has been devoted to the earth's geophysical attributes. This aspect of geodesy is important to surveyors because the geoid—the mean sea level to which elevations are referenced—is actually defined by an equipotential surface. Since the distance between equipotential surfaces is defined in terms of the work required to move a unit mass from one to the other, the perpendicular distance between two level surfaces is not constant but varies from the equator to the pole. Precise differential-leveling computations must accommodate that difference.

It is not possible or practical to cover all aspects of geodesy in one chapter of this surveying handbook. Therefore, the remainder of this chapter will be devoted to (1) a brief history of geodesy, (2) artful applications of

Figure 13-2. Status of horizontal control in the United States (Courtesy of National Geodetic Survey, NOAA/National Ocean Survey.)

geometrical geodesy, and (3) one brief section each on physical and satellite geodesy.

13-4. HISTORY OF GEODESY

Who first pondered the extent of the earth beyond the horizon? Who first realized inferences about out planet could be drawn from star observations? Although answers to these questions can only be conjectured, it is known that Pythagoras (b. 582 B.C.) declared the earth to be a globe, and Aristotle (384–322 B.C.) concluded that the earth must be spherical. However. an Alexandrine scientist named Eratosthenes (276–195 .c) is given credit for first determining the earth's size; admittedly, his measurements were crude by today's standard, but the method correct for this assumption of a spherical earth. The length be obtained for the earth's circumference was only about 16% too large.

Little was recorded about geodesy from the time of Eratosthenes until after the Middle Ages. However, a new epoch of geodesy began in the early 1600s with the invention of telescopes, publication of 14-place logarithms, and applications of triangulation to arc measurement. Later developments include the theory of gravity, differential and integral calculus, standardization of lengths and techniques of least-squares adjustment.

In 1615, a Dutchman, Willebrord Snellius, measured an arc over 80 mi long with a series of 33 triangles. The distance he obtained for the earth's radius was too small by about 3.4%. Next a Frenchman, Jean Picard, measured an arc on the meridian through Paris in 1669–70 and obtained a length for the earth's radius too large by only 0.7%.

Later, Picard's work was extended north to Dunkirk and south to Collioure by the Cassini brothers. The total latitude difference from Dunkirk to Collioure is 8°20′, but the arc was completed in two segments: the parts north and south of Paris. The length of one degree of latitude—and subsequently the earth's radius—for the northern part was found to be shorter than for the southern one. Hence, based on the triangulation arc through Paris, the Cassini brothers concluded, and even insisted, that the earth is not a sphere but elongated at the poles.

In 1687, Issac Newton published his law of gravitation, in which he stated that the earth is flattened at the poles; Figure 13-3 illustrates his logic. The force of gravity experienced by a plumb bob near the earth's surface is the vector sum of gravitational attraction and centrifugal force due to the earth's rotation. A level surface (sea level) is always perpendicular to the direction of the gravity vector, and a plumb bob points toward the earth's center only if the observer is standing at the equator

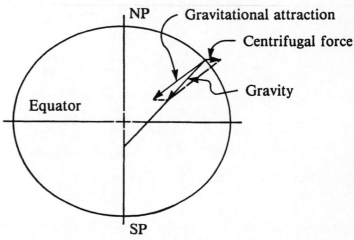

Figure 13-3. Gravity, the vector sum of gravitational attraction and centrifugal force.

or poles. Realizing this, Newton concluded that the earth must be a rotational ellipsoid flattened at the poles.

An ensuing dispute between the British (followers of Newton) and French (Cassini followers) regarding the earth's true shape was settled by two geodetic surveying expeditions sponsored by the French Academy of Science. In 1735, the first party went to the equatorial region of Peru (present-day Ecuador) to make arc measurements. In 1736, the second party went to the northern latitude of Lapland (present-day Finland). Results of the expeditions showed quite conclusively that the earth is flattened at the poles as stated by Newton (see Figure 13-4).

The lack of a universal length standard was a problem that plagued early geodesists and still affects modern interpretation of early efforts. In the late 1700s, two Frenchmen, Delambre and Mechain, were charged with determining the meridian arc distance, equator to pole, as accurately as possible. That distance was then set as *10 million meters*, the length

standard now accepted worldwide. Table 13-1 shows the values for early measurements of the earth's size and shape, ending with results obtained by Delambre and Mechain.

Since the early 1800s, there have been numerous determinations of the earth's size and shape, and some are still in use (Table 13-2). If the earth was truly a homogeneous rotating fluid as postulated by Newton, one would expect the numbers in Table 13-2 to agree better than they do. However, since the internal density of the earth is not uniformly distributed, a "best-fitting" ellipsoid for any area of the earth will not necessarily be best fitting elsewhere. Consequently, practical applications of geodesy to survey control networks have been based on different ellipsoids, depending on the part of the world (or continent) in question. The Clarke Spheroid of 1866 was used as the reference ellipsoid for geodetic datums in the United States from 1879 to 1983 (Table 13-3).

With the advent of satellite triangulation and Doppler point-positioning, it has become possible to obtain parameters of an earth-

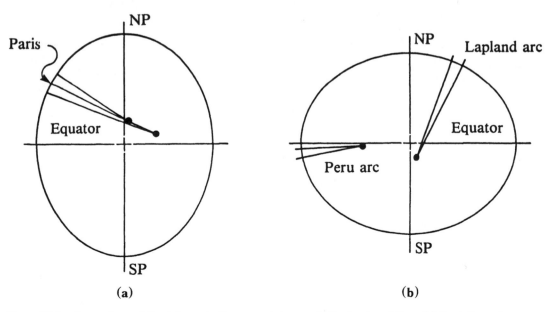

Figure 13-4. Comparison of Cassini arcs in France and the arcs in Lapland and Peru. (a) Cassini's prolate spheroid. (b) Newton's oblate spheroid.

Table 13-1.   Early determinations of the earth's size and shape

Investigator	Approximate Date	Meridian Quadrant Arc Length (m)	Flattening
Eratosthenes	200BC	11,562,500	—
Willebrord Snellius	AD 1615	9,660,000	—
Jean Picard	1670	10,009,081	—
Cassini Brothers	1700	10,042,652	$-1:66$
French Academy of Science	1750	10,000,157	$1:310.3$
Delambre and Méchain	1800	10,000,000	$1:334$

Source: I. I. Mueller and K. H. Ramsayer, 1979, *Introduction to Surveying*, Frederick Ungar Publishing Co., New York, p. 148.

Table 13-2.   Ellipsoids and area where used

Name	Semimajor Axis (m)	$1/f$	Used in
Everest 1830	6 377 276.345	300.801 7	India
Bessel 1841	6 377 397.155	299.152 8	China, Japan, Germany
Clarke 1866	6 378 206.4	294.978 7	North and Central America
Modified Clarke 1880	6 378 249.145	293.465	Africa
International 1924	6 378 388	297	Europe
Krasovskiy 1942	6 378 245	298.3	Former Soviet Republic and adjacent countries
Australian National 1965	6 378 160	298.25	Australia
South American 1969	6 378 160	298.25	South America

Source: I. I. Mueller and K. H. Ramsayer, 1979, *Introduction to Surveying*, Frederick Ungar Publishing Co., New York, p. 148.

Table 13-3.   Geodetic datums used in the United States

Datum Name	Year Adopted	Reference Ellipsoid	Remarks
New England Datum	1879	Clarke 1866 $a = 6,378.206.4$ m $b = 6,356,583.5$ m	First official U.S. Datum, datum origin: station Principo in Maryland
U.S. Standard Datum	1901	Clarke 1866	Datum origin moved to Meades Ranch in Kansas
North American Datum	1913	Clarke 1866	A name change only to reflect adoption by Canada and Mexico
North American Datum of 1927	1927	Clarke 1866	A general readjustment holding location of station Meades Ranch
North American Datum of 1983	1983	Geodetic Reference System of 1980 $a = 6378137.000$ m $1/f = 298.257222101$	An extensive readjustment on a new reference ellipsoid having its origin at the earth's center of mass

centered ellipsoid based on a global best fit. The *North American Datum of 1983* is a comprehensive readjustment of the North American continent geodetic horizontal-control networks. The various national systems are tied to a worldwide geometric satellite network computed on the new ellipsoid, the Geodetic Reference System of 1980, adopted by the 17th General Assembly of the International Union of Geodesy and Geophysics Meeting in Canberra, Australia, December 1979. The adoption and use of an earth-centered ellipsoid make accurate global mapping possible. Additionally, the geodetic position of any point is fixed, working "from the whole to the part" on a global scale.

13-5. GEOMETRICAL GEODESY

13-5-1. Geometry of the Ellipsoid

As stated in Section 13-3, the ellipsoid is obtained by rotating an ellipse about its minor axis. The minor axis coincides with the earth's spin axis; the major axis sweeps out an equatorial plane as the ellipse is rotated. Any cross section of the ellipsoid containing both poles is a *meridian section* and shows the form of the original ellipse as illustrated in Figure 13-5a. The meridian section through Greenwich, England, Figure 13-6, is taken as the reference meridian. All other meridians are tied to this *prime meridian* by their *longitude*, the angular difference between meridian sections Figure 13-5b. Longitude starts with 0° at the prime meridian and increases eastward to 360° for a complete revolution. However, it is common practice in the Western Hemisphere to use longitude increasing *westward* from Greenwich to 180°W at the international date line. A word of caution: If west longitude is employed, it should be so noted to avoid confusion with the higher practice of east longitude 0 to 360°.

A position on a meridian is defined by its *geodetic latitude*, the angular distance north or south of the equator. The geodetic latitude goes from 90°S ($-90°$) at the South Pole to 90°N ($+90°$) at the North Pole. As shown in Figure 13-5a, the *normal* is perpendicular to the ellipse tangent and goes from a point of tangency to the spin axis. The angle ϕ that a

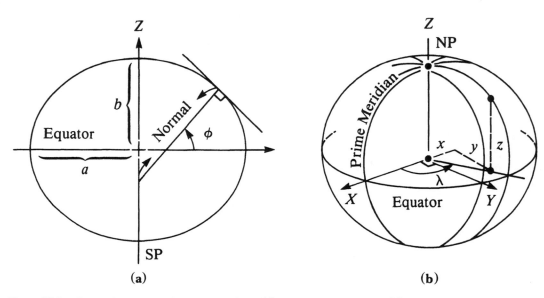

Figure 13-5. Comparison of meridian section ellipse (a) and rotational ellipsoid (b).

Figure 13-6. Astraddle prime meridian in Greenwich, England, (Courtesy of Engineering Surveys Ltd., West Byfleet, England.)

normal makes with the equatorial plane is the point's geodetic latitude.

There is also an orthogonal three-dimensional coordinate system associated with the ellipsoid (Figure 13-5b). The origin is at the intersection of the spin axis and equatorial plane. A two-dimensional XY-plane lies in the equatorial plane with the positive X-axis pointing toward the prime meridian. The Y-axis is at longitude 90°E and the positive Z-axis points toward the North Pole. For this reason, the following discussion of a two-dimensional ellipse will be in terms of the XZ prime-meridian plane instead of an XY-plane commonly used for two-dimensional coordinates.

13-5-2. The Two-Dimensional Ellipse

A two-dimensional ellipse is a conic section described as the path of a point moving in a plane, so the total distance to two points, called foci, remains constant. The equation of an ellipse in the XZ-plane is given by

$$\frac{X^2}{a^2} + \frac{Z^2}{b^2} = 1 \qquad (13\text{-}1)$$

where a is the ellipse semimajor axis and b the semiminor axis.

Flattening of the ellipse and its eccentricity are defined in terms of a and b as

$$\text{flattening } f = \frac{(a-b)}{a} = 1 - \frac{b}{a} \quad (13\text{-}2)$$

$$\text{eccentricity } e = \frac{\sqrt{a^2 - b^2}}{a} \quad (13\text{-}3)$$

$$\text{second eccentricity } e' = \frac{\sqrt{a^2 - b^2}}{b} \quad (13\text{-}4)$$

Two parameters are required to define an ellipse. Previously, the semimajor axis a and semiminor axis b have been used; however, an ellipse can be defined equally well by a and e, e', or f. Current practice defines the ellipsoid size and shape with the semimajor axis a and reciprocal flattening $1/f$. Given these two parameters, the eccentricity and semiminor axis are

$$e^2 = 2f - f^2, \quad e = \sqrt{e^2} \quad (13\text{-}5)$$

$$b = a(1 - f) = a\sqrt{1 - e^2} \quad (13\text{-}6)$$

Equations (13-5) and (13-6) are obtained by substitution and the algebraic manipulation of (13-2) and (13-3).

Construction of an Ellipse

There are three ways to construct an ellipse of any size and shape. The first method is a mechanical one using a piece of string and a pencil. Since the sum of the distances from each focus to a point on the ellipse is constant, a curve can be drawn by anchoring the string ends at the foci, taking up the slack with a pencil, and tracing one-half of the ellipse while keeping the string taut. The second half is drawn by taking up the slack in the opposite direction, rather than wrapping the string around one focus. The following items are illustrated in Figure 13-7:

1. Total string length is twice the ellipse semimajor axis is shown by the pencil being positioned at point P_1.

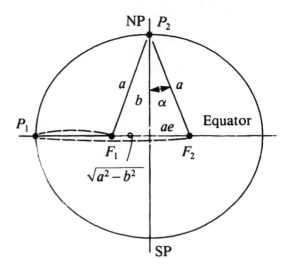

Figure 13-7. Mechanical construction of ellipse: String method.

2. Separation of the foci from the origin determines the minor axis length. If both foci are at the origin, the semiminor and semimajor axes are identical, so the ellipse becomes a circle ($e = 0$). If the two foci are separated so the string becomes taut, the semiminor axis length goes to zero and the ellipse degenerates to a straight line ($e = 1$).

3. Stopping the pencil at point P_2 on the minor axis produces a symmetrical figure, and the distance from each focus to P_2 is a, one-half the total string length. The resulting right triangle is solved for the distance focus to origin as $\sqrt{a^2 - b^2}$.

$$\text{Sin } \alpha = \frac{\sqrt{a^2 - b^2}}{a}$$

from Figure 13-7 is the same as e, the eccentricity (α is the *angular eccentricity*). *Note:* The distance between each focus and the origin is given by the product ae.

The second method of constructing an ellipse is a graphical one. First, two circles are drawn. The radius of the outer circle is a and the inner one b. Next, any number of radial lines are drawn as shown in Figure 13-8. Finally, lines are drawn parallel to the X-axis from the intersection of the radial line and

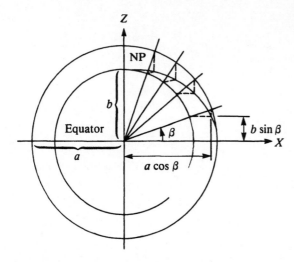

Figure 13-8. Graphical construction of ellipse showing parametric (reduced) latitude.

inner circle and parallel to the Z-axis from the intersection of the radial line with the outer circle. The intersection of these two lines from the same radial line falls on an ellipse, which is formed by plotting a sufficient number of points and "connecting the dots." Two ellipse properties illustrated in Figure 13-8 are as follows:

1. The angle between the equatorial plane (X-axis) and radial line is the *parametric latitude* β. In some geodetic literature it is called "reduced latitude."

2. The X- and Z-coordinates of a point on an ellipse are given, respectively, by

$$X = a \cos \beta \quad \text{and} \quad Z = b \sin \beta \quad (13\text{-}7)$$

The third and perhaps the most efficient method of constructing an ellipse is to compute X- and Z-coordinates for a sufficient number of points and plot them directly. Coordinates are computed using a, the semimajor ellipse axis; e^2, the eccentricity squared; and ϕ, the geodetic latitude of a point. The

following are equations for X and Z:

$$X = \frac{a \cos \phi}{(1 - e^2 \sin \phi)^{1/2}} \quad (13\text{-}8)$$

$$Z = \frac{a(1 - e^2) \sin \phi}{(1 - e^2 \sin^2 \phi)^{1/2}} \quad (13\text{-}9)$$

Important ellipse properties in Figure 13-9 are as follows:

1. It is doubly symmetrical. Coordinates need to be computed for one quadrant only.
2. N, the normal, goes from the ellipse to the spin axis.
3. The normal length is $X/\cos \phi$. Using Equation (13-8), we can write it as

$$N = \frac{a}{(1 - e^2 \sin^2 \phi)^{1/2}} \quad (13\text{-}10)$$

Three Types of Latitude

Three types of latitude are routinely encountered in geometrical geodesy. Geodetic latitude and parametric latitude have already been discussed. The third one is *geocentric latitude*. ψ, the angle between the equatorial plane and a line from the ellipse center to a surface point (Figure 13-10a). Tan ψ is obtained directly as Z/X. The geocentric latitude, geodetic latitude, and parametric latitude are related by substituting values of X and Z as contained in Equations (13-7), (13-8), and (13-9).

$$\text{Tan } \psi = \frac{Z}{X} = \frac{b \sin \beta}{a \cos \beta}$$

$$= \frac{b}{a} \tan \beta = (1 - e^2)^{1/2} \tan \beta \quad (13\text{-}11)$$

$$\text{Tan } \psi = \frac{Z}{X}$$

$$= \frac{a(1 - e^2)\sin \phi / (1 - e^2 \sin^2 \phi)^{1/2}}{a \cos \phi / (1 - e^2 \sin^2 \phi)^{1/2}}$$

$$= (1 - e^2)\tan \phi \quad (13\text{-}12)$$

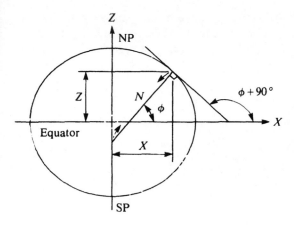

Figure 13-9. Mathematical construction of ellipse.

Equations (13-11) and (13-12) can be summarized as

$$\text{Tan } \psi = (1 - e^2)^{1/2} \tan \beta$$

$$= (1 - e^2)\tan \phi \qquad (11\text{-}13)$$

Comparison of Equation (13-13) with Figure 13-10b shows that the three latitudes are identical at the equator and poles. Between the equator and poles, the geocentric latitude is smaller than the parametric and geodetic latitudes, whereas the geodetic latitude is larger than either the geocentric or parametric latitudes. The maximum difference between geodetic and geocentric latitude occurs when the former is more than 45°, while the latter is smaller than 45°.

Radius of Curvature in Meridian Section

If the meridian section was spherical—i.e., if *e* equals zero—the ellipsoid would be a sphere and the radius of curvature the same at any point. Since the meridian section is an ellipse, its radius of curvature is not constant, but changes with increasing latitude. The *radius of curvature M* at any point in the meridian section is obtained by taking the first and second derivatives of Equation (13-1) and substituting those expressions in the general equation for radius of curvature given in Equation (13-14).

$$M = \left[1 + \left(\frac{dZ}{dX}\right)^2\right]^{3/2} \bigg/ \frac{d^2Z}{dX^2} \qquad (13\text{-}14)$$

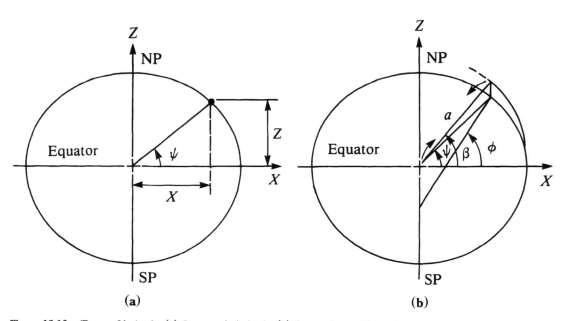

Figure 13-10. Types of latitude. (a) Geocentric latitude. (b) Comparison of latitudes.

The result is Equation (13-15), which gives the meridian radius of curvature for a specific ellipse at any geodetic latitude.

$$M = \frac{a(1 - e^2)}{(1 - e^2 \sin^2 \phi)^{3/2}} \qquad (13\text{-}15)$$

Length of Meridian Arc

Early determinations of the earth's size were made by measuring a portion of a meridian arc and comparing that length to the angle subtended at the earth's center. (The angle was usually found by astronomical observations.) Different values for the length of one degree of arc at various latitudes implied that a meridian section of the earth was ellipsoidal. Having selected an ellipsoid as a model for the earth, we compute the arc length by integrating the differential geometry elements in Figure 13-11, where the arc length of differential elements dS equals the instantaneous radius of curvature M times the differential change in geodetic latitude $d\phi$ ($d\phi$ in rad).

$$dS = M\,d\phi \qquad (13\text{-}16)$$

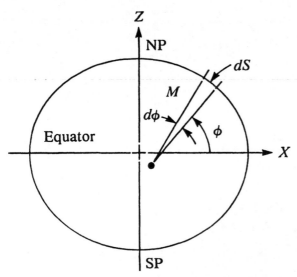

Figure 13-11. Differential elements of meridian arc length.

Meridian arc length from one latitude to another is obtained by integrating Equation (13-17) between selected limits.

$$S_{\phi_1 \to \phi_2} = \int_{\phi_1}^{\phi_2} M\,d\phi \qquad (13\text{-}17)$$

The value of M from Equation (13-15) is substituted in Equation (13-17) and the constant portion moved outside the integral to get

$$S_{\phi_1 \to \phi_2}$$
$$= a(1 - e^2)\int_{\phi_1}^{\phi_2}(1 - e^2 \sin^2 \phi)^{-3/2}\,d\phi \quad (13\text{-}18)$$

Equation (13-18) is an elliptical integral and cannot be integrated in closed form—i.e., the expression inside the integral must be expressed in a series expansion containing ever-smaller terms that can then be integrated individually. A solution is obtained by including all those terms that make a difference in the answer to the accuracy desired. Terms beyond those are dropped.

Evaluation of the series expansion of an elliptical integral gives the meridian arc length of a specified ellipsoid as

$$\begin{aligned}
S_{\phi_1 \to \phi_2} = a(1 - e^2)\Big[& A(\phi_2 - \phi_1) \\
& - \frac{B}{2}(\sin 2\phi_2 - \sin 2\phi_1) \\
& + \frac{C}{4}(\sin 4\phi_2 - \sin 4\phi_1) \\
& - \frac{D}{6}(\sin 6\phi_2 - \sin 6\phi_1) \\
& + \frac{E}{8}(\sin 8\phi_2 - \sin 8\phi_1) \\
& - \frac{F}{10}(\sin 10\phi_2 - \sin 10\phi_1)\Big] \quad (13\text{-}19)
\end{aligned}$$

where

$$A = 1 + \frac{3}{4}e^2 + \frac{45}{64}e^4 + \frac{175}{256}e^6 + \frac{11,025}{16,384}e^8 + \frac{43,659}{65,536}e^{10} + \cdots$$

$$B = \frac{3}{4}e^2 + \frac{15}{16}e^4 + \frac{525}{512}e^6 + \frac{2205}{2048}e^8 + \frac{72,765}{65,536}e^{10} + \cdots$$

$$C = \frac{15}{64}e^4 + \frac{105}{256}e^6 + \frac{2205}{4096}e^8 + \frac{10,395}{16,384}e^{10} + \cdots$$

$$D = \frac{35}{512}e^6 + \frac{315}{2048}e^8 + \frac{31,185}{131,072}e^{10} + \cdots$$

$$E = \frac{315}{16,384}e^8 + \frac{3465}{65,536}e^{10} + \cdots$$

$$F = \frac{693}{131,072}e^{10} + \cdots$$

Note: In Equation (13-19), the latitude difference in the A coefficient term is in radian units. Also, if limits 0 and 90° are chosen for ϕ_1 and ϕ_2, the length of a meridian quadrant is

$$S_{0° \rightarrow 90°} = a(1 - e^2)\left[A\frac{\pi}{2} \right] \qquad (13\text{-}20)$$

Other methods for computing meridian arc length expand the elliptical integral in terms of e'^2 instead of e^2 as listed here.

13-5-3. The Three-Dimensional Ellipsoid

Elements of a two-dimensional ellipse have been discussed, but the earth is three-dimensional. Point position on an ellipsoid surface is defined by three-dimensional coordinates or by latitude and longitude (Figure 12-5b). Given a point on an ellipsoid surface, there are several three-dimensional elements to be considered.

The Normal Section

A *normal section* is created by intersecting a plane containing the normal at a point and the ellipsoid. The plane, which can be oriented in any azimuth, is sometimes illustrated as the vertical plane that rotates about a theodolite's standing axis. (This is correct if the deflection of the vertical is zero. The standing axis of a theodolite is perpendicular to the geoid, but the normal is perpendicular to the ellipsoid. The difference is the *deflection of the vertical*.) A normal section having an azimuth of 90° (or 270°) at a point defines the *prime vertical* plane through it. The "normal" computed by Equation (13-10) is the prime vertical instantaneous radius of curvature. The normal section radius of curvature at a point on the ellipsoid in any azimuth is given by Euler's theorem, as follows:

$$R_\alpha = \frac{MN}{M \sin^2 \alpha + N \cos^2 \alpha}$$

$$\text{or} \quad \frac{1}{R_\alpha} = \frac{\cos^2 \alpha}{M} + \frac{\sin^2 \alpha}{N} \qquad (13\text{-}21)$$

where M is the radius of curvature in the meridian section, N the radius of curvature in the prime vertical, and α the normal section azimuth.

Radius of curvature properties for normal sections in various azimuths include:

1. $R_\alpha = M$ for a normal section in azimuth 0°.
2. $R_\alpha = N$ for a normal section in azimuth 90°.

3. $R_{30°} = R_{150°} = R_{210°} = R_{330°}$ due to symmetry. (Values of R_α repeat mirror-fashion with respect to both axes throughout all four quadrants.)

4. Values of R_α will always be greater than M and smaller than N.

As the geodetic latitude changes, the lengths of M and N also change. Note what happens to values of M and N at the equator and poles. For the equator, substitute $0°$ in Equations (13-10) and (13-15).

$$N_{0°} = \frac{a}{(1 - e^2 \sin^2 0°)^{1/2}} = a$$

$$M_{0°} = \frac{a(1 - e^2)}{(1 - e^2 \sin^2 0°)^{3/2}} = a(1 - e^2)$$

At a pole, substitute $\pm 90°$ in Equations (13-10) and (13-15).

$$N_{90°} = \frac{a}{(1 - e^2 \sin^2 90°)^{1/2}}$$

$$= \frac{a}{(1 - e^2)^{1/2}} = \frac{a^2}{b}$$

$$M_{90°} = \frac{a(1 - e^2)}{(1 - e^2 \sin^2 90°)^{3/2}}$$

$$= \frac{a}{(1 - e^2)^{1/2}} = \frac{a^2}{b}$$

It is obvious that M and N are equal at the poles, because the prime vertical of a given meridian is itself a meridian section. The ellipsoid radius of curvature at the poles is the same in all azimuths.

$$c = \frac{a^2}{b} = \text{the polar radius of curvature} \quad (13\text{-}22)$$

Length of a Parallel

A *parallel* of constant latitude on the ellipsoid describes a small cicle, as opposed to a great circle, whose plane is parallel to the

equatorial plane. A parallel crosses all meridians at a 90° angle (Figure 13-12a) and is a circle whose radius equals $N \cos \phi$ (Figure 13-12b). Since a parallel is a circle, its length is simply $2\pi r$. Partial length of parallel L_p can be computed as a proportionate part of the total circumference or calculated directly as a product of the radius ($r = N \cos \phi$) times the subtended angle in radians. The subtended angle is the longitude difference between meridian sections.

$$L_p = (\lambda_2 - \lambda_1)N \cos \phi = \Delta \lambda N \cos \phi \quad (13\text{-}23)$$

Ellipsoid Surface Area

Ellipsoid surface area is computed by integrating the differential-area elements in figure 13-13. The differential area dA is the product of the differential meridian length times the differential parallel length. Area is obtained by performing a double integration of Equation (13-24).

$$dA = (M d\phi)(N \cos \phi \, d\lambda) \quad \text{or}$$

$$\text{Area} = \int_{\lambda_1}^{\lambda_2}\int_{\phi_1}^{\phi_2} MN \cos \phi \, d\phi \, d\lambda \quad (13\text{-}24)$$

Previous expressions for M and N, Equations (13-15) and (13-10), are substituted into Equation (13-24) and a double integration performed to obtain

$$\text{Area} = \frac{(\lambda_2 - \lambda_1)a^2(1 - e^2)}{2}$$

$$\times \left[\frac{\sin \phi}{(1 - e^2 \sin^2 \phi)} \right.$$

$$\left. + \frac{1}{2e} \ln\left(\frac{1 + e \sin \phi}{1 - e \sin \phi} \right) \right]_{\phi_1}^{\phi_2} \quad (13\text{-}25)$$

Equation (13-25) will give the ellipsoid surface area for any block defined by latitude and longitude limits. The entire ellipsoid surface area is computed by choosing limits of longitude from 0 to 2π radians and latitude limits

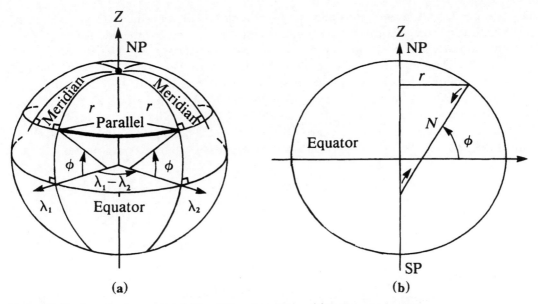

Figure 13-12. Length of a parallel. (a) Parallels and meridians. (b) Radius of a parallel.

from $-90°$ at the South Pole to $+90°$ at the North Pole

$$\text{Total area} = 2\pi a^2(1 - e^2)$$
$$\times \left[\frac{1}{1 - e^2} + \frac{1}{2e}\ln\left(\frac{1 + e}{1 - e}\right) \right]$$

$$(13\text{-}26)$$

The Geodetic Line

The shortest distance between any two surface points on an ellipsoid is the *geodetic line*, also known as the *geodesic*. The geodetic line on the ellipsoid surface is analogous to a great circle arc on a sphere. When a geodetic line is drawn on a rectangular graticule of meridians and parallels, it appears as a curved line similar to a great circle.

Starting on the equator and traversing a geodetic line to the *antipole* (the point 180° from the beginning point), the route would go across either the North or South Pole and follow a meridian exactly. If the terminal point is several kilometers east or west of the an-

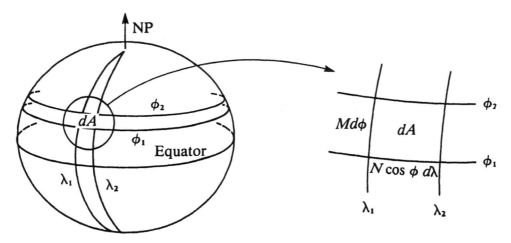

Figure 13-13. Computation of ellipsoid surface area.

tipole, the geodetic line will not pass over either pole, but will reach some maximum (or minimum) latitude, where it will cross the meridian at 90°. Note also that if the geodetic line does not follow a meridian, it will cross the equator at some azimuth other than 0°. As the azimuth at the equator increases, the maximum latitude reached decreases (Figure 13-14), until the geodetic line stays on the equator.

Given a point on the equator, the shortest distance to the antipole is a geodetic line over a pole. The shortest distance to the *lift-off point* is along the equator (Figure 13-14). Any point between the antipole and lift-off point is reached following a geodetic line crossing the equator between 0 and 90°C; for the earth, the lift-off point is approximately 34 km from the antipole. Except for the geodetic line following a meridian or equator, the azimuth changes continuously with respect to the meridian it crosses. The azimuth of a geodetic line can be determined at any point of known latitude by using *Clairaut's constant*, defined as

$$N \cos \phi \sin \alpha = K \quad \text{(Clairaut's constant)} \quad (13\text{-}27)$$

Given the latitude of a point and geodetic line azimuth at the point, Clairaut's constant is

computed. That result is then used at other latitudes to solve for the geodetic line azimuth there. For example, in Figure 13-15, the latitude of point A is 42°15′28″.17621, point B is 42°20′16″.96171, and the geodetic line azimuth at point A is 55°16′28″.12. With the GRS 1980 ellipsoid (a = 6378137.0 m, e^2 = 0.006694380023), and Equation (13-10) for N, Clairaut's constant is computed as

$$\frac{a}{\left(1 - e^2 \sin \phi_A\right)^{1/2}} \cos \phi_A \sin \alpha_A$$

$$= \frac{3,879,837.711 \text{ m}}{0.9984852096} = 3,885,723.768 \text{ m}$$

The geodetic line azimuth at point B is obtained by rewriting Equation (13-27) as

$$\sin \alpha_B = \frac{\text{Clairaut's constant}}{N_B \cos \phi}$$

$$= \frac{3,885,723.768 \text{ m}}{4,721,791,697 \text{ m}} = 0.8229341778$$

The azimuth at point B is 55°22′46″.57. The geodetic line azimuth difference between points A and B is due to convergence of the meridians. Therefore, Clairaut's constant and

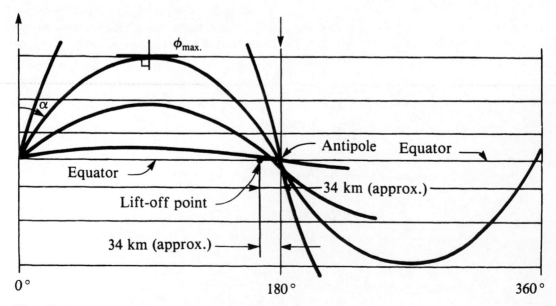

Figure 13-14.   Geodetic lines around the earth.

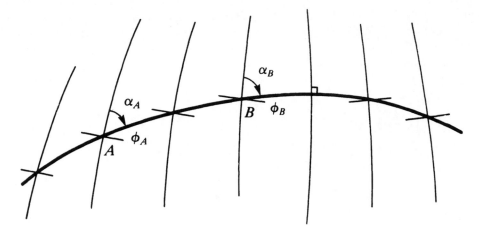

Figure 13-15. Geodetic line crossing meridians.

the geodetic line can be used to compute convergence between points.

$$\text{Covergence } (A \rightarrow B)$$

$$= \alpha_B - \alpha_A = 000°06'18''.45$$

Equations for convergence typically involve an approximation using the midlatitude of the line between points A and B. Clairaut's constant and a geodetic line provide a closed-formula method for determining convergence between points.

The maximum latitude reached by a geodetic line occurs where it crosses a meridian at 90°. Since we know the value of Clairaut's constant for the line and that $\sin 90° = 1.0$, it is possible to solve for the maximum latitude reached by a given geodetic line through writing Equation (13-27) as follows:

$$N_{\max} \cos \phi_{\max}(1.0)$$

$$= K \quad \text{(Clairaut's constant)}$$

from which—with considerable manipulation —the following can be written:

$$\cos \phi_{\max} = \frac{K(1 - e^2)^{1/2}}{(a^2 - K^2 e^2)^{1/2}}$$

$$= \frac{K}{(c^2 - K^2 e'^2)^{1/2}} \quad (13\text{-}28)$$

Clairaut's constant is not a unique property of a geodetic line. The constant remains unchanged along a parallel of latitude, although a parallel is not the shortest distance between two ellipsoid points.

Comparison of Geodetic Line and Normal Section

Due to a difference in direction of the normals at points A and B in Figure 13-16a, the normal section trace on the ellipsoid from A to B is different from the trace from B to A. The geodetic line between points A and B is not the trace of either normal section between the points. The geodetic line shown in Figure 14 when reverses its curvature only when it crosses the equator. However, when comparing the geodetic line with normal section traces between the points, it is impossible to show the geodetic line without giving it a double curvature (Figure 12-16c). The difference between a geodetic line azimuth A_g and the normal section azimuth A_n is given to a close approximation by

$$A_n - A_g = \frac{e^2}{12} \frac{S^2}{N_A^2} \cos^2 \phi_m \sin 2A_n \quad (13\text{-}29)$$

where e is the eccentricity of the ellipsoid, S the distance from point A to B, ϕ_m the mean latitude of line, and N_A the normal at point A.

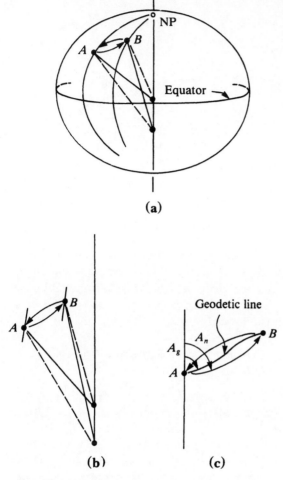

(a)

(b)   **(c)**

Figure 13-16. Comparison of normal plane curves and geodetic line. (a) Normal plane curves on the reference ellipsoid. (b) Normal plane curves: from A to B and B to A. (c) The geodetic line between A and B.

13-5-4. Geodetic Position Computation

A geodetic line can be used to compute the latitude and longitude of a point, given the distance and direction from a known point. The procedure is a *geodetic direct* computation. *Geodetic inverse* is employed to compute the distance and direction between points when the latitude and longitude of both are given. Details for performing both the "direct" and "inverse" geodetic position computations are presented by Jank and Kivioja in the Septem-

ber 1980 issue of *Surveying and Mapping*.[1] Their method utilizes the numerical integration of differential geometry elements shown in Figure 13-17. Clairaut's constant is used to determine the correct azimuth of each geodetic line element.

Although numerical integration may not be as quick as other methods for short lines, it is superior because any desired level of accuracy can be obtained regardless of line length. This is achieved by choosing a sufficiently small differential length element and programming a calculator or computer to do the repetitive calculations. According to Jank and Kivioja, centimeter accuracy can be expected for length elements up to 2 km long. If length elements are kept smaller than 200 m, millimeter accuracy can be attained.

The following Puissant Coast and Geodetic Survey formulas, used for geodetic direct and inverse computations, are quite accurate for lines up to 60 mi long.

Geodetic Direct

Given the latitude and longitude (ϕ_1, ϕ_2) of point 1, a geodetic line azimuth from north through point 1, α_1 and the distance S in meters along the geodetic line to point 2, find the latitude, longitude, and azimuth of the geodetic line at point 2.

$$\phi_2 = \phi_1 + \Delta\phi$$

$$\Delta\phi'' = SB\cos\alpha_1 - S^2C\sin^2\alpha_1$$

$$- D(\Delta\phi'')^2 - hS^2E\sin^2\alpha_1 \quad (13\text{-}30)$$

where

$B = \rho/M_1$ (sec per m)

$h = SB\cos\alpha_1$ (sec)

$C = \dfrac{\rho\tan\phi_1}{2M_1N_1}$ (sec per m^2)

$D = \dfrac{3e^2\sin\phi_1\cos\phi_1}{2\rho(1-e^2\sin^2\phi_1)}$ (per sec)

$E = \dfrac{(1+3\tan^2\phi_1)(1-e^2\sin^2\phi_1)}{6a^2}$ (per m^2)

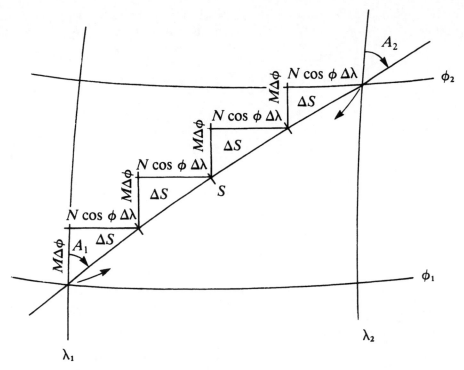

Figure 13-17. Differential elements of geodetic line computation.

The constants h, B, C, D, and E are computed using $\rho = 206264.8062470964$ sec per rad, M [Equation (13-15)] is the radius of curvature in the meridian, N [Equation (13-10)] the normal, and e^2 the ellipsoid eccentricity squared. Note that $\Delta\phi''$ appears on both sides of Equation (13-30), requiring an iterative solution. Use $\Delta\phi'' = $ zero for the first iteration.

With the latitude of point 2 known, the longitude is calculated.

$$\lambda_2 = \lambda_1 + \Delta\lambda \quad \text{(east longitude)}$$

$$\Delta\lambda'' = \frac{S\rho \sin \alpha_1}{N_2 \cos \phi_2} \tag{13-31}$$

The azimuth from point 2 back to point 1 can be computed using Clairaut's constant; however, the Puissant formulas use the following:

$$\alpha_2 = \alpha_1 + \Delta\alpha + 180°$$

$$\Delta\alpha'' = \frac{\Delta\lambda'' \sin \phi_m}{\cos\left(\dfrac{\Delta\phi}{2}\right)}$$

$$+ (\Delta\lambda'')^3 \frac{\sin \phi_m \cos^2 \phi_m}{\rho} \tag{13-32}$$

where

$$\phi_m = \left(\phi_1 + \frac{\phi_2}{2}\right)$$

Geodetic Inverse

Given the latitude and longitude of two points, it is required to find the geodetic line azimuth at each point and the distance from one to the other.

$$\Delta\phi = \phi_2 - \phi_1 \quad \text{and} \quad \Delta\lambda$$

$$= \lambda_2 - \lambda_1 \quad \text{(east longitude)}$$

$$x = \frac{\Delta\lambda'' N_2 \cos^2\phi_2}{\rho} = S \sin\alpha_1$$

$$y = \frac{1}{B}[\Delta\phi'' + Cx^2 + D(\Delta\phi'')^2$$

$$+ E(\Delta\phi'')x^2] = S \cos\alpha_1$$

$$\tan\alpha_1 = \frac{S\sin\alpha_1}{S\cos\alpha_1} = \frac{x}{y} \quad \text{(from north)} \quad (13.33)$$

$$S = \sqrt{x^2 + y^2} \quad \text{(in meters)} \quad (13.34)$$

13-6. GEODETIC DATA TRANSFORMATIONS

13-6-1. Use of a Model

This section discusses how geodetic directions and distances on the ellipsoid are obtained from field measurements. Since the ellipsoid is an abstract mathematical model, the data must be transformed from the actual measurement configuration to its equivalent representation on a model. The distance transformations shown in Figure 13-18 is an example. The measured slope distance must be transformed to an equivalent distance on the reference ellipsoid (the model) before it is used in a geodetic position computation.

13-6-2. Target Height Correction

When a target is sighted through a theodolite, the direction to it is the normal section azimuth from instrument to the target. If the target is not on the ellipsoid, there will be a difference in directions to it and to the station on the ellipsoid. The difference occurs because the normals at the target and theodolite are not parallel (Figure 13-19). The situation is analogous to sighting the top of a range pole

that is not held plumb over a point. A correction to the observed direction can be computed from Equation (13-35), as follows

$$\alpha = \alpha_h + \Delta\alpha$$

$$\Delta\alpha'' = \frac{\rho h e^2 \cos^2\phi_1}{2N_1(1-e^2)}\left(\sin 2\alpha_h - \frac{S}{N_1}\sin\alpha_h \tan\phi_1\right)$$

$$(13\text{-}35)$$

where $\rho = 206264.8062470964$ sec of arc per rad, h is the target height above the ellipsoid, e^2 the ellipsoid eccentricity squared, ϕ the geodetic latitude of instrument station, α the normal section direction from point 1 to point 2, and α_h the normal section direction from point 1 to the target elevated above point 2.

Note the following items regarding the use of Equation (13-35):

1. The sign of the correction $\Delta\alpha''$ is determined by $\sin 2\alpha_h$. It is positive if the target is in the NE or SW quadrants and negative for the SE and NW quadrants.
2. Theodolite elevation is immaterial because the standing axis contains the vertical, and the instrument measures the dihedral angle.
3. The correction for target height is quite small ($\pm 0''.5$) for elevations under 4000 m, but could be significant on precise surveys if targets are located on high mountaintops.

13-6-3. Deflection-of-the-Vertical Correction

As shown in Figure 13-20b, a normal is perpendicular to an ellipsoid, but a vertical is perpendicular to the geoid. Deflection of the vertical (also called deviation of the vertical) is the difference in directions of the normal and vertical. Deflection of the vertical at a point can be scaled from an accurate geoid map showing contours of the geoid for a given area. Figure 13-20a shows part of the *Geoid Contour Map of the North American Datum of 1927*, printed by the U.S. Army Map Service.[2] The geoid is below the ellipsoid and slopes upward to the northeast, across central Ore-

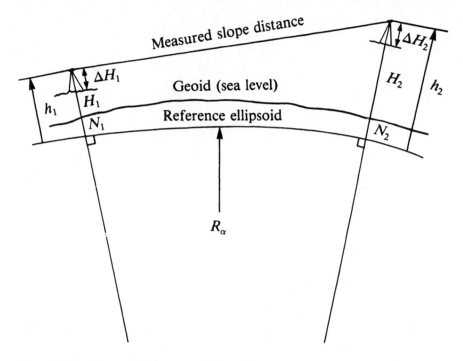

Figure 13-18. Elements of distance transformation.

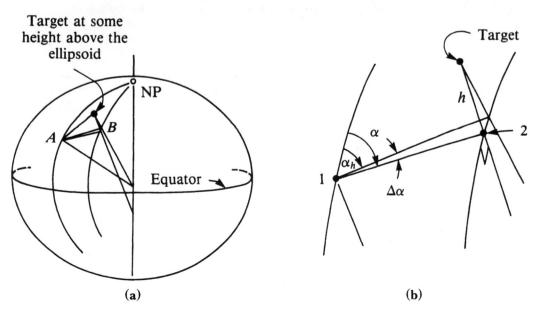

Figure 13-19. Difference in direction due to height of target. (a) Target height above ellipsoid. (b) Difference in observed azimuth due to height of the target.

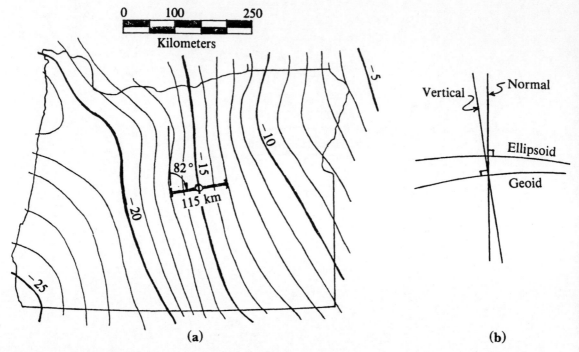

Figure 13-20. (a) Slope of geoid across the state of Oregon. (b) Deflection of the vertical. (Geoid Contours in North America; 1967 Washington D.C.; U.S. Army Map Service.)

gon. It rises 4 m in a scaled distance of 115 km. The azimuth of a line perpendicular to the geoid contours—the direction water would flow—scales 262°. From this data, approximately three significant figures, the total deflection of the vertical is 7.17 sec, and the NS and EW components are determined.

$$\text{NS: } xi = \xi = 7''.17\cos(262°) = -1''.00$$

$$\text{EW: } eta = \eta = 7''.17\sin(262°) = -7''.10$$

By convention, if the vertical above a theodolite is deflected into the northeast quadrant, both components are positive. Stated differently, if the geoid slopes upward to the north and east as in Figure 12-20b, both components are negative.

Deflection of the vertical is actually defined and determined using gravity measurements and physical geodesy techniques. The geoid map is compiled from aggregate deflection-of-the-vertical data and gravity measurements over the entire geodetic control network.

Geodetic scientists compile geoid maps and geodetic surveyors use geoid maps to relate geographic positions determined from astronomical observations to the point's geodetic position.

Equations (13-36), (13-37), and (13-38) give the relationships between astronomical and geodetic latitude, longitude, and azimuth, as follows:

$$\phi = \Phi - \xi \tag{13-36}$$

$$\lambda = \Lambda - \frac{\eta}{\cos\phi} \tag{13-37}$$

$$\alpha = A - \eta\tan\phi \tag{13-38}$$

where ϕ is the geodetic latitude, λ the geodetic longitude, α the geodetic azimuth, Φ the astronomical latitude, Λ the astronomical longitude, A the astronomical azimuth, ξ the NS component of deflection of the vertical, and η the EW component. Thus, if astronomical observations are made for the geographic position of a point, the geodetic latitude and longitude can be determined using equations (13-36) and (13-37). Additionally, the geodetic azimuth of a normal section from theodolite to target can be obtained from the line's astronomical azimuth using equation (13-38).

The steps required to transform an observed astronomical azimuth of a normal section to the corresponding geodetic line azimuth are the following:

1. Convert from the astronomical azimuth to the geodetic azimuth of the normal section, Equation (13-38).

2. Correct the normal section azimuth for the height of target above the ellipsoid, Equation (13-35).

3. Compute the geodetic line azimuth from the normal section azimuth, Equation (13-29).

13-6-4. EDMI Distance Transformation to the Ellipsoid

As shown in Figure 13-21, the slope distance measured at some elevation must be transformed to its equivalent ellipsoid distance before being used in geodetic position computations. The method and formulas shown for distance transformation are adapted from Appendix I of Fronczek's "Use of Calibration Base Lines." [3]

The ray path of an electronic distance-measuring instrument (EDMI) is not exactly a straight line. Due to electromagnetic wave refraction by the atmosphere, the distance measured by an EDMI must be corrected for the ray-path curvature to obtain a straight-line chord distance before it is transformed to the ellipsoid. Symbols used in the distance transformation are as follows:

a = semi major axis of reference ellipsoid
e^2 = eccentricity squared of reference ellipsoid
α = mean azimuth of line, referenced to N or S
ϕ = mean latitude of line
H_i = elevation of station 1 or 2 above the geoid
ΔH_i = theodolite or reflector height above the station mark

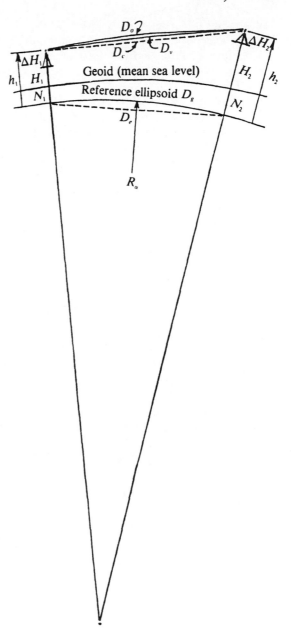

Figure 13-21. Transformation of measured distance to reference ellipsoid.

N_i = geoid height at station 1 or 2 (*Note:* In geodetic literature there is an unfortunate duplication in using the capital letter N to denote both the geoid height and radius of curvature in the prime vertical.)
h_i = instrument or reflector height above the ellipsoid
M = radius of curvature in the meridian

N = radius of curvature in the prime vertical

R_α = ellipsoid radius of curvature in azimuth α

k = index of refraction (for light-wave instruments $k = 0.18$, for microwave instruments $k = 0.25$)

D_o = observed slope distance corrected for temperature, pressure, eccentricity, reflector constant, and electrical center of EDMI

$D_v = D_o + I$ − second velocity correction

D_c = chord distance at instrument elevations

D_e = chord distance at ellipsoid surface

D_g = geodetic arc distance on the ellipsoid

The algorithm to transform an EDMI distance to geodetic arc distance is

$$M = \frac{a(1 - e^2)}{(1 - e^2 \sin^2 \phi)^{3/2}} \qquad (13\text{-}15)$$

$$N = \frac{a}{(1 - e^2 \sin^2 \phi)^{1/2}}$$

$$(N = \text{radius of curvature}) \qquad (13\text{-}10)$$

$$R_\alpha = \frac{MN}{M \sin^2 \alpha + N \cos^2 \alpha} \qquad (13\text{-}21)$$

$$D_v = \frac{D_o - (k - k^2)D_o^3}{12R_\alpha^2} \qquad (13\text{-}39)$$

$$D_c = 2\,\frac{R_\alpha}{k} \sin\left[\frac{D_1 k}{2R_\alpha}\left(\frac{180}{\pi}\right)\right] \qquad (13\text{-}40)$$

$$h_1 = H_1 + \Delta H_1 + N_1$$

$$(N = \text{geoid height}) \qquad (13\text{-}41)$$

$$h_2 = H_2 + \Delta H_2 + N_2$$

$$(N = \text{geoid height}) \qquad (13\text{-}42)$$

$$D_e = \left[\frac{\left(D_c^2 - (h_2 - h_1)^2\right)}{\left(1 + \dfrac{h_1}{R_\alpha}\right)\left(1 + \dfrac{h_2}{R_\alpha}\right)}\right]^{1/2} \qquad (13\text{-}43)$$

$$D_g = 2R_\alpha\left[\sin^{-1}\left(\frac{D_e}{2R_\alpha}\right)\right]\left(\frac{\pi}{180}\right) \qquad (13\text{-}44)$$

13-7. GEODETIC DATUMS

13-7-1. Regional Geodetic Datums

A geodetic datum is a mathematical model of the earth's figure on which geodetic computations are based. A *regional geodetic datum* is one that "fits" and is intended to be used in a specific area. Mitchell[4] defines a geodetic datum using geometrical geodesy concepts and the following five elements:

1. a = semimajor axis of the ellipsoid
2. b = semiminor axis of the ellipsoid
3. ϕ = latitude of the initial point
4. λ = longitude of the initial point
5. α = azimuth from initial point to another point

Unstated in the definition are assumptions that the ellipsoid and geoid are coincident at the initial point, and the earth's spin axis is parallel to the ellipsoid minor axis. The *North American Datum of 1927*, with its initial point at Meades Ranch in Kansas and based on the Clarke Spheroid of 1866, is an example of such a datum.

Further development and sophistication are reflected in Ewing and Mitchell[5] in which a regional geodetic datum is defined by:

1. a = semimajor axis of the reference ellipsoid
2. f = flattening of the reference ellipsoid
3. ξ_o = deflection of the vertical in the meridian at the datum origin, $\xi_o = \Phi_o - \phi_o$
4. η_o = deflection of the vertical in the prime vertical at the datum origin,

$$\eta_o = (\Lambda_o - \lambda_o)\cos \Phi$$

5. α_o = geodetic azimuth from the origin along an initial line of the network, $\alpha_o = A_o - \eta_o \tan \Phi$
6. N_o = geoid height at the datum origin—i.e., the distance between the reference and geoid
7. The condition that the ellipsoid minor axis be parallel to the earth's spin axis

The *North American Datum of 1927* also fits this definition of a regional geodetic datum, with one exception—i.e., the initial azimuth from station Meades Ranch to station Waldo is published as 75°28′09″.64, but orientation throughout the network is controlled by geodetic azimuth determinations utilizing astronomical azimuths and deflection of the vertical at numerous stations (such a station is *Laplace station*).

The geoid height at station Meades Ranch was assumed to be zero in 1927, but subsequent observations and refinements in the network have shown residual components of deflection of the vertical exist there. Hence, the ellipsoid and geoid coincide at station Meades Ranch ($N_o = 0$), but the two surfaces are not tangent there.

13-7-2. Global Geodetic Datums

A regional geodetic datum has an initial point on the ellipsoid surface and is chosen for its approximation to the earth's shape for a particular area or continental land mass, but a *global geodetic datum* has its datum point at the earth's center of mass and a reference ellipsoid chosen on the basis of a global "best fit." Thus, points on any continent can be accurately related to any other points throughout the world that are tied to the same global datum.

Global and regional datums are not defined the same way. They both use a reference ellipsoid, but since the earth's shape is actually determined by forces of gravitational attraction and centrifugal acceleration, those physical geodesy elements, along with others, are employed to define a global geodetic datum. Moritz[6] gives the following as defining parameters of a global geodetic datum:

1. The datum origin is located at the earth's center of mass

2. The Z-axis is the direction of the Conventional International Origin (CIO) defining a mean North Pole

3. The X-axis is parallel to the zero meridian adopted by the Bureau International De L'Heure (BIH) and known as the Greenwich mean astronomical meridian

4. A reference ellipsoid is defined by
   (a) a = the semimajor axis
   (b) GM = the geometric gravitational constant (the Newtonian constant G times the mass M of the earth, including the atmosphere
   (c) J_2 = zonal spherical harmonic coefficient of second degree
   (d) ω = earth's angular velocity

Given the four physical geodesy parameters of a reference ellipsoid, the ellipsoid eccentricity is computed using Equation (13-45), as follows:

$$e^2 = 3J_2 + \frac{4}{15}\frac{\omega^2 a^3}{GM}\frac{e^3}{2q_o} \qquad (13\text{-}45)$$

where

$$2q_o = \left(1 + \frac{3}{e'^2}\right)\arctan e' - \frac{3}{e'}; \; e'^2 = \frac{e^2}{(1-e^2)}$$

Equation (13-45) has e, the eccentricity, on both sides of the "equals" signs, which means it must be solved iteratively even though it is in chosen form.

13-7-3. Parameters of Selected Regional Geodetic Datums

1. *North American Datum of 1927: Clarke Spheroid of 1866*

a = 6,378,206.4 m	b = 6,356,538.8 m
ϕ_o = 39°13′26″.686 N	Origin: Meades Ranch, Kansas, USA
λ_o = 98°32′30.″506 W	
α_o = 65″28′09″.64 from origin to station Waldo	

2. *European Datum: International Ellipsoid of 1924*

$$a = 6{,}378{,}388.0 \text{ m} \qquad 1/f = 297.00$$
$$\phi_o = 52°22'51''.45 \text{ N} \qquad \text{Origin: Helmert Tower, Potsdam, Germany}$$
$$\lambda_o = 13°03'58''74 \text{ E}$$

3. *Pulkovo Datum: Krassovski Ellipsoid of 1942*

$$a = 6{,}378{,}245 \text{ m} \qquad 1/f = 298.3$$
$$\phi_o = 59°46'18''.55 \text{ N} \qquad \text{Origin: Pulkovo Observatory, Leningrad, former Soviet republic}$$
$$\lambda_o = 30°19'42''.09 \text{ E}$$

4. *Tokyo Datum: Bessel Elipsoid of 1841*

$$a = 6{,}377{,}397.155 \text{ m} \qquad 1/f = 299.1528$$
$$\phi_o = 35°39'17''.51 \text{ N} \qquad \text{Origin: Tokyo Observatory, Japan}$$
$$\lambda_o = 193°44'40''.50 \text{ E}$$

13-7-4. Parameters of Selected Global Geodetic Datums

1. *Geodetic Reference System of 1967*

$$a = 6{,}378{,}135 \text{ m}$$
$$GM = 3.98603*10^{14} \text{ m}^3/\text{sec}^2$$
$$J_2 = 0.0010827$$
$$\omega = 7.2921151467*10^{-5} \text{ rad/sec}$$
$$1/f = 298.247 \quad \text{(computed and rounded)}$$
exact

2. *World Geodetic System of 1972*

$$a = 6{,}378{,}135 \text{ m}$$
$$GM = 3.986005*10^{14} \text{ m}^3/\text{sec}^2$$
$$J_2 = 0.001082616$$
$$\omega = 7.2921151467*10^{-5} \text{ rad/sec}$$
$$1/f = 298.26 \quad \text{(computed and rounded)}$$
exact

3. *Geodetic Reference System of 1980*

$$a = 6{,}378{,}137 \text{ m}$$
$$GM = 3.986005*10^{14} \text{ m}^3/\text{sec}^2$$
$$J_2 = 0.00108263$$
$$\omega = 7.29115*10^{-5} \text{ rad/sec}$$
exact

$$e^2 = 0.006694380022903416$$
$$1/f = 298.2572221008827$$
$$e'^2 = 0.006739496775481622$$
$$f = 0.003352810681183637$$
$$b = 6{,}356{,}752.314140347$$
$$c = 6{,}399{,}593.625864032$$
(computed to 16 significant figures)

13-8. PHYSICAL GEODESY

13-8-1. Gravity and Leveling

The study of physical geodesy concerns the earth's gravity field and spacing of equipotential surfaces. Due to the earth's eccentricity and gravity-field irregularities, vertical spacing between equipotential surfaces varies from point to point (Figure 13-22a).

The equipotential surface most commonly known and understood is mean sea level, which serves as a reference datum for the elevation of points on or near the earth's surface. The vertical distance from mean sea level to each bench mark is its *orthometric height*, determined by applying an "orthometric height correction" to the differences in elevation observed along a given line of precise levels. It has been said that differential levels is the simplest surveying concept to teach or understand, but the most difficult when considering long lines, high precision, and large differences in elevation. The orthometric height correction is one subtle concept that makes the statement true.

It is easy to accept orthometric height as a vertical distance from mean sea level to the equipotential surface, until one considers that the orthometric height of the water surface at the south end of Lake Huron is 5 cm higher than the same equipotential surface at the north end. How can the same water surface have two heights, when a loop of precise levels around the lake shows no difference in elevation? Additional clarification is required.

Another way to visualize the apparent discrepancy is to observe, in Figure 13-22b, the difference in elevation between points 1 and 2 by following low route A that is greater than along high route B. A loop from point 1 to point 2, along route A and back along route B, will fail to close because equipotential surfaces are not parallel.

For the Lake Huron example, an orthometric-height correction must be computed and applied to obtain a geopotential number that is the same for the entire lake surface. For a precise level line, the orthometric height cor-

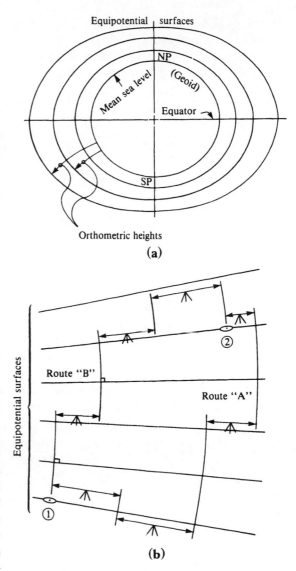

Orthometric heights

(a)

(b)

Figure 13-22. Orthometric heights related to precise leveling. (a) Equipotential surfaces. (b) Route-dependent leveling.

rection must be computed and applied to the observed differences in elevation to close the loop. This means that the correct difference in orthometric height between points 1 and 2 can be determined, irrespective of the route taken between the two points.

Orthometric height correction is a function of the force of gravity that, in turn, is related to the altitude, latitude, and longitude of a point. Hence, the further study of gravity and physical geodesy is vitally important to control surveyors and geodesists concerned with precise elevations over large areas. Adequate treatment of the topic is beyond this chapter's scope.

13-9. SATELLITE GEODESY

13-9-1. Geodetic Positioning

Although satellite geodesy is primarily concerned with orbits of satellites and other spacecraft, the use of satellite signals and space-age technology for geodetic positioning has revolutionized geodetic surveying practice. In the past, the line of sight between points was required to make triangulation measurements. However, with the launching of the Echo I satellite, intervisibility ceased to be critical because the satellite was photographed against a star background, simultaneously from two stations. Photographic images of the satellite were then analyzed to obtain a geometri-

cal tie between the two stations. A worldwide geometrical satellite triangulation net was completed using the Wild BC-4 ballistic camera system and Pageos satellites, but the program was discontinued in favor of better all-weather positioning systems.

Doppler Positioning with the Transit System

The next satellite positioning system to enjoy worldwide application was the Doppler positioning system, utilizing signals from a group of five satellites in polar orbits having approximately 107-min periods. The satellites broadcast two very stable frequencies (150 and 400 MHz), with orbital parameters and timing data phase-modulated on the signal. A stationary receiver on the ground (one type is shown in (Figure 13-23) receives a higher or lower frequency, depending on whether the satellite is moving toward or away from the receiver. This observed change in frequency (the *Doppler shift*) from numerous satellite passes is analyzed to determine a control point's geodetic

Figure 13-23.   Magnavox MX 1502 geoceiver/satellite surveyor. Portable, battery-operated precise point-positioning and translocation system (Courtesy of Magnavox Advanced Products and Systems Co.)

position. If two Doppler receivers are used in pairs, with one receiver positioned over a known control point, the relative positions of unknown points can be determined to submeter accuracy, with data from approximately 25 acceptable satellite passes.

The Doppler positioning (transit) system was developed by the U.S. Navy for global navigation, and it became operational in January 1964. Since released for civilian use in July 1967, the transit system has proved to be very reliable for worldwide geodetic point positioning. Line of sight is no longer required between geodetic control points, and unlike satellite triangulation it can be used day or night, rain or shine. Additionally, Doppler receivers are quite portable, making it possible to establish geodetic positions in remote locations with minimum logistical support.

The GPS (NAVSTAR) Positioning System

Despite the functional success and heavy use of the transit system, another satellite positioning system has been developed. The Geodetic Positioning System (GPS) will involve a group of 18 satellites with 12-hr periods and provide 24-hr global coverage. Chapter 15 is devoted entirely to this subject.

NOTES

1. W. Jank, and L. A. Kivioja. 1980. Solution of the direct and inverse problems of reference ellipsoids by point-by-point integration using programmable pocket calculators. *Surveying and Mapping Journal* 40 (3).
2. Geoid Contour Map of North American Datum of 1927; Washington D.C., U.S. Army Map Service.
3. C. J. Fronczek. 1977. Use of calibration baselines. NOAA Technical Memorandum NOS NGS-10
4. H. C. Mitchell. 1948. Definition of terms used in geodetic and other surveys. USC & GS Special Publication No. 242, Washington, D.C..
5. E. Ewing, and M. M. Mitchell. 1970. *Introduction to Geodesy*. New York: Elsevier North-Holland.
6. H. Moritz. 1978. The definition of a geodetic datum. 2nd International Symposium on Re-

definition of North American Geodetic Networks. Arlington, VA, April 24–28.

REFERENCES

BOMFORD, G. 1971. *Geodesy*, 3rd ed. London: Oxford University Press.

COLLINS, J. 1982–1983. The global positioning for surveying—today. *P.O.B Magazine* 8 (2).

FRONCEZK, C. J. 1967. Geoid contours in North America—from astrogeodetic deflections, 1927 North American datum. U.S. Army Map Service, Washington, D.C.

HEISKANEN, W. A., and F. A. VENNING MEINESZ. 1958. *The Earth and Its Gravity Field*. New York: McGraw-Hill.

HEISKANEN, W. A., and H. MORITZ. 1967. *Physical Geodesy*. San Francisco and London: W. H. Freeman.

HOAR, G. J. 1982. *Satellite Surveying—Theory, Geodesy, Map Projections*. Torrance, CA: Magnavox Advanced Products and Systems.

JORDAN, W., and EGGERT, O. 1962. *Jordan's Handbook of Geodesy*. (Translated into English by M. W. Carta). Washington D.C.: U.S. Army Map Service.

MORITZ, H. 1980. Geodetic reference system 1980. *Bulletin Geodesique* (*Paris*) 54 (3).

MUELLER, I. 1964. *Introduction to Satellite Geodesy*. New York: Frederick Ungar Publishing Company.

SEPPELIN, T. O. 1974. The Department of Defense world geodetic system 1972. Proceedings of the International Symposium on Problems Related to the Redefinition of North American Networks. Frederiction, New Brunswick, Canada, May.

SMITH, J. R. 1988. *Basic Geodesy—An introduction to the History and Concepts of Modern Geodesy Without Mathematics*. Rancho Cordora, CA: Landmark Enterprises.

STANSELL, T. A. 1978. The TRANSIT navigation satellite system—status, theory, performance, applications. Magnavox Government and Industrial Electronics Company, Report 5933, Torrance, CA, Oct.

TORGE, W. 1991. *Geodesy*. Berlin–New York: Walter DeGruyter.

VANIECK, P., and E. KRAKIWSKY, 1982. *Geodesy: The Concepts*. Amsterdam–New York–Oxford: North-Holland Publishing Company.

14

Inertial and Satellite Positioning Surveys

David F. Mezera and Larry D. Hothem

14-1. INTRODUCTION

Godetic control networks have classically been divided into two distinct categories: (1) horizontal and (2) vertical. Each network has its own respective set of monumented points—i.e., horizontal control (or "triangulation") stations and bench marks in horizontal and vertical networks, respectively. Similarly, classical control surveying methods are divided into two nearly independent categories: (1) horizontal methods that include traversing (Chapter 9), triangulation (Chapter 11), trilateration (Chapter 12), and combinations of the three; and (2) vertical methods that consist of differential and trigonometric leveling (Chapter 6).

More recently, methods have developed that overlap the two classical categories, enabling direct determinations of three-dimensional geodetic positions or position differences. The computer revolution, space-age spin-offs, and related technology have fostered the development of nonconventional geodetic surveying instruments, systems and techniques such as *electronic tachymeters* or *"total station" instruments*; *photogrammetric geodesy*; *laser ranging*; *very long base-line interferometry* (VLBI); and *inertial and satellite positioning*. Some of these are most

appropriate for scientific applications—e.g., earth crustal-movement studies. However, geodetic positioning (*geopositioning*) surveys by inertial and satellite methods have proved feasible for many applications previously restricted by default to classical geodetic control surveying methods. Future developments will probably expand the number of potential applications, so reliance in these two nonconventional control surveying methods will increase rapidly.

14-1-1. Inertial Positioning

Inertial surveying involves determining position changes from acceleration and time measurements and sensing the earth's rotation and local vertical direction. Adapted and modified to meet surveying practice requirements, *inertial positioning systems* are relatively new evolutions. The first commercial unit became available for nonmilitary use in 1975. Despite the short period of time since its introduction, the inertial surveying system (ISS) has been quite widely accepted. Applicability to surveying, mapping, geodesy, and engineering projects has been proven by numerous successful applications. The U.S. Bureau of Land Management's Cadastral Survey Division employs several ISS units for extensive original subdivisions of public lands in Alaska. They have been used on many federal mapping control projects in Canada. Inertial methods have also provided control for various geophysical prospecting projects.

An ISS features a number of attractive advantages when compared with more conventional surveying instrumentation. The equipment is typically mounted in a vehicle such as a van or helicopter, so the measurement work proceeds rapidly, with only brief stops required at survey stations of interest and a few selected intermediate points. Clear sight lines between adjacent stations are not required as in conventional traverse and triangulation surveys. Therefore, an ISS is especially well-adapted to surveys that involve numerous points, long distances between them, or areas where sight-line lengths are restricted by structures, vegetation, or rugged terrain. Considerable savings in time and labor costs, key features of an ISS, can be realized for such projects.

14-1-2. Satellite Positioning

Geodetic surveying techniques dependent on artificial satellites began in the early 1960s. Soon after the first Sputnik satellite was launched in 1957, results of satellite tracking and orbit-determination activities led to investigations of possible satellite-aided navigation systems. These, in turn, quickly demonstrated the potential for accurate satellite-based systems for geopositioning surveys. Since that time, numerous different schemes have been proposed, and several systems have reached operational status. *Satellite triangulation*, *SECOR*, and *Doppler satellite positioning* were three of the earliest systems to be investigated and developed. Satellite triangulation involved the use of precise metric cameras set up at widely spaced stations to simultaneously photograph illuminated satellites against their respective star backgrounds. Geodetic positions were determined by methods similar to photogrammetric aerotriangulation. Sequential

collation of ranges (SECOR) was essentially a trilateration system, with distances (ranges) from ground stations to satellites measured electronically using radio signals. Distances from three known satellite positions to an unknown ground point were sufficient to determine its geodetic position.

Doppler satellite positioning methods superseded both satellite triangulation and SECOR by the early 1970s, when portable geodetic Doppler receivers became available. Improved accuracy was also possible using the Doppler method. The design and development of a new-generation satellite global positioning system (GPS) began in the 1970s and continued in the 1980s.

Like inertial systems, satellite positioning does not require optical line of sight between adjacent survey points. However, a satellite receiver cannot operate if nearby obstructions block the satellite's signals from reaching the antenna. Both Doppler and GPS receivers can be operated in a point (absolute) or relative positioning mode. Demonstrations based on developing GPS technology have indicated a potential for the relative positioning of points to subcentimeter accuracy, with just a few minutes of observation.

14-2. DEVELOPMENT OF INERTIAL SURVEYING SYSTEMS

Navigation systems based on inertial technology continue to be used extensively in ships, aircraft, and space vehicles. The marine gyrocompass was invented in 1906, providing a nonmagnetic navigation aid for ships. The horizon and direction gyro was introduced during World War I, enabling pilots to orient aircraft with respect to the earth. The capability to derive reliable travel distances from measured accelerations and time intervals was developed during World War II; combined with advanced gyroscope technology, it provided the basis for today's modern navigation systems. Inertial surveying systems have evolved from these.

14-2-1. Military Initiatives

In the early 1960s, the U.S. Army Engineer Topographic Laboratories (ETL) contracted with General Electric to study the development of an artillery surveying system based on inertial navigation concepts. Between 1965 and 1972, ETL contracts with Litton Systems, Inc. resulted in a successful working model called the position and azimuth determining system (PADS). Test accuracies of ± 10 m in latitude, longitude, and height over 210-km open traverses strongly indicated that the PADS might be modified for geodetic surveying applications. By 1975, successive refinements by Litton for the U.S. Defense Mapping Agency (DMA) had produced, first, the inertial positioning system and, later, the rapid geodetic surveying system. The latter model was capable of position and elevation accuracies of ± 1 m when used in a closed traverse mode. This conclusively demonstrated the feasibility of inertial accuracies sufficient for many surveying purposes. The capability of inertial systems for determination of gravity and deflections of the vertical was also verified by the Litton and DMA tests.

14-2-2. Commercial Development

By early 1975, Litton had developed, concurrently with its work for the DMA, the Auto-Surveyor, a commercial version of the inertial positioning system. Since then, ISS units have been used for production work by several government agencies and numerous private enterprises in the United States, Canada, and elsewhere. In addition to the Auto-Surveyor, systems have been developed and are currently marketed by Ferranti Ltd. inertial land surveyor (FILS), and Honeywell, Inc. (GEO-SPIN).

The purchase of an ISS requires a large investment, with current costs in excess of half a million dollars for complete systems. Despite this, they have proven to be cost-effective for organizations with work volumes large enough to efficiently utilize their capabilities. Inertial surveying services are now provided on a contract basis by several private consulting companies. Because of high purchase costs, most inertial survey users will subcontract to one of these organizations for services or lease of equipment.

Reliabilities of early ISS versions were somewhat low, a common shortcoming of any complex, new hardware development. As newer versions were introduced, reliabilities have gone up. There have been reported average ISS "downtimes" as low as 5%.

14-3. INERTIAL POSITIONING THEORY

An in-depth knowledge of the operational theory and detailed inner workings of an ISS are not needed to successfully use inertial technology for surveying. However, it is helpful to know the basic theory and general function of its principal components rather than simply regarding the device as a "black box." Such knowledge will improve understanding of the capabilities and limitations of an ISS, leading to a well-informed choice between available surveying methods. If an inertial survey is selected, this background information will help to optimize survey design and organization. This section, therefore, introduces the basic concepts of inertial positioning.

Accelerations and times are the fundamental quantities measured by an ISS. Acceleration is a vector quantity with magnitude and direction, both of which typically vary continuously as the system moves from one point to another along the survey. At each precisely measured time interval, acceleration components are measured in directions parallel to three mutually orthogonal axes.

Acceleration is the rate of velocity change with respect to time. If a constant acceleration is maintained for a short time, the incremental velocity during that interval equals the product of acceleration and time. From the example in Figure 14-1, an acceleration of 2 m/sec^2 occurring over the 1-sec interval $a - b$ produces a velocity increase of $2 \times 1 = 2$ m/sec. Repetition of such a numerical integration process yields the terminal velocity for each time interval. Acting over the 1-sec interval $d - e$ in Figure 14-1, a constant velocity of 5 m/sec produces an incremental travel distance of $5 \times 1 = 5$ m.

In an ISS, the three acceleration components are measured at very small time intervals —on the order of 60 times per second—so they can be numerically integrated to yield accurate travel-distance components. When the ISS stops at a point of interest, each of three orthogonal components of distance traveled to that point equals the sum of incremen-

tal components in that particular direction since the previous stop. The ISS is a relative positioning system—i.e., distance components are measured from an initial known reference position, and new points are located relative to that control point.

14-4. INERTIAL SYSTEM CONFIGURATION

An ISS is a complex instrument consisting of precisely made mechanical parts, electronic circuitry, control devices, and computer hardware and software. However, to assist in the general understanding of the system's operation, a brief description of its principal components will be given, with emphasis on each component's function rather than its physical construction.

14-4-1. Accelerometers

The devices that measure accelerations are called accelerometers. Each may be visualized as a small pendulum equipped with a feedback system (Figure 14-2). When the accelerometer is at rest or moving at a uniform velocity, the pendulum hangs at its rest position. The pendulum's inertia resists changes in velocity, so it will try to swing away from its rest position when acceleration occurs. A detector senses the pendulum motion's beginning and sends an amplified electrical signal to a forcing (or torquing) system that, in turn, directs application of a force (or torque) just sufficient to

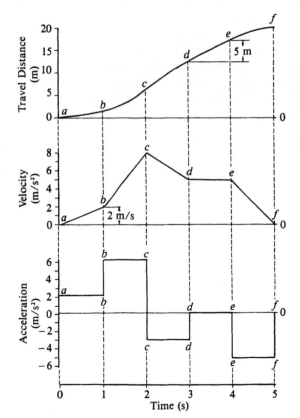

Figure 14-1. Acceleration, velocity, and travel distance as a function of time.

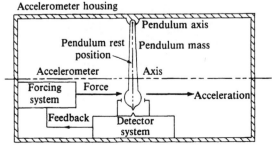

Figure 14-2. Simple accelerometer.

keep the pendulum from swinging. The feed-back signal to the torquing system is a direct measure of applied force and, thus, accelera-tion also, since the two quantities are propor-tional according to Newton's first law, which states that force equals mass times accelera-tion. The electrical signal is scaled and digi-tized, then multiplied by the corresponding time interval to obtain the incremental veloc-ity component.

Each accelerometer is fully sensitive only along its axis, the direction normal to both the pendulum's swing axis and its rest position axis (Figure 14-2). If acceleration is perpendic-ular to this axis, the instrument will detect nothing. If it occurs in any other direction, only that component parallel to the axis will be detected. Three orthogonally mounted ac-celerometers are thus required to measure the three constituent components of the total ac-celeration vector (Figure 14-3).

Without control or restraint or isolation from vehicle motions, the orthogonal ac-celerometer triad's orientation will vary with respect to inertial space as the ISS moves from point to point during a survey. To correctly combine incremental distance components between points for ISS position-change deter-minations, these orientation variations must be controlled and monitored.

14-4-2. Gyroscopes

Gryoscopic control is employed to stabilize the accelerometer triad in inertial space and provide a reference for detecting its attitude variations. A gyroscope, or gyro, has a symmet-rical rotor—a wheel or ball of dense metal made to spin rapidly about its axis on bearings located in a gimbal housing. The rotor devel-ops a large angular momentum and resists any forces that would tend to change its spatial orientation. Once a gyro has aligned itself in inertial space, a stable reference is provided that permits translations but resists rotations (Figure 14-4a).

Each gimbal also has a rotation axis at right angles to the rotor axis. It permits the spin-

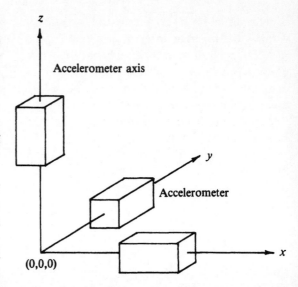

Figure 14-3. Orthogonal accelerometer triad.

ning rotor to retain its original angular orien-tation, despite gimbal rotations about either the rotor or gimbal axis. This gyro is said to have a "single degree of freedom" since gim-bal rotations about a third axis, mutually per-pendicular to the other two, will change the rotor orientation. For a gyro with "two de-grees of freedom," the gimbal axis itself ro-tates in bearings set in an outer gimbal, with a rotation axis perpendicular to both inner gim-bal and rotor axes. This permits the rotor to retain its angular orientation during outer gimbal rotations in any direction (Figure 14-4b).

The gimbal rotation axes have detector-feedback-torquer devices, similar in function to those on the accelerometers, that serve to counteract unwanted rotations. The torquers may also be used to drive the gyros to a desired orientation.

14-4-3. Inertial Measuring Unit

The accelerometer triad and associated gy-roscopes are mounted on an *inertial platform*, the heart of the inertial measuring unit (IMU). If single-degree-of-freedom gyros are used, there must be one corresponding to each ac-

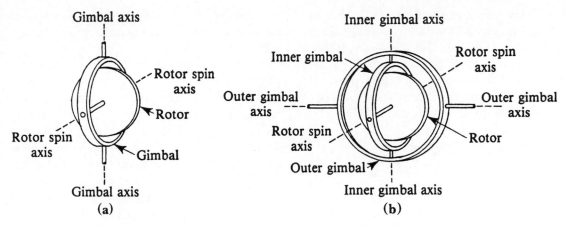

Figure 14-4. Simple gyroscopes. (a) Gyro with single degree of freedom. (b) Gyro with two degrees of freedom.

celerometer axis (Figure 14-5). Two gyros are sufficient if they each have two degrees of freedom. The accelerometer triad axes define the inertial platform axes, denoted either x-y-z or east-north-vertical. The whole inertial platform, of course, must also be gimbal-mounted with two degrees of freedom to provide for maintenance of its angular orientation in space, as the survey vehicle moves from one point to another. These gimbal rotation axes are also equipped with detector-torquer devices to monitor platform orientation varia-

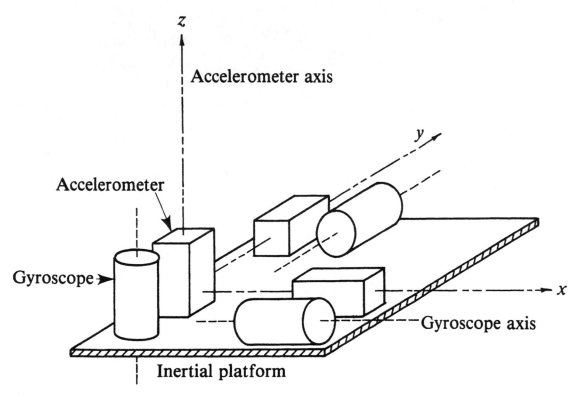

Figure 14-5. Inertial platform schematic.

tions and drive the platform to desired atti-
tudes. An actual inertial platform, discon-
nected from the other IMU hardware, is shown
in Figure 14-6. The IMU and associated parts
are housed in a closed case that provides a
thermally stabilized environment as well as
protection from damage (Figure 14-7).

14-4-4. Control and Data-Handling Components

An *on-board computer* is an integral component of an ISS. The final computation and
adjustment of the data are usually done on a
larger off-line computer, but the on-board unit
may perform the initial position calculation in
"real time" as the survey proceeds. The computer controls and monitors system operation,
and it may check and filter the raw measurement data before they are recorded for later

processing. It may compute misclosures when
an inertial traverse closes on a control point,
and it also directs the calibration and alignment of the platform at the initial point of a
survey.

A *control and display unit*—typically mounted
on or near the instrument panel of a survey
vehicle—controls operations of the IMU and
data-recording device. It allows communication with the on-board computer, initialization
of calibration and measurement sequences, input of externally collected data, and visual
readout of preliminary coordinates, closure
errors, and other operational data (Figure 14-
8).

A *data-recording unit*—typically a magnetic
tape cassette recorder— is used to capture the
raw or filtered measurement data for later
processing and analysis. It must be a well-designed unit to ensure reliable operation in the
survey vehicle's variable environment. Simi-

Figure 14-6. Inertial platform for Ferranti
ISS. (Courtesy of Shell Canada Resources,
Ltd.)

Figure 14-7. Complete Ferranti FILS II ISS. (Courtesy of Shell Canada Resources, Ltd.)

Figure 14-8. Control and display unit for Litton Auto-Surveyor.

larly, high-quality data tapes must be used for recording and preserving measurement information.

14-4-5.   Survey Vehicle and Power Supply

An ISS may be transported in a variety of different vehicles. For ground operations, a van-type light-duty truck is suitable (Figure 14-9), but helicopters are also frequently used, especially in remote areas where ground travel may be difficult or impossible (Figure 14-10). For a ground vehicle, the main requirement is installation space for the necessary equipment. In addition to seats for the driver and operator, there must be room inside the vehicle for the inertial measuring unit, computer, recorder, batteries, and the control and display unit, plus space in the engine compartment for the power supply alternator. Four-wheel-drive capability or even a tracked vehicle may be required if off-road travel is necessary. A helicopter must be of sufficient size and power to safely accommodate both the equipment and personnel. For some applica-

tions, the driver/pilot may also serve as the operator.

A power supply system must be provided to support operation of the IMU and its thermally stabilized environment, the on-board computer, control and display unit, and data recorder. For a truck-type survey vehicle, this is typically a 24-V battery system with recharge provided by a special heavy-duty alternator powered by the truck's engine. For helicopter use, the ISS may be operated from the aircraft's 28V power system.

15-5.   INERTIAL SURVEYING OPERATIONS

There are two primary system types used for inertial surveying: (1) space-stabilized and (2) local level. If an ISS is "space-stabilized," its inertial platform orientation is held fixed with respect to inertial space for the survey duration. With a local-level (or local-vertical, local-north) system, two platform axes are aligned parallel to local-vertical and north directions, respectively, so the third axis points eastward.

Figure 14-9.   Truck-mounted Litton DASH II ISS. (Courtesy of International Technology, Ltd.)

Figure 14-10. Ferranti FILS II inertial system mounted in Bell Jet Ranger helicopter. (Courtesy of Shell Canada Resources, Ltd.)

This local orientation is maintained and updated at each new point throughout the course of a survey. Much of the following description of ISS operation refers primarily to a local-level system.

14-5-1. System Calibration

Before beginning each day's operation, or each time the ISS is turned on anew, the system must be warmed up and calibrated. This calibration or alignment is mainly an automated sequence of self-tests initiated by the operator, but thereafter directed by the on-board computer under system software control. The calibration procedure may require an hour or more to complete. During this time, the IMU components are monitored for operational stability, and the gyro and accelerometer parameters tested to see if they fall within acceptable ranges. These parameters are recorded for later use during filtering and adjustment. The calibration sequence is repeated, perhaps several times, if the parameters do not meet specifications. Continued failure to successfully calibrate may necessitate system adjustment or repair.

In addition to daily calibrations, it is desirable to conduct periodic dynamic calibrations. Ideally, this would consist of inertial traversing over an L-shaped course, with traverse legs along cardinal directions and precise coordinates at the angle and terminal points. Analysis of the results will yield accelerometer and gyro parameter information.

14-5-2. Initial Orientation of the Inertial Platform

If the maximum sensitivity direction of a stationary accelerometer is perpendicular to the direction of local gravity, it will indicate zero acceleration. At the initial point of a survey with a local-level system, the platform leveling is accomplished by orienting the X- and Y-accelerometers to produce zero accelerations and aligning the Z-axis of the system along the local plumb line. If deflection of the vertical is known, platform orientation with respect to the geodetic reference is also defined.

A gyrocompassing procedure is used to align the sensitive direction of the Y-accelerometer toward local north. The earth's rotation causes a torque on the corresponding gyroscope axis, and the resulting precession concludes with it aligned in the plane of the earth's rotation axis (Figure 14-11a).

14-5-3. Compensation for Platform Movements

Without continuous correction of the gyroscope orientations, the local-level alignment of a platform is lost immediately because of the earth's rotation, even though the ISS remains stationary at the initial point. Compensation to maintain alignment with the geodetic reference is accomplished by the gyroscope torquers, as directed by the on-board computer (Figure 14-11b). The angular orientation correction is based on the earth's mean rotation rate and elapsed time since the initial orientation or previous correction.

When the platform is moved to another point, a similar loss of alignment occurs if the computer does not signal for orientation correction. But the required compensation is now composed of two parts: (1) allowance for earth rotation during the elapsed travel time and (2) compensation for the distance traveled (Figure 14-11c). With a space-stabilized ISS, compensations for both components of platform movement are computed and applied during the data processing, rather than torquing the gyros to reorient the platform during a field observation period.

Periodically during an inertial survey, it is necessary to stop the vehicle and initiate a control instruction sequence to inform the computer that the ISS is stationary, and thus all three current velocity values should be zero. This procedure is called a *zero velocity update* or ZUPT. Each accelerometer and gyro output is monitored and recorded as an indication of that component's drift since its initial calibration state. A ZUPT requires a stop duration of a minute or shorter.

14-5-4. Position Measurements

When the ISS reaches a point where a geodetic position is desired, a *station mark* is initiated. This position measurement procedure is similar to the ZUPT, except that in addition to the accelerometer and gyro data, preliminary coordinates are also recorded.

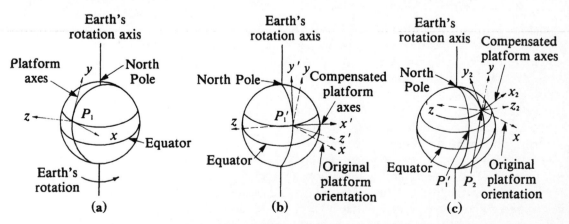

Figure 14-11. Compensation for inertial platform movements. (a) Local-level platform orientation at point P_1. (b) Compensation for rotation of point P_2 to P_1. (c) Compensation for rotation to point P_1 and platform travel to point P_2.

Since the basic point of reference for an ISS measurement is at the inertial platform's center and thus inaccessible, an offset reference mark is set on the exterior of the IMU by the instrument manufacturer. The ISS computer software is designed to account for components of this offset, in all position computations.

During field operations, offset measurements must be made since the reference mark cannot be set directly on the survey points. However, in most cases, the offsets are not measured directly to the reference mark on the IMU but for convenience and speed to a secondary reference point instead. This *measuring mark* may consist of a simple sighting

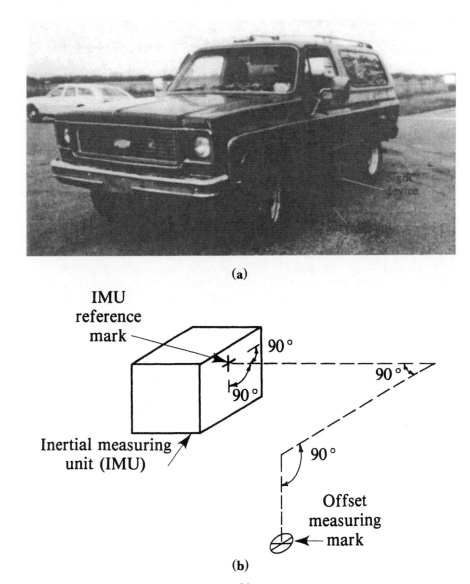

(a)

(b)

Figure 14-12. ISS offset measuring mark. (a) Offset measuring-mark sighting device beneath driver's door. (b) Measurement of offset components from IMU to measuring mark.

device mounted on the vehicle exterior—truck door or helicopter skid—for convenient use by the driver (see Figure 13-12a). For points that are not so accessible, offset direction, distance, and elevation difference may be measured from a protractor mounted on the front or rear of the vehicle (see Figure 14-13). For greater accuracy, especially for larger offsets, these components may be measured with an electronic tachymeter or total station instrument from a measuring mark located inside the vehicle (see Figure 14-14).

The offset components to the measuring mark may be determined using conventional surveying instruments (see Figure 14-12b). These values are input to the computer as constants to be applied to each position determined. Any additional offset from the measuring mark to a particular survey point also must be input, so the total offset from an inertial platform to the point will be included in the position computation. Regardless of the technique employed, care must be used for all offset measurements to ensure that desired positional accuracies are maintained.

For some surveys with less stringent accuracy requirements, both ZUPTs and station marks have been accomplished with a nearly stationary ISS is in airborne helicopter hovering over the point. The U.S. BLM has investigated the use of a similar technique for meandering water boundaries in Alaska.

14-6. INERTIAL DATA PROCESSING AND ADJUSTMENTS

Like all surveying measurements, those made with an ISS are subject to errors from instrumental, natural, and personal sources. Many procedures included in the calibration, measurement, and data-reduction/adjustment phases are designed to limit error magnitudes and reduce their effects on final survey results. The measurements are also subject to mistakes or blunders due to confusion or carelessness, but these should be discovered and eliminated through the use of repeated measurements and independent checks.

14-6-1. Error Sources and Characteristics

Nonorthogonality of the three accelerometer axes may cause the measured acceleration components to be incorrect. The daily ISS calibration prior to field observations can virtually eliminate this error if it remains con-

Figure 14-13. Measuring offset from vehicle measuring mark to survey point. (Courtesy of Shell Canada Resources, Ltd.)

Figure 14-14.   Offset measurement to survey point using electronic distance-measuring instrument. (Courtesy of Shell Canada Resources, Ltd.)

stant during the day's survey. The electronic signals are scaled and digitized, so if scaling is incorrect, double integrations of resultant erroneous accelerations result in incorrect travel-distance components. Errors in elapsed time measurements also result in incorrect travel distances, but time-measurement accuracy is very high, so these errors are very small.

Gyroscopes slowly drift away from their original alignments, thus allowing the inertial platform to drift. Most of this drift is a linear function of time that can be modeled and, therefore, eliminated during data reduction, but the small irregular component is more difficult to eliminate, and random position errors may result. Accelerometers also have generally linear drifts for which corrections can be made, but small, irregular drifts may contaminate accelerations, thus affecting computed distance components. Vehicle vibrations caused by the engine or strong winds may also cause accelerometer errors during a stop for a ZUPT or position determination.

Setting the measuring mark over a survey point or measuring the mark-to-point offset are processes subject to all the observation errors inherent in conventional surveying methods. The same is true of the procedure for finding the measuring-mark offset from the IMU reference mark. The ISS control unit operations, including data input, always provide opportunities for human mistakes or blunders, as do the data-reduction procedures. The computer software may be designed to automatically detect some large discrepancies, but as in all surveys, good practice will include careful and thorough checking of all input, output, and computed values.

14-6-2. Data Filtering

The on-board computer of an ISS may operate software that statistically filters raw measurement data as it is collected. The best estimates are obtained by a *Kalman filter*, an algorithm that weights each new data value based on the characteristics of previous measurements and prior knowledge of the nature and usual magnitudes of errors. Initial estimates of gyro and accelerometer error parameters, such as drift rates and biases, are supplied to the Kalman filter after the daily calibration procedure. As the survey proceeds, these parameters are updated at each stop for a ZUPT or station mark. This statistical filtering process significantly retards error growth in preliminary coordinates calculated by the on-board computer, so these real-time position estimates may be accurate enough for some purposes without further adjustment. The position error accumulates so rapidly in an ISS without a Kalman filter that rough field coordinates have little value. In such cases, it is necessary to wait for off-line postprocessing of the raw data.

14-6-3. Horizontal and Vertical Positions

For each inertial traverse, coordinates of the beginning control station are input to the ISS. Following recommended practice, a survey concludes at a second control point with known coordinates, forming a closed traverse. The reliability of an open-ended traverse is always uncertain and should be avoided regardless of the accuracy sought.

At each desired survey point along a traverse, the ISS accumulates incremental travel-distance components from the previous point to obtain three-dimensional geodetic coordinates of the current point, relative to the initial control station. Coordinates are computed in terms of latitudes, longitudes, and elevations—although for display purposes and later use, latitudes and longitudes may be transformed into a local plane coordinate system such as state plane coordinates (i.e., output in terms of X, Y, elevation).

The vertical accelerometer measures accelerations due to the combined effects of vehicle motion and the local gravity vector. These two components must be separated to determine elevation differences between surveyed points. A gravity-effect estimate can be re-

moved from the vertical acceleration at each integration step. This estimate is derived from a mathematical model of the earth's gravity field or measured gravity value at the initial station. Elevation differences are computed using residual vertical accelerations.

For a space-stabilized ISS, each of the three accelerometers will detect a component of vertical acceleration, since platform orientation is not held fixed with respect to local vertical. In this case, the computer software must enable determination of the net vertical acceleration and separation of the gravity-field component, during either on-line processing by the on-board computer or off-line postprocessing.

14-6-4. Data Adjustments

In following good practice, an inertial traverse begins at a known control station and ends at another. Closing a traverse on a known point provides a reliability check of intermediate survey points, since preliminary closing point coordinates—inertially derived—can be compared with known values. Any coordinate misclosures should be within both the tolerances expected for the ISS utilized and ranges specified for the particular survey. If either test fails, the traverse must be resurveyed.

If misclosures are within allowable limits, preliminary coordinates should be adjusted (or *smoothed*) by distributing accumulated errors of closure in proportion to elapsed travel time, travel distance (analogous to the compass rule traverse adjustment), or a time-distance combination. Smoothed coordinates may be determined by the on-board computer immediately on conclusion of the traverse run and recorded on the data tape. Alternatively, smoothed coordinates can be obtained during off-line postprocessing of the data.

If desired, detailed mathematical algorithms may be employed to rigorously model ISS error parameters and thus produce more accurate distributions of accumulated closure errors. Despite various attempts to improve modeling of the inertial measurement process, experience has shown that smoothed coordinate values typically exhibit sizable systematic error effects. To help reduce such systematic errors and provide desirable checks, each traverse should be run in both forward and reverse directions—e.g., from control point *A* to control point *B*, then from *B* back to *A*. Hereafter, a traverse run in this fashion will be referred to as a "single inertial traverse." This method is analogous to the common procedure of measuring with a theodolite in both direct and inverted telescope positions to cancel systematic instrumental errors from the mean angle value. Adjusted survey point coordinates can be taken as the weighted mean of smoothed values from forward and reverse traverse runs.

Uncertainities of adjusted survey point coordinates can be reduced somewhat by additional repetitions in the forward and back directions, just as any observed value may be statistically improved by repeated measurements. In any case, numbers of forward and reverse runs should be equal along any traverse line.

14-7. INERTIAL SURVEY DESIGN AND TYPICAL RESULTS

Because of the unique operational characteristics of an ISS, careful preparation and detailed organization are extremely important, both before and during an inertial survey.

14-7-1. Logistics

Since inertial system hardware is usually mounted semipermanently in a survey vehicle, accessibility of points to be positioned is a primary consideration during initial planning and design. When an ordinary ground vehicle is used, survey points must be either on or near traversable roadways or trails, depending on the offset measurement method utilized and terrain difficulty. Availability of a four-wheel-drive or all-terrain survey vehicle to transport the ISS allows more freedom in selecting survey point locations. Helicopters provide even greater flexibility, especially in remote areas where ground vehicle travel is difficult or impossible. However, the relatively large space required for helicopter landing may generate additional logistical problems, particularly in heavily forested areas.

Selecting and marking desired survey points are tasks best completed before ISS measurements commence. Since errors tend to build up as elapsed time increases, unnecessary delays should be avoided during inertial traversing. Because the number of points positioned by an inertial survey is usually large, work involved with reconnaissance, selection, marking, referencing and describing desired station locations comprises a significant portion of the total project effort. If many of the points must be permanently monumented survey stations, the labor, materials, and transportation required can make this project phase very costly in terms of both time and money.

Whether points are temporary or permanent, they must be clearly marked or flagged, so an ISS can be transported directly to each one without delay. In some cases, it may be necessary to run a "scout vehicle" in advance of the ISS unit to guide it to consecutive survey station locations. If approximate coordinates of a station are known, the navigational capability of an ISS may be used to locate the point marker. The display unit can be directed to indicate direction and distance to a desired point as the ISS approaches it. In congested areas, traffic-control measures may be necessary to ensure safety for the ISS vehicle and its occupants. Inertial surveys are sometimes conducted during late night and early morning hours to avoid peak traffic conditions.

14-7-2. Other Design Considerations

In addition to logistics, other factors must be considered in the design of an inertial survey. An ISS is a relative-positioning device; thus, two known geodetic control stations must be available to serve as inertial traverse beginning and ending points, respectively. Open-ended traverses and closed-loop traverses with only one control station should be avoided, since no satisfactory check or systematic error-correction technique exists for either of these survey types. In fact, if the purpose is to densify an area's existing geodetic control network, the U.S. Federal Geodetic Control Committee (FGCC) recommends that each inertial traverse tie into a minimum of four existing control points rather than just two. Additionally, each traverse should be run in both forward and reverse directions as described in Section 14-6-4. Accuracies of inertially derived positions cannot exceed those of control stations utilized for the survey and generally will be somewhat lower. Unchecked or suspect geodetic control should be avoided since erroneous control values incorporated into an inertial survey will degrade the accuracy of ISS results.

The direct relationship between elapsed time and ISS positional error accumulation constrains allowable traverse length. The desired quality of results limits allowable elapsed time intervals between both ISS stops at terminal control stations and consecutive ZUPT or station mark stop-points. Typical ZUPT intervals vary between about 1 and 8 min, depending on the accuracy level desired. Total elapsed time for a single inertial traverse is commonly limited to 1 or 2 hr or shorter for most applications. But, depending on the ISS vehicle type, inertial traverse lengths can easily reach 10 to 100 km or more.

An ISS produces the best results when vehicle travel approximates a straight-line path connecting the terminal control stations, desired intermediate points, and ZUPT stop-points. Point spacings and vehicle travel rate should each be nearly uniform and sharp horizontal direction changes ("doglegs") avoided. Normal BLM guidelines restrict any desired point along an inertial traverse to be within a corridor extending no more than 0.8 km either side of a straight line connecting the two terminal control stations. The FGCC recommends a maximum angular deviation from a straight-line path of 20 to 35°, depending on the desired accuracy.

For greatest reliability, each important survey point is linked to control stations by more than a single inertial traverse. Such a point should be connected to additional control stations by a crossing inertial traverse (see Figure 14-15a). Ideally, for a control densification or similar area-wide survey, known stations are

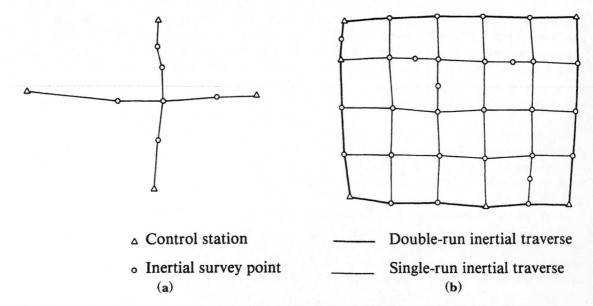

    △ Control station   —— Double-run inertial traverse

    ○ Inertial survey point   ——— Single-run inertial traverse

          (a)   (b)

Figure 14-15. Inertial traverse survey configurations. (a) Intersecting inertial traverses. (b) Inertial grid network.

located around the area perimeter, and desired survey point coordinates established using a series of crossing and interconnected inertial traverses surveyed in a uniform grid pattern (see Figure 14-15b). In practice, it is very difficult to design such perfect survey networks, especially for projects employing ground vehicles traveling on established roadways. However, to achieve maximum ISS accuracy, the network design process must pursue this ideal configuration, and inertial traversing must be done with great care, with a least-squares network adjustment used to obtain the most probable coordinate values. As with any type of survey, it is desirable to verify the inertial results on a sample of project points, making independent checks using another method.

14-7-3. Typical Inertial Survey Results

In a discussion of obtainable ISS accuracies, it is important to describe the survey conditions under which the results were produced. The possibility exists that uncorrected systematic errors remain even after a rigorous network survey and adjustment; it most especially exists for a single traverse with no cross-connecting inertial runs, where accuracy may be affected by such factors as ZUPT intervals, quality of existing control stations, and traverse shape, length, and duration.

For a Litton Auto-Surveyor ISS, a single, generally straight-line traverse of 2-hr duration can be expected to produce positional standard errors for points along the traverse path of about 0.40 m horizontally and 0.30 m vertically. For a similar but L-shaped traverse, the corresponding values are about 1.00 and 0.30 m, respectively. These results assume reasonable care in survey conduct, uniform ZUPT intervals of 3 to 5 min, and use of manufacturer's software. For a grid network of criss-crossing traverses (see Figure 14-15b), the corresponding positional standard errors are both expected to be about 0.15 m for a Litton ISS. This assumes adequate area perimeter control,

traverse durations each shorter than $1\frac{1}{2}$ hr, and a rigorous least-squares network adjustment. Some users have reported relative positional errors below the 0.10 m level under near-ideal or strictly controlled conditions.

When using a Ferranti FILS ISS in a helicopter with ZUPT intervals of 3 to 5 min was used, mean horizontal and vertical errors (relative to convetional survey positions) of about 0.25 and 0.15 m, respectively, have been obtained for generally straight-line traverse of no more than $1\frac{1}{2}$-hr duration and 30-km length. But with a ground vehicle operating in undulating or rugged terrain that necessitated circuitous and L-shaped traverses, the Ferranti ISS produced corresponding mean errors of 0.75 and 0.25 m, respectively—values quite similar to the expected errors for a Litton ISS. Likewise, a Ferranti FILS system should produce grid network accuracies comparable to those for an Auto-Surveyor, or about 0.15 m both horizontally and vertically.

For a Honeywell GEO-SPIN ISS in a ground vehicle, expected accuracies are about 0.15 m horizontally and 0.30 m vertically over a generally straight-line 51-km traverse. These numbers degrade to about 0.50 and 0.35 m, respectively, for a 39-km L-shaped traverse. For two intersecting traverses (as in Figure 14-15a) with lengths of 51 and 26 km, corresponding accuracies of 0.10 and 0.20 m can be expected. In one test, several nearly straight-line traverses were run with ZUPT intervals of 1 min and the results compared with positions obtained by conventional control surveying methods. For six traverses, each 4 km in length with survey mark points at about 0.4-km intervals, root-mean-square (RMS) errors were below 0.06 m, both horizontally and vertically. On two double-run 2-km traverses, corresponding RMS errors were 0.04 m or smaller. Thus, the potential of a GEO-SPIN ISS for establishing closely spaced control stations in a grid network (Figure 14-15b) is excellent. If traverses of such high quality are combined with a grid pattern and subjected to a rigorous adjustment, comparable accuracies should be attainable over much longer traverse lengths.

An ISS also has the potential to yield information about the earth's gravity field. Although not ordinarily needed for most survey purposes, gravity anomalies and deflections of the vertical may be derived from the observed data, provided that appropriate software configurations are available for both the on-board computer and postprocessing hardware. This gravity information may be a very important component of certain geodetic, geophysical, and geological surveys. Under good conditions, accuracies of 1.0 mGal and 0.5″ have been obtained for gravity anomalies and deflections of the vertical, respectively.

14-8.   INERTIAL SURVEYING APPLICATIONS

Many different types of surveys have been successfully executed using inertial technology. In many instances, ISS usage has proved to be an economically viable alternative to conventional surveying methods, particularly for larger projects requiring geopositioning of many points over large areas.

14-8-1.   Application Examples and Projections

As previously implied, most ISS projects could be characterized as control surveys. In such cases, an inertial survey provides a spatial framework of geodetically positioned points to support another activity, such as a more detailed survey utilizing a different technique.

An example of such a situation is a topographic mapping project utilizing photogrammetry (see Chapter 20). Such a project might be necessary for mapping a utility transmission-line corridor, a highway or railway right-of-way, or a new dam and reservoir site. An ISS can provide the required photo ground control points, positioned with respect to the area's existing geodetic control network. ISS accuracies would normally be adequate to support photogrammetric mapping. In remote areas,

the helicopter mode probably would be most efficient, but a ground vehicle could be used if a good road network exists.

The navigation capability of an ISS can be relied on to guide the survey vehicle to a preselected control point location. For example, as an aerial photography flight is planned, desirable photo control locations may be identified on a map and coordinates scaled. These values can be input to the on-board computer and used to direct the ISS to a desired point's immediate vicinity. A suitable point location can then be selected after a first-hand inspection of the area. While accurate position is observed with the ISS, this point is staked and flagged for a separate crew that will place the photo targets.

Cadastral surveying is another major application area for which an ISS is well-suited, particularly for large land parcels in remote areas. The BLM Cadastral Survey Division utilizes helicopter-borne Litton Auto-Surveyors to establish original U.S. public land survey system (PLSS) boundaries in Alaska. Inertial surveys have also been used to determine coordinate positions of remonumented PLSS section and quarter-section corners in several other states, such as Illinois and Wisconsin, so the geodetic and legal (real property) networks could be integrated.

Other extensive applications of inertial methods have included engineering, geophysical, and construction surveys. Seismic and gravity surveys for geophysical prospecting have utilized ISS geopositioning, including some surveys in offshore areas and positioning for drilling platforms. Inertial systems have also been used in helicopters in conjunction with laser profiling of ground terrain.

Research and development have been conducted using an ISS to determine the position and orientation of an aircraft and camera during an aerial photography flight; this system, when perfected, will greatly reduce photogrammetry ground control requirements. Some ISS use has developed in conjunction with mapping boundaries of natural resources, such as wetlands and soil types; similar inertial surveys could be employed on floodplain and land use/land cover mapping.

When the special characteristics of an ISS are considered, it seems likely there are numerous additional applications yet to be exploited. This is a self-contained system that has no line-of-sight restrictions, can move rapidly from point to point in a ground vehicle or helicopter, and produces three-dimensional geodetic positions with an accuracy adequate for many purposes. Better understanding of its capabilities should lead to increased future use.

14-8-2. Economic Considerations

Because of its high cost, the purchase of an ISS is only viable for an organization that has an applications volume sufficient to utilize its very high productivity. Measured in terms of output point positions, it may be as much as 5 to 20 (or even more) times that of conventional survey methods. However, alternatives to purchase exist, including inertial surveying services via contract and inertial equipment leasing.

Several factors may influence the cost of an inertial project. A significant part of the total expenditure may be the mobilization cost—the expense to ship equipment and transport personnel to the project site, install the ISS in a survey vehicle, and other overhead factors that are relatively fixed. The project area's geographical location with respect to the source of the inertial equipment or services can greatly influence these mobilization costs. Of course, the number and accuracy of survey points desired and project area size are important factors: The costs per point generally decrease rapidly as point numbers increase. Helicopter transportation, on a daily basis, will be much more expensive than use of a ground vehicle; however, increased productivity could make helicopter use much more economical. Advance preparation and planning by the client, as well as active participation during the actual inertial survey, will help keep down

the project cost. Because of the many expense variables involved, it is difficult to make generalized statements about inertial survey outlays. However, there have been many inertial projects reported to have a cost per point one-half to one-quarter (or even below) that of an equivalent conventional survey. Corresponding survey duration comparisons appear to be even more dramatic: There have been reports of as much as a 20-to-1 reduction compared with time estimates for corresponding conventional surveys.

14-9. DEVELOPMENT OF SATELLITE POSITIONING SYSTEMS

Shortly after the former Soviet republic launched its first Sputnik satellite in October 1957, its orbit was accurately determined by measuring and analyzing the Doppler frequency curve from a single satellite pass. Shortly thereafter, the U.S. Navy navigation satellite system (NNSS) was conceived. Early research and development was conducted at the Johns Hopkins University Applied Physics Laboratory. The U.S. Navy supported development of the NNSS, commonly referred to as the transit system, to provide accurate, all-weather, passive, worldwide navigation capability for its Polaris submarine fleet.

The first United States satellite to be used extensively for geodetic purposes, the Vanguard, was launched in 1958. The original navigation satellite and another designed to broadcast precise information on its own orbital position were orbited in 1960 and 1961, respectively. Since 1964, there has been at least one transit-equipped satellite continuously operational to provide Polaris navigation.

In addition to the transit system, several other satellite positioning systems developed concurrently during the 1960s. Photogrammetric satellite triangulation was used to establish a North American reference network of 21 widely spaced stations and a 45-station worldwide net that interconnected several previously unrelated areas. At several stations, calibrated metric cameras were synchronized to simultaneously photograph illuminated satellites against their star backgrounds. Some satellites carried flashing lights, others utilized sun illumination. Photographic coordinate measurements of satellite and star images were used to compute directions to the satellites, and geodetic positions of camera stations then found by classical astronomy techniques. Required field equipment was bulky and transportable only by large vans; data-reduction procedures long and complex; and positional accuracy limited to about 5 m, so satellite triangulation was replaced by other methods in the early 1970s.

The ranging method (satellite trilateration) was also utilized in the 1960s, and a phase-shift measurement technique, similar to that employed in EDM surveying instruments (see Chapter 5), used with the SECOR system. A ground station transmitted a phase-modulated electromagnetic signal to a satellite equipped with a SECOR transponder. The signal was received and retransmitted to the ground station, where its returned signal phase could be measured relative to the transmitted signal phase. The observed phase shift is a function of signal travel distance. If one unknown station and three known stations simultaneously measure the ranges to at least three satellite positions, the unknown station coordinates can be determined. Repeatability (precision) using SECOR was about 6 m, but for widely separated stations, relative accuracies as high as 1 : 100,000 were obtained by comparison with conventionally surveyed positions. Other methods also superseded SECOR because of its accuracy limitations and low mobility due to complex bulky ground station antenna and receiver equipment.

Another trilateration method developed during the 1960s, satellite laser ranging (SLR), became operational in the early 1970s, and continues in use today. Distances to satellites equipped with retroreflectors can be measured with 1-cm accuracies. Despite these ex-

cellent results, SLR use is limited primarily to crustal motion studies owing to the complexity of instrumentation and data-processing requirements.

The Doppler NNSS became available for civilian use in 1967, when details of receiver equipment and computation requirements were made public. Until then, usage had been restricted primarily to the navigation of U.S. Navy vessels. Although the control surveying potential of the transit system had been demonstrated, applicable Doppler receiver equipment was bulky with complex field operation and data-reduction procedures. But portable geodetic receivers were developed and available for testing by the early 1970s (see Figure 14-16 and 14-17). Initial results indi-

Figure 14-16. JMR-1 Doppler survey set in field operation. (Courtesy of JMR Instruments, Inc.)

cated that accuracies better than 1.5 m were possible, thus making many geodetic applications feasible. Improvements in hardware, software, and operational procedures have resulted in achievable accuracies of 0.5 to 1.0 m for point positioning and 15 to 30 cm for relative positioning. Today, the NNSS is used routinely for both navigation and geodetic control surveying purposes in the private sector.

In 1973, the success and future promise of transit led the U.S. Department of Defense to initiate design and development work on the navigation satellite timing and ranging (NAVSTAR). The first NAVSTAR satellites were orbited in 1978 to support testing and development of both the system's space and ground control segments and receivers for navigation and geodetic purposes.

14-10. SATELLITE ORBITS AND COORDINATE SYSTEMS

All satellite-based positioning systems require that several satellite coordinate positions either be known a priori or observed as an integral part of the measurement scheme. In either case, it is necessary to know the parameters used to define a satellite's orbit and elements of the coordinate system in which its positions will be expressed.

14-10-1. Satellite Orbits

The path around the earth of an orbiting artificial satellite is approximately elliptical (see Figure 14-18a). An ellipse is the locus of points with a constant sum of distances from the two fixed focal points. For a satellite orbit, one

Figure 14-17. Magnavox ANPRR-14 geoceiver. (Courtesy of Magnavox Advanced Products and Systems Co.)

focal point is at the geocenter (the earth's center of gravity). Ellipse size and shape are fixed by two parameters such as semimajor axis and eccentricity. The perigree point marks the satellite's closest approach to earth while the apogee is farthest away. If extended, a line connecting the perigee and apogee, the *line of apsides*, also passes through the ellipse's foci and coincides with the x_s-axis of an orbital coordinate system. This is a right-handed, three-dimensional Cartesian system with its y_s- and z_s-axes in and perpendicular to, respectively, the elliptical plane.

The orbital ellipse orientation, with respect to an astronomical or celestial coordinate system (see Chapter 17), may be defined by four

parameters: (1) the true anomaly, (2) argument of perigee, (3) inclination, and (4) the right ascension of the ascending node (see Figure 14-18b). These four parameters, plus the semimajor axis and eccentricity, define a smooth nonvarying elliptical path and are sometimes referred to as the *Keplerian orbital elements*. However, an orbit would follow such a path exactly only if the earth was a uniform-density sphere, and no disturbing forces acted on a satellite other than gravitational attraction toward the geocenter.

Of course, these conditions are not satisfied, and disturbing factors such as the following must be taken into account: (1) the earth's irregular gravity field; (2) attractions of the

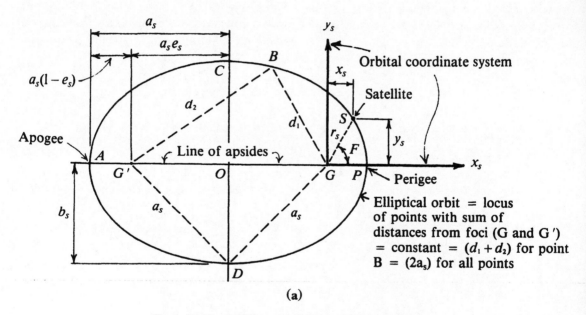

(a)

Figure 14-18. Satellite orbit definition and orientation.
(a) Elliptical satellite orbit and orbital coordinate system.
(b) Astronomical orientation of elliptical satellite orbit.

O = center of ellipse

G, G' = foci of ellipse

G = center of gravity of earth (geocenter) and origin of orbital coordinate system

a_s = semi-major axis = $\overline{OA}$ = $\overline{OP}$

b = semi-minor axis = $\overline{OC}$ = $\overline{OD}$

e_s = eccentricity = $\{(a_s^2 - b_s^2)^{1/2}/a_s\}$

F = true anomaly

r_s = distance from geocenter (G) to satellite

r_s = vector from geocenter (G) to satellite

$$r_s = \begin{bmatrix} x_s \\ y_s \\ z_s \end{bmatrix} = \begin{bmatrix} r_s \cos F \\ r_s \sin F \\ O \end{bmatrix}$$

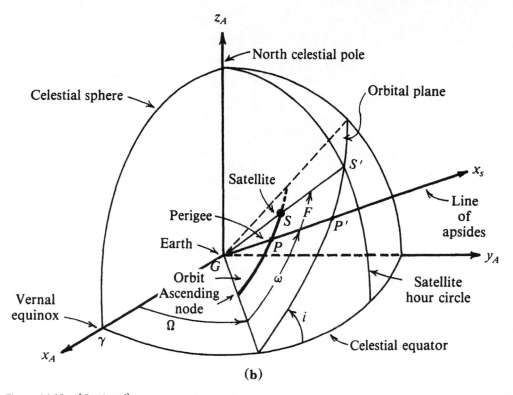

Figure 14-18. (*Continued*).

F = true anomaly

ω = argument of perigee

i = inclination of orbital plane

Ω = right ascension of ascending node

S' = satellite position projected on celestial sphere

sun, moon, and planets; (3) atmospheric drag; and (4) the sun's radiation pressure. The effects of these factors must be included in any mathematical model utilized to accurately compute or predict an orbit. The actual satellite locations, disturbed from an ideal elliptical path, may be expressed as a closely spaced time series of three-dimensional coordinate positions. Alternatively, locations may be defined by parameters for a mean elliptical orbit plus a time-variant set of corrections to be applied to obtain actual orbit positions.

14-10-2. Coordinate Systems

A satellite's orbital position must eventually be expressed in the same coordinate system used for a ground receiver station. To accomplish this, positions in the satellite orbital coordinate system (see Figure 14-18a) are transformed to an astronomical coordinate system (see Figure 14-18b) by rotating through three angles: (1) the argument of the perigee, (2) inclination, and (3) the right ascension of the ascending node. One additional rotation is required to transform satellite coordinates to a mean terrestrial system (see Figure 14-19). The rotation angle is Greenwich apparent sidereal time, a function of time and the earth's rotation rate; it is the angle between the vernal equinox and mean Greenwich meridian.

The mean terrestrial coordinate system has its origin at the earth's center of gravity and is

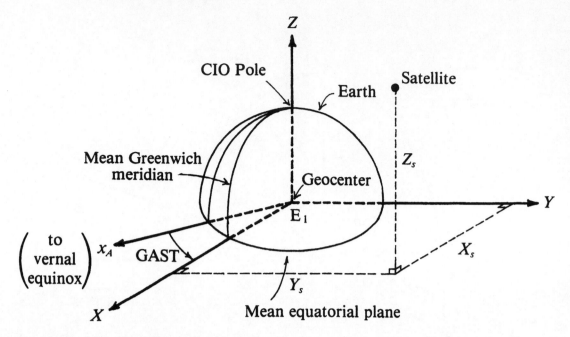

Figure 14-19.   Mean terrestrial coordinate system.

thus a geocentric system. The Z-axis passes through the Conventional International Origin (CIO) pole, a reference north pole that differs from the instantaneous pole because of "polar motion" (see Chapter 13). The X-axis is oriented in the meridian that passes through the mean Greenwich Observatory in England.

Geocentric Cartesian coordinates in the mean terrestrial system may be converted to geodetic latitudes, longitudes, and heights if a reference ellipsoid is selected. In most instances, however, the geodetic coordinates desired are those defined relative to the local geodetic datum, a nongeocentric system (see Chapter 13). For the general case, a three-dimensional coordinate transformation is required to convert receiver station coordinates from the mean terrestrial system to a geodetic datum.

14-11.   NAVY NAVIGATION SATELLITE SYSTEM

As of 1986, six operational satellites were in near-circular polar orbits, with small eccentric-

ity values and inclinations near 90°. The satellites orbit at approximately 1100-km altitudes and have orbital periods of about 107 min. For such a satellite, the earth rotates "beneath it" about 27° of longitude per orbit. Thus, each satellite passes within the line of sight of a particular ground station at least four times per 24 hrs, two each as that satellite passes from north-to-south and south-to-north. The orbits are usually spaced so that the interval between available satellites varies from about 35 to 100 min as a function of receiver latitude, with the longest interval at the equator.

Each satellite continuously broadcasts two coherent carrier frequencies, nominally 400 and 150 MHz, with both signals derived from the same ultrastable oscillator. Actual transmitted frequencies are about 80 ppm below nominal values, or 32 and 12 KHz below 400 and 150 MHz, respectively. The transmitter frequency must be stable, since the basic measurement made by a ground station receiver is a function of small frequency shifts in the received signal caused by the Doppler effect.

A satellite position is required for each observation, so orbital parameters must be known and observations accurately timed. A navigation message that defines the satellite's position is phase-modulated onto the 400-MHz signal, beginning and ending every even minute. This *broadcast ephemeris* message includes so-called fixed and variable parameters. Fixed parameters, included in each 2-min message, define a smooth, precessing elliptical orbit for a 12-hr interval. Variable parameters, updated every 2 min, are used to correct the elliptical path and find actual orbital coordinates at those times. Therefore, a transit satellite's signal consists of two stable carrier frequencies, precise timing marks, and a broadcast ephemeris message that defines its orbital coordinates as a function of time.

The basic NNSS ground support system consists of four tracking stations (Maine, Minnesota, California, and Hawaii) plus two computing and injection stations (Minnesota and California). Each tracking station observes Doppler data for signals transmitted from all satellites. The computing centers use the data for a particular satellite, plus historical tracking data, to compute parameters of the observed orbit and predict its orbit for the next 12 hr. Predicted orbit information is then transmitted from an injection station to update that satellite's memory.

14-12. GEODETIC DOPPLER RECEIVERS

Portable Doppler receivers have been developed and marketed by several firms, including Magnavox Corporation, JMR Instruments, Inc., Canadian Marconi Company, Decca Navigation Company, and Motorola, Inc. Doppler receiver system components include a receiver, antenna, data recorder, and power supply unit. The weight of a typical receiver system is in the 25- to 40-kg range, making it feasible to hand-carry or backpack the unit to locations inaccessible to vehicles (see Figures 14-20a and 14-21). A stable receiver oscillator generates a reference signal from which 400-

and 150-MHz frequencies are derived; thus, these receiver references are approximately 80 ppm higher than the respective satellite frequencies. This ocsillator also drives a clock that can be synchronized to time marks in a satellite ephemeris message.

Antenna design is important because the received signal strength is low. It must also be lightweight and transportable (see Figure 14-20b). After an antenna is set up and leveled, offsets—both horizontal and vertical—are carefully measured from the station mark to the antenna's effective center (see Figure 14-22).

Some early receiver models recorded observed data on punched paper tape (see Figure 14-17). Data recording and storage for most current models are accomplished by the use of magnetic tape cassettes (see Figure 14-23). At least one receiver utilizes a bubble memory cartridge for data collection. Receivers are designed for all-weather operation and may be left unattended in an automatic data-acquisition mode. Current systems operate on 12-V battery power and include a microprocessor that controls receiver operations, tests for equipment malfunctions, verifies recorded data, and even provides on-site position computations (see Figures 14-24 and 14-25). Modular construction provides for easy field repairs.

14-13. DOPPLER SATELLITE POSITIONING THEORY

Due to relative motion of the receiver antenna and satellite transmitter, the satellite's constant transmitted frequency—as a consequence of *Doppler shifts*—is no longer constant when received at a ground station antenna (see Figure 14-26). As a satellite approaches a receiver station, received Doppler frequency is greater than satellite-transmitted frequency. As the satellite passes and the receiver-satellite distance, or satellite range, begins to increase, the received frequency drops below the transmitted value. This Doppler shift is a function of signal propagation velocity and the time rate of change of satellite range, whereas the received Doppler-frequency-curve slope is a measure of satellite range.

Corresponding received Doppler and receiver reference frequencies are differenced to produce lower-frequency signals. The *beat frequency* for the 400-MHz signal is nominally 32 KHz, with a maximum variation of ± 9 KHz due to the Doppler effect. The receiver includes a counter to observe the number of beat-frequency cycles that occur during successive time intervals. Selected time intervals may be 2 min or shorter; commonly, nominal

30-sec intervals are used. During a typical satellite pass, 20 or more "30-sec Doppler counts" may be observed, the number depending on how long the satellite remains above the horizon. The maximum number of counts, about 36, would result from horizon-to-horizon tracking of a satellite that passed directly over the receiver. The beat-frequency cycle counts constitute the basic observables of a Doppler receiver.

Each Doppler count is a measure of the satellite-range change during a corresponding time interval—e.g., range difference $(r_2 - r_1)$ that occurs during time interval $(t_2 - t_1)$ in Figure 14-26a. Each range difference can be

(a)

Figure 14-20. (a) JMR-2000 global surveyor carried in standard backpack frame. (b) JMR-2000 global surveyor showing antenna storage. (Courtesy of JMR Instruments, Inc.)

(b)

Figure 14-20. (*Continued*)

expressed mathematically as a function of electromagnetic wave propagation velocity, receiver reference frequency, Doppler count, Doppler-frequency shift, and time interval (see Equation 1 in Figure 14-26c).

A range difference is also a function of satellite coordinates and receiver antenna coordinates, since an individual range can be expressed geometrically in terms of end-point coordinates. The antenna position is the desired unknown, so estimated coordinates are substituted initially. Satellite coordinates are computed from the orbital parameters defined by broadcast ephemeris data. If 30-sec Doppler counts are used, required satellite

positions corresponding to those intermediate times are interpolated from positions defined at 2-min time marks. By combining these two range-difference expressions (Equations 1 and 2 in Figure 14-26c), a nonlinear observation equation is formed (Equation 3 in Figure 14-26c). Each Doppler count corresponds to a similar equation, with antenna coordinates as only unknowns. These equations are linearized and solved by the least-squares method (see Chapter 16) to find corrections to the estimated coordinates. The solution is iterated until a three-dimensional set of antenna coordinates is found that best fits observed Doppler counts.

Figure 14-21. Motorola Mini-Ranger satellite survey system in carrying cases. (Courtesy of Motorola, Government Electronics Group.)

14-14. DOPPLER DATA-REDUCTION METHODS

Doppler surveys may be performed in two different modes: (1) point positioning or (2) relative positioning. Single point positioning can be done with one receiver any place on earth. A point's three-dimensional geodetic position in the satellite coordinate system can be computed with data from as few as two satellite passes. Relative positioning requires simultaneous operation of a minimum of two

Figure 14-22. Magnavox MX 1502 geoceiver satellite surveyor field site setup. (Courtesy of Magnavox Advanced Products and Systems Co.)

Figure 14-23. Data tape cassette for Motorola Mini-Ranger satellite survey system. (Courtesy of Motorola, Government Electronics Group.)

receivers: One occupies a station with known geodetic coordinates, the other is placed at an unknown point. The data-reduction process involves solving for the coordinate differences between the two stations and then applying them to the known station coordinates to find those of the unknown point. A relative-positioning Doppler survey may utilize more than two receivers if desired, thereby gaining additional reliability from the network configuration produced.

A two-dimensional "horizontal" position can be determined by using data collected during a single satellite pass, but a position so obtained may have up to a 30-m uncertainty. This figure can be greatly reduced by collecting and reducing the Doppler data for multiple satellite passes. Three-dimensional repeatability has been reported to be about 9 m for 10-pass solutions, 5 m for 50-pass solutions, and 2 m for 100-pass solutions. These results were obtained using orbital parameters from

Figure 14-24. Receiver control panel for Magnavox MX 1502 geoceiver satellite surveyor. (Courtesy of Magnavox Advanced Products and Systems Co.)

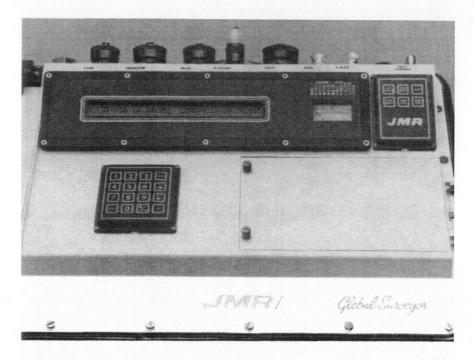

Figure 14-25. Control panel for JMR-2000 global surveyor in operation. (Courtesy of JMR Instruments, Inc.)

the *broadcast ephemeris* and predicted values used on tracking data that is a minimum of 6 hr old at the time of the satellite pass. Errors in satellite orbital positions have a large effect on computed receiver station coordinates.

A postcomputed orbit, based on tracking data taken at about the same time as the observed satellite pass, will yield better results. Computed and distributed by the U.S. Department of Defense, this *precise ephemeris* is based on two days of tracking data obtained at over 20 stations distributed around the world. Point-positioning repeatability obtained using the precise ephemeris has been reported to be about 1 m or smaller for 10- to 33-pass solutions, whereas accuracies better than 1 m have been reported for multiple-pass solutions compared to external references such as the high-precision geodimeter traverse. The precise ephemeris has several disadvantages: (1) It is *not* generally available for public use; (2) it is only available for one or two of the satellites; and (3) receipt of ephemeris data, for autho-

rized users, is delayed as much as a week or more, so position computation cannot be done in "real time."

Doppler satellite relative positioning, in its simplest form, is referred to as *translocation.* Relative-positioning methods provide improved accuracy because they take advantage of highly correlated position errors for two or more receiver stations located in the same general area. These errors included effects of atmospheric propagation delays as well as orbit inaccuracies. Translocation uses data from each satellite pass observed simultaneously by two or more receivers. The *short-arc* method, another relative-positioning data-reduction technique, is similar to translocation, except that additional unknown parameters are included in the solution to permit adjusting the satellite orbit. The *semi short-arc* method likewise provides orbit adjustment, but it utilizes fewer additional parameters and yields a slightly less rigorous solution. Accuracies of 50 cm or better have been reported for multiple-

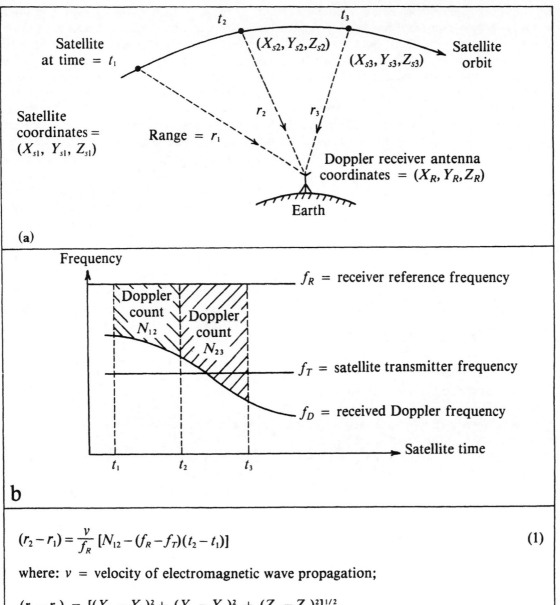

(a)

b

$$(r_2 - r_1) = \frac{v}{f_R}\left[N_{12} - (f_R - f_T)(t_2 - t_1)\right] \qquad (1)$$

where: v = velocity of electromagnetic wave propagation;

$$(r_2 - r_1) = \left[(X_{s2} - X_R)^2 + (Y_{s2} - Y_R)^2 + (Z_{s2} - Z_R)^2\right]^{1/2}$$

$$- \left[(X_{s1} - X_R)^2 + (Y_{s1} - Y_R)^2 + (Z_{s1} - Z_R)^2\right]^{1/2} \qquad (2)$$

$$\frac{v}{f_R}\left[N_{12} - (f_R - f_T)(t_2 - t_1)\right] = \left[\Delta X_2^2 + \Delta Y_2^2 + \Delta Z_2^2\right]^{1/2} - \left[\Delta X_1^2 + \Delta Y_1^2 + \Delta Z_1^2\right]^{1/2} \quad (3)$$

where: $\Delta X_i = (X_{si} - X_R)$; $\Delta Y_i = (Y_{si} - Y_R)$; $\Delta Z_i = (Z_{si} - Z_R)$.

(c)

Figure 14-26. Doppler satellite positioning theory. (a) Satellite-transmitter/receiver-antenna geometry. (b) Receiver Doppler count measurements. (c) Doppler positioning mathematical model.

pass solutions using translocation, and short-arc multiple-pass computations have yielded reported accuracies within 10 cm.

Although nominally a known value, the difference between ground reference and satellite-transmitted frequencies $(f_R - f_t)$ is treated as an unknown in a position solution. Small frequency drifts of each oscillator may be accounted for in this manner.

14-15. GLOBAL POSITIONING SYSTEM

The GPS is a worldwide system of navigation satellites implemented by the U.S. Department of Defense (see Chapter 15).

14-16. GPS EXPERIENCE AT THE NGS

The NGS's first experience with the GPS was in January 1983, as a participant of the FGCC's test of the Macrometer V-1000 survey system. In March 1983, the NGS acquired two Macrometer V-1000 GPS receivers. Test surveys were conducted from April through July; in September, the first of several operational control survey projects was begun, and in October 1983, a third Macrometer V-1000 was acquired.

In cooperation with the U.S. Defense Mapping Agency, the NGS conducted its first GPS satellite survey with Texas Instruments TI-4100 geodetic receivers in June 1984. In May 1985, two of seven TI-4100 GPS satellite receivers acquired by NGS were delivered; tests began that month, and the first operational survey started in June. Two more receivers were delivered by December 1985 and the remaining three units scheduled or use by the summer of 1986.

The NGS experience in GPS satellite surveying has been extensive, with projects carried out in various regions of the conterminous United States and Alaska. Over 50 GPS survey projects have been carried out since early 1983, involving the occupation of more than 500 stations. These GPS survey projects were executed to meet a wide range of control requirements or special survey investigations. The extensive experience gained from these surveys has clearly demonstrated that observations of GPS satellite signals yield very accurate three-dimensional relative-position data. Depending on spacing between stations, centimeter- to decimeter-level relative-position accuracies have been achieved. Because uncertainty levels in the order of centimeters can be obtained at a significant reduction in costs compared with conventional methods, the NGS has adopted GPS satellite surveying technology as the primary way to establish geodetic control. Consequently, this acceptance of GPS as a primary tool is affecting everyone involved in control, land, and engineering survey practices. Chapter 15 provides current practical survey techniques using GPS.

14-17. CONCLUSIONS

Inertial and satellite geopositioning technologies are revolutionary developments that have already had a significant impact on surveying practices. Inertial surveying methods are particularly appropriate for large projects that require rapid repositioning of widely scattered points or a high density of points in a smaller area. A truck- or helicopter-mounted ISS has the potential for high productivity in moderate-to-high precision applications. Because of its absolute point positioning capability, the Doppler satellite method is well suited to the establishment of control networks in remote areas. Doppler surveys are also appropriate for relative positioning of widely separated control stations.

The impact of GPS surveying technology is being felt and will continue to advance rapidly. Surveyors need to be aware of its potential and learn as much as possible about the principles of GPS surveying practice. Although by no means a panacea, GPS satellite surveying technology is proving to be a very useful tool in satisfying general and special control surveying needs. In particular, it has been or is potentially applicable in the following:

1. Strengthening the existing national geodetic reference system (NGRS)
2. Establishing new horizontal and vertical control connected to the NGRS
3. Providing data to improve estimates of geoidal undulations
4. Determining reliable estimates for geoid height differences and useful values for orthometric heights at points not connected by differential leveling to the national vertical control network
5. Greater efficiency in establishing network control
6. Complementing and thereby enhancing the usefulness of other precise survey systems, such as the inertial survey and total station methods
7. Making high-precision measurements at the cm- to few-cm-level for purposes such as measuring horizontal motion, monitoring land subsidence and uplift and special engineering surveys for deformation studies, etc.

The GPS surveying technology is a revolutionary tool expected to dominate the state of the art for control and geodetic surveying through the end of this century.

REFERENCES

ASCE. 1983. Special issue on advanced surveying hardware. *Journal of Surveying Engineering 109 (2)*.

HOAR, G. J. 1982. *Satellite Surveying*. Torrance, CA: Magnavox Advanced Products and Systems.

The Institute of Navigation. 1980, 1984, 1987. *Global Positioning System*. Washington,D.C., Vols. I–III.

KING, R. W., E. G. MASTERS, C. RIZOS, A. STOLZ, AND J. COLLINS. 1985. *Surveying with GPS*. Kensington, Australia: School of Surveying, University of N.S.W.

Positioning with GPS-1985. Proceedings of the 1st International Symposium on Precise Positioning with the Global Positioning System, Vols. I and II. 1985. Rockville, MD.

Proceedings of the International Geodetic Symposium on Satellite Positioning, Vols. 1 and 2. 1976. Las Cruces, NM.

Proceedings of the 1st International Symposium on Inertial Technology for Surveying and Geodesy. 1977. Canadian Institute of Surveying, Ottawa, Canada.

Proceedings of the 2nd International Geodetic Symposium on Satellite Doppler Positioning, Vols. 1 and 2. 1979. Austin, TX.

Proceedings of the 2nd International Symposium on Inertial Technology for Surveying and Geodesy. 1981. Banff, Canada.

Proceedings of the 3rd International Geodetic Symposium on Satellite Doppler Positioning, Vols. 1 and 2. 1982. Las Cruces, NM.

Proceedings of the 3rd International Symposium on Inertial Technology for Surveying and Geodesy, Vols. 1 and 2. 1985. Banff, Canada.

Proceedings of the 4th International Geodetic Symposium on Satellite Positioning, Vols. 1 and 2. 1986. Austin, TX.

SCHERRER, R. 1985. *The WM GPS Primer*. Norcross, GA: WM Satellite Survey, Wild Heerbrugg Instruments.

STANSELL, T. A. 1978. *The Transit Navigation Satellite System*. Torrance, CA: Magnavox Government and Industrial Electronics.

Surveying Engineering Division of the ASCE 1985. Advanced positioning systems. *Engineering Surveying Manual*. New York, Chap. 20.

WELLS, D., ET AL. 1989. *Guide to GPS Positioning*. Fredericton, Canada: Canadian GPS Associates.

15

Global Positioning System Surveying (GPS)

Bryant N. Sturgess and Ellis R. Veatch II

15-1. INTRODUCTION

This chapter is not intended to be a text on global positioning surveying. The bibliography contains many other sources that will greatly expand one's knowledge of the topic to any depth or breadth required. The objective of the chapter is to provide information to eliminate problems commonly experienced by first time users of GPS technology. Some related topics are triangulation (Chapter 11), trilateration (Chapter 12), geodesy (Chapter 13), and inertial and satellite surveys (Chapter 14).

The navigation satellite time and ranging (NAVSTAR) was developed by the Department of Defense (DOD) as a three-dimensional, satellite-based, 24-h, global, all-weather navigational system.

The use of differential GPS as a surveying tool was not considered in its initial development and early satellite launches that began in February 1978. Surveying and other nonnavigational applications are added bonuses developed primarily by the private sector. GPS has eclipsed and will ultimately transcend the use of classical and conventional surveying instrumentation and techniques such as triangulation, trilateration, and traversing, while providing a highly accurate, precise, and cost-effective technology for control extension, control densification, airborne and terrestrial

photo control, deformation monitoring, hydrographic positioning, and a host of other applications. GPS has pushed many applications of conventional instrumentation into obsolescence. With internal system precision of better than 1 ppm, GPS is often several orders superior to the basic primary control on which the subsequent constrained adjustment of GPS observations is based. The National Geodetic Survey (NGS) now uses GPS exclusively for horizontal control work.

The planned constellation of 21 satellites and three spares in 12-h orbits at an altitude of 20,180 km will provide visibility of five or more satellites, 24 h a day to users over the globe. The satellites are arranged four to an orbital plain, with six circular planes inclines at 55° to the equator and skewed 60° between each other. The mandate is for mission control to maintain 21 operational satellites at all times, providing three-dimensional positioning around the clock as shown in Figure 15-1. The system will have three working spares to help ensure the continuation of this level of coverage.

This constellation array has been delayed, but by the end of 1991, the system contained 16 operating satellites with five prototype block I's and 11 production model block II's providing virtual 24-h satellite visibility sufficient for two-dimensional position, and 18-h three-di-

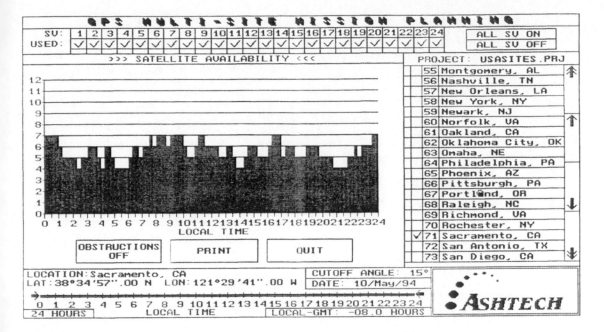

Figure 15-1. Projected satellite availability as of May 10, 1994, 15° cutoff.

mensional position in latitude, longitude, and height with a 10° cutoff angle above the horizon. With a 15° cutoff angle, the available time is reduced by $2\frac{1}{2}$ hr to a $15\frac{1}{2}$-hr three-dimensional observation period as in Figure 15-2.

Figures 15-1 and 15-2 are taken from typical mission planning software that shows satellite availability. Figure 15-1 demonstrates satellite availability for May 10, 1992 at Sacramento, CA when three-dimensional GPS measurements were possible using four or more satellites. Also shown are periods when only two-dimensional (x, y) measurements were possible with three satellites, and times shown with black shading when the constellation was not sufficient for surveying purposes.

15-2. GPS BASICS

Position computations with GPS are done by ranging to the satellites. The satellite positions are known—i.e., they are computed from the orbit parameters in the broadcast ephemeris.

By intersecting the ranges (analogous to distances) from multiple satellites, the position of the receiver antenna can be calculated, in effect, performing a classic resection in which the known stations are the satellites.

At least four satellites are required to compute a three-dimensional position. The ranges from three satellites would provide a good position if the timing were perfect. However, two clocks are being used: the satellite clock and receiver clock. If the receiver clock is different when compared to the satellite clock, and it always is, the computed position will be wrong by the amount the receiver clock is offset from the satellite clock. A fourth range is required to solve for the receiver clock offset to compute a position.

The GPS system is based on time. Precise atomic clocks control the frequency of the carrier signals and timing of code and message modulations. The code itself does not convey information; it is simply a set of unique patterns that are used to identify the satellites and pick the signal from the background noise.

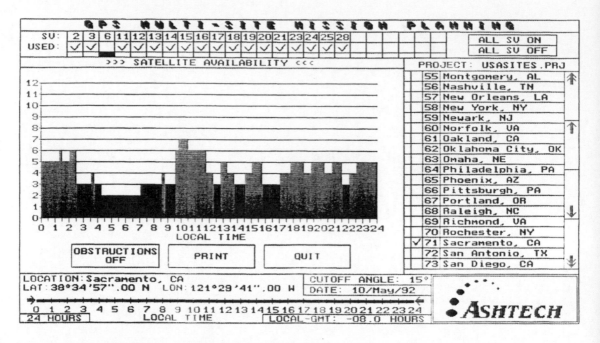

Figure 15-2. Satellite availability as of May 10, 1992, 15° cutoff.

Each satellite has its own code and signature. The range to a satellite is measured by correlating the code generated by the satellite with the same code generated by the receiver to determine the time shift between the two codes, or the transmission time of the satellite code, as illustrated in Figure 15-3.

The transmission time multiplied by the speed of light provides the distance, or range, to the satellite. This measurement contains several sources of error and is therefore termed a pseudorange, meaning a false range or range with error. These errors are primarily clock errors, but also include orbit error and signal delays caused by the ionosphere and troposphere. These errors limit the accuracy of GPS in computing point positions.

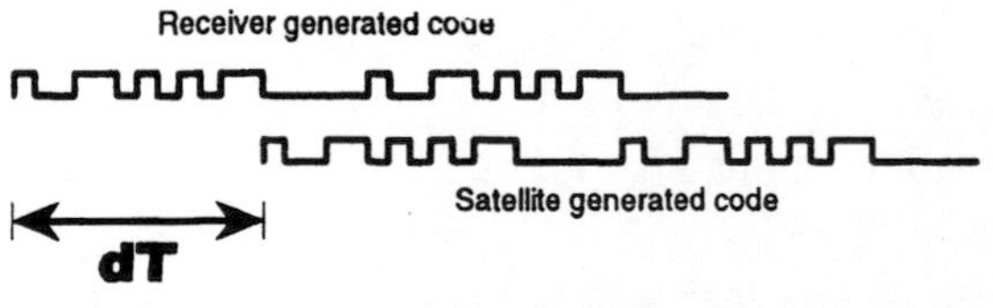

By using two receivers, many of the errors in the system can be cancelled. In differential navigation or *code-phase ranging*, the differencing is done by computing range corrections derived by comparing the observed ranges with the expected ranges based on the known position of the base receiver. The range corrections are then broadcast and applied by the remote receiver in its position computations. The resulting positions are accurate to 1 to 10 m as opposed to the 10- to 100-m accuracy of a standalone receiver. Although providing an improvement in position accuracy, pseudorange corrections are still quite inadequate for surveying use. There is, however, a second kind of ranging available.

Carrier-phase ranging, using the wavelength of the underlying carrier frequency, is employed for surveying measurements. Whereas code-phase ranging is straightforward (time of transmission multiplied by speed of light equals the distance to the satellite), carrier-phase

Figure 15-3. Code correlation. Time shift (dT) between the receiver-generated code and satellite-generated code.

ranging introduces an ambiguity into the observations. The receiver can measure the phase of a carrier wavelength very accurately, but it cannot directly measure the whole number of wavelengths associated with the first measurement of phase. This unknown number of cycles is referred to as the phase ambiguity or integer bias. The combination of the phase measurement and integer bias equals the initial range to satellites. The elements of a carrier base range are illustrated in Figure 15-4.

This situation is somewhat analogous to measuring a long distance with a surveyor's canyon chain when the head chainperson is out of voice contact with the rear chainperson. The hundreths are recorded precisely at the head end of the chain, but the exact number of feet are not known until the head and rear chaining notes are later combined. By knowing the orbits of the satellites and keeping track of the change in ranges throughout the observation session, the integer biases can be computed, thereby determining which foot the hypothetical rear chainperson was actually reading at each satellite.

By refining the ruler and differencing the carrier-phase measurement made with two receivers, not only are common errors removed, but accuracy is substantially increased. Millimeter measurements are made possible using a system originally designed for several-meter accuracy.

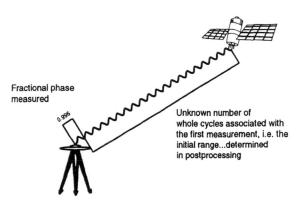

Fractional phase measured

0.996

Unknown number of whole cycles associated with the first measurement, i.e. the initial range...determined in postprocessing

Figure 15-4. The carrier-phase range.

Three-dimensional base lines, or vectors, are the result of these efforts. Any GPS survey project is simply a collection of three-dimensional (spatial) vectors.

15-3. BASIC NETWORK DESIGN

U.S. Geological Survey quadrangle maps at a scale of $1:24,000$ ($7\frac{1}{2}$ min) remain a most useful tool for preplanning. However, other types of maps and resources, such as commercial road maps, county atlases, state highway maps, county assessors' maps, and aerial photography, are all valuable for providing additional information in those areas where the quadrangles are either not topical, or outdated. In any event, select the most current map and plot all available (existing) horizontal and vertical control of the highest order available in the area and obtain network diagrams, station descriptions, and to-reaches from the appropriate government agencies such as the NGS, county surveyor, or local highway department. Plot the location of the horizontal and vertical control on the quad and then add the locations for which GPS positions are desired. So far, this is no different than the preplanning formerly required for conventional instrumentation for triangulation (Chapter 11) or trilateration (Chapter 12).

The location and relative distribution of the existing control are not as rigorous with GPS as with earlier conventional terrestrial technology. GPS is very sensitive to satellite geometry and less sensitive to network geometry. An example of good satellite geometry would be, say, four or more satellites spread out over the sky in different quadrants, as compared to the same satellites bunched together near the zenith.

GPS does not require station intervisibility, except in those cases where a check azimuth is required on an adjacent control net, or for purposes of establishing an eccentric (ECC) to an existing monument that is not possible to occupy with GPS due to local obstructions screening the satellites. In those instances where an ECC is necessary, two intervisible GPS stations are required in order to provide for the necessary backsight for conventional measurements. The minimum acceptable distance between the GPS stations for subsequent conventional terrestrial work is dependent on the accuracy required for the resulting azimuth. Table 15-1 provides a handy reference for determining station spacing.

There are factors that must be considered in designing the network. For example, the following conditions need to be given consideration before any network design can begin:

1. Standards and specifications for project: What minimum accuracy is required to accomplish the needs of the project while providing for necessary redundancy.

2. Distribution of existing horizontal and vertical control.

3. Minimum and maximum existing network control.

4. Minimum and maximum secondary station spacing.

It is a good idea to include as many existing network horizontal control points in the GPS schedule as economically possible. The number is independent on the relative size of the GPS project and funding envelope. Commonly a minimum of three existing NAD 83 horizontal control stations of 1 : 100,000 are included.

If NAD 83 stations are not available, then select the highest order in the area. On larger projects such as shown in Figure 15-5, existing horizontal control stations were used to control the network, providing maximum geometrical redundancy with good success. It is best to err on the side of generosity in the selection of horizontal and vertical project control. Place as much as can be afforded.

The heights produced from GPS surveys are referenced to the ellipsoid, not the geoid. Constantly, it is necessary to include bench marks referenced to NGVD in the GPS observation scheme as well. The selection of at least three bench marks spaced somewhat evenly in the area might be adequate on smaller projects of approximately 15 to 20 km. However, in projects where the separation between the ellipsoid and geoid are irregular, as could be expected in foothills or mountainous areas, more benches will be necessary to allow for accurate modeling should GPS be used to generate orthometric elevations (NGVD) for photo control or other engineering projects. It is important that the selected bench marks used to control the GPS survey have not been disturbed, and the elevation of the BM is consistent with that of adjacent BMs in terms of epoch and loop adjustment. It may be necessary to validate the elevation with a conventional level run if either stability or elevation is in doubt.

Network design can be a very individualized process. Probably the most common method is to simply link stations by sessions, leap frog-

Table 15-1. Station Spacing

	Azimuth Accuracy Arc Seconds (one sigma)				
	1	2	4	6	10
Station Spacing (m)	+ −	mm		Confidence	
100	—	—	—	3	5
200	—	2	4	6	10
300	—	3	6	9	14
400	2	4	8	12	19

ging from session to session until all stations have been connected. This tends to be a random method and often results in networks with one or more unnecessary sessions. The following design process is an attempt to be more orderly. It is based on, but does not rigorously follow, NGS guidelines. These guidelines can be found in *Geometric Geodetic Accuracy Standards and Specifications for Using GPS Relative Positioning Techniques*.[1]

The first step in designing an efficient network is to connect the stations into loops of nontrivial base lines as shown in Figure 15-6. This provides the framework of the network and allows loop closures according to NGS specification. In Figure 15-6, the network has been connected into three loops. The loops should contain no more than 10 base lines and not exceed 100 km in perimeter.

When creating the loops, try to keep them as "boxy" as possible. Avoiding a series of parallel loops will allow more flexibility in the second step.

The second step connects the nontrivial lines together in sessions. Each session should contain $N - 1$ (where N is the number of receivers being used) of the nontrivial loop base lines. In Figure 15-7, four receivers are being used and each session contains three loop base lines. The network is complete in seven sessions with no extra work; therefore, there are 21 nontrivial base lines in our network. Conversely, if you know the number of nontrivial base lines in your network, you know exactly how many sessions you need to complete the observations.

Try several approaches to each network. Networks do not always work out as cleanly as the example above. The addition of one other station in the network above would create two more nontrivial base lines and necessitate another session.

A good way to practice is to put a group of dots on a sheet of paper and make several copies. Try different loops and session groups. The example in Figure 15-7 is based on using four receivers; try laying out the same network

for three receivers (two nontrivial lines per session instead of three). Practice leads to efficiency.

The term nontrivial as used above to describe the base lines in the loop framework is somewhat misleading and has been misused. As used here, it denotes the minimum number $(N - 1)$ of base lines necessary to connect all stations in a session. Whether or not there are any trivial base lines is dependent on the processing method. The trivial base lines in Figure 15-7 are shown as dashed lines. These lines are only trivial if the processing software considers the data from all receivers at the same time, i.e., multi-base-line processing, resulting in true zero loop closures. If the base lines were processed independently, as is usually the case, the resulting lines $[(N*(N - 1))/2]$ are correlated, but they are not trivial. A recommended method is to process all base lines independently and use them in the network. In redundancy is strength. A fundamental rule in GPS surveying is, "Think network, not traverse."

A summary of recommendations for controlling a GPS network follows:

1. *Horizontal.* Technically two stations will provide an azimuth and a scale. However, there would be no redundancy for either azimuth or scale. It is recommended that the project be bracketed with NAD 83 horizontal control, three stations minimum, whereas four or more horizontal stations are recommended for larger control projects.

2. *Vertical.* Three stations are sufficient to solve for the two tilt biases: east-west and north-south. A fourth vertical provides a necessary redundancy. The tilt defines the slope of the geoid in relation to the ellipsoid and may be sufficient to provide reliable elevations in small areas where the geoid is fairly consistent. In areas where gravity anomalies are suspected or in projects over large areas, using a geoid model will be necessary to provide good elevations. The geoid model may be improved by comparing elevation differences and ellipsoid height differences

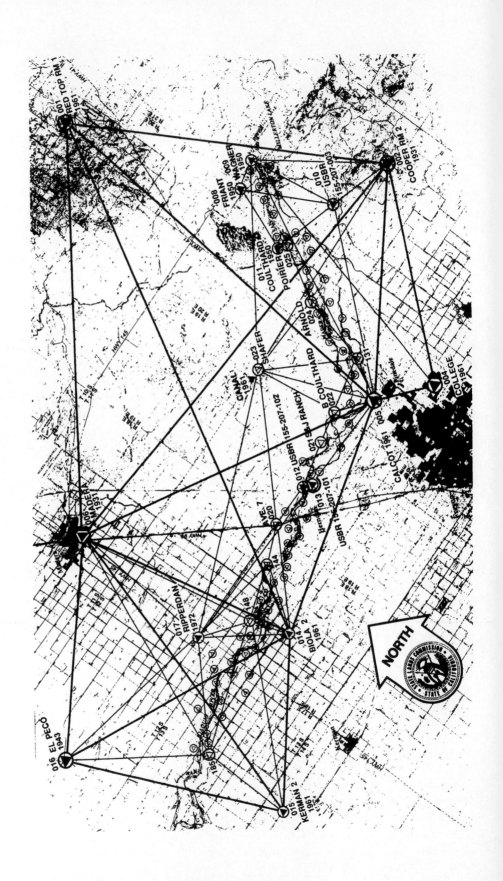

LEGEND

▲ First Order NGS Primary Horizontal Control.

▲ Second Order NGS Horizontal Control.

△ Horizontal Control Established by GPS This Survey.

● Horizontal and Vertical position established by Conventional Survey Instrumentation (CSI).

⊛ ⊘ Horizontal Control with NGVD Elevation.

000 GPS Identifier.

☐ Bench Mark.

▬▬▬ Fixed Baseline, Validated by GPS, Held Published Terminal Position.

───── GPS Measured Baseline, Established New Terminal Position.

– – – Single Direction Line, Measured by Conventional Survey Instrumentation (CSI).

Figure 15-5. Sample control project.

SURVEYOR'S STATEMENT

I, Bryant N. Sturgess, Licensed Surveyor No. 4333, certify that this survey and these plats were completed under my direction and that all monuments shown hereon existed as of the date of this survey during the months of February and March 1989, and that their positions are correctly shown.

Signed and sealed

For Descriptions, Coordinates, Geodetic Positions, and Elevations See DATA SHEETS 8, 9 and 10 of 17.

For Pertinent Notes Relating To This Plat See Sheet 2 of 17.

STATE of CALIFORNIA
STATE LANDS COMMISSION

PLAT OF GPS SURVEY

FOR

PHOTO CONTROL

IN THE

COUNTIES OF FRESNO AND MADERA

ALONG

THE SAN JOAQUIN RIVER

FROM

FRIANT DAM TO GRAVELLY FORD
W23104

SCALE 1" = 2 miles   SHEET 1 of 17

JUNE 1989 . REVISED SEPTEMBER 1989

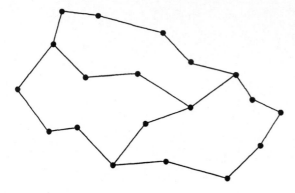

Figure 15-6.   Nontrivial base line loops.

between known bench marks. Terrain changes can provide clues to the existence of gravity anomalies. For example, a gravity anomaly might be indicated in an area marked by the transition of valley floor to foothills, or by foothills to mountains. Use caution in mountainous regions by providing a generous number of known elevations. NGS specifications outline the frequency of multiple occupations required for a typical network. For example, 10% of the stations require triple occupation, whereas 5% of the base lines require double measurement loops. Thirty% of the new stations, 100% of the vertical control, and 25% of the horizontal control, any azimuth pairs, all require double station occupation. Designing a network to meet NGS specifications often results in more sessions and is, consequently,

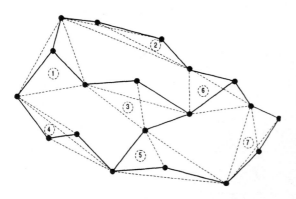

Figure 15-7.   Trivial base lines (dashed) and nontrivial base lines (solid).

more expensive. Current NGS specifications may be obtained by directly contacting the agency. See Appendix 2 for its address.

15-4.   RECONNAISSANCE

GPS reconnaissance in many respects is similar to conventional reconnaissance with one major exception: Usually, GPS does not require stations to be intervisible as in conventional terrestrial surveying. There are times, however, when intervisibility is a requirement. One such occasion would be when an independent validation of either the GPS system or primary control net is included for purposes of furnishing an independent check on the GPS measurements and/or adjustment. Until GPS technology is a universally accepted and recognized technique by the courts, this might serve as a method of convincing either the judge or jury that the results are what they are claimed to be. As time goes on and the use of GPS becomes a surveying technique used by more surveying organizations, this idea of validation will become of lesser and lesser importance. Another possible situation where intervisibility is a requirement occurs when a backsight is necessary in a GPS network to provide an azimuth for use by conventional terrestrial survey measurements to position a point that is not possible to occupy with a GPS antenna due to obstructions restricting line of sight to the satellites.

It is essential that all primary control be validated at some time during the GPS campaign to ensure no disturbance of the station. Usually, the checking of published measurements from the reference monuments (RMs) to the station will satisfy at least a part of the validation relating to site stability or possible disturbance. There is another validation that should take place in any primary network before published primary control positions are accepted in the final constrained adjustment of GPS base lines. That is the direct measurement between primary network control stations using GPS receivers. This may seem to be an unnecessary luxury, or perhaps overkill. However, the direct measurement between primary control stations serves as the best evidence of regional stability and fit with the control possible.

Pre–NAD 83 (NAD 27, e.g.), it was always a good idea to determine if the primary control being used for the campaign was entirely within the same NGS network adjustment. Post–NAD 83, this practice is no longer as rigid. Most NAD 83 positions have been validated to better than 1 part in 100,000. Nevertheless, sufficient occupations should be planned to verify the adjustment before the control is used. Usually, the GPS campaign will result in so-called free (unconstrained) adjustments that are superior to the constrained (fixed) ones. This is a situation rarely, if ever, experienced with conventional terrestrial surveys except those of the highest order.

15-4-1. Public Contact Strategy

More often than not, existing primary control necessary for referencing the GPS campaign, as well as monumentation that must be established during the campaign, is located on private (or government) property belonging to someone other than the client. In these cases, proper permission or notification for entry should be obtained. Securing permission of the property owner to enter private property, is not only good business, but it is polite, and in some locales, it will keep the surveyor out of legal complications. A diligent attempt must be exercised to secure proper permission. Failing this, the owner could be sent a letter of intent to enter the property to a survey with some explanation of the survey's purpose.

15-4-2. Site Selection

GPS sites must be selected free of significant obstructions and multipath conditions. Multipath is caused by an object or a media

that causes signal reflection. Multipath conditions can result from metallic objects located above the antenna plane such as buildings, signs, semitrailers, tankers, chain-link fences, all of these can be a common source of multipath. In short, avoid sites where signal-reflecting material is above the plane of the antenna. Figure 15-8 is an actual NGS primary control station site where a metal-sided fire lookout tower was a source of not only severe multipath, but also potential satellite signal blockage covering nearly 40° in azimuth and nearly 45° in altitude (see Figure 15-9).

Objects interfering with a direct signal from the satellite during the planned observation periods cause cycle slips. When a satellite signal is obstructed in any way, tracking is interrupted. However, when lock to the satellite is regained, the fractional part of the measured phase is restored. The lost integer count will be repaired during postprocessing. The signals propagate from the satellite and are transmitted to the receiver antenna along the line of sight. These signals cannot penetrate water, soils, walls, trees, buildings, or other obstacles. Standing between the receiver antenna and satellite can interrupt signals. Consequently, while the data are being collected, the operator should stay away from the antenna to avoid excessive blockages of signals resulting in cycle slips. Cycle slips are a loss of count and could be one cycle or a billion cycles (1 billion cycles represents 1 sec of signal transmission). Occasional cycle slips are not a problem and will be automatically repaired by the postprocessing software. Excessive cycle slips or blockages should be avoided. Figure 15-9 shows a polar plot of the terrain depicted in Figure 15-8, with the track of the satellites superimposed.

In this example, it can be seen that at approximately 0030 hours, satellite no. 11 will disappear behind the fire lookout structure. Accordingly, should GPS observations be planned between 2320 and 0020 hours, satellite no. 11 will be visible. However, multipath will still remain as a significant problem due to the metal slab-sided structure and positions of

Figure 15-8. NGS station "Red top" showing a metal-sided building, the source of severe multipath conditions.

the balance of satellites from azimuth 90 to 320°. An eccentric station was necessary at this location to eliminate potential multipath conditions.

The adage "monuments are where you find them" remains true today. Established monuments are either clear for GPS measurements or they are not. Often, the latter is the common case, as in the example shown in Figure 15-8. In the event the sky is obstructed, a location must be selected for the setting of an eccentric station that has a clear view of the sky from the mask limit of, say, 15 to 20° to the zenith.

Site selection for new monumentation can be determined after the consideration of a few basic elements. It is not a complex operation and in many ways has similarities to conventional field methods. The site must be con-

ducive to the installation of the type of monumentation specified for the campaign. Future access to the site must be considered as well as monument survivability. If the location forms a part of the overall GPS network, then consideration must be given to its long-term preservation for subsequent use by others. It is only fair that the property owner be advised that other surveyors might require the use of this new station in the future with the possible result that the initial setting of the station might establish a precedence for continued usage.

If it is likely that the new station will be needed in the future for other possible activities, anticipate the effect tree growth will have on satellite visibility, or the relative safety of the site from construction activities, vandalism, or road widening, e.g. It is not possible to plan

California State Lands Commission
Survey Unit

GPS VISIBILITY CHART

Station: RED TOP   Latitude: 37° 07' 41"
Date: 18 JAN 90   Longitude: 119° 46' 36"
By: B.N.S.   Elevation: 1,840'

18-Jan-90 (Day 18)

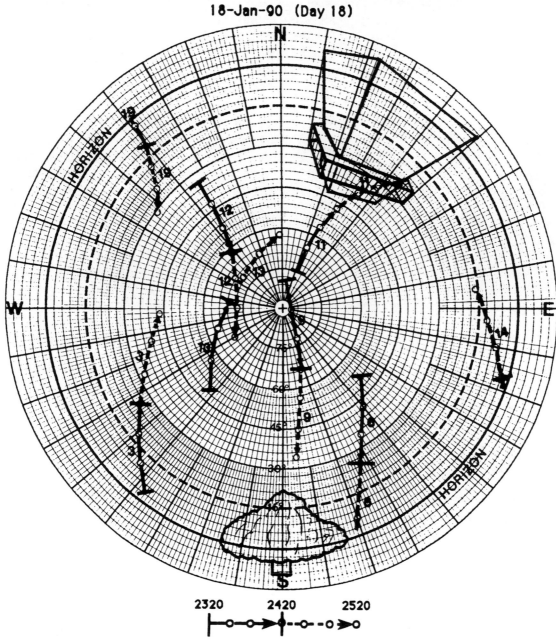

Figure 15-9. Polar plot of NGS "Red top."

for all contingencies, but a greater percentage of new monumentation will survive the perils of development, agricultural activities, etc. and be usable for the next surveying job if some though is given to site selection.

Since GPS allows a high rate of productivity, easy station accessibility for the receiver operators is an essential element, no matter which technique for GPS measurement is selected, be it rapid-static, static, pseudokinematic, or kinematic. It is necessary for the receiver operators to get to the station in the least amount of time possible in order for the project to proceed with the smallest operational cost and highest productivity. The selection of GPS sites must be maximized for easy setups, while avoiding unnecessary walking to the station when a slight change of location would result in a drive-to-station. Avoid locations that require the excessive unlocking and locking of gates if possible. If access by key is a requirement, it would be advisable to have a duplicate set made for *each* receiver operator. By taking this precaution, no access complications will occur since all operators can access any locked area. This is one less problem to address and allows full flexibility in station assignments during the GPS measurement phase of the campaign.

Give consideration to the time of year when the actual observations will be scheduled. Actually, the reconnaissance should be followed as closely as possible by the GPS schedule with a minimum of delay because conditions rapidly change in the field. Reconnaissance done in the spring of the year might indicate a particular station and site with its resulting polar plot to be clear of obstruction, but its GPS occupation 2 or 3 wk later could fail due to the growth of arboreal foliage, or perhaps construction activities.

If the campaign is scheduled during the rainy season, determine if the site might be subject to ponding by water, or made inaccessible due to mud, and that vehicular access will be possible along the planned access route under these conditions. As late as 1990, due to limited satellites and inadequate geometry, it

was a given that at least part of the year GPS operations would be conducted during nighttime, or part day and part night depending on the time of the year. In 1992, yearround daytime use was possible in the midlatitudes due to the increased numbers of satellites made available during 1990 and 1991 providing 24-hr two-dimensional coverage 18-hr three-dimensional position in latitude, longitude, and height with a 10° cutoff angle. With a 15° cutoff, the available time is reduced by $2\frac{1}{2}$ hr to a $15\frac{1}{2}$-hr observational period. What this means is that at the present and for the future, nighttime GPS will be necessary only for certain applications. One such application where such a schedule would be advantageous is in an area of high daytime traffic, or a hazardous location that if occupied at night would make the site less perilous for the receiver operator. Another application of nighttime operations would be the case of maximizing productivity by incorporating both daytime and nighttime missions when time constraints are a factor and maximum productivity is paramount.

With proper site selection and planning, nighttime observations can approach productivity levels associated with daylight. If proper consideration has not been given to site selection, access, safety, or logistical considerations, nighttime operations, as well as daytime GPS operations, will not be productive.

15-4-3. Monumentation

The criteria used for the selection of sites and setting of existing control monuments during the era when control extension utilized conventional instrumentation were different than presently is the case for GPS. Earlier networks required line-of-sight visibility between adjacent network stations, sometimes only obtainable from the erection of tall inner and outer observation towers placed over the stations. Generally, clear visibility from 10° above the horizon to the zenith was not a requirement except in cases where a Laplace observation or an astronomic azimuth was a requirement. GPS, on the other hand, requires no station intervisibility, but line of sight to the satellites.

In some ways, GPS is no different than conventional control densification or extension in the selection of the type of monumentation for installation. Depending on the intended use of the new station, the monumentation may be of temporary design for short-term use such as ECC to a property corner or photo control point that cannot be occupied due to satellite signal interference caused by the proximity of buildings, trees, or other obstructions. The selection of monumentation in this example could be typically a hub, nail and tin, pipe, or rebar. For those stations requiring

some degree of permanence, consideration should be given to one or a combination of commercial ultrastable three-dimensional types, as in Figure 15-10, constructed of aluminum or stainless steel rods driven to refusal, then capped, collared, and covered for identification and protection. In some cases, due to local site requirements, it might be necessary to bury the monument to ensure survivability. Generally, the use of poured-in-place monuments has fallen into disfavor as permanent GPS stations or even bench marks, for that matter, due to possible vertical instability, so careful consideration should be given to the actual planned use of the monument. A per-

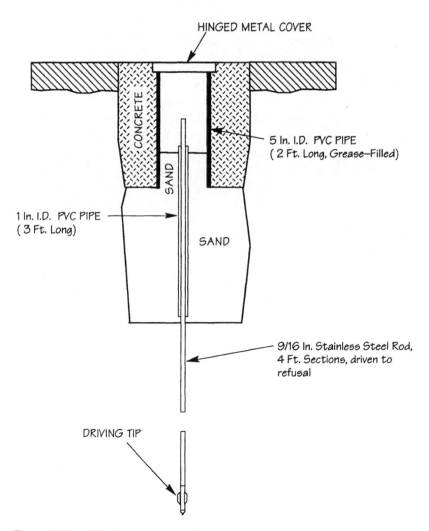

Figure 15-10. NGS three-dimensional monument.

manent GPS control station must be stable in three dimensions. Epoxying or grouting a monument in bedrock or a rock outcrop is a desirable and stable installation. The difference in actual cost between a temporary-type monument and one of more long-lasting design is negligible. Consider the future possible use of the station and remember that GPS is a precise three-dimensional measurement technology that must have vertical as well as horizontal stability.

Contrary view is that with the ease and accuracy of establishing high-precision points with GPS, it is not cost-effective to spend the time and resources in establishing superpermanent marks.

14-4-4. Recovery Notes

Recovery descriptions remain an invaluable resource to the GPS campaign. They had their beginning in the earliest days of triangulation and, as is the case with any good idea, they will be with us into the future. As long as there is a need to search for and recover control stations, there will be a need for some form of document describing access to the monument, its location and description. The recovery note (Figure 15-11) is an essential tool to the mission planner and receiver operator who must get to the station on time for the scheduled sessions to be a success. The recovery description is a narrative describing the station and gives particulars about its site. It contains specific instructions guiding the operator to the station, usually from a common starting point with other such stations on the project. It should contain specifics such as street names, landmarks, and cardinal directions and distances associated with the instructions for left, right, or straight-ahead movements. Additionally, it ought to include hazard warnings such as aggressive dogs, livestock, gate combinations, instructions for the closing gates, dust on crops, possible driving hazards, or anything that will allow the receiver operator to reach the station site in the least amount of time with as little risk as possible, while observing

safety and any special requests by the property owner. If property owers or caretakers were contacted en route, these references with phone numbers, addresses, or other appropriate comments should be noted.

If the survey work is to be conducted during darkness, the recovery description should reflect instructions that give consideration to the special difficulties of nighttime operations. There ought to be references to land falls, other factors while en route that will confirm the observer is on the right track. An aid for nighttime operations is adding accessories such as highway reflectors, reflective tape, reflective paint, etc. on fence posts, trees, gates, etc. to guide the operator to the station; these again should be noted in the recovery description. If nighttime operators are necessary, driving times should be buffered if the reconnaissance and to-reach preparation were done during the daytime. The GPS receiver operator probably will be unfamiliar with the site and station. The accuracy and completeness of the to-reach will determine the success of the GPS receiver operator in locating the assigned site and station.

15-4-5. Station Notes

The station noteform, as in Figure 15-12, used to document all recoveries; to-reaches; contacts with the public; names, addresses, and phone numbers; and all contacts resulting from recovery efforts. This information (along with the to-reach description) is essential and is entered into the database for later use during mission planning for GPS. This field noteform can also be used for all station recoveries and descriptions. The station notes and recovery descriptions are used by the mission planner in his or her tour of the project for purposes of familiarization with the primary control and new GPS stations. The receiver operator also uses these same documents to find the station under severe time constraints during the GPS measurement phase. It is important that the to-reach be clearly written and the station recovery notes be sufficiently

STATE OF CALIFORNIA

STATE LANDS COMMISSION

STATE LANDS DIVISION

SURVEY NOTES

PARTY CHIEF	RECORDER	RODMAN	CHAINMEN
BN STURGESS	BNS		HEAD
			REAR

REMARKS	INSTRUMENTS	WEATHER
		OVERCAST, COLD

RECOVERY NOTES "BIOLA-2, 1961"

STATION "TO REACH"

TO REACH THE STATION FROM THE NEW POST OFFICE IN BIOLA GO SOUTH ON BIOLA STREET FOR 0.5 MILES TO THE INTERSECTION OF "G" STREET. CROSS THE INTERSECTION PASSING THE FIRE STATION, THE OLD RR. RIGHT-OF-WAY, AND GO 0.15 MILE TO THE CURVE TO THE LEFT ("I" STREET) AND THE STATION.

VISIBILITY TO THE NORTH IS TOTALLY OBSCURED BY BUILDINGS. THE WATER TANK IS NOT VISIBLE. IN ORDER FOR BIOLA-2 TO BE UTILIZED BY CONVENTIONAL GEODETIC MEASUREMENTS AT LEAST 50 FEET OF BULKY IS NEEDED. BIOLA-2 AZIMUTH MARK COULD BE USED AS AN ECC. 10 TO 15 FEET AT THE AZIMUTH WOULD MAKE MOST OF MAIN SCRIBE VISIBLE, HOWEVER 10 FEET WOULD STILL BE NEEDED AT BIOLA-2.

BIOLA 2

W.O. 21005	COUNTY FRESNO	VICINITY BIOLA	DATE 15 DEC 87	CHECKED	DATE	PAGE 3 OF 3

FORM 40.20 (4-75)

Figure 15-11. A Recovery Description.

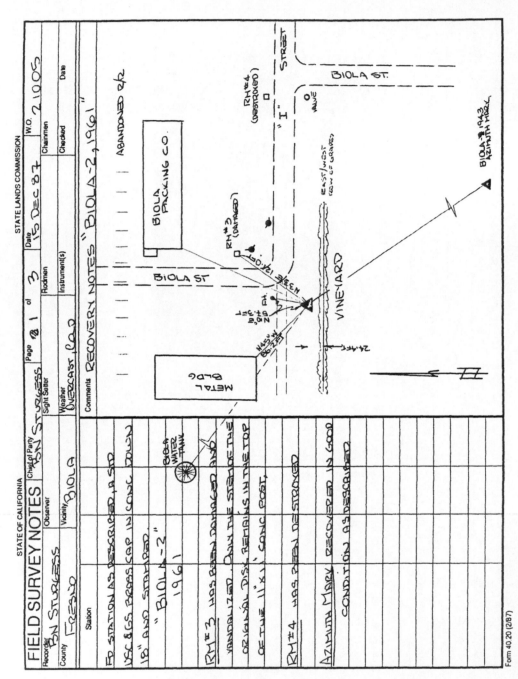

Form 40 20 (2/87)

Figure 15-12. Station notes.

350

detailed to represent the site of the monument so as to be recognizable to a person unfamiliar with the location. At this time, station notes have a benefit. As a practical consideration, station notes often are combined with the recovery description on the same sheet or page to provide a concise document for recovery uses.

During recovery, the station site should be marked in a distinctive manner in order to be instantly recognizable by the receiver operator. Plastic flagging if eaten by livestock can be hazardous; consequently, its use should be considered carefully.

15-4-6. Polar Plot

The polar plot (see Figure 15-9) shows the sky line around the prospective GPS station in graphic form. The plot details the height of obstructions in degrees above the horizon and the azimuth of obstructions between the station and sky. It is essential that a plot be produced when possible obstructions are unavoidable as part of the reconnaissance agenda in order that mission planning proceed.

At new GPS stations being established, it is desirable to select sites free of obstructions that could interrupt satellite signals. It is not always possible to select a GPS site absolutely clear of obstructions from the horizon to the zenith clear around the compass. It follows also that few existing monuments for GPS positioning are completely free from possible obstructions to the satellites. If the sites are clear, there is not need whatsoever for a polar plot.

If the reconnaissance person knows when the GPS observations are planned to be made, the predicted locations of the satellites can be noted on the polar plot in the field and a site selected that allows those sectors of the sky to be clear even though the balance of the sky is occluded. The use of a conventional transit equipped with a plate compass is the preferred instrument to make the necessary measurements. Other options are acceptable, such as a Brunton compass, or perhaps an older-style vane compass on a Jacob staff or a reconnaissance compass used in conjunction with an Abney hand level. Do not forget to allow for declination because the compass measurements must be a "true" bearing. The measurements need not be any more precise than + or $-1°$. Remember that even obstructions such as leaves on a tree can cause a loss of lock, so be sure to outline the entire outside edge of the tree when taking measurements for the polar plot. The plot need not go below $10°$ above the horizon. Make certain to measure the azimuth to both sides of power poles and signs as well as the tops. The instrument should be set as high as possible above 5 ft to more closely represent the height of the receiver antenna during the data collection phase of the campaign. The use of a contrasting color is recommended for plotting the obstructions. Typical polar plot field notes should contain provisions for the following information:

1. Station name
2. GPS ID number (this is an ID number assigned for the individual campaign)
3. Observer
4. HI of instrument
5. Declination set into the instrument

In the event polar data will be later entered into a mission planning software program, a time-saving alternative to the preparation of the polar plot diagram is a set of field notes detailing the same written information as would be represented on the diagram. The field notes should have provisions for entering the following information:

1. Bearing to object
2. Vertical angle to the object
3. Description of the object or portion of the object being observed
4. GPS ID number or name
5. Name of observer
6. HI of instrument
7. Declination set into the instrument

This method is quicker for those times when the observer is the only person available at the site and is a time-saver when it comes to the task of data entry later in the office. Unless this type of summary is prepared, the data entry will involve individually inspecting and scaling each point detailed on the polar plot for the bearing and vertical angle and subsequent computer entry.

The required precision of all horizontal and vertical angle measurements is + or − 1°; field notes recording measured values are not required. Any feature that may obstruct satellite radio transmissions must be shown. This may include such factors as trees, power poles, transmission lines or towers, structures, terrain (hills), etc. In the case of power poles, be sure to include a measurement to the top of the pole. Do not go overboard in the number of observations to outline a feature: This wastes time. For economy, use the minimum number to outline an object.

The information provided by the polar plot or field notes can either be used directly by the mission planner as in the past, or it can be entered into a database that is part of the mission planning software provided by the manufacturer. Once entered, an image can be assembled on the screen that duplicates the mask measured in the field, while the satellite ephemeris provides a template of the available satellites on the julian date and time selected. This allows the mission planner to select times when satellites will not be obscured by features such as trees, power poles, signs, structures, etc.

15-4-7. Photos and Rubbings

During the reconnaissance phase, it is desirable for documentation purposes to photograph the station site, the monument either found or set, and any other feature at the location that may help to recover the station in the future. Station notes as shown in Figure 15-13 are a diagram prepared that documents the site and shows a planimetric view of the location together with any reference points

(RPs) set or reference measurements to aid in either the resetting of the station should it be destroyed or disturbed, or recovery at a future date. The photos and station notes form a very powerful team for the preservation and perpetuation of monumentation. However, as time goes on, the future value of such precautions will become questionable as the ease of reestablishing positions by GPS evolves into the realm of real time.

The technique of taking rubbings consists of laying a page of blank field notes over a monument such as a brass cap and rubbing the surface of the disk with a soft lead pencil or lumber crayon, thus producing an exact image of the monument. The use of this technique can occur at the time of the preparation of the station notes, but is primarily useful when the station is visited on mission day by the receiver operator. This rubbing provides proof that the receiver was on the station designated for the session and not a reference mark. Even the most careful of surveying assistants manage, on occasion, to set up on a reference monument instead of the designated station. Of course, this method is not practical, e.g., when the monument is a galvanized pipe or rebar without any characteristic that serves to make this monument distinctively different from others. It is recommended that rubbings be used at least on the primary control stations.

15-5. GPS FIELD OPERATIONS

All GPS surveying is dependent on the measurement of the carrier phase and the solution of the ambiguities utilizing the basic carrier that is a 19-cm sine wave. A phase measurement is a portion of this basic carrier. A fractional phase can be directly determined, but the number of complete cycles (integer ambiguity) from the satellite is not directly measured. This is determined over a long observation period in static measurements, or in pseudokinematics by occupying a station twice dur-

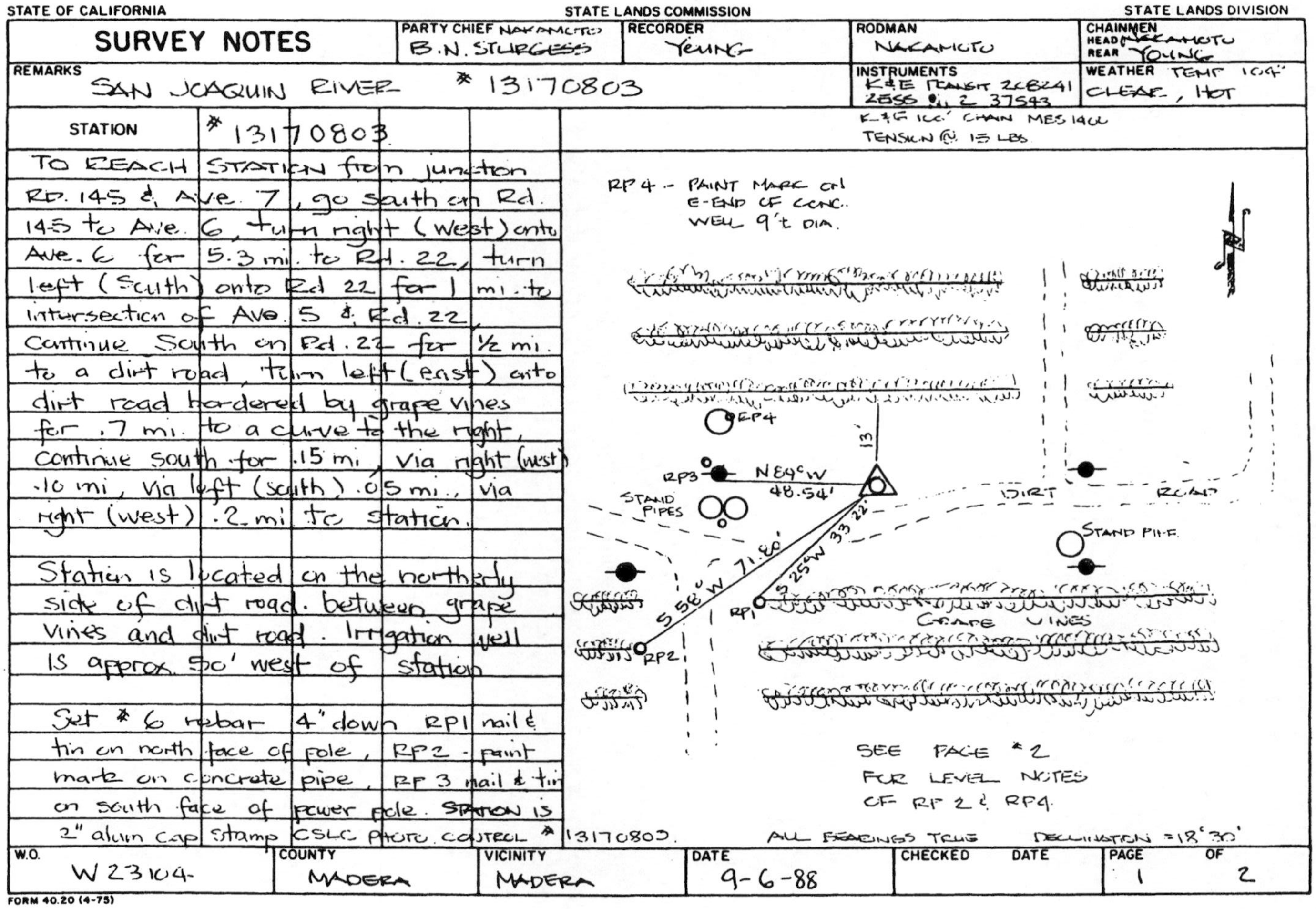

Figure 15-13. Sample GPS field notes, long-form.

ing the same session spaced by 1 hr for 30 to 60 epochs (about 5 to 10 min). In kinematics, the ambiguities are solved by initializing on a known base line that may have been created by an antenna swap. This allows for the instantaneous solution of the rover position.

15-5-1. Statistics

The term *static* was adopted because this technique requires receivers to remain stationary on a monument for a session length approaching an observing time of about 45 min (135 epochs) to 90 min (270 epochs) or more, as required depending on the standards and specifications for the campaign. The overall length of the static occupation is further dependent on the charging geometry of the position of the satellites with respect to the receivers. During times of rapidly changing *positional dilution of precision* (PDOP), shorter observation times can yield the same accuracy as longer occupation times with low, relatively unchanging levels of PDOP.

At the culimination of a session, the usual process is for one receiver to remain in place while the remaining receivers all move to a new location. The "fixed" receiver provides the hinge or pivot point for extending the control into the new session.

Typical uses of the static GPS technique might be in the areas of surveys of the highest order for such purposes as deformation, crustal motion studies, or high-order statewide GPS reference control nets, where the precision of 1 part in 1,000,000 or better is common. The static technique might also be employed in those cases where the spacing of the stations is so geographically distant that multiple sessions are not a practical consideration.

The number of sessions possible in a mission day is governed by the number of four (or more) satellite windows with acceptable PDOP, the length of satellite windows with acceptable PDOP, and travel time between the stations. The productivity of static procedures is also dependent on other variables such as weather, traffic, and luck.

The static technique is a common choice for the first-time user of GPS. It is the easiest in terms of training staff, or mission planning, but is the lowest in productivity of all possible techniques. For those networks requiring long occupation times because of high-precision specifications, or stations spaced far apart making a second session impossible during the same satellite window because of lengthy drive times, this is the only choice available. Statics can be accomplished with either dual- or single-frequency receivers.

The standards and specifications control productivity as well, by impacting the length of time necessary for an occupation. The higher the order, the more epochs are required. The same condition exists for long base lines in which extended time on station is a requirement.

The sample GPS field notes shown in Figure 15-13 are typical for static observations. This noteform would not be practical for other GPS field techniques in which high productivity would be impaired by the 2 or 3 min necessary to complete this form. However, for static measurements the observer is on station for at least 45 min so the time liability is not a problem in this case.

15-5-2. Rapid-static

Rapid-static is not a technique in itself. Rapid-static is simply an improvement in the software algorithms that currently use the additional information available with P-code to resolve the phase ambiguities faster. Simply stated, rapid-static is static observations for a shorter span of time. Occupations of 5 to 10 min take the place of occupations for 45 to 90 min. Current implementations utilize the P-code, but it is possible to perform rapid-static measurements with codeless dual-frequency data. Future software improvements may even make it possible to perform short sessions with single-frequency receivers. Rapid-static is certainly more productive than static or simultaneous rover pseudokinematic and twice as productive as pseudokinematic, because it requires only a single observation and half the total observation time. Rapid-static rivals kinematics in productivity and does not require that lock be maintained. The current implementations of rapid-static rely on the P-code, and one must realize that stated DOD policy is to employ antispoofing (AS) when the GPS system is fully operational. Antispoofing may or may not be turned on at all times. However, selective availability (SA) is turned on all the time, and it is reasonable to assume that the DOD will turn on antispoofing all the time. When AS is turned on, the P-code will essentially be useless to civilians without the information to decrypt the signal. Therefore, it is reasonable to assume that purchasing dual-frequency P-code receivers at this time should only be done if the GPS project has sufficient numbers of stations to amortize the cost differential prior to the implementation of AS.

15-5-3. Pseudokinematics

Pseudokinematic (PK), also known as false kinematic, pseudostatic, broken-static etc., can be the equal in accuracy and precision to 1-hr static GPS. PK is only a field concept and actually is a modified static method. There is no difference in the processing when compared to statics. The mechanics of PK is closely related to static in not only the postprocessing of the data, but also the common reliance on the changing of satellite geometry over a span of time. Static employs an unbroken period of observations of an hour or more to solve for the integer ambiguities. Pseudokinematics, on the other hand, uses short intervals of data separated by changes in the geometry of the satellites over the span of approximately 1 hr. In effect, PK is simply a static session in which the middle has been removed. PK does not require that lock be maintained between GPS setups, is not as productive as K or rapid-static, but can be used in locations where kinematics cannot. PK does not require dual frequency, as does rapid-static.

One variant of the technique employs a receiver on a known station (fixed station) and

one or more rover receiver operators visiting a series of unknown stations twice. The use of a base station guarantees common time measurements whenever the rover is observing on a point. One occupation is done initially on each station and then each station is revisited in the same original order about an hour after the first occupation. The technique is based on the changing geometry of the satellites during an interval of 1 h (or more) that solves for integer ambiguities. It is not necessary to maintain lock on the satellites when moving between stations. However, it is recommended that the receiver remain on to avoid unnecessary warm-up time. The receiver on the base station remains in place during the complete session, never being shut down and collecting data during the entire period. The rover receivers occupy different unknown points twice for 30 to 60 epochs at 10-sec epoch intervals (about 5 to 10 min) at each station, with a return revolution through the same stations about 1 h later during the same session, collecting data for another 30 to 60 epochs as before. During the second tour of stations, each site should have a minimum of four satellites common with the first revolution at that same station. Each individual receiver antenna height must be the same on the second tour as it was on the first. This requires the use of fixed-height antenna poles!

The recommended epoch interval of 10 sec with 6 observations per minute for 5 min is subject to some variation, depending on data quality and postprocessing results. In some cases, it might be advisable to increase the total time on station during each visit by several minutes.

The sample GPS field notes shown in Figure 15-14 are typical for high-production-type observations such as rapid-statics, pseudokinematics, kinematics, and other methods. This noteform contains considerably less data than that shown in Figure 15-13, which could pose a problem during postcampaign analysis in attempting to isolate an error.

Another variation of pseudokinematics called simultaneous rover pseudokinematics

FIELD DATA SHEET
GLOBAL POSITIONING SYSTEM OBSERVATION
California State Lands Commission

Survey: SAN JOAQUIN RIVER
W.O.: 24073
page 1 of 1

Station Name: USBR 155-207-102
TRSQ:
Local Julian Day: 167
Start Date: 16 JUNE 91

GPS Station Number: 0012
Session #: A
Observers: BOB LEA

Latitude:
Longitude:
Elevation:

Receiver #: 003
Serial #: ASHTECH
Micro Antenna #: 003
Cable Length: 10 meters
Epoch Cycle: SET 10 sec

Power Battery: X
Other:
Weather: CLEAR, COOL, CALM

Leave Time: 10
FIELD OFFICE @ 0000
Refpos:
Vehicle:

Mission-Session	SV's	Time PST PDST	Meteorology				Antenna Height		
			Wet Bulb Temp.	Dry Bulb Temp.	Humidity%	Pressure	Slant Ft.- meters		H I
BEGIN A91.167	19,02,14,11, 6,16,18	0845	59°F	72°F	–	Ft. 270 Hg° 29.71	5' 10 5/16"	1.786m	
END						Ft. Hg°			
						Ft. Hg°			
A91.167	19,02,11,06, 15,16,18	1120	88°F	101°F	–	Ft. 270 Hg° 29.71	CHECK EPO SESSION 5' 10 5/16"	1.786m	

Description of Monument and Remarks:
FD. CSLC BRASS CAP IN CONCRETE STAMPED
"USBR 155-207-102, 1961"
RESET 1988 LS 4333

* LOST LOCK ON SV 14 & 06 DURING TRAIN PASSAGE – BOTH SATS WERE
LOW ON HORIZON @ 4° & 9° RESPECTIVELY

Plumb Bob Check ☒ YES

Disk	Other Stations If Named	Other Stations Number TRSQ	GPS
SESSION A91.167	STATIC = GPS 0012 & 0138		
	PSEUDO KINEMATIC ROVERS = 0268, 0260, 026A, 028B		

JULIAN DAY FIRST DAY OF THE MONTH
Jan-001, Feb-032, Mar-060, Apr-091, May-121, June-152, July-182

26 Jan 89

Figure 15-14. Sample GPS field notes, short-form.

(SRPK), or no master pseudokinematics, is a very productive adaptation of pseudokinematics. The project referenced by Figure 15-18 partially utilized this field technique that resulted in the completion of a very large GPS control and cadastral survey project in less than one-half the time necessary for the usual static GPS procedures. SRPK is very effective with multiple rovers, three or more, each making the scheduled station occupations simultaneously. This not only results in the measured base lines between the rovers and reference station, but creates additional base lines between each of the rovers simultaneously. SRPK requires communication between all field personnel, either by cellular telephone or two-way radio, to coordinate the beginning and end of the common epochs. SRPK is a mission planning and session coordinating nightmare. This GPS technique requires that all rovers be on station, on time (to the minute), with setup times, tear-down times, travel times to multiple stations, etc., all with timing carefully orchestrated in order for the session to be a success. When the field work clicks, this technique is really worth the effort and complicated planning required. Production in terms of completed base lines is higher than with PK, but less than with K and rapid-statics.

In PK and SRPK, the antenna height on each rover antenna must be the same on the first circuit as it is on the return circuit at each point. The rover antennas can be of different heights, but they must not change between the first and second revolution. The use of fixed-height poles instead of tripods is strongly recommended. In the event that the rovers are shut down (not recommended) between moves, an additional 2 min, or more, must be allowed for warm-up to reestablish good satellite lock.

Figure 15-15 demonstrates the productivity possible with SRPK, even though in this example two of the five receivers remain fixed in location throughout the session. During this session, the four individual occupations resulted in the measurement of 37 base lines including both trivial and nontrivial vectors.

Had the session been PK, then only 25 baselines would have resulted during the same satellite window.

15-5-4. Kinematics

Kinematics (K) can approach the same accuracy and precision as 1-hr static GPS. Kinematics is based on solving the integer ambiguities (one ambiguity for each satellite) up front through a rapid initialization on a known base line. This base line can be known from a previous GPS survey, or it can be determined prior to a kinematic operation by an antenna swap, or a static observation. Once the integer ambiguities have been solved and tracking maintained to four or more satellites, the differential position of the rover antenna(s) relative to the fixed receiver antenna can be computed instantaneously.

The difference between pseudokinematic (fake kinematic) and kinematic GPS is the method of ambiguity resolution. Pseudokinematic is simply static processing of an hour of data with the middle portion of data missing. The ambiguities are solved during the processing just as if the receivers had been on station for the full time period. In kinematic processing, the ambiguities are solved up front by fixing a known base line. Once the ambiguities have been solved, they are carried forward throughout the survey. This is what requires the maintaining of lock to four satellites at all times. Without four satellites, the equations fall apart and the ambiguities cannot be carried forward. When this happens, the system must be reinitialized on another known base line, typically the last surveyed point.

The antenna swap in kinematics two receivers as shown in Figure 15-16. One receiver antenna setup is made on the base or reference station, the other is set up on an arbitrary point established at a convenient distance of approximately 2 to 10 m away (within a cable length of the reference station). The receivers record a specified number of epochs, usually numbering approximately 6 to 10; then

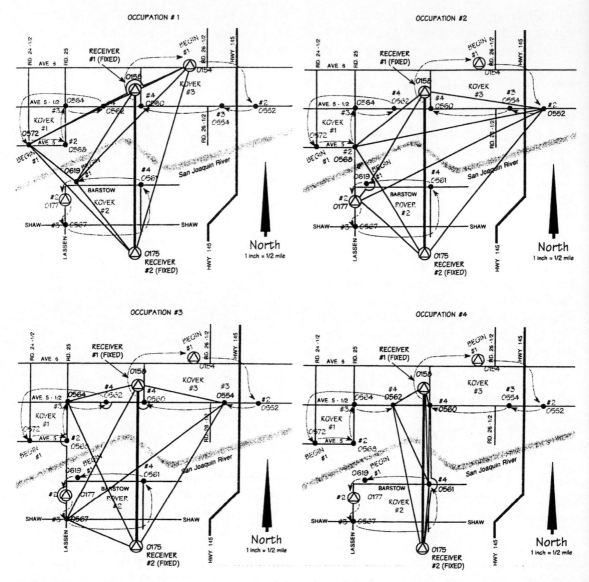

Figure 15-15.   Session employing simultaneous rover pseudokinematics.

each antenna is moved to the opposite station to record the same number of epochs. This concludes an antenna swap.

Normally, two antenna swaps are performed as a safety measure. After the antenna swaps are completed, the antennas are returned to the home-leg position to initialize the kinematic survey. It is important to note that the only purpose of the antenna swap is to provide the required known base line. The antenna swap is not part of the kinematic

survey. The kinematic survey is initialized by observing 6 epochs on the now known base line with the antenna of the stationary receiver on the base station. After the initialization, the roving antenna is removed to a fixed-height pole and placed on a vehicle to commence the kinematic survey proper. At this time, any other roving receivers would initialize on the known base line for 6 epochs at the swap point prior to being placed on their fixed-height poles and going mobile.

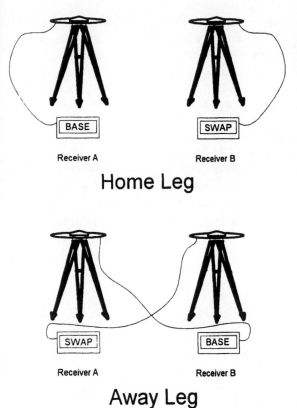

Home Leg

Receiver A   Receiver B

Away Leg

Receiver A   Receiver B

Figure 15-16. Antenna swap.

Should a loss of lock occur during kinematics, the rover simply returns to the last good station for additional measurements consisting of another 6 epochs, then continues as before, avoiding the problem that caused the loss of lock. It is not necessary to return to the location of the antenna swap to reinitialize. At the end of the session, each rover must return to the initializing point to close out the session.

An antenna swap may also be done at the end of a session as well as the beginning. This not necessary, as the check in observation on the known base line is quite sufficient to initialize the survey for reverse processing. If an antenna swap is performed at the end of a session, it must be done after all rovers have checked in on the known base line. If an antenna swap is done at the end, it may be performed between the base and any one of the rovers.

During the entire antenna swap and kinematic survey, the receivers must maintain lock on four satellites, or more. The receiver operators must be extremely careful when moving the antennas to keep them oriented to the sky, and out of the way of obstructions. Both power and antenna connections must be maintained throughout the session. It is a good precaution to connect two power sources to each receiver to ensure an unbroken power supply. Also, the antenna height of the base receiver and roving receiver must be the same when occupying the same point. The best way to perform the swap is to set up two tripods and tribrachs, attaching identical tribrach adapters to each of the antennas, When swapping, move only the antenna and adapter, thus ensuring that the antenna heights will remain constant at each point. In all kinematic GPS operations, it highly recommended that fixed-height poles be used on which to mount the antennas. Fixed-height poles will eliminate one of the most common causes of error, incorrect antenna height measurement.

If kinematic lock is lost, i.e., if any of the roving receivers should track fewer than four satellites, the kinematic survey can be reinitialized on any point that is well known in relation to the base station. Well known is defined as a point or station having been measured directly by GPS methods. Typically, this is the last station occupied prior to losing lock.

During the antenna swapping process, keep clear of the top of the antenna or loss of lock will occur. The antenna heights at the reference station and swap station must not change during either the swap or entire session. Antenna swaps yield very precise azimuths good to + or − 2 arc seconds. The receivers must maintain lock at all times; consequently, all receivers are left on during the entire session until after the last antenna swap is completed at the end of the session.

It is assumed that the tracking of four or more satellites will be maintained throughout the session without a loss of lock. Therefore, it is necessary to plan for a five-plus satellite window to allow for the occasional cycle slip.

When the operator anticipates a situation in which a loss of lock might occur, a temporary point such as a PK nail or hub and tack is set and a reading made at that location. This provides a point in the proximity to which the receiver operator can return should loss of lock occur.

Kinematic GPS, although a highly productive technique, is not in common usage because of the requirement for maintaining lock to at least four satellites at all times. For the best results, five or more satellites should be a requirement in mission planning. Kinematic GPS is second in productivity only to rapid-static. It is most demanding of the mission planner and receiver operators, and is very sensitive to route selection between stations due to the required avoidance of satellite signal interruptions from hazards to data collection such as trees and overhead obstructions. The avoidance of loss of lock is very difficult and virtually impossible in urban, wooded, or rough-terrain areas.

Work is proceeding on instantaneous ambiguity resolution that will allow reinitialization "on the fly" and provide robust kinematic operations in both real time as well as postprocessing. Results have been reported, and when the method is successfully implemented, kinematics will be a viable tool for not only the collection of data, but also real-time stakeout tasks

Refer to Table 15-2 for a comparison of the different GPS measurement techniques.

15-5-5. GPS Survey Party Staffing

GPS data must be processed on a daily basis. Neither the size of the project, nor the number of receivers really matter; postprocessing and network adjustment must proceed as the project progresses in order to spot possible areas of poor results or errors. On-site* post-

processing and network adjustment, mission and session planning, should all be a capability of the GPS survey party. Consequently, these requirements mandate participation in the survey by more than just receiver operators. A representative party size for four receivers on a campaign employing mostly static procedures, e.g., might be shown as in Figure 15-17.

Figure 15-17 does not necessarily suggest that four GPS receivers require eight persons. Some responsibilities can be combined to maximize cost efficiency without impacting productivity. For example, the project director, chief of party, and mission planner could be the same person. The tasks of the computer operator and those of the mission planner could be combined and made the responsibility of one person. This will only be successful provided that no complications or problems with either computations or mission planning arise requiring additional attention. Given the previous example of Figure 15-17 with mostly static procedures, this would (or could be) a workable solution. Computer operation and mission planning are two areas on which the progress of the campaign is crucially dependent.

Mission planning is done in near real time, preferably several hours before the actual mission day is scheduled to begin. With the evolution of more userfriendly, more automated postprocessing software in use at this time, the work of the computer operator has been reduced compared with several years ago. This is offset, however, by greater field productivity resulting from the use of rapid-statics, pseudokinematics, and kinematics, all doubling and redoubling the work of the mission planner.

Combining the responsibilities of computer operation and mission planning is a sensible solution for efficient and economical staffing for schedules and conditions where static procedures predominate, and two sessions per mission day are the norm. This combination

---

*For the purposes of discussion, "on-site" means in the proximity of the survey location, such as the field office or motel.

Table 15-2.   A table of methods

Method	Observing Time	Accuracy	Productivity*	Required SVs	Notes
Static	45 min to 7 hr	cm, mm	9 new points	3 or more, recommend 4 plus	More time higher accuracy
Rapid-static	5–10 min	cm	27 new points	4 or more, the more the better	Typically uses P-code
Pseudo kinematic or pseudostatic	2 repeated 5–10 min obs.; total 10–20 min	cm	18 new points	4 or more, 4 common between repeated observations	Complete loss of lock is ok
Kinematic	Instantaneous epoch by epoch solutions; 1 min of obs. typically averaged for survey points	cm, highly subject to satellite geometry with only four satellites	54 new points (10,800 1-sec epoch per trajectory)	Must maintain lock to at least 4 SV's at all times, min. of 5 satellites for practical use	Most productive but most demanding, needs open terrain
Stop & go kinematic	1-min obs.; trajectory epochs ignored	Same as kinematic	54 new points	Same as kinematic	Same as kinematic

*Rough estimates if we assume a 3-hr window, 4 receivers, open terrain, 5 or more satellites above 15°, 1 base or "swing" station, normal static sessions, and accessible points spaced 1–2 km.

works for as many as possibly five or six receivers, but no more. Increases in productivity that result from using either rapid-static pseudokinematic, or kinematic procedures will quickly overload the person who must keep up with both computer operation and mission planning. The use of five or more receivers with mostly nonstatic procedures requires that the functions of mission planning and computer operation be separated and the responsibility of two different persons.

For high-productivity techniques using five or more receivers on a large project, combin-

1 project director
1 chief of party
1 mission planner
1 computer operator*
4 receiver operators**

*Daily postprocessing and network adjustment.
**Note that safety, productivity, time constraints, or other factors might require more than one operator per receiver.
Figure 15-17.   Proposed GPS party configuration for a medium-size network.

ing static procedures with kinematic and pseudokinematic procedures, a different staffing strategy is more appropriate and necessary as the example in Figure 15-18 indicates. The best efficiency is realized here if the responsibilities of the project director and mission planner are combined. The computer operation is vested in one individual. The role of chief of party would fall on the most experienced of the receiver operators. This position would be a dual role. Although justified by special circumstances such as the level of experience of the operators, safety considerations, or unique problems in the field, it is necessary for the chief of party to be free to go wherever assistance is necessary. This requires another receiver operator. The survey from which Figure 15-18 was taken was of this nature. The party configuration and other details were as given there.

15-5-6.   Vehicle Considerations

The selection of vehicles is less specialized and design-specific than in conventional terrestrial surveys. GPS does not require the transport of an inordinate amount of equipment inventory and, in most cases, would not exceed the capacity of the trunk of an automobile. In fact, many GPS surveys are successfully conducted with the receiver operators working out of ordinary rental sedans. Naturally, in these circumstances, the survey should not require off-road driving, or rapid GPS production techniques, unless special modifications are made to the vehicles. Use what is available and best-suited for the job at hand.

In the event that either rapid-statics, kinematics, or pseudokinematics are being employed, it is necessary to be as time-efficient as possible. It is not productive for the receiver operator to be required, due to space or transport limitations imposed by the type of vehicle, to assembly the array of equipment arrival at the site, then to disassemble the same equipment when the collection of data is complete. It is better to have a vehicle that allows for the transport of equipment as preassembled as possible to save time in setting up and

Project Details

Geographical area:	200 square miles, combination of city, suburban, rural, and remote
Terrain:	Rolling hills, river valley, river obstruction with limited locations for crossing
Ground cover:	Mature trees with large canopies over roads and access locations such as oaks, sycamore, conifer, agricultural tree crops such as almond, pistachio, citrus, etc.
Purpose of survey:	NAD 83 positioning of cadastral monuments, property corners, various BMs, and historical monuments
Minimum precision:	+ or − 1 cm or 1 part in 50,000, whichever is greater
Special considerations:	The majority of cadastral monuments are located within public rights of way, often in the pavement of busy roads and highways, requiring traffic control, signing, and traffic cone protection of the receiver operators
Number of stations:	Approximately 200
Average station spacing:	Approx. 0.75 m
Measurement techniques:	Static GPS 35%
	Simultaneous, rover pseudokinematic GPS 59%
	Conventional survey 5%
	Kinematics 1% due to tree canopies, obstructions, etc.
Campaign duration:	2 wk
Number of receivers:	5 single-frequency, 12-channel
Number of vehicles:	8 total (7 ea. 4 × 4's, all 2-way FM radio-equipped and 1 sedan)

GPS Party Configuration

1 project director:	Project supervision, mission planning, mission briefings and debriefings, public contact, public relations
1 computer operator:	assistant project director
1 chief of party:	Receiver 1 : 2 persons*
	Receiver 2 : 2 persons*
	Receiver 3 : 2 persons*
	Receiver 4 : 2 persons* (occasionally 1 person)
	Receiver 5 : 1 person
TOTAL STAFFING:	12 persons

*Two persons required due to safety considerations in high traffic areas.
Figure 15-18. Details of a large GPS campaign.

tearing down. The choice of two- or four-wheel capability is terrain- and weather-dependent.

The selection of most 4 × 4 types such as van, station wagon, utility body pickups, and other enclosed utility-type vehicles, all winch-equipped if possible, is the most common and universally successful for any type of GPS survey and practically any location where vehicle access to the station site is possible. These vehicles will be successful wherever the project is located. Some locations still require access by packing-in or airlift.

It is axiomatic that the vehicle must get the receiver operator to the station. Various functions of the vehicle, such as the electrical system, e.g., will be taxed, especially in cases where rapid GPS production techniques are employed. Proper tires for the terrain, a spare tire for the same application, accessories such as windshield wipers, a heater, and even air conditioning must all be in proper working order. In short, using an unreliable, poorly equipped vehicle will be counterproductive when the receiver operator must be on site, on station, and ready to begin data collection at a specific time.

The convenient and secure storage of GPS equipment is a prime requirement of the vehicle. There are two types of storage required. The first is for general transportation to and from the job site or from the main office to the field office. This type does not generally take productivity into account and the equipment is usually safely stored in shipping containers or cases. The second type is storage designed for the rapid deployment of the receiver gear from the vehicle to the station on site. This requires a different consideration. Time is crucial; often, the session planning and coordination allow only minutes for the receiver operator to unpack, set up the equipment, log on, and begin collecting data. This kind of scheduling mandates that maximum preassembly be utilized to save valuable time. There are commercial accessories available that allow the mounting of a GPS bipod or tripod on a universal roof assembly that can be transferred between different vehicle types.

Other custom variants might allow for the mounting of a GPS bipod or tripod rod on the front bumper of a vehicle. This type is a preference by many because the assembly can be viewed from the driver's seat while the vehicle is en route to the station. There are magnetic roof mounts available that permit the antenna to be transported on the roof between stations to help avoid the loss of lock during kinematic measurements. Keep these magnetic mounts away from the computer disks.

In the example shown in Figure 15-19, a 2.5-in. pipe flange has been bolted to the winch bumper. A galvanized pipe is screwed into the flange. The pipe is used as the GPS bipod (tripod) holder, and the GPS rod is inserted within for transport. The pipe is sleeved to provide padding for the protection of the GPS rod from vibration and abrasion during travel from station to station. Note that the antenna is attached to the GPS rod for transport, and the antenna cable to the receiver for quick deployment. The receiver, battery, and coiled antenna cable are housed

Figure 15-19. Vehicle modification for high productivity.

within a foam-padded amo can bolted to the winch bumper. This assembly proved to be very safe and time-efficient. During the survey pictured in Figure 15-19, four of the five receiver vehicles were staffed by two field personnel due to the large geographical area being surveyed during this campaign. It proved to be more efficient, cost-effective, safer, and more productive to assign an extra person to assist in directing the driver to the station site while reading the to-reach, since it was not always possible for each receiver operator to memorize the route and on-site details of approximately 200 such stations. Time constraints and production schedules did not leave time for the receiver operator to visit the station prior to the days' mission. Once at the station, the extra person assisted in setting up and completing the GPS data log form. The typical elapsed time from arrival to log-on and beginning data collection was 2 min.

15-6. MISSION PLANNING

The term *mission* as used in GPS generally signifies a GPS work day. A mission is broken into several smaller parts called *sessions*. From GPS beginnings in the late 1970s, mission planning for GPS surveys has consisted of the same general and primary steps. The technology of field data collection has evolved from static procedures through several innovative techniques to improve efficiency such as kinematic, pseudokinematic procedures, to the latest, rapid-static. From the smallest to the very largest GPS survey, proper mission planning has not changed significantly, but has become more compacted due to more efficient GPS field techniques. The importance of proper planning cannot be overstressed, and it will help to guarantee the success of the project.

Geometry concerns have rotated to the zenith. Satellite geometry is more important than ground geometry in a well-configured network. Without good satellite geometry, the most ideally configured network would fail.

Measurements on the ground are indirect measurements made by differencing measurements to the satellites. Compared to doing a terrestrial resection, the measurements to the satellites are subject to the same geometrical constraints.

The satellites available for the session must not be bunched together in the sky, but be spread, ideally in four quadrants. This is not always possible, but using mission planning software available from several sources enables the mission planner to monitor an expression called the positional dilution of precision (PDOP). Generally, PDOP values of six or less are sufficient for GPS relative positioning techniques. Do not overlook the use of periods identified by the mission planning software in which the PDOP indicates a spike. This could indicate the rapidly changing geometry that can result in shorter occupation times, yielding the same accuracy as would result with longer occupation times.

The mission planner must have an intimate knowledge of the terrain. This requires a visit to each site and assurance that the recovery notes are still sufficiently accurate for the receiver operator to get to the station site and recover the station.

The mission planner must know the driving time required to access the station from any location in the project area. In the event packing-in is required, then the time necessary to reach the end of truck travel must be known, as well as the walking time necessary to reach the station. Most of the foregoing is information that should be available from recovery notes and to-reaches resulting from a thorough reconnaissance. Packing-in requires an awareness of the physical abilities of the receiver operator and the knowledge of possible dangers en route to the station before personnel assignments are made for the session. The personal strengths and weaknesses of each staff member must be given consideration when making session assignments.

Figure 15-20 shows a key route index map that was taken from an actual campaign; it details specified main access routes for the receiver operators and associated driving times between road intersections defining the geographical limits of the GPS network. The information was extracted from recovery notes prepared for the campaign, as well as other sources to aid in realistic mission and session planning for multiple sessions on the same mission day.

The selection of GPS field technique for the campaign can only be done after the reconnaissance is completed. If a station-to-station spacing represents, say, 5 min of travel or less, then rapid-statics, SRPK, PK, or K would be a strong consideration for high productivity.

The selection of GPS field `methodology involves an intimate knowledge of the geography and transportation routes of the campaign area, as well as the capabilities of the various GPS field techniques, so that technique can be matched with area. The mission planner must have personal and first-hand information of the stations to be visited by the GPS crew, in addition to various access routes to the stations. Such an awareness does not result from the recovery notes alone, but from a station-by-station tour of the project for the purposes of familiarization, and formulation of the most favorable GPS technique for these specific locations.

After the agenda for a mission day has been designed by the mission planner, especially in those cases where high-productivity methods are to be employed, it is advisable to have the monuments in the schedule freshened before the session begins. This requires an available member of the party to proceed from station to station ahead of the GPS crew, uncovering the markers should they be buried, or opening well covers and bailing out water if necessary. Additionally, where monuments are located in a roadway, the setting of necessary traffic signs and cones in preparation for the arrival of the GPS observer is a significant time-saver.

As an aid to rapid recovery during the GPS measurement phase of the campaign, it would be very helpful if, during reconnaissance, the station is well lathed and flagged. If the monument is located in an asphalt roadway, then a unique GPS identity number can be painted on the pavement in a large enough size to be easily visible to the GPS observer from a vehicle approaching the station.

Initial advance preplanning should be for the sole purpose of estimating general campaign costs and scheduling, while providing sufficient padding to allow for contingencies. Actual mission planning should occur in near real time in order to be realistic enough to reflect the realities of present field conditions, allowing for flexibility in immediately responding and adapting to field, satellite, weather, personnel, equipment, and access difficulties that might exist. A survey will not be successful if inflexible mission planning was done 2 mon in advance of the project back in the home office.

Figure 15-21 shows an actual plan for an SRPK session prepared for the A session (there was also a B and C session on the same mission day) of Julian day 161 during 1991. The plan details a sketch of the geographical area, station assignments for each rover, start-up and shut-down times, move times, station identities, and more. This document was prepared at the end of the previous mission day (91.160), together with two other session plans for B91.161 and C91.161 (no figures shown). Prior to the beginning of mission day 91.161 the GPS field crew was briefed on the details of A91.161 (Figure 15-21), and the subsequent B and C sessions of the same day.

Although the example shown was for an SRPK session, it is recommended that this type of document be drawn regardless of the GPS technique selected. Figure 15-15 is a detailed breakdown of the four moves (setups) contained within session A91.161; it demonstrates the high-productivity potential of SRPK.

15-6-1. Premission Briefing

The premission briefing is an essential meeting between the mission planner and all GPS crew members and should occur immedi-

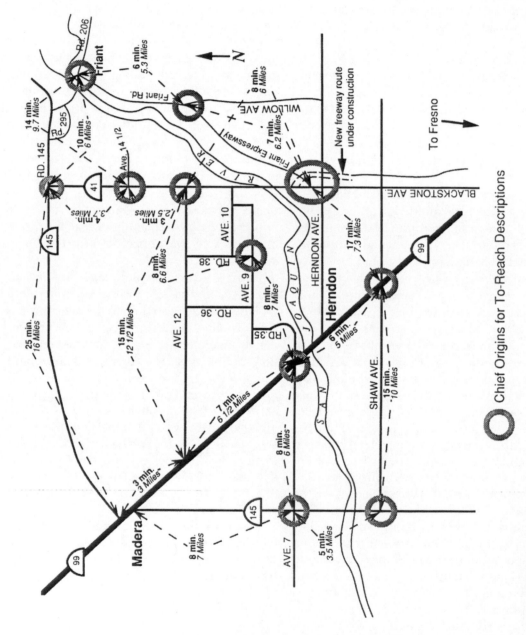

Chief Origins for To-Reach Descriptions

Figure 15-20. Key route index map.

ately before the crew takes to the field. The briefing covers individual equipment assignments, vehicles, station assignments in each session, access to stations, observation schedules, receiver start-up and shut-down times, route changes that may differ from the to-reach, and any changes in conditions since reconnaissance that may impact access to station sites. If there is new information to be added to the recovery notes, this is the time to do it. Other factors to discuss include any new information such as new locks and keys, or combinations, bad dogs, new restrictions, or conditions imposed by property owners, and other practical information. This is the opportunity to comprehensively discuss the start-up and shut-down times as well as anything crucial to the success of the mission. Whatever has gone wrong during the last session should be discussed and critiqued by the receiver operators so that each can learn from the mistakes of others. Perhaps a better time for such input occurs at the time of the postsession debriefing. Feedback to the mission planner is crucial. For example, if insufficient time is being factored in for the moves in pseudokinematics, statics, or rapid-statics, or, on the other hand, the allotment of time is too generous, this would be the time to bring this matter to the attention of the mission planner so that adjustments can be made to the scheduled start-up and shut-down times to provide greater productivity.

Session plan distribution should be made to all receiver operators at the time of the premission briefing to ensure the success of the session. Additionally, it is further recommended that each operator be given a folder for each station assignment that contains a copy of the following:

1. The to-reach
2. Station notes
3. Local map
4. Assessor's parcel map
5. Photo of station
6. Blank notes for the GPS observation, either long- or short-form.

15-7. SPECIAL EQUIPMENT

The importance of two-way radio communication can never be overstressed or overemphasized. At the very least, the receiver operators should be equipped with CB radios for inter-site communication. Ideally, the receiver vehicles should have two-way FM radios and walkies, and at least one of the vehicles in the field should have a cellular telephone if at all possible. The resulting close communication enables the coordination of start-up and shut-down times during the sessions and allows for variables in arrival times or setup times at the station sites due to wrong turns, delays, equipment malfunction, or similar misfortunes. It also allows the flexibility to extend data-collection times if necessary, or for earlier start-up time than scheduled by the mission planner, provided that proper satellite geometry is present. In the event of an emergency or any situation where assistance is necessary, communication is the key to response and should *never* be ignored.

15-7-1. Tripods and Tribrachs

Except for GPS hardware such as antenna, cables, receiver, etc., the normal equipment and accessories usually found in the inventory of most surveying offices are sufficient. GPS antenna assemblies are designed to fit on practically every type of tribrach. A tribrach and tripod are all that is required. Naturally, the calibration of the tribrach must be validated and the tripod in good repair. But why use a tripod and tribrach? The tribrach is expensive, subject to chronic centering (collimation) errors, is fragile, is not weatherproof, and with its sophisticated optics and precision machining, perhaps too refined and complex for the job at hand and maybe not essential, or even practical, for production-oriented GPS.

15-7-2. GPS Bipod-Tripod

As an alternative to the conventional tripod and tribrach, many governmental agencies, es-

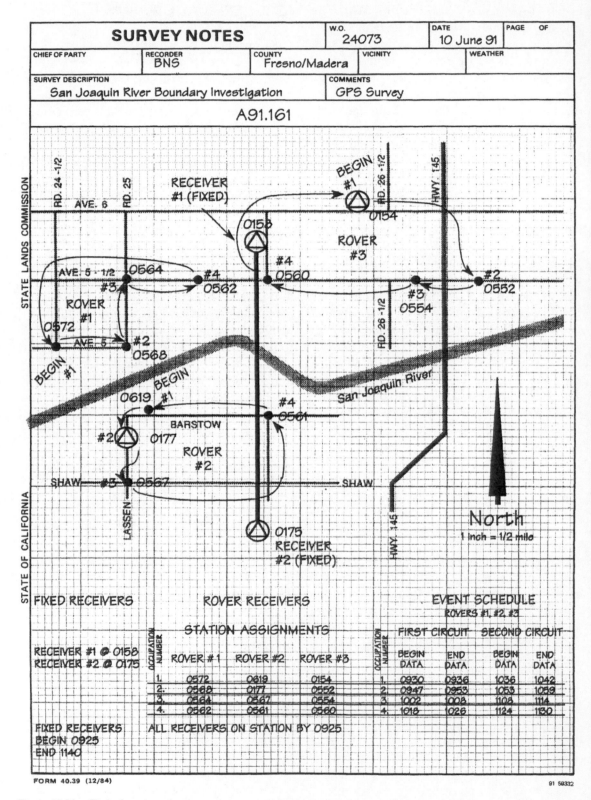

Figure 15-21. Typical session plan.

pecially federal and state, and private industry as well are switching to GPS bipod-type stake-out rods in preference to the conventional tripod and tribrach. For the price of a better-quality conventional tripod, a GPS stakeout rod can be purchased that is easier to set up, more suitable for GPS measurements, less subject to damage than the conventional tribrach, and about one-third the cost of a tripod and tribrach. Additionally, at least 2 to 3 min are cut from the setup time when a GPS bipod- or tripod-type stakeout rod is used. In conventional work, this may not seem like much, but when the session plan calls for a GPS receiver operator to break down the setup, pack up, and travel to a new station, where the equipment is again set up and the receiver logged on before data collection begins again, all in a time frame of 10 min, the shaving off of 2, 3, or 4 min from total operational time becomes significant. Time is of the essence; time saved in any GPS operation will yield greater productivity no matter what GPS field operation technique is utilized. Time is saved in setting up the antenna and time is saved during HI measurements, since it is only measured once, and in recording the data. The GPS rod assembly shown in Figure 15-22 is lighter in weight, more suitable for packing into a station, and occupies considerably less space, which is beneficial in those situations where storage capacity in a vehicle or aircraft is at a premium. Higher-precision-level bubbles are available that can yield centering of the 1-mm level. These assemblies make the GPS crew more mobile and flexible to individual site requirements and cut down on a bulky and unnecessary inventory. With the addition of screw-in rod extensions, the height of the antenna can be elevated several times beyond that of a tripod and tribrach, thus solving the problems associated with difficult setup locations where obstructions would have made observations impossible.

The GPS bipod or tripod is similar to the conventional stakeout rod, but with larger diameter for greater stability and strength. The length of the rod is adjustable by the selection

Figure 15-22. GPS rod.

of appropriate lengths of screw-in sections that are available in various lengths, generally of 1, 2, or 3 feet. Once assembled, the rod is left in whatever length configuration is selected for the mission. The use of an adjustable-length (telescoping) rod is not a recommended alternative due to problems inherent with HI measurements. The goal is to reduce unnecessary field operations and eliminate the incidence of HI errors. A telescoping rod is subject to a change in the antenna height if the friction collar slips. Should this occur and go undetected, any GPS measurements made will be invalid. Leave it in the equipment room to keep company with the tripods and tribrachs. The use of uniform length should be a rule during the campaign. This does not mean that all the rods need to be exactly the same; they do not. However, to avoid errors in the HI measurement, this consistency should be maintained. The mismeasurement of HI is one of the more common mistakes made with GPS observations. If each GPS rod is premeasured for HI, this is one less task that the receiver operator must accomplish in his or her busy schedule and one less possible source of error.

All that needs to be recorded in the GPS field notes is the rod identification number. Since the rod is premeasured, the HI is already known to the mission planner and computer operator.

The choice between the two- or three-leg GPS stakeout rod is a matter of preference. However, there are advantages and disadvantages with either selection.

The bipod is more efficient to set up. How much more time-efficient is a matter of practice and technique but should not amount to more than 10 sec. The bipod is naturally somewhat lighter in weight than the tripod since it has less hardware and does not have that extra leg which sometimes gets in the way. The bipod will stand by itself, provided there are no gusts of wind or other conditions that could upset the balance. Consequently, the bipod must be attended throughout the occupation to prevent it from falling over.

The tripod version of the GPS rod is more stable in traffic, windy or gusty conditions and is recommended for most applications requiring unattended operations. Naturally, adequate carrying cases are advised to protect the GPS rod for shipment.

15-7-3. Multiple GPS Rod Bubbles

Centering (or plumbing) over the mark is accomplished by using the level bubble attached to the rod. GPS rods can be purchased with more than one bull's-eye level. It is recommended that an array of three such levels with a sensitivity of 10 min be installed on each rod. These rod bubbles shown in Figure 15-23 are arrayed at 120°, a convenient distance above the ground so as not to interfere with the operation of the bipod or tripod legs of the GPS rod. Make certain that they are in exact adjustment and firmly attached and mounted to the rod as the level bubbles may be subjected to unavoidably harsh handling. Why three bubbles? If only one bubble is used on the GPS rod assembly and it should go out of adjustment, the condition would probably

Figure 15-23. GPS rod with three bubbles.

go unnoticed by the receiver operator and might result in good observations to the antenna, but bad observations relative to the mark since the antenna phase center is in a different plane. Two bubbles installed on the GPS rod and adjusted during plumbing calibration serve as a check on each other. Should one of the pair of bubbles go out of adjustment, the operator would not know which of the two is suspect, and would need to abort his or her session schedule until the problem is cured. The use of three bubbles is the practical limit and provides for a situation in which one of the levels has been knocked out of adjustment; the remaining two should have the same reading when the rod is set over the mark, thus reassuring the observer that collimation is still good. Thus, three bubbles would allow the session to proceed without interruption.

15-7-4. GPS Rod Collimation Adjustment: A Shortcut

Rod collimation can be accomplished using several different procedures. One of the simplest methods that does not require elaborate equipment can be done in any doorway. Simply suspend a plumb bob from a small nail driven securely into the top underside of a doorway. Mark the plumb point on the floor. Remove the plumb bob and set up the rod with the point of the rod on the plumbed point and the top of the rod plumbed underneath the nail (the plumb bob might be needed here) and secure the legs. Once this is accomplished, the rod is now vertical and in collimation with both points. Now adjust the rod bubbles so that they are centered. Check by slowly rotating the rod through 360°, noting any movement of the bubbles and any movement of the top of the rod away from its plumbed position under the plumb bob hanging from the nail. There should be no movement in either the bubbles or top of the rod. If such is not the case, the rod is either bent or the rod bubbles are at fault. Replace and/or repair and readjust as required.

15-8. HAZARDS TO SUCCESSFUL DATA COLLECTION

The interruption of satellite signals can be caused by a multitude of sources. Rain or condensation causing moisture on the electrical connections, or the head, hands, hard hat, etc., while servicing or maintaining the receiver and antenna can cause interference or blocking of the signal. Basically, anything that passes between the antenna and satellite can interrupt the signal. Keep away from the antenna while collecting data. The body can block signals from the satellite just as surely as tree limbs, power poles, and other inanimate objects. Avoid placing anything between the antenna and satellites or the signal will be blocked and a loss of lock will occur. Another classic is the "experienced" field hand who, during kinematics, automatically shoulders the GPS rod or tripod when picking up. This, of course, is guaranteed to cause a loss of lock and trip back to the last known position for reinitialization.

15-8-1. Multipath

Multipath is a condition in which a satellite signal arrives at the receiver antenna by way of several different paths. It is caused by reflected, indirect, signals from a satellite and can originate from a multitude of sources. Figure 15-9 is a good example. Bodies of water, structures, nearby vehicles (especially slab-sided semitrailers, or vehicles or similar design), freeway signs, chain-link fences, or similar reflective objects can cause a condition of multipath.

Sometimes, the use of a larger-accessory ground plain can reduce or eliminate this condition. Some antennas do not even have an integral ground plane. These antenna types should be avoided except for special applications. Avoid the problem whenever possible by locating the station away from possible interference. If the station is part of the primary net, then an eccentric might be considered, or possibly a higher setup to clear the possible reflection. Multipath is not detectable until post-data-reduction is done, so be observant. Proper reconnaissance will identify stations where this problem exists and precautions can be taken to minimize or eliminate the problem. During data collection, the survey vehicle must be parked far enough away from, or below, the antenna to eliminate the possibility of multipath from that source. Good site selection during reconnaissance will either minimize or eliminate problems due to multipath.

15-8-2. Electronic Interference

Consult with the manufacturer regarding possible sources of electronic interference that could corrupt the signals of the satellites. The following electromagnetic energy sources will not necessarily affect all receivers. Some receivers are adequately shielded; however, it would be best to err on the side of caution if there is any doubt. In the design of any electronic circuit, one important factor is to separate the input signal from unwanted signals and amplify it in the required way without producing distortion beyond an acceptable degree. Effective passive filters are installed in the GPS receivers to filter out unwanted signals; however, if the unwanted signals are powerful enough, the filters are ineffective. Powerful signals cause the amplification to be adversely affected in the form of distortion and this can affect the GPS receiver performance. Some of the sources that may cause interference with GPS units are:

1. Vehicle detectors for actuating traffic lights
2. Portable transceivers for radio communication
3. Signals emitted from antennas generated by radio and television stations
4. Amateur ham band and citizen band transceivers
5. Microwave antennas and transmitters for equipment such as telephones
6. Radar installations

15-8-3. Traffic Sensor Devices

Vehicle detectors come in two varieties: below-ground (loop detector) and above-ground (pole-mounted). The ground plane antenna on GPS units would shield the signal from the below-ground, or loop, detectors, that have a range of approximately 10 ft straight up. Therefore, they would present little risk of interference to the GPS receivers. The pole mounts have two operating frequencies: microwave (10.525 GHz) and ultrasonic. The microwave vehicle detectors have a greater possibility of affecting GPS receivers because of the closeness of operating frequencies; also they are located above the GPS receiver antenna. On the other hand, the GPS receiver antenna would have to be relatively close to a vehicle detector because the low power (2.5 to 6.0 W) of the vehicle detector and limited range (60 ft more or less). Ultrasonic vehicle detectors have little or no effect on GPS receivers because of the difference in operating signal frequencies between the two units.

15-8-4. Two-Way Radios

Portable transceivers such as handheld and mobile radios should be operated as little as possible and as far from the GPS unit as is convenient. Transmission is not recommended during data collection except with the manufacturer's OK. The concentrated signal in the vicinity of a portable transceiver can resemble a much stronger signal and possibly corrupt the satellite signal. Cellular phone transmissions, when done in the close proximity of the receiver, could interfere with satellite transmission reception as well.

15-8-5. Radio, Television, Microwave Antennas

Radio, television, radar, and microwave antennas radiate a powerful signal and generally the antennas occupy high points such as the tops of mountains and buildings. These signals can also present a problem for GPS units. The signal from these antennas is so strong that it can affect a GPS signal a considerable distance away. Satellite dishes such as those for home use are for receiving only and present no problem unless the satellite dish is in the GPS signal path to the receiver antenna. The satellite dishes for commercial television stations both transmit and receive; however, they are very directional and so also should present no problem. The configuration is not necessarily an indication of the function of the antenna since antennas can be used for both transmitting and receiving.

Airports typically have more than a fair share of potential electronic interference.

Large airports always have a greater abundance of exotic electronic devices than any other location, most of which are capable of interference.

Other possible sources of problems such as high kv power transmission lines can cause problems during certain atmospheric conditions. Any time that buzzing or arcing can be heard, satellite signals could be subject to possible interference. Thunderstorms, even though miles away, are reported to be disruptive.

15-8-6. Geomagnetic Disturbances

Geomagnetic disturbances are caused by solar flares, solar storms, and other similar natural solar phenomena. These solar disturbances release large amounts of energy in the ionosphere, an electrically conductive series of layers of the earth's upper atmosphere extending from 50 to 400 km above the surface.

Flares and the resulting geomagnetic storms that sometimes accompany them can disrupt low-frequency systems such as Loran C and GPS. Communication systems like television, radio, microwave, and short-wave ones are also impacted, sometimes to the point of total disruption. The current solar cycle is believed to be one of the highest ever and should continue into the mid–1990s.

Geomagnetic storms are sometimes accompanied by the appearance of the Aurora Borealis or northern lights, even in the lower mid-latitudes. In a time of major geomagnetic storms, such as the flares of March 1989, an aurora was visible in the Gulf states. The Aurora Borealis during this event was seen as far south in California as 35° north latitude.

The use of dual-frequency GPS receivers will minimize the effects of geomagnetic disturbance. The majority of receivers in use at this time are of the single-frequency variety that are susceptible to cycle slips caused by the electromagnetic noise associated with these disturbances.

The appearance of the aurora should be a visual warning to the project manager and mission planner to carefully view the post-data-reduction for signs of noisy data and possible cycle slips. Mission planning should include the monitoring of all geomagnetic and solar advisories that are available on the joint USAF/NOAA solar region summary bulletin board service. This bulletin provides values for the A-index and K-index and predicted values for A and K that can be used as an indication of solar activity levels. Generally, A-index values greater than 20 and K-index values greater than 5 are indicators of high geomagnetic disturbances, possibly contributing detrimentally to the collected data.

With single-frequency receivers, a few measures could be employed to a limited extent in an attempt to salvage a few mission days in a campaign unfortunate enough to be accidentally scheduled within a period of high solar activity. If there are short base lines in the project of 2 km or less, these short lines would be the preferred measurements to attempt. Longer base lines are typically out of the question and likely to fail. The order of preference for measurement techniques during high solar activity periods is

1. Statics
2. Kinematics
3. Pseudokinematics or SRPK

It is very possible that cycle slips will go unnoticed with the pseudokinematic technique during periods of high solar activity; increasing the data-collection time by 50% or more might help. It is reported that nighttime levels are somewhat lower than daytime levels and midday levels lower than early AM or late afternoon levels. Watch the resulting residual statistics with caution and modify the mission plans to include redundant vectors and independent checks on the observed stations. If unrepairable cycle ships occur in spite of these precautions, either obtain dual-frequency receivers or cancel the field operations until the geomagnetic disturbances subside.

15-8-7. Power Source

Any system that utilizes battery power is subject to a multitude of energy-related problems. Do not believe the sales representative when a claim is made that a brand A receiver will run all day on an AAA battery. Never underestimate the power consumption of the receiver system. Always allow for a comfortable, if not generous, safety margin. It does not matter how high-tech and exotic surveying technology becomes. The simple truth remains: We are slaves to our battery power source. The battery and two-bit battery connections remain the two most common causes of receiver failures. Always provide the receiver operators with backup battery reserves and extra cables and connectors, and insist that all batteries be topped off between missions.

The selection of battery capacity is dependent on the power requirements of the unique receiver, which can vary from manufacturer to manufacturer. A good guideline would be to select a battery with sufficient reserves for possibly 150% or more of the required power consumption for approximately two missions —i.e., two data-collection days or perhaps 10 hr of actual data collection, whichever is the greater. This would allow for the loss of battery efficiency from low temperatures or possibly a failure of the operator to recharge to a full 100% level. Naturally, pack-in situations require lightweight as well as ample reserves so allowances should be made to mix battery types—i.e., weight and capacity according to the specific session and station requirement —to cover long-term continuous data-collection sessions as well as situations in which a lightweight and free mobility are important.

15-8-8. Antenna Height Measurement

The importance of this measurement is often overlooked. GPS is a three-dimensional system and requires an HI to compute the position at the mark. Antenna height measurement is a common source of problems in the field. Without reliable antenna height measurements, the system will not accurately compute final position and elevation at the station. The position will be to some nonrepeatable point in space. The common error is either failure to measure the height or erroneous measurement of an HI. As a check against possibly flawed measurements, a good technique is to independently measure in two different units of length such as feet and meters. The use of fixed-height antennas is the greatest elixir for either bad or missing HI measurements. However, be advised in otherwise identical GPS rods that there can be variations in antenna height measurements. It is a good idea to individually identify each GPS rod with a unique name or number and HI. This identity is then incorporated as data entry on the GPS field notes and entered into the GPS receiver at the site. When conventional tripods and tribrachs are being employed, extra care must be exercised to ensure that the HI being measured is error-free. Once again, the best method involves measurement in two different systems, such as feet and meters, with an independent conversion of feet to meters or from meters to feet done in the field by the receiver operator as a check. It should be standard operational policy for field personnel to make this conversion as part of the data-collection process.

The use of a GPS rod, either bipod or tripod, practically eliminates this error due to missing HIs since the setups are done at a uniform height above the mark. The length of the rod from the point to the antenna needs to be measured and checked only once.

15-9. TYPICAL PROBLEMS ENCOUNTERED

No matter how detailed, complete, and intuitive the planning and preparations are, something will always occur to disrupt the survey.

GPS is subject to a host of familiar maladies common to conventional terrestrial survey operations and a few that are not. Most problems associated with delays and downtime are vehicle- and human-error-related. Table 15-3 is a tabulation of various things that went wrong during two actual campaigns, each survey consisting of approximately 200 stations. The first campaign employed static procedures with two, or more, nighttime sessions per mission during a 4-wk period in February and March 1989. The second survey employed a combination of static, pseudokinematic, and kinematic procedures with two or more daytime sessions per mission during a 2-wk period in June of 1991. During each survey, premission briefings and postmission debriefings were conducted by the mission planner, with the chief of party, computer operator, and GPS receiver operators in attendance.

Items 1 through 9, and perhaps item 19, in Table 15-3 are, to the greatest extent, classified as human blunders and mostly preventable. Perhaps more emphasis can be placed on high-repeat problems at the time of the postmission debriefing or next premission

Table 15-3. Problems Encountered on Two Typical Jobs

Malady	Number of Instances
1. Receiver operator getting lost en route, wrong turn en route to a GPS station	6
2. Receiver operator on wrong GPS station	4
3. Bad setup (tribrach not leveled, GPS rod not plumb, etc.)	4
4. Receiver operator forgets to load an equipment component necessary for session	6
5. Operator error: Wrong data keyed, wrong epoch interval, etc.	4
6. Bad HI measurement	3
7. Cycle slip: Loss of lock due to carelessness	2
8. Vehicle stuck	2
9. Vehicle out of gas or lack of fuel delays or constrains planning	3
10. Vehicle will not start	4
11. Vehicle breakdown	4
12. Vehicle battery dead	2
13. Access to station not possible: Receiver operator locked out, new lock installed, no key, key not working, etc.	3
14. Traffic delays	4
15. Receiver battery problems: Battery failure, bad connections, etc.	11
16. GPS equipment failure, either receiver or antenna	2
17. Geomagnetic storm (causing unrepairable cycle slips): Dual-frequency receivers would not have been affected:	6
18. Bad weather: Thunder, lightning, snow, rain, etc.	1
19. Ground swing. Multipath (semitrailer parked near station, survey vehicle parked too close to the antenna, etc.)	2
SUMMARY of PROBLEMS by TYPE	
Human error, blunder-related	34
Vehicle, traffic, access	17
Unavoidable, would possibly occur again	9
GPS hardware	2
GPS power source, connectors	11
TOTAL	73
Summary	
GPS hardware-related	02%
GPS power source, connectors	15%
Unavoidable, usual delays	13%
Human error, non–GPS equip.-related	47%
Vehicle, traffic, access	23%

briefing. Some problems are due to operator fatigue; others are preventable by the application of good judgement, and some are going to occur no matter what preventative measures are taken.

Items 10 through 14 are vehicular or access-related problems: Some are preventable and some are not. Vehicle breakdowns are going to occur, but can be minimized with good maintenance and careful, nonabusive driving practices. Access problems can be reduced by revisiting the station (premission) to update the to-reach if time permits. Should there be sufficient time before a session begins, a luxury that does not occur very often, the receiver operator could tour assigned stations to become familiar with the routes to them. With good planning, traffic delays can be factored into most session plans.

Items 15 and 16 are GPS hardware-related. Item 15 is a famililar foe and will continue to be as long as a battery is a part of any technology. Batteries need to be recharged and connections are necessary to feed the energy to the electronics. This is nothing new as battery and connector difficulties have always been a problem with conventional survey instrumentation and will continue with GPS as well. With the implementation of standard procedural policy for field personnel, energy-related malfunctions can be minimized. Carrying spare batteries is essential; keeping both the main and spare battery fully charged is paramount. Policy should be that the receiver operator is responsible for topping off these power sources immediately after the mission day. If insufficient time exists between missions to completely recharge, then additional batteries must be issued to permit the cycling and charging of the battery inventory. Power interruptions due to faulty connectors are intolerable during data collection. Good manufacturer design should be insisted on; if problems occur, complain loudly. Keep the connections tight and tape them if necessary to prevent movement and moisture.

Item 16 is not supposed to happen, but does in spite of the best of design and manufacturing practices. The best advice is to buy reliable, robust equipment from manufacturers or firms with established postsale support and good performance history.

Items 17, 18, and 19 are random. Geomagnetic storms or ionospheric disturbances in the earth's upper atmosphere caused by solar disturbances are not predictable with any reliability. These disturbances can cause unrepairable cycle slips in single-frequency GPS receivers when levels are high. The use of dual-frequency receivers virtually eliminates this problem. Both the surveys referenced by Table 15-3 were impacted by solar disturbances that aborted several mission days. Had dual-frequency receivers been employed, the campaigns could likely have continued without interruption and costly crew downtime.

GPS is universally represented as an "all-weather system." This is not always true. GPS is subject to most of the same influences as conventional instrumentation, except those relating to line of sight. Thunderstorms and lightening can cause cycle slips, and the hazards of a strike from being near the tripod or GPS rod during such events are a reality. Rain can cause shorts, or open circuits in the power leads and connectors. Snow collecting on the antenna can cause a loss of lock, and wind can topple a tripod or GPS rod, possibly irreparably damaging equipment and causing downtime for repair or replacement. It would be advisable to carry as many spares as the budget and conditions warrant. An extra set of cables and power leads should be issued to each receiver operator. Rain, ice, snow, and mud can cause gross differences in the access times required to reach stations; this could impact mission planning to a substantial degree. Weather should always be factored in with all mission plans. Should bad weather set in, wait it out if economically possible. GPS is the most productive and efficient technology ever to impact the surveying profession. Lost time can be made up. Inclement weather is hard on equipment, results in undue hardship for field personnel, and because of hazardous driving and field conditions places the staff in possible danger.

15-10. NETWORK ADJUSTMENT

Network adjustments are performed for two reasons: (1) To detect and remove blunders, and (2) provide a best fit into the local datum. Blunder detection adjustments should be performed on the daily results throughout the project. Each day the daily results should be added to the previous vectors and a daily subtotal adjustment performed to verify data continuity. At the end of the project, it will then be a simple task to fix the control and perform a final constrained adjustment, fitting the data to the local control.

There are several excellent least-squares adjustment packages on the market for GPS observations. Most of these packages allow the inclusion of conventional data and use the information contained in the base-line processing solution files for weighting the lines. However, one commonly used package does not allow the mixing of conventional and GPS data and its expects the user to input a priori error estimates for weighting instead of using the solution information. There are pros and cons to either approach, both as to the inclusion of conventional data use of the base-line solution statistics. Either approach is correct and will produce excellent results if used properly. Following is a description of the process using least-squares adjustments. The method is applicable to either of the least-squares weighting schemes, the use of the base-line statistics or a priori error estimates, and based on realistic error estimates regardless of the weighting scheme used.

Most of the work in a network adjustment is performed as a *free adjustment*. A free adjustment is a minimally constrained adjustment where the latitude, longitude, and height components of a single station are held fixed. A free adjustment results in the adjustment of the data on itself without constraints as to the known positions of other stations or rotation angles between the GPS system and local datum. Free adjustments are for blunder detection and eliminating bad measurements. Once the data have been cleaned up, the control

station values can be held fixed and the final constrained adjustment performed.

Starting directly with a constrained adjustment results in greater difficulty in isolating any problems. If there are any problems in the adjustment, it would be extremely difficult to determine if the problem is actually in the GPS data or control itself. By cleaning up the data before fixing the control, it is easy to determine that the problem lies with the control. A properly performed network adjustment consists of daily free adjustments, daily subtotal free adjustments, a total free adjustment, and a total constrained adjustment. Each step of the adjustments, especially the total constrained adjustment, may take several iterations when vectors are deleted or deweighted. Daily processing and adjusting cannot be overstressed. It is important to find any errors as soon as possible so that they can be corrected immediately. Reobservation is a dirty word only if spoken at the wrong time.

15-10-1. Statistical Indicators

Before the steps in an adjustment are detailed, look at the apriori and a posteriori indicators that will determine the adjustment quality. These indicators are the *apriori error estimates*, either entered by the operator or taken from the base-line solution statistics; the *residuals*; the *normalized residuals* (standardized residuals); and the *standard error of unit weight* (variance of unit weight). To be accurate, all these indicators depend on realistic estimates of apriori error.

Apriori error estimates are generally created from the standard errors of the base-line components (sigma X, sigma Y, and sigma Z) and the correlation matrix. These are combined into the covariance matrix that is used for weighting the vectors. The apriori errors can also be empirically derived and specified manually as a base error (mm or cm) plus an allowable additional part-per-million error. Either way, it is important that the estimates of error be realistic.

The residuals are the amounts by which the adjusted vectors have been shifted in the adjustment. They are the differences between the ΔX, ΔY, and ΔZ components of the observed vector and adjusted vector. They are generally reported in meter units.

The normalized residuals are the residual divided by the apriori standard error. The normalized residual indicates outliers. A normalized residual of 2.0 indicates that the residual is twice as large as it should be based on the apriori errors. If the apriori errors are realistic and the standard error of unit weight is close to 1, statistical outliers will have a value of 3 or more.

The standard error of unit weight indicates the degree with which the data and apriori errors agree. The ideal, a standard error of unit weight equal to 1.0, indicates that the quality of the data exactly fits the model. The standard error of unit weight should approach 1.0 or less. Exceeding 1.0 by more than a very small amount indicates problems with the data, or overly optimistic error estimates—i.e., the apriori errors are too small. On the other hand, a very small standard of unit weight indicates pessimistic apriori errors, or extremely good data. If the estimates are pessimistic—i.e., if they can be realistically reduced—they should be. Pessimistic apriori errors can hide problem data. When the apriori errors are too large, the normalized residuals will be smaller than they should be, and data that might be an outlier with realistic error estimates can be hidden.

15-10-2. Network Adjustment Procedure

Starting with the first mission-day observations, a daily free adjustment is performed. The daily free adjustment is where the majority of work is done. It is at this stage that blunders are detected and removed, and final decisions as to the fixed or float solution selection are made.

Most processing software packages have a filter program that selects the best solution for each vector. This filter selects either the *float double difference solution* or *fixed double difference*

solution, depending on its particular algorithm looking at the base-line processing statistics in the solution files.

A suggested practice is to perform two adjustments if any float solutions are selected. One of the adjustments will contain the combination of fixed and float solutions selected by the filter, and the other all fixed solutions. A quick comparison of the standard error of unit weight will usually determine whether the fixed or float solutions are best.

Sometimes, it is necessary to go further and return one of the fixed solutions back to a float solution and perform a tired adjustment. Decisions like this are based on the SE of unit weight and values of the normalized residuals. The results of the daily free adjustment(s) should be the vectors in the adjustment with the lowest SE of unit weight. These would be the best fitting vectors and those to pass along as we build the total network.

Each day, the vectors of the final daily adjustment are added to the previously selected vectors, and a subtotal adjustment is performed. This is to verify that the day's sessions agree with those of other previous days. Although it is not common, a session not agreeing with others happens on occasion, due probably to bad broadcast ephemeris information. To find this out in a timely fashion, perform daily subtotal adjustments. Generally, the broadcast ephemeris is satisfactory for surveys up to a few parts per million. For surveys of 1 part per million or better, the use of precise orbits is mandatory.

The use of daily subtotal adjustments also provides an easy way of building the network using the best vector solutions available. It is virtually assured that the final free adjustment will be simply a matter of adding the final daily adjustment vectors to the subtotal and performing the final free adjustment.

When the final free adjustment has been completed, blunders have been removed and the data has been validated. At this point, the first task of the network adjustment has been accomplished: the removal of blunders. It is now time to constrain our free adjustment by the control and perform the final task, fitting to the local datum.

Generally, the free adjustments are performed using the WGS 84 latitude, longitude, and ellipsoidal heights derived from the baseline processing solutions. One of these positions is simply held in all three components for the free adjustments. To constrain the network, specify the datum and reference ellipsoid to which the free adjustment will be molded and fix the control stations by entering their positions relative to the datum selected.

For example, if it is desired to perform a constrained NAD 83 adjustment, make sure the adjustment program will use the GRS 80 reference ellipsoid parameters, then supply NAD 83 latitude and longitude, and ellipsoid eights (elevation plus geoidal separation) for the fixed positions. If NAD 27 positions are desired, make sure the Clark 1866 reference ellipsoid is used, and NAD 27 latitude ad longitude, and orthometric heights (sea-level elevations) for the control stations are entered for the fixed stations.

Once the control has been fixed, perform a constrained adjustment. It may be a minimally constrained adjustment or fully constrained adjustment. A minimally constrained adjustment has only enough control fixed to solve uniquely for scale and rotation biases between the GPS system and local datum. For example, two horizontal stations will provide one and only one solution for scale and azimuth rotation. Three vertical stations will provide one and only one solution for the gamma X and gamma Y rotations—i.e., the vertical tilts in the east-west and north-south directions. A fully constrained adjustment will provide redundant solutions for these bias parameters,—e.g., three or more horizontal control stations and four or more vertical control stations.

If the control values are good and estimates of error accurate, a standard error of unit weight close to 1.0 should result. If the standard error of unit weight is larger than 1.0, it indicates one of two things: Either the estimates of error are too optimistic, i.e., too

small, or the control does not fit well. Most commonly in an NAD 27 adjustment, it will be the control causing the problem. Unless there is a single station that can be proven to be in error, it will be necessary to mold the data to fit the local datum. This may take several iterations and involve deweighting of vector components and constraining, instead of fixing, some control values.

Avoid any temptation to leave the adjustment in the free adjustment stage, i.e., radial surveys. The data will fit best on itself, but will not fit the real world or local control. Any ties from the free adjustment to other national network stations could easily result in substandard closure statistics because the rotation biases were not solved and applied.

15-10-3. Dual Heights

Elevation, as measured with a level, is the height above the geoid, an irregular surface of equipotential gravity commonly associated with the mean sea level. Ellipsoid height is height above the reference ellipsoid, a smooth mathematical surface. GPS measures this ellipsoid height. The ellipsoid height used in the NAD 83 adjustments may be broken down into two elements: the *orthometric height* (commonly referred to as sea-level elevation) and *geoid height* (the height of the geoid above or below the ellipsoid). Normally, the surveyor does not know the ellipsoid height of a station, but will know the elevation. The geoid height of any station can be interpolated from a tabular data set that has been created using a model such as the GEOID 90 model available from NGS. By adding the elevation and geoid height (a negative value in the continental United States), the ellipsoid height of the vertical control stations can be computed, and conversely the adjusted elevations of unknown stations by subtracting the geoid heights from the adjusted ellipsoid heights.

In Figure 15-24, the relationship between the geoid and ellipsoid in world terms can be seen. The WGS 84 or GRS 80 ellipsoids are mathematical surfaces defined to best fit the

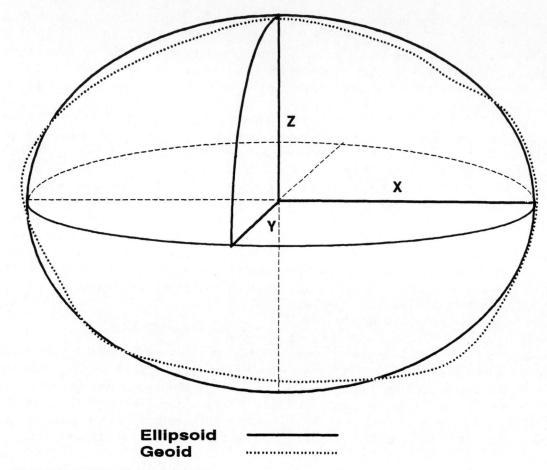

Ellipsoid ──────
Geoid ·····················

Figure 15-24.   Ellipsoid and global geoid.

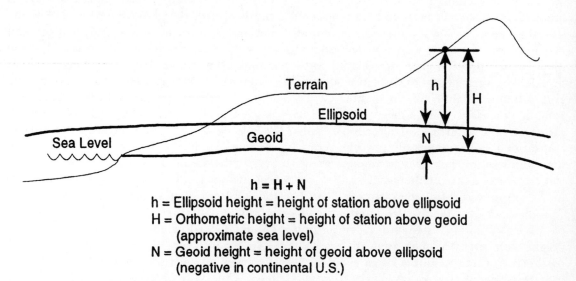

$$h = H + N$$

h = Ellipsoid height = height of station above ellipsoid
H = Orthometric height = height of station above geoid
      (approximate sea level)
N = Geoid height = height of geoid above ellipsoid
      (negative in continental U.S.)

Figure 15-25.   Elements of height.

shape of the earth. The WGS 84 was intended to be the same ellipsoid as the GRS 80, but a mistake with precision caused a change in the inverse of flattening parameter.

In Figure 15-25, a detail section showing the relationship between the ellipsoid and geoid can be viewed. The calculation of the ellipsoid height *h* is shown as the addition of the elevation *H* and geoid height *N*. In the example shown, *N* is a negative number (the geoid is below the ellipsoid), resulting in *h* being smaller than *H*. This is the case in all of the continental United States. The term geoid height is somewhat confusing. Unlike the term ellipsoid height, or orthometric height, it does not refer to the station height in relation to the datum. It refers instead to the height of the geoid in relation to the ellipsoid. A more appropriate term would be geoid separation.

A good adjustment package will allow easy use of modeled geoid separations and easy conversion between ellipsoid heights and elevations. Ellipsoid heights are not of much use to the surveyor; water does not flow based on differences in ellipsoid height.

15-11. GPS AND THE FUTURE

A blending and blurring of differential GPS surveying and navigation will result. Work is being done to develop on-the-fly ambiguity resolution allowing for cm level for ships, aircraft, and most significantly the surveyor, and aircraft camera positioning to eliminate the need for costly ground control. On-the-fly ambiguity resolution will have other applications such as profiling tasks, and bathymetry without the need for static initialization and a host of others. For the average surveyor, the name of the game will be rapid-statics.

REFERENCES

ANANDA, M. P., ET AL. 1988. Global positioning system (GPS) autonomous user system. *Navigation, Journal of the Institute of Navigation 35*(2), Summer 1986.

BLACKWELL, E. G. 1985. Overview of differential GPS methods. *Navigation Journal of the Institute of Navigation 32*(2), Summer 1985.

CANNON, M. E. 1990. High-accuracy GPS semikinematic positioning modeling and results. *Navigation, Journal of the Institute of Navigation 37*(1), 26 Spring. 1990.

FERGUSON, K. E., ET AL. 1989. Kinematic and pseudo-kinematic surveying with the ASHTECH XII. Proceedings of ION GPS 1989. Colorado Springs, CO, Institute of Navigation, Sept. 27–29, pp. 35–37.

FERGUSON, K. E., and E. R. VEATCH. 1990. Centimeter level surveying in real time. GPS90, Ottawa, Canada, Sept. 3–7.

FREI, E., and G. BEUTLER. 1989. Some considerations concerning an adaptive optimized technique to resolve the initial phase ambiguities. Proceedings of the 5th International Geodetic Symposium on Satellite Positioning. DOD/DMA, NOAA/NGS, Las Cruces, NM, March 13–17, New Mexico State University, Vol. 2, pp. 671–686.

FREI, E., and G. BEUTLER 1988. Rapid static positioning based upon fast ambiguity resolution approach "FARA": Theory and results. *Manuscripta Geodaetica 15*, 325–356.

GEORGIADOU, Y., and A. KLEUSBERG. 1988. On the effect of ionospheric delay on deodetic relative GPS positioning. *Manuscripta Geodaetica 13*(1).

GPSIC. 1990. *GPS Information Center Users Manual.* U.S. Department of Transportation, U.S. Coast Guard.

Global positioning system papers. 1980, 1984, 1986. *Navigation*, Vols. I–3. Washington, D.C. Institute of Navigation.

GREEN, COLONEL G. B., ET AL. 1989. The GPS 21 primary satellite constellation. *Navigation, Journal of the institute of Navigation 36*(1), Spring.

GUNTER, H. W., ET AL. 1989. High-precision kinematic GPS differential positioning and integration of GPS with a ring laser strapdown inertial system. *Navigation, Journal of the Institute of Navigation 36*(1), Spring.

HATCH, R. 1989. Ambiguity resolution in the fast lane. Proceedings of ION GPS 1989. Institute of Navigation, Colorado Springs, CO, Sept. 27–29, pp. 45–50.

HATCH, R. 1990. Instantaneous ambiguity resolution. KIS Symposium, Baniff, Canada, September 11.

HOTHEM, L. D., C. C. GOAD, and B. W. REMONDI. GPS satellite surveying: A Practical aspects. *The Canadian Surveyor 38*(3), Autumn.

HOTHEM, L. D., and G. E. STRANGE. 1985. Factors to be considered in the development of specifications for geodetic surveys using relative positioning GPS techniques. Proceedings 1st International Symposium on Precise Positioning with the Global Positioning System, Vol. II. pg 87 Rockville, MD, April.

HWANG, P. Y. C. 1991. Kinematic GPS for differential positioning: Resolving integer ambiguities on the fly. *Navigation, Journal of the Institute of Navigation 38*(1), Spring.

KREMER, G. T. 1990. The effect of selective availability on differential GPS corrections. *Navigation, Journal of the institute of navigation 37*(1), Spring.

JORGENSEN, P. S. 1989. An assessment of ionospheric effects on the GPS user. *Navigation, Journal of the Institute of Navigation 36*(2), Summer.

KING, R. W., ET AL. 1985. Surveying with GPS. Monograph No. 9, School of Surveying, University of New South Wales, Kensington, NES Australia, Nov.

KUNCHES, J. M., and J. W. HIRMAN. 1990. Predicted Solar flare activity for the 1990's: Possible effects on navigational systems. *Navigation, Journal of the Institute of Navigation 37*(2), Spring.

LACHAPELLE, G., ET AL. 1988. Shipborne GPS kinematic positioning for hydrographic applications. *Navigation, Journal of the Institute of Navigation 35*(1), Spring.

LAPINE, COMMANDER L. A. 1990. Practical photogrammetric control by kinematic GPS. *GPS World 1*(3), May–June.

LEICK, A. 1990. *GPS Satellite Surveying*. New York: John Wiley & Sons.

Minutes of FGCC Meeting. 1988. Rockville, MD, May 27.

REILLY J. P. 1990. *Practical Surveying with GPS*. POB Publishing Company, Feb. Canton MI.

REMONDI, B. W. 1985. Performing centimeter accuracy relative surveys in seconds using carrier phase. Proceedings of the 1st International Symposium on Precise Positioning with the Global Positioning System. National Geodetic Information Center, NOAA, Rockville, MD, April 15–19, pp. 789–797.

REMONDI, B. W. 1985. Performing centimeter-level surveys in seconds with GPS carrier phase, initial results. NOAA Technical Memorandum NOS NGS-43, National Geodetic Information Center, NOAA, Rockville, MD.

REMONDI, B. W. 1988. Kinematic and pseudo-kinematic GPS. Proceedings of the Satellite Divisions International Technical Meeting. Colorado Springs, CO, Institute of Navigation, Sept. 19–23, pp. 115–121.

REMONDI, B. W. 1991. Pseudo-kinematic GPS results using the ambiguity function method. *Navigation, Journal of the Institute of Navigation 38*(1), Spring.

ROEBER, J. F. 1986–1987. Where in the world are we? *Navigation, Journal of the Institute of Navigation 33*(4), Winter.

TALBOT, N. C. 1991. High-precision real-time GPS positioning concepts: Modeling and results. *Navigation, Journal of the Institute of Navigation 38*(2), Summer.

VEATCH II, E. R., and J. OSWALD. 1989. The kinematic GPS revolution: Surveying on the move. ASPRS/ACSM Annual Convention, Baltimore, MD, Vol. 5, pp. 288–297.

WELLS, D., ET AL. 1987. *Guide to GPS Positioning*. Fredericton, New Brunswick, Canada: Canadian GPS Associates, May.

16

Survey Measurement Adjustments by Least Squares

Paul R. Wolf and Charles Ghilani

16-1. INTRODUCTION

The general subject of errors in measurement was discussed in Chapter 3, and the two classes of errors, *systematic* and *random* (or accidental), were defined. It was noted that systematic errors follow physical laws, and that if the conditions producing them are measured, corrections to eliminate these can be computed and applied; however, random errors will still exist in all observed values.

As explained in Chapter 3, experience has shown that random errors in surveying follow the mathematical laws of probability, and that any group of measurements will contain random errors conforming to a "normal distribution" as illustrated in Figure 3-5. With reference to that figure, it can be seen that random errors have the following characteristics: (1) small errors occur more frequently than large ones, (2) positive and negative errors of the same size occur with equal frequency, and (3) very large errors seldom occur. They must be avoided through alertness and careful checking of all measured values.

If proper procedures are used in surveying work—after eliminating mistakes and making corrections for systematic errors—the presence of remaining random errors generally

should be evident. In leveling, e.g., as discussed in Chapter 7, circuits should be closed on either the starting bench mark or another of equal or higher reliability. Any misclosure in the circuit can then be computed, providing an indication of random errors that remain. Similarly, in an angle measurement as described in Chapter 6, the sum of all angles measured around the horizon at a point should equal 360°, and in plane surveying the sum of the angles in any closed polygon should equal $(n - 2)\,180°$, where n is the number of angles in the figure. Also, as discussed in Chapter 9, the algebraic sums of the latitudes and departures of a closed-polygon traverse must equal zero. After eliminating mistakes and correcting for systematic errors, any remaining deviations (misclosures) from these required conditions indicate the presence of random errors in the measured values.

In surveying, adjustments are applied to measured values to distribute misclosure errors and produce mathematically perfect geometric conditions; various procedures are used. Some simply apply corrections of the same size to all measured values, where each correction equals the total misclosure divided by the number of measurements. Others introduce corrections of varying size to certain values on the basis of their suspected errors. Still others employ rules of thumb—e.g., the compass rule for adjusting latitudes and departures of closed traverses.

Because random errors in surveying are "normally distributed" and conform to the mathematical laws of probability, it follows logically that for the most rigorous adjustment procedure, corrections should be computed in accordance with that theory. The method of *least squares* is based on the laws of probability.

In the sections of this chapter that follow, the fundamental condition that is enforced in least-squares adjustment is described and an elementary example given. Then, systematic procedures for forming and solving least-squares equations are given, including the use of matrix methods. Following this, specific procedures for adjusting level nets, trilateration, triangulation, and traverses are described, and example problems solved.

16-2.   THE FUNDAMENTAL CONDITION OF LEAST SQUARES

Making adjustments of measured values by the method of least squares is not new. It was done by the German mathematician Karl Gauss as early as the latter part of the 18th century. Until the advent of computers, however, least-squares techniques were seldom employed because of the lengthy calculations involved. Now the procedures are routinely performed.

Least squares is applicable for adjusting any of the basic measurements made in surveying, including observed differences in elevation, horizontal distances, and horizontal and vertical angles. Applying least squares for adjusting these observations in the commonly employed surveying procedures if leveling, trilateration, triangulation, and traversing is the thrust of this chapter. Least squares is also applied in photogrammetric, inertial, and GPS surveys, but these procedures are not described here.

For a group of equally weighted observations, the fundamental condition that is enforced in least-squares adjustment is that the sum of the squares of the residuals is minimized. This condition, which has been developed from the equation for the normal distribution curve (see Section 3-10), provides most probable values for the adjustment quantities. Suppose a group of m equally weighted measurements were taken having residuals $v_1, v_2, v_3, \ldots, v_m$. Then, in equation form, the fundamental condition of least squares is expressed as follows

$$\sum_{i=1}^{m} (v_i)^2 = (v_1)^2 + (v_3)^2 + \cdots$$
$$+ (v_m)^2 = \text{minimum} \qquad (16\text{-}1)$$

If measured values are weighted (see Section 3-22) in least-squares adjustment, then the fundamental condition enforced is that the sum of the weights p times their corresponding squared residuals is minimized, or in the following equation form:

$$\sum_{i=1}^{m} p_i (v_i)^2 = p_1 (v_1)^2 + p_2 (v_2)^2 + p_3 (v_3)^2 + \cdots$$
$$+ p_m (v_m)^2 = \text{minimum} \qquad (16\text{-}2)$$

Some basic assumptions underlying least-squares theory are that (1) mistakes and systematic errors have been eliminated, so only random errors remain; (2) the number of observations being adjusted is large; and (3) as stated earlier, the frequency distribution of the errors is normal. Although these basic assumptions are not always met, least-squares adjustment still provides the most rigorous error treatment available, and hence it has become very popular and important in modern surveying. Besides yielding most probable values for the unknowns, least-squares adjustment also enables (1) determining precisions of adjusted qualities, (2) revealing the presence of large errors and mistakes so steps can be taken to eliminate them, and (3) making possible the optimum design of survey procedures in the office before going into the field to take measurements. The latter topic is beyond the scope of this discussion but can be found in references cited at the end of this chapter.

16-3. LEAST-SQUARES ADJUSTMENT BY THE OBSERVATION-EQUATION METHOD

There are two basic methods of employing least squares in survey adjustments: (1) the *observation-equation* method and (2) *condition-equation* approach. The following discussion in this chapter concentrates on the former procedure.

In the observation-equation method, observation equations are written relating measured values to their residual errors and the unknown parameters. One observation equation is written for each measurement. For a unique solution, the number of equations must equal the number of unknowns. If redundant observations are made, then more observation equations can be written than are needed for a unique solution, and most probable values of the unknowns can be determined by the method of least squares. For a group of equally weighted observations, an equation for each residual error is obtained from each observation equation. The residuals are squared and added to obtain the function expressed in Equation (16-1).

To minimize the function in accordance with equation (16-1), partial derivatives of this expression are taken with respect to each unknown variable and set equal to zero. This yields a set of so-called *normal equations*, which are equal in number to the number of unknowns. The normal equations are solved to obtain the most probable values for the unknowns.

Example 16-1. As an elementary example illustrating the method of least-squares adjustment by the observation-equation method, adjust the following three equally weighted distance measurements taken between points *A*, *B*, and *C* of Figure 16-1:

$$AC = 431.71$$
$$AB = 211.52$$
$$BC = 220.10$$

In terms of unknown distances x, and y, the following three equations can be written:

$$x + y = 431.71 \text{ ft}$$
$$x = 211.52 \text{ ft}$$
$$y = 220.10 \text{ ft}$$

These equations relate unknowns x and y to the observations. Values for x and y could be obtained from any two of these equations so that the remaining equation is redundant. Notice, however, that values obtained for x and y will differ, depending on which two equations are solved. It is therefore apparent

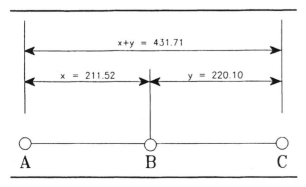

Figure 16-1. Measurements for least-squares adjustment of Example 16-1.

386 *Survey Measurement Adjustments by Least Squares*

that the measurements contain errors. The equations may be rewritten as observation equations by including residual errors as follows:

$$x + y = 431.71 + v_1$$
$$x = 211.52 + v_2$$
$$y = 220.10 + v_3$$

To obtain the least-squares solution, the observation equations are rearranged to obtain expressions for the residuals; these are squared and added to form the function

$$\sum_{i=1}^{m} (v_i)^2$$

as follows:

$$\sum_{i=1}^{m} (v_i)^2 = (x + y - 431.71)^2$$
$$+ (x - 221.52)^2 + (y - 220.10)^2$$

This function is minimized, enforcing the condition of least squares, by taking partial derivatives with respect to each unknown and setting them equal to zero. This yields the following two equations:

$$\frac{\partial \Sigma v^2}{\partial x} = 0 = 2(x + y - 431.71)$$
$$+ 2(x - 211.52)$$
$$\frac{\partial \Sigma v^2}{\partial y} = 0 = 2(x + y - 431.71)$$
$$+ 2(y - 220.10)$$

These are normal equations. The reduced normal equations are as follows:

$$2x + y = 643.23$$
$$x + 2y = 651.81$$

Solving the reduced normal equations simultaneously yields $x = 211.55$ and $y = 220.13$. According to the theory of least squares, these values have the highest proba-

bility. Having the most probable values for the unknowns, the residuals can be calculated by substitution in the original observation equations, or

$$v_1 = 211.55 + 220.13 - 431.71 = -0.03$$
$$v_2 = 211.55 - 211.52 = +0.03$$
$$v_3 = 220.13 - 220.10 = +0.03$$

By substituting these adjusted values for x and y into the original observation equations, the following adjusted measurements result:

$$x + y = 431.71 - 0.03 = 431.68 \text{ ft} = AC$$
$$x = 211.52 + 0.03 = 211.55 \text{ ft} = AB$$
$$y = 220.10 + 0.03 = 220.13 \text{ ft} = BC$$

Note that the adjusted values are now consistent—i.e., $x + y = 431.68$—no matter which measurements are used. Whereas other adjustments could be made to achieve consistency, there is no other combination of residuals possible that will render the sum of their squares a smaller value. Thus, the condition of least squares is realized.

This simple example serves to illustrate the method of least squares without complicating the mathematics. Least-squares adjustment of large systems of observation equations is performed in the same manner.

16-4. SYSTEMATIC FORMULATION OF NORMAL EQUATIONS

In large systems of observation equations, it is helpful to utilize systematic procedures to formulate normal equations. Consider the following system of m linear observation equations of equal weight containing n unknowns:

$$a_1 A + b_1 B + c_1 C + \cdots + n_1 N - L_1 = v_1$$
$$a_2 A + b_2 B + c_2 C + \cdots + n_2 N - L_2 = v_2$$

$$(16\text{-}3)$$

. . . .

$$a_m A + b_m B + c_m C + \cdots + n_m N - L_m = v_m$$

In Equation (16-3), the a's, b's, c's, etc are coefficients of unknowns A, B, C, etc.; the L's constants and the v's residuals. By squaring the residuals and summing them, the function Σv^2 is formed. Taking partial derivatives of Σv^2 with respect to each unknown A, B, C, etc. yields n normal equations, After reducing and factoring the normal equations, we can obtain the following generalized system for expressing normal equations:

$$[aa]A + [ab]B + [ac]C + \cdots + [an]N = [aL]$$

$$[ba]A + [bb]B + [bc]C + \cdots + [bn]N = [bL]$$

$$[ca]A + [cb]B + [cc]C + \cdots + [cn]N = [cL]$$

. . . .

$$[na]A + [nb]B + [nc]C + \cdots + [nn]N = [nL]$$

$$(16\text{-}4)$$

In Equation (16-4), the symbol [ ] signifies the sum of the products; e.g., $[aa] = a_1 a_1 = a_2 a_2 + a_3 a_3 + \cdots + a_m a_m$; $[ab] = a_1 b_1 + a_2 b_2 + a_3 b_3 + \cdots + a_m b_m$; etc.

It can be similarly shown that normal equations may be systematically formed from weighted observation equations in the following manner:

$$[paa]A + [pab]B$$

$$+[pac]C + \cdots +[pan]N = [paL]$$

$$[pba]A + [pbb]B + [pbc]C + \cdots +[pbn]N = [pbL]$$

$$[pca]A + [pcb]B + [pcc]C + \cdots [pcn]N = [pcL]$$

. . . .

$$[pna]A + [pnb]B$$

$$+[pnc]C + \cdots +[pnn]N = [pnL] \quad (16\text{-}5)$$

In Equation (16-5), the terms are as described previously, except that the p's are the relative weights of the individual observations. Examples of the bracket terms are $[paa] = p_1 a_1 a_1 + p_2 a_2 a_2 + \cdots +p_m a_m a_m$; $[pbL] = p_1 b_1 L_1 + p_2 b_2 L_2 + \cdots +p_m b_m L_m$; etc.

16-5. MATRIX METHODS IN LEAST-SQUARES ADJUSTMENT

It has been previously mentioned that least-squares computations are quite lengthy and therefore best performed on a computer. The algebraic approach—Equations (16-4) or (16-5)—for forming normal equations and obtaining their simultaneous solution can be programmed for computer solution; however, the procedure is much more easily adapted to matrix methods.

In developing matrix equations for least-squares computations, analogy will be made to the algebraic approach in Section 16-4. First, observation Equation (16-3) may be prepresented in matrix form as follows:

$$_m\mathbf{A}_n {}_n\mathbf{X}_1 = {}_m\mathbf{L}_1 + {}_m\mathbf{V}_1 \qquad (16\text{-}6)$$

where

$$\mathbf{A} = \begin{bmatrix} a_1 & b_1 & c_1 & \cdots & n_1 \\ a_2 & b_2 & c_2 & \cdots & n_2 \\ \cdot & \cdot & \cdot & \cdots & \cdots \\ \cdot & \cdot & \cdot & \cdots & \cdots \\ a_m & b_m & c_m & \cdots & n_m \end{bmatrix}, \quad \mathbf{X} = \begin{bmatrix} A \\ B \\ C \\ \cdot \\ \cdot \\ \cdot \\ N \end{bmatrix}^1$$

$$\mathbf{L} = \begin{bmatrix} L_1 \\ L_2 \\ L_3 \\ \cdot \\ \cdot \\ L_m \end{bmatrix}^1, \quad \mathbf{V} = \begin{bmatrix} v_1 \\ v_2 \\ v_3 \\ \cdot \\ \cdot \\ v_m \end{bmatrix}^1$$

On studying the following matrix representation, it will be realized that it exactly produces normal Equation (16-4):

$$\mathbf{A}^T\mathbf{A}\mathbf{X} = \mathbf{A}^T\mathbf{L} \qquad (16\text{-}7)$$

In this equation, $\mathbf{A}^T\mathbf{A}$ is the matrix of normal equation coefficients for the unknowns.

Premultiplying both sides of Equation (16-7) by $(A^TA)^{-1}$ and reducing, we obtain

$$(A^TA)^{-1}(A^TA)X = (A^TA)^{-1}A^TL$$

$$IX = (A^TA)^{-1}A^TL \qquad (16\text{-}8)$$

$$X = (A^TA)^{-1}A^TL$$

In this reduction, I is the identity matrix. Equation (16-8) is the basic least-squares matrix equation for equally weighted observations. The matrix X consists of most probable values for unknowns $A, B, C, \ldots, N$. For a system of weighted observations, the following matrix equation provides the X matrix:

$$X = (A^T P A)^{-1} A^T P L \qquad (16\text{-}9)$$

In Equation (16-9), the matrices are identical to those of the equally weighted equations, except that P is a diagonal matrix of weights defined as follows:

$$P = \begin{bmatrix} p_1 & & & & & \\ & p_2 & & & & \\ & & p_3 & & & \\ & & & \cdot & & \\ & & & & \cdot & \\ & & & & & p_m \end{bmatrix}^m$$

In the P matrix, all off-diagonal elements are zeros. This is proper when the individual observations are independent and uncorrelated—e.g., they are not related to each other. This is usually the case in surveying.

If the observations in an adjustment are all of equal weight, Equation (16-9) can still be used, but the P matrix becomes an identity matrix with ones for all diagonal elements. It therefore reduces exactly to Equation (16-8). Thus, Equation (16-9) is a general one that can be used for both unweighted and weighted adjustments. It is readily programmed for computer solution.

Example 16-2. Solve example 16-1 using matrix methods.

1. The observation equations of Example 16-1 can be expressed in matrix form as follows:

$$_3A_2 \; _2X_1 = {_3}L_1 + {_3}V_1$$

where

$$A = \begin{bmatrix} 1 & 1 \\ 1 & 0 \\ 0 & 1 \end{bmatrix}^2, \qquad X = \begin{bmatrix} x \\ y \end{bmatrix}^1,$$

$$L = \begin{bmatrix} 431.71 \\ 211.52 \\ 220.10 \end{bmatrix}^1, \qquad V = \begin{bmatrix} v_1 \\ v_2 \\ v_3 \end{bmatrix}^1$$

2. Solving matrix Equation (16-7), yields

$$(A^TA) = \begin{bmatrix} 1 & 1 & 0 \\ 1 & 0 & 1 \end{bmatrix}\begin{bmatrix} 1 & 1 \\ 1 & 0 \\ 0 & 1 \end{bmatrix} = \begin{bmatrix} 2 & 1 \\ 1 & 2 \end{bmatrix}$$

$$(A^TA)^{-1} = \frac{1}{3}\begin{bmatrix} 2 & -1 \\ -1 & 2 \end{bmatrix}, \quad A^TL = \begin{bmatrix} 643.23 \\ 651.81 \end{bmatrix}$$

$$X + (A^TA)^{-1}A^TL = \frac{1}{3}\begin{bmatrix} 2 & -1 \\ -1 & 2 \end{bmatrix}$$

$$\begin{bmatrix} 643.23 \\ 651.81 \end{bmatrix} = \begin{bmatrix} 211.55 \\ 220.13 \end{bmatrix}$$

Note that this solution yields $x = 211.55$ and $y = 220.13$, which are exactly the same values obtained through the algebraic approach of Example 16-1.

As previously stated, digital computers are normally used in least-squares adjustments due to the relatively lengthy nature of the calculations. Because the equations are so conveniently programmed and solved using matrix algebra, the balance of this chapter will stress this approach.

16-6. MATRIX EQUATIONS FOR PRECISIONS OF ADJUSTED QUANTITIES

The matrix equation for calculating residuals after adjustment, whether the adjustment is weighted or not, is

$$V = AX - L \qquad (16\text{-}10)$$

The standard deviation of unit weight for an unweighted adjustment is

$$S_0 = \sqrt{\frac{(V^T V)}{r}} \qquad (16\text{-}11)$$

The standard deviation of unit weight for a weighted adjustment is

$$S_0 = \sqrt{\frac{(V^T P V)}{r}} \qquad (16\text{-}12)$$

In Equations (16-11) and (16-12), r is the number of degrees of freedom in an adjustment and equals the number of observations equations minus the number of unknowns, or $r = m - n$.

Standard deviations of the individual adjusted quantities are as follows:

$$S_{x_i} = S_0 \sqrt{Q_{x_i x_i}} \qquad (16\text{-}13)$$

In Equation (16-13), S_{x_i} is the standard deviation of the ith adjusted quantity—e.g., the quantity in the ith row of the **X** matrix; S_0 the standard deviation of unit weight as calculated by Equation (16-11) or (16-12); and $Q_{x_i x_i}$ the diagonal element in the ith row and the ith column of the matrix $(A^T A)^{-1}$ in the unweighted case or the matrix $(A^T P A)^{-1}$ in the weighted case. The $(A^T A)^{-1}$ and $(A^T P A)^{-1}$ matrices are the so-called *covariance* matrices.

Example 16-3. Calculate the standard deviation of unit weight and standard deviations

of the adjusted quantities x and y for the unweighted problem of Example 16-2.

1. By Equation (16-10), the residuals are as follows:

$$V = \begin{bmatrix} 1 & 1 \\ 1 & 0 \\ 0 & 1 \end{bmatrix} \begin{bmatrix} 211.55 \\ 220.13 \end{bmatrix}$$

$$- \begin{bmatrix} 431.71 \\ 211.52 \\ 220.10 \end{bmatrix} = \begin{bmatrix} -0.03 \\ 0.03 \\ 0.03 \end{bmatrix}$$

2. By Equation (16-11), the standard deviation of unit weight is

$$V^T V = \begin{bmatrix} -0.03 & 0.03 & 0.03 \end{bmatrix} \begin{bmatrix} -0.03 \\ 0.03 \\ 0.03 \end{bmatrix}$$

$$= 0.0027$$

$$S_0 = \sqrt{\frac{0.0027}{3-2}} = \pm 0.052$$

3. With Equation (16-13), the standard deviations of the adjusted values for x and y are

$$S_x = \pm 0.052 \sqrt{\frac{2}{3}} = \pm 0.042$$

$$S_y = \pm 0.052 \sqrt{\frac{2}{3}} = \pm 0.042$$

In part 3, the numbers 2/3 under the radicals are the 1,1 and 2,2 elements of the $(A^T A)^{-1}$ matrix of Example 16-2. The interpretation of the standard deviations computed under part 3 is that a *68%* probability exists the adjusted values for x and y are within ± 0.042 of their true values. Note that for this simple example, the three residuals calculated in part 1 were equal, and the standard deviations of x and y were equal in part 3. This is due to the symmetric nature of this particular problem (illustrated in Figure 16-1), but it is not generally the case with more complex problems.

16-7. ADJUSTMENT OF LEVELING CIRCUITS

The method of least squares is extremely valuable as a means of adjusting leveling circuits, especially those consisting of two or more interconnected loops that form networks. A simple example is illustrated in Figure 16-2. Here the objective was to determine elevations of *A*, *B*, and *C*, which were to serve as temporary project bench marks to control construction of a highway through the cross-hatched corridor. Obviously, it would have been possible to obtain elevations for *A*, *B*, and *C* by beginning at *BMX* and running a single closed loop consisting of only courses 1, 5, 7, and 4. Alternatively, a single closed loop could have been initiated at *BMY* and consist of courses 2, 5, 7, and 3. However, by running all seven courses, redundancy is achieved that enables checks to be made, blunders isolated, and precision increased.

Now that we have run all seven courses of Figure 16-2, it would be possible to compute the adjusted elevation of *B*, e.g., using several different single closed circuits. Loops 1-5-6, 2-5-6, 3-7-6, and 4-7-6 could each be used, but it is almost certain that each would yield a

different elevation for *B*. A more logical approach, that will produce only one adjusted value for *B*—its most probable one—is to use all seven courses in a simultaneous least-squares adjustment.

In adjusting level networks, the observed difference in elevation for each course is treated as one observation containing a single random error. This single random error is the total of the individual random errors in backsight and foresight readings for the entire course. The table of Figure 16-2 lists the total difference in elevation observed for each course. In the figure, the arrows indicate the direction of leveling. Thus, for course number 1, leveling proceeded from *BMX* to *A* and the observed elevation difference was +5.10 ft.

Example 16-4. Adjust the level net of Figure 16-2 by least squares. Consider all observations to be equally weighted. Compute the precisions of the adjusted elevations.

1. First, observation equations are written relating each measurement of the difference in elevation of a line to the most probable values for unknown elevations *A*, *B*, and *C*

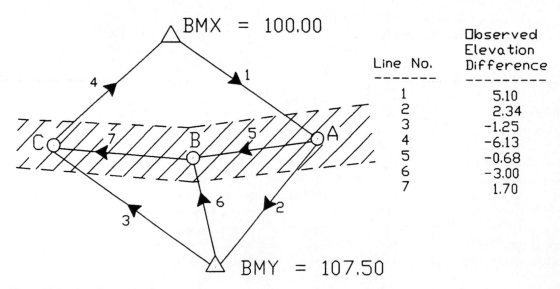

Line No.	Observed Elevation Difference
1	5.10
2	2.34
3	−1.25
4	−6.13
5	−0.68
6	−3.00
7	1.70

Figure 16-2. Leveling network.

and residual errors in the measurements, as follows:

$$A = BMX + 5.10 + v_1$$

$$BMY = A + 2.34 + v_2$$

$$C = BMY - 1.25 + v_3$$

$$BMX = C - 6.13 + v_4 \qquad (16\text{-}14)$$

$$B = A - 0.68 + v_5$$

$$B = BMY - 3.00 + v_6$$

$$C = B + 1.70 + v_7$$

2. Introducing the elevations of *BMX* and *BMY*, reducing, and rewriting observation Equations (16-14) in a form compatible with Equation (16-6) give the following:

$$A = 105.10 + v_1$$

$$-A = -105.16 + v_2$$

$$C = 106.25 + v_3$$

$$-C = -106.13 + v_4$$

$$-A + B = -0.68 + v_5$$

$$B = 104.50 + v_6$$

$$-B + C = 1.70 + v_7$$

3. The observation equations expressed in matrix form are as follows:

$$_7\mathbf{A}_{33}\mathbf{X}_1 = {}_7\mathbf{L}_1 + {}_7\mathbf{V}_1$$

where

$$\mathbf{A}^T\mathbf{A} = \begin{bmatrix} 1 & 0 & 0 \\ -1 & 0 & 0 \\ 0 & 0 & 1 \\ 0 & 0 & -1 \\ -1 & 1 & 0 \\ 0 & 1 & 0 \\ 0 & -1 & 1 \end{bmatrix}, \quad \mathbf{X} = \begin{bmatrix} A \\ B \\ C \end{bmatrix},$$

$$\mathbf{L} = \begin{bmatrix} 105.10 \\ -105.16 \\ 106.25 \\ -106.13 \\ -0.68 \\ 104.50 \\ 1.70 \end{bmatrix}, \quad \mathbf{V} = \begin{bmatrix} v_1 \\ v_2 \\ v_3 \\ v_4 \\ v_5 \\ v_6 \\ v_7 \end{bmatrix}$$

4. The matrix solution for most probable values is

$$\mathbf{A}^T\mathbf{A} = \begin{bmatrix} 1 & -1 & 0 & 0 & -1 & 0 & 0 \\ 0 & 0 & 0 & 0 & 1 & 1 & -1 \\ 0 & 0 & 1 & -1 & 0 & 0 & 1 \end{bmatrix}$$

$$\times \begin{bmatrix} 1 & 0 & 0 \\ -1 & 0 & 0 \\ 0 & 0 & 1 \\ 0 & 0 & -1 \\ -1 & 1 & 0 \\ 0 & 1 & 0 \\ 0 & -1 & 1 \end{bmatrix}$$

$$= \begin{bmatrix} 3 & -1 & 0 \\ -1 & 3 & -1 \\ 0 & -1 & 3 \end{bmatrix}$$

$$(\mathbf{A}^T\mathbf{A})^{-1} = \frac{1}{21}\begin{bmatrix} 8 & 3 & 1 \\ 3 & 9 & 3 \\ 1 & 3 & 8 \end{bmatrix}$$

$$\mathbf{X} = (\mathbf{A}^T\mathbf{A})^{-1}\mathbf{A}^T\mathbf{L} = \frac{1}{21}\begin{bmatrix} 8 & 3 & 1 \\ 3 & 9 & 3 \\ 1 & 3 & 8 \end{bmatrix}$$

$$\times \begin{bmatrix} 210.94 \\ 102.12 \\ 214.08 \end{bmatrix}$$

$$= \begin{bmatrix} 105.14 \\ 104.48 \\ 106.19 \end{bmatrix}$$

$$\mathbf{A}^T\mathbf{L} = \begin{bmatrix} 1 & -1 & 0 & 0 & -1 & 0 & 0 \\ 0 & 0 & 0 & 0 & 1 & 1 & -1 \\ 0 & 0 & 1 & -1 & 0 & 0 & 1 \end{bmatrix}$$

$$\begin{bmatrix} 105.10 \\ -105.16 \\ 106.25 \\ -106.13 \\ -0.68 \\ 104.50 \\ 1.70 \end{bmatrix} = \begin{bmatrix} 210.94 \\ 102.12 \\ 214.08 \end{bmatrix}$$

Thus, the adjusted bench-mark elevations are $A = 105.14$, $B = 104.48$, and $C = 106.19$.

5. The residuals by Equation (16-10) are as follows:

$$\mathbf{AX} = \begin{bmatrix} 1 & 0 & 0 \\ -1 & 0 & 0 \\ 0 & 0 & 1 \\ 0 & 0 & -1 \\ -1 & 1 & 0 \\ 0 & 1 & 0 \\ 0 & -1 & 1 \end{bmatrix} \begin{bmatrix} 105.14 \\ 104.48 \\ 106.19 \end{bmatrix}$$

$$= \begin{bmatrix} 105.14 \\ -105.14 \\ 106.19 \\ -106.19 \\ -0.66 \\ 104.48 \\ 1.71 \end{bmatrix}$$

$$\mathbf{V} = \mathbf{AX} - \mathbf{L} = \begin{bmatrix} 105.14 \\ -105.14 \\ 106.19 \\ -106.19 \\ -0.66 \\ 104.48 \\ 1.71 \end{bmatrix}$$

$$- \begin{bmatrix} 105.10 \\ -105.10 \\ 106.25 \\ -106.13 \\ -0.68 \\ 104.50 \\ 1.70 \end{bmatrix}$$

$$= \begin{bmatrix} +0.04 \\ +0.02 \\ -0.06 \\ -0.06 \\ +0.02 \\ -0.02 \\ +0.01 \end{bmatrix}$$

6. Utilizing Equation (16-11), we obtain the estimated standard deviation of unit weight as

$$S_0 = \sqrt{\frac{0.0101}{7-3}} = \pm 0.050 \text{ ft}$$

where

$$\mathbf{V}^T\mathbf{V} = (0.04)^2 + (0.02)^2 + (-0.06)^2 + (-0.06)^2 + (0.02)^2 + (-0.02)^2 + (0.01)^2 = 0.0101.$$

7. By Equation (16-13), the estimated standard deviations of unknown elevations of A, B, and C are

$$S_A = S_0\sqrt{Q_{AA}} = \pm(0.05)\sqrt{8/21} = \pm 0.031 \text{ ft}$$
$$S_B = S_0\sqrt{Q_{BB}} = \pm(0.05)\sqrt{9/21} = \pm 0.033 \text{ ft}$$
$$S_C = S_0\sqrt{Q_{CC}} = \pm(0.05)\sqrt{8/21} = \pm 0.031 \text{ ft}$$

Note in these calculations that terms in the radicals are diagonal elements of the $(\mathbf{A}^T\mathbf{A})^{-1}$ matrix.

Example 16-5. Adjust the level net of Figure 16-2 by the method of weighted least squares. Use weights that are inversely proportional to the course lengths of 4, 3, 2, 3, 2, 2, and 2 m, respectively.

1. The **A**, **X**, **L**, and **V** matrices are exactly the same as for Example 16-4. The diagonal elements of the P matrix are $1/4, 1/3, 1/2, 1/3, 1/2, 1/2,$ and $1/2$, respectively, and after we multiply each by 12, the **P** matrix becomes

$$\mathbf{P} = \begin{bmatrix} 3 & & & & & & \\ & 4 & & & & & \\ & & 6 & & & & \\ & & & 4 & & & \\ & & & & 6 & & \\ & & & & & 6 & \\ & & & & & & 6 \end{bmatrix}$$

2. The matrix solution of Equation (16-9) is

$$(\mathbf{A}^T\mathbf{PA})^{-1} = \begin{bmatrix} 0.0933 & 0.0355 & 0.0133 \\ 0.0355 & 0.0770 & 0.0289 \\ 0.0133 & 0.0289 & 0.0733 \end{bmatrix}$$

$$\mathbf{A}^T\mathbf{P} = \begin{bmatrix} 1 & -1 & 0 & 0 & -1 & 0 & 0 \\ 0 & 0 & 0 & 0 & 1 & 1 & -1 \\ 0 & 0 & 1 & -1 & 0 & 0 & 1 \end{bmatrix}$$

$$\times \begin{bmatrix} 3 & & & & & & \\ & 4 & & & & & \\ & & 6 & & & & \\ & & & 4 & & & \\ & & & & 6 & & \\ & & & & & 6 & \\ & & & & & & 6 \end{bmatrix}$$

$$= \begin{bmatrix} 3 & -4 & 0 & 0 & -6 & 0 & 0 \\ 0 & 0 & 0 & 0 & 6 & 6 & -6 \\ 0 & 0 & 6 & -4 & 0 & 0 & 6 \end{bmatrix}$$

$$A^T PA = \begin{bmatrix} 3 & -4 & 0 & 0 & -6 & 0 & 0 \\ 0 & 0 & 0 & 0 & 6 & 6 & -6 \\ 0 & 0 & 6 & -4 & 0 & 0 & 6 \end{bmatrix}$$

$$\times \begin{bmatrix} 1 & 0 & 0 \\ -1 & 0 & 0 \\ 0 & 0 & 1 \\ 0 & 0 & -1 \\ -1 & 1 & 0 \\ 0 & 1 & 0 \\ 0 & -1 & 1 \end{bmatrix}$$

$$= \begin{bmatrix} 13 & -6 & 0 \\ -6 & 18 & -6 \\ 0 & -6 & 16 \end{bmatrix}$$

$$A^T PL = \begin{bmatrix} 3 & -4 & 0 & 0 & -6 & 0 & 0 \\ 0 & 0 & 0 & 0 & 6 & 6 & -6 \\ 0 & 0 & 6 & -4 & 0 & 0 & 6 \end{bmatrix}$$

$$\times \begin{bmatrix} 105.10 \\ -105.16 \\ 106.25 \\ -106.13 \\ -0.68 \\ 104.50 \\ 1.70 \end{bmatrix}$$

$$= \begin{bmatrix} 740.02 \\ 612.72 \\ 1072.22 \end{bmatrix}$$

$$X = (A^T PA)^{-1} A^T PL$$

$$= \begin{bmatrix} 0.0933 & 0.0355 & 0.0133 \\ 0.0355 & 0.0770 & 0.0289 \\ 0.0133 & 0.0289 & 0.0733 \end{bmatrix}$$

$$\times \begin{bmatrix} 740.02 \\ 612.72 \\ 1072.22 \end{bmatrix}$$

$$= \begin{bmatrix} 105.15 \\ 104.49 \\ 106.20 \end{bmatrix}$$

In summary, the adjusted elevations for *A*, *B*, and *C* are 105.15, 104.49, and 106.20, respectively. Note that these differ slightly from the unweighted results of Example 16-4, as they should.

3. By equation (16-10), the residuals are as follows:

$$AX = \begin{bmatrix} 1 & 0 & 0 \\ -1 & 0 & 0 \\ 0 & 0 & 1 \\ 0 & 0 & -1 \\ -1 & 1 & 0 \\ 0 & 1 & 0 \\ 0 & -1 & 1 \end{bmatrix} \begin{bmatrix} 105.15 \\ 104.49 \\ 106.20 \end{bmatrix}$$

$$= \begin{bmatrix} 105.15 \\ -105.15 \\ 106.20 \\ -106.20 \\ -0.66 \\ 104.49 \\ 1.71 \end{bmatrix}$$

$$V = AX - L = \begin{bmatrix} 105.15 \\ -105.15 \\ 106.20 \\ -106.20 \\ -0.66 \\ 104.49 \\ 1.71 \end{bmatrix} - \begin{bmatrix} 105.10 \\ -105.16 \\ 106.25 \\ -106.13 \\ -0.68 \\ 104.50 \\ 1.70 \end{bmatrix}$$

$$= \begin{bmatrix} +0.05 \\ +0.01 \\ -0.05 \\ -0.07 \\ +0.02 \\ -0.01 \\ +0.01 \end{bmatrix}$$

4. The estimated standard deviation of unit weight by Equation (16-12) is

$$S_0 = \pm \sqrt{\frac{0.046}{7-3}} = \pm 0.107$$

where

$$V^T PV = 3(0.05)^2 + 4(0.01)^2$$
$$+ 6(-0.05)^2 + 4(-0.07)^2$$
$$+ 6(0.02)^2 + 6(-0.01)^2 + 6(0.01)^2$$
$$= 0.0461.$$

5. By Equation (16-13), the estimated standard deviations in elevations of bench marks A, B, and C are as follows:

$$S_A = S_0\sqrt{Q_{AA}} = \pm(0.107)\sqrt{(0.0933)}$$

$$= \pm0.033 \text{ ft}$$

$$S_B = S_0\sqrt{Q_{BB}} = \pm(0.107)\sqrt{(0.0770)}$$

$$= \pm0.030 \text{ ft}$$

$$S_C = S_0\sqrt{Q_{CC}} = \pm(0.107)\sqrt{(0.0733)}$$

$$= \pm0.029 \text{ ft}$$

Note that values under the radicals in the above calculations are diagonal elements of the $(\mathbf{A}^T\mathbf{PA})^{-1}$ matrix.

16-8. ADJUSTMENT OF HORIZONTAL SURVEYS

In addition to level nets, which can be referred to as vertical surveys, another basic class that can be adjusted by least squares is *horizontal surveys*. They are run for the purpose of establishing horizontal positions of points, expressed either in terms of X and Y plane coordinates (usually state plane systems), or as geodetic latitudes and longitudes. These surveys are often executed as one of three specific types: (1) *trilateration*, (2) *triangulation*, or (3) *traversing*. The methods of running these surveys are described in Chapters 9, 11, and 12, respectively.

Trilateration consists exclusively of distance measurements, triangulation principally involves angle measurement with some base-line distances observed, and traverses contain both distance and angle measurements. Therefore, to perform least-squares adjustments of horizontal surveys by the method of observation equations, it is necessary to write observation equations for these two types of measurements. The equations are nonlinear, so they are first linearized using Taylor's theorem and then solved iteratively.

16-9. THE DISTANCE OBSERVATION EQUATION

Distance observation equations relate measured lengths and their inherent random errors to the most probable coordinates of their end points. This procedure is often referred to as the method of the *variation of coordinates*. In this section, X and Y plane coordinates will be used. Referring to Figure 16-3, we may write the following distance observation equation for any line IJ:

$$L_{ij} + V_{L_{ij}} = \sqrt{(X_j - X_i)^2 + (Y_j - Y_i)^2} \quad (16\text{-}15)$$

In Equation (16-15), L_{ij} is the observed length of line IJ; V_{Lij} is the residual error in the observation; and X_i, Y_i, X_j, and Y_j are the most probable coordinates of points I and J. The right side of the equation is a nonlinear function of unknown variables X_i, Y_i, X_j, and Y_j. The equation may be rewritten as

$$L_{ij} + V_{L_{ij}} = F(X_j, Y_j, X_j, Y_j) \quad (16\text{-}16)$$

where

$$F(X_i, Y_i, X_j, Y_j) = \sqrt{(X_j - X_i)^2 + (Y_j - Y_i)^2}.$$

With the Taylor series, linearization of the function F, after we drop as negligible all

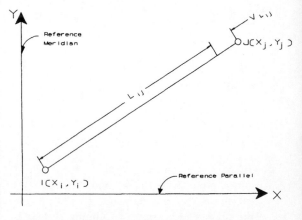

Figure 16-3. Geometry of distance observation equation.

terms of order two or higher, takes the following form:

$$F(X_i, Y_i, X_j, Y_j) = F(X_{i_0}, Y_{i_0}, X_{j_0}, Y_{j_0})$$

$$+ \left[\frac{\partial F}{\partial X_j}\right]_0 dX_i + \left[\frac{\partial F}{\partial Y_i}\right]_0 dY_i + \left[\frac{\partial F}{\partial X_j}\right]_0 dX_j$$

$$+ \left[\frac{\partial F}{\partial Y_j}\right]_0 dY_j \qquad (16\text{-}17)$$

In Equation (16-17), X_{i_0}, Y_{i_0}, X_{j_0}, Y_{j_0} are initial approximations of unknowns X_i, Y_i, X_j, and Y_j; $(\partial F/\partial X_j)_0$ is the partial derivative of F with respect to X_i evaluated at the initial approximations, etc.; and dX_i, dY_i, dX_j, and dY_j are corrections to be applied to the initial approximations such that

$$X_i = X_{i_0} + dX_i, \qquad X_j = X_{j_0} + dX_j$$

$$Y_i = Y_{i_0} + dY_i, \qquad Y_j = Y_{j_0} + dY_j$$

Evaluating partial derivatives of the function F, substituting them into Equation (16-17), and then in turn substituting into Equation (16-16) and rearrange, we arrive at the following linearized observation equation for distance measurements:

$$K_{L_{ij}} + V_{L_{ij}} = \left[\frac{X_{i_0} - X_{j_0}}{IJ_0}\right] dX_i + \left[\frac{Y_{i_0} - Y_{j_0}}{IJ_0}\right] dY_i$$

$$+ \left[\frac{X_{j_0} - X_{i_0}}{IJ_0}\right] dX_j +$$

$$\left[\frac{Y_{j_0} - Y_{i_0}}{IJ_0}\right] dY_j \qquad (16\text{-}18)$$

where $K_{L_{ij}} = L_{ij} - (IJ_0)$, and

$$(IJ_0) = \sqrt{(X_{j_0} - X_{i_0})^2 + (Y_{j_0} - Y_{i_0})^2}$$

Evaluation of partial derivatives in the previous development is quite straightforward. However, to illustrate the procedure, the partial of F with respect to X_i is demonstrated. Having done this, we may easily visualize the

remaining partials without actually performing the steps.

$$F = \left[(X_j - X_i)^2 + (Y_j - Y_i)^2\right]^{1/2}$$

$$\frac{\partial F}{\partial X_i} = \frac{1}{2}\left[(X_j - X_i)^2\right.$$

$$\left. + (Y_j - Y_i)^2\right]^{-1/2}[2(X_j - X_i)(-1)]$$

Reducing gives

$$\frac{\partial F}{\partial X_i} = \frac{-X_j + X_i}{\left[(X_j - X_i)^2 + (Y_j - Y_i)^2\right]^{1/2}}$$

Rearranging and evaluating at initial approximations yield the following:

$$\frac{\partial F}{\partial X_i} = \frac{X_{j_0} - X_{j_0}}{(IJ_0)}$$

16-10. THE ANGLE OBSERVATION EQUATION

Angle observation equations relate measured angles and their inherent random errors to the most probable coordinates of the occupied station, backsight station, and foresight station. This is also termed the variation of coordinates method. Again, in this treatment X and Y plane coordinates are used. Referring to Figure 16-4, we may write the following observation equation for the measured angle at I between points J and K:

$$\theta_{jik} + V_{\theta_{jik}} = \text{azimuth}_{ik} - \text{azimuth}_{ij}$$

$$= \tan^{-1}\left(\frac{X_k - X_i}{Y_k - Y_i}\right)$$

$$- \tan^{-1}\left(\frac{X_j - X_i}{Y_j - Y_i}\right) + C \quad (16\text{-}19)$$

Equation (16-19) relates observed angle θ_{jik} and its residual error $V_{\theta_{jik}}$ to the most proba-

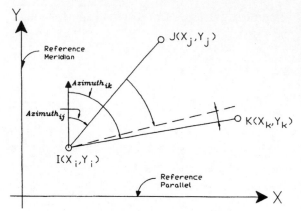

Figure 16-4. Geometry of angle observation equation.

ble angle in terms of unknown variables X_j, Y_j, X_i, Y_i, X_k, and Y_k, the most probable coordinates of the points involved. The term C makes the equation general and accounts for the fact that the azimuths of lines IK and IJ can be in any directions. Figure 16-5 shows the 12 possi-

ble quadrant locations for stations J, I, and K for angles under 180°. For the six cases a through f, $C = 0$; for the six cases g through 1, $C = 180°$.

Equation (16-19) is also nonlinear and may be linearized using the Taylor series as follows:

$$\theta_{jik} + V_{\theta_{jik}} = U(X_j, Y_j, X_i, Y_i, X_k, Y_k) \quad (16\text{-}20)$$

where

$$U(X_j, Y_j, X_i, Y_i, X_k, Y_k)$$

$$= \tan^{-1}\left(\frac{X_k - X_i}{Y_k - Y_i}\right) - \tan^{-1}\left(\frac{X_j - X_i}{Y_j - Y_i}\right) + C$$

The Taylor series approximation for function U, after we drop as negligible all terms of

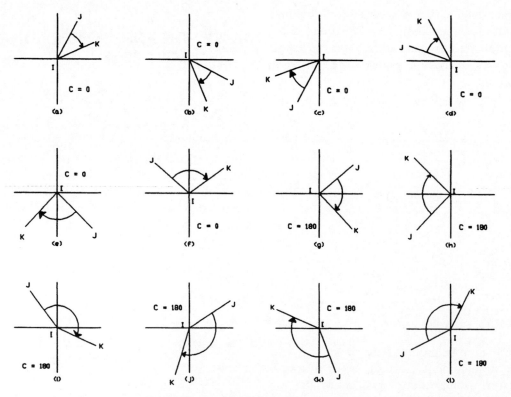

Figure 16-5. Twelve different quadrant locations (a) through (l) for stations J, I, and K in angle measurement.

order two and higher, is

$$U(X_j, Y_j, X_i, Y_i, X_k, Y_k)$$

$$= U(X_{j_0}, Y_{j_0}, X_{i_0}, Y_{i_0}, X_{k_0}, Y_{k_0})$$

$$+ \left[\frac{\partial U}{\partial X_j}\right]_0 dX_j + \left[\frac{\partial U}{\partial Y_j}\right]_0 dY_j$$

$$+ \left[\frac{\partial U}{\partial X_i}\right]_0 dX_i + \left[\frac{\partial U}{\partial Y_i}\right]_0 dY_i$$

$$+ \left[\frac{\partial U}{\partial X_k}\right]_0 dX_k + \left[\frac{\partial U}{\partial Y_k}\right]_0 dY_k \quad (16\text{-}21)$$

In Equation (16-21), terms are defined as in Equation (16-17). Evaluating partial derivatives of the function U, substituting them into equation (16-21), and then in turn substituting into equation (16-20) and rearranging, we get the following linearized observation equation for angle measurements:

$$K_{\theta_{jik}} + V_{\theta_{jik}} = \left[\frac{Y_{i_0} - Y_{j_0}}{(IJ_0)^2}\right] dX_j + \left[\frac{X_{j_0} - X_{i_0}}{(IJ_0)^2}\right] dY_j$$

$$+ \left[\frac{Y_{j_0} - Y_{i_0}}{(IJ_0)^2} - \frac{Y_{k_0} - Y_{i_0}}{(IK_0)^2}\right] dX_i$$

$$+ \left[\frac{X_{i_0} - X_{j_0}}{(IJ_0)^2} - \frac{X_{i_0} - X_{k_0}}{(IK_0)^2}\right] dY_i$$

$$+ \left[\frac{Y_{k_0} - Y_{i_0}}{(IK_0)^2}\right] dX_k$$

$$+ \left[\frac{X_{i_0} - X_{k_0}}{(IK_0)^2}\right] dY_k \quad (16\text{-}22)$$

where $K_{\theta_{jik}} = \theta_{jik} - \theta_{jik_0}$, and

$$\theta_{jik_0} = \tan^{-1}\left(\frac{X_{k_0} - X_{i_0}}{Y_{k_0} - X_{i_0}}\right)$$

$$- \tan^{-1}\left(\frac{X_{j_0} - X_{i_0}}{Y_{j_0} - Y_{i_0}}\right) + C$$

$$(IJ_0) = \sqrt{(X_{j_0} - X_{i_0})^2 + (Y_{j_0} - Y_{i_0})^2}$$

$$(IK_0) = \sqrt{(X_{k_0} - X_{i_0})^2 + (Y_{k_0} - Y_{i_0})^2}$$

Equations (16-18) and (16-22) are linearized distance and angle observation equation, respectively, and a system of these may be formed and manipulated conveniently by matrix methods for adjusting horizontal surveys.

In equation (16-22), $K_{\theta_{jik}}$ and $V_{\theta_{jik}}$ are in radian measure. Since it is more common to work in the sexagesimal system in the United States and because the magnitudes of the angle residuals are generally in the seconds range, the equation's units may be converted to seconds by multiplying the right-hand side by ρ (rho) the number of seconds per radian, which is $206{,}264.8''/\text{rad}$.

16-11.   TRILATERATION ADJUSTMENT

As noted in Section 16-8, trilateration surveys consist of only distance measurements. The geometric figures used are many and varied. All are equally adaptable to the observation-equation method of adjustment, although each different geometric configuration poses a specific adjustment problem. Consider, e.g., the adjustment of simple Figure 16-6. Points A, B, and C are horizontal control points whose X- and Y-coordinates are known and fixed. The position of point U is to be established from the measurement of distances AU, BU, and CU. Obviously, any two of these distances would

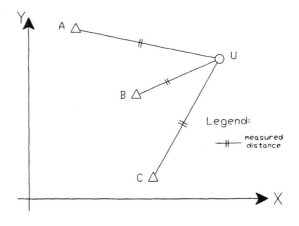

Figure 16-6.   Simple trilateration example.

be sufficient to establish X- and Y-coordinates of U. The third distance is, therefore, redundant and makes possible the adjustment and calculations of most probable values of X_u and Y_u.

The observation equations are developed by substituting into prototype Equation (16-18). The equation for measured line AU, e.g., may be formed by interchanging subscript i with a, and j with u. For line BU the i and j subscripts of Equation (16-18) are replaced by b and u, respectively, ad for line CU, the subscripts i and j are replaced by c and u, respectively. It is to be noted that if one of the end points of a measured line is a control point, then its coordinates are invariant, and hence these terms drop out of the prototype observation equation. By using the procedures, described, the following three linearized observation equations result:

$$(L_{au} - AU_0) + V_{L_{au}} = \left[\frac{X_{u_0} - X_a}{AU_0}\right] dX_u$$

$$+ \left[\frac{Y_{u_0} - Y_a}{AU_0}\right] dY_u$$

$$(L_{bu} - BU_0) + V_{L_{bu}} = \left[\frac{X_{u_0} - X_b}{BU_0}\right] dX_u$$

$$+ \left[\frac{Y_{u_0} + Y_b}{BU_0}\right] dY_u$$

$$(L_{cu} - CU_0) + V_{L_{cu}} = \left[\frac{X_{u_0} - X_c}{CU_0}\right] dX_u$$

$$+ \left[\frac{Y_{u_0} - Y_c}{CU_0}\right] dY_u \quad (16\text{-}23)$$

where

$$AU_0 = \sqrt{(X_{u_0} - X_a)^2 + (Y_{u_0} - Y_a)^2}$$

$$BU_0 = \sqrt{(X_{u_0} - X_b)^2 + (Y_{u_0} - Y_b)^2}$$

$$CU_0 = \sqrt{(X_{u_0} - X_c)^2 + (Y_{u_0} - Y_c)^2}$$

Also, L_{au}, L_{bu}, and L_{cu} are the observed distances, and X_{U_0} and Y_{U_0} initial approxima-

tions of coordinates of point U, which can be obtained from a scaled diagram or computed from two of the observed distances.

This system of linear observation equations may be expressed in matrix form as follows:

$$_3A_2\ _2X_1 = {}_3K_1 + {}_3V_1$$

where $\mathbf{A}$ is the matrix of coefficients of unknown; $\mathbf{X}$ the matrix of unknown corrections dX_u and dY_u; $\mathbf{K}$ the matrix of constants, i.e., measured lengths minus lengths computed from initial approximate coordinates, and $\mathbf{V}$ the matrix of residuals in the measured lengths. Most probable corrections dX_u and dY_u, and hence the most probable coordinates X_u and Y_u, may be calculated by applying the least-squares matrix equation. If we consider equal weights for the observations the equation is

$$_2X_1 = ({}_2A_3^T\ _3A_2)^{-1}\ _2A_3^T\ _3K_1$$

The $\mathbf{X}$ matrix consists of corrections to be added to the initial approximations of the coordinates. Because Taylor series linearization drops terms higher than order one, an iterative solution is required. For the second iteration, the corrected coordinates are used to reformulate the $\mathbf{A}$ and $\mathbf{K}$ matrices. Then, these are used in another least-squares solution to obtain a new $\mathbf{X}$ matrix consisting of corrections. The process is repeated until the values of the computed $\mathbf{X}$ matrix become small enough to be considered negligible.

Example 16-6. Adjust the example of Figure 16-6 by least squares if measured distances AU, BU, and CU are 6049.00, 4736.83, and 5446.49 ft, respectively, and coordinates of the control points are as follows:

$X_a = 865.40,$ $X_b = 2432.55,$ $X_c = 2865.22$

$Y_a = 4527.15,$ $Y_b = 2047.25,$ $Y_c = 27.15$

A. FIRST ITERATION.

1. Calculate initial approximates for X_{u_0} and Y_{u_0}.
   (a) Calculate azimuth AB.

$$\text{Azimuth}_{AB} = \tan^{-1}\left(\frac{X_b - X_a}{Y_b - Y_a}\right) + 180°$$

$$\text{Azimuth}_{AB} = \tan^{-1}\left[\frac{2432.55 - 865.40}{2047.25 - 4527.15}\right]$$

$$+ 180° = 147°43'34''$$

(b) Calculate length AB.

$$AB = \sqrt{(X_b - X_a)^2 + (Y_b - Y_a)^2}$$

$$= \sqrt{\begin{array}{c}(2432.55 - 865.20)^2 \\ + (2047.25 - 4527.15)^2\end{array}}$$

$$= 2933.58 \text{ ft}$$

(c) Calculate azimuth AU_0, (from cosine law: $c^2 = a^2 + b^2 - 2ab\cos C$).

$$\cos(\angle UAB)$$

$$= \frac{6049.00^2 - 4736.83^2 + 2933.58^2}{2(6049.00)(2933.58)}$$

$$\angle UAB = 50°06'50''$$

$$\text{Az}_{AU_0} = 147°42'34'' - 50°06'50'' = 97°35'44''$$

(d) Calculate X_{u_0} and Y_{u_0}

$$X_{u_0} = 865.40 + 6049.00\sin(97°35'44'')$$

$$= 6861.324 \text{ ft}$$

$$Y_{u_0} = 4527.15 + 6049.00\cos(97°35'44'')$$

$$= 3727.587 \text{ ft}$$

2. Calculate AU_0, BU_0, and CU_0. For this first iteration, AU_0 and BU_0 are exactly equal to their respective measured distances because X_{u_0} and Y_{u_0} were calculated from these

measured values. Therefore,

$$AU_0 = 6049.00, \qquad BU_0 = 4736.83$$

$$CU_0 = [(6861.32 - 2865.22)^2$$

$$+ (3727.59 - 27.15)^2]^{1/2} = 5446.29 \text{ ft}$$

3. Formulate the matrices. (a) The **A** matrix. Observation Equations (16-18) may be simplified as follows:

$$a_{11}\,dX_u + a_{12}\,dY_u = k_1 + v_1$$

$$a_{21}\,dX_u + a_{22}\,dY_u = k_2 + v_2$$

$$a_{31}\,dX_u + a_{32}\,dY_u = k_3 + v_3$$

where

$$a_{11} = \frac{6861.32 - 865.40}{6049.00} = 0.991$$

$$a_{12} = \frac{3727.59 - 4527.15}{6049.00} = -0.132$$

$$a_{21} = \frac{6861.32 - 2432.55}{4736.83} = 0.935$$

$$a_{22} = \frac{3727.59 - 2047.25}{4736.83} = 0.355$$

$$a_{31} = \frac{6861.32 - 2865.22}{5446.29} = 0.734$$

$$a_{32} = \frac{3727.59 - 27.15}{5446.29} = 0.679$$

(b) The **K** matrix.

$$k_1 = 6049.00 - 6049.00 = 0.00$$

$$k_2 = 4736.83 - 4736.83 = 0.00$$

$$k_3 = 5446.49 - 5446.29 = 0.20$$

(c) The **X** and **V** matrices.

$$\mathbf{X} = \begin{bmatrix} dX_u \\ dY_u \end{bmatrix}, \qquad \mathbf{V} = \begin{bmatrix} v_{au} \\ v_{bu} \\ v_{cu} \end{bmatrix}$$

4. The matrix solution using unweighted least-squares Equation (16-7) is

$$\mathbf{X} = (\mathbf{A}^T\mathbf{A})^{-1}\mathbf{A}^T\mathbf{K}$$

$$\mathbf{A}^T\mathbf{A} = \begin{bmatrix} 0.991 & 0.935 & 0.734 \\ -0.132 & 0.355 & 0.679 \end{bmatrix}$$

$$\times \begin{bmatrix} 0.991 & -0.132 \\ 0.935 & 0.355 \\ 0.735 & 0.679 \end{bmatrix}$$

$$= \begin{bmatrix} 2.395 & 0.699 \\ 0.699 & 2.395 \end{bmatrix}$$

$$(\mathbf{A}^T\mathbf{A})^{-1} = \frac{1}{0.960} \begin{bmatrix} 0.605 & -0.699 \\ -0.699 & 2.395 \end{bmatrix}$$

$$\mathbf{A}^T\mathbf{K} = \begin{bmatrix} 0.991 & 0.935 & 0.734 \\ -0.132 & 0.355 & 0.679 \end{bmatrix}$$

$$\begin{bmatrix} 0.000 \\ 0.000 \\ 0.200 \end{bmatrix} = \begin{bmatrix} 0.144 \\ 0.135 \end{bmatrix}$$

$$\mathbf{X} = \frac{1}{0.960} \begin{bmatrix} 0.605 & -0.699 \\ -0.699 & 2.395 \end{bmatrix}$$

$$\begin{bmatrix} 0.144 \\ 0.135 \end{bmatrix} = \begin{bmatrix} -0.007 \\ +0.232 \end{bmatrix}$$

The revised coordinates of U are as follows:

$$X_u = 6861.324 - 0.007 = 6861.317$$

$$Y_u = 3727.587 + 0.232 = 3727.819$$

B. Second Iteration.

1. Calculate AU_0, BU_0, and CU_0.

$$AU_0 = \sqrt{\begin{aligned} (6861.73 - 865.40)^2 \\ + (3727.819 - 4527.15)^2 \end{aligned}}$$

$$= 6048.963$$

$$BU_0 = \sqrt{\begin{aligned} (6861.317 - 865.40)^2 \\ + (3727.819 - 2047.25)^2 \end{aligned}}$$

$$= 4736.907$$

$$CU_0 = \sqrt{\begin{aligned} (6861.317 - 2865.22)^2 \\ + (3727.819 - 27.15)^2 \end{aligned}}$$

$$= 5446.443$$

2. Formulate the matrices. With these minor changes in the lengths, the **A** matrix (to three places) does not change. Hence, $(\mathbf{A}^T\mathbf{A})^{-1}$ does not change either. The **K** matrix does not change, however, as shown by the following computations:

$$k_1 = 6049.00 - 6048.963 = 0.037$$

$$k_2 = 4736.83 - 4376.907 = -0.077$$

$$k_3 = 5446.49 - 5446.443 = 0.047$$

3. The matrix solution.

$$\mathbf{A}^T\mathbf{K} = \begin{bmatrix} 0.991 & 0.935 & 0.734 \\ -0.132 & 0.355 & 0.679 \end{bmatrix}$$

$$\times \begin{bmatrix} 0.037 \\ -0.077 \\ 0.047 \end{bmatrix}$$

$$= \begin{bmatrix} -0.0008 \\ -0.0003 \end{bmatrix}$$

$$\mathbf{X} = \frac{1}{0.960} \begin{bmatrix} 0.605 & -0.699 \\ -0.699 & 2.395 \end{bmatrix}$$

$$\times \begin{bmatrix} -0.0008 \\ -0.0003 \end{bmatrix}$$

$$= \begin{bmatrix} -0.0003 \\ -0.0002 \end{bmatrix}$$

The revised coordinates of U are

$$X_u = 6861.317 - 0.0003 = 6861.317$$

$$Y_u = 3727.819 - 0.0004 = 3727.819$$

Satisfactory convergence is indicated by the very small size of the corrections computed in the second iteration. Having calculated most probable coordinates, we may then calculate residuals using Equation (16-10), followed by calculations for the standard deviation of unit weight and the standard deviations of adjusted coordinates using Equations (16-11) and (16-13), respectively. Of course, the measured lengths could be weighted, in which case the appropriate weighted least-squares equations would be used.

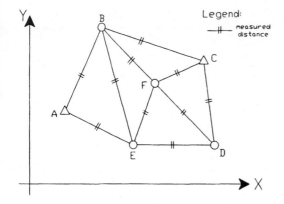

Figure 16-7. More complex trilateration network.

are readily programmed for computer solution.

16-12. TRIANGULATION ADJUSTMENT

Triangulation is a method of horizontal control extension in which angle measurements are the basic observations. Positions of widely spaced points are then computed based on these angles and only a minimum number of measured distanced called *base lines*. As in trilateration, the geometric figures used are many and varied.

Least-squares triangulation adjustment may be performed by condition equations, direction observation equations, or angle observation equations. In this section, the angle observation equation method is presented. This procedure is relatively simple and adaptable to any combination of angle measurements and geometric figures.

In formulating angle observation equations I is always assigned to the station of the angle's vertex. If we consider the angles turned clockwise, J is the backsight station and K the foresight station. This designation of stations must be strictly adhered to in forming observation equations from prototype Equation (16-22). The procedure is demonstrated in numerical examples in this chapter.

16-12-1. Adjustment of Intersections

Intersection is one of the simplest and sometimes most practical methods for locating the horizontal position of an occasional isolated point, if the point is visible from two or more existing horizontal control stations. Intersection is especially well-adapted over inaccessible terrain. For a unique position computation, the method requires that at least two horizontal angles be measured from two con-

As was demonstrated in Example 16-6, X- and Y-coordinates of control stations can readily be held fixed in an adjustment. This is accomplished by assigning values of zero to their dX and dY terms, and hence those terms drop out of the equations. Note that in each observation equation of the example, only two unknowns appear, since in each case one end of those lines was a control station and thus was held fixed. Directions of lines can also be held fixed, and methods of doing so are described in references cited at the end of this chapter.

The procedure for adjusting larger trilateration networks, such as that shown in Figure 16-7, follows the same approach as that used for the small figure of Example 16-6. The basic difference is that the sizes of the matrices are larger, but the individual elements are still formed using the same prototype equation. Suppose in Figure 16-7 that A and C are control points whose coordinates are fixed. Thus, there are ten observations and eight unknowns. Points A and C in the network are held fixed by giving the terms dX_a, dY_a, dX_c, and dY_c zero coefficients. Hence, these terms drop out of the solution. The **A** matrix formulated from Equation (16-18) would have nonzero elements as indicated by the X's in Table 16-1.

Formulation of the matrices and their least-squares solution in an iterative manner

Table 16-1. Nonzero elements for **A** matrix from Equation (16-18)

Unknowns Distance	dY_b	dX_b	dY_d	dX_d	dY_e	dX_e	dY_f	dX_f
AB	X	X	0	0	0	0	0	0
AE	0	0	0	0	X	X	0	0
BC	X	X	0	0	0	0	0	0
BF	X	X	0	0	0	0	X	X
BE	X	X	0	0	X	X	0	0
CD	0	0	X	X	0	0	0	0
CF	0	0	0	0	0	0	X	X
DF	0	0	X	X	0	0	X	X
DE	0	0	X	X	X	X	0	0
EF	0	0	0	0	X	X	X	X

trol points, as angles θ_1 and θ_2 measured at control points A and B of Figure 16-8. If additional control is available, the position computation for unknown point U can be strengthened by measuring redundant angles θ_3 and θ_4 in the figure. When redundant measurements are taken, most probable coordinates of the point U may be calculated by the least-squares procedure.

Example 16-7. Adjust the example of Figure 16-8 by least squares if the measured angles (equally weighted) and coordinates of control points are the following:

$$\theta_1 = 50°06'50''$$

$$\theta_3 = 98°41'17''$$

$$\theta_2 = 101°30'47''$$

$$\theta_4 = 59°17'01''$$

$$X_a = 865.40$$

$$X_b = 2432.55$$

$$X_c = 2865.22$$

$$Y_a = 4527.15$$

$$Y_b = 2047.25$$

$$Y_c = 27.15$$

1. Calculate initial approximations for AU_0, BU_0, CU_0, X_{u_0}, and Y_{u_0} as follows:

$$AB = \sqrt{\begin{array}{c}(2432.55 - 865.40)^2 \\ + (4527.15 - 2047.25)^2\end{array}}$$

$$= 2933.58 \text{ ft}$$

$$AU_0 = \frac{AB \sin \theta_2}{\sin(180° - \theta_1 - \theta_2)}$$

$$= \frac{2933.58 \sin 101°30'47''}{\sin 28°27'23''}$$

$$= 6049.00 \text{ ft}$$

$$\text{Azimuth } AB = \tan^{-1}\left(\frac{X_b - X_a}{Y_b - Y_a}\right) + 180°$$

$$= \tan^{-1}\left(\frac{2432.55 - 865.40}{2047.25 - 4527.15}\right)$$

$$+ 180° = 147°42'34''$$

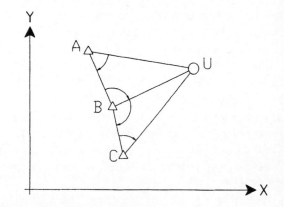

Figure 16-8. Intersection example.

Azimuth $AU_0 = 147°42'34'' - 50°06'50''$

$$= 97°35'44''$$

$X_{u_0} = X_a + AU_0 \sin(\text{azimuth } AU_0)$

$$= 865.40 + (6049.00)\sin 97°35'44''$$

$$= 6861.35$$

$Y_{u_0} = Y_a + AU_0 \cos(\text{azimuth } AU_0)$

$$= 4527.15 + (6049.00)\cos 97°35'44''$$

$$= 3727.59$$

$$BU_0 = \sqrt{\begin{array}{c}(6861.35 - 2432.55)^2 \\ + (3727.59 - 2047.25)^2\end{array}}$$

$$= 4736.83$$

$$CU_0 = \sqrt{\begin{array}{c}(6861.35 - 2865.22)^2 \\ + (3727.59 - 27.15)^2\end{array}}$$

$$= 5446.29$$

2. Formulation of the matrices. As in trilateration adjustment, a control station's coordinate can be held fixed by assigning zeros to its dX and dY values, so that these terms drop out of the equations. In Figure 16-8, the vertex points of all angles are control stations; thus, their dX and dY values are zeros, and these terms do not appear in the observation equations. In forming the observation equations, I, J, and K are assigned as previously described. Thus for angle θ_1, i, j, and k are placed in prototype Equation (16-22) by a, u, and b, respectively. Angle θ_1 conforms to Figure 14-4b.

Referring to Equation (16-22), we may write the following observation equations for the four observed angles:

$$\left[\frac{Y_a - Y_{u_0}}{(AU_0)^2}\right](dX_u) + \left[\frac{X_{u_0} - X_a}{(AU_0)^2}\right](dY_u)$$

$$= \theta_1 - \left\{\tan^{-1}\left(\frac{X_b - X_a}{Y_b - Y_a}\right)\right.$$

$$\left. -\tan^{-1}\left(\frac{X_{u_0} - X_a}{Y_{u_0} - Y_a}\right) + 0°\right\} + v_1$$

$$\times\left[\frac{Y_{u_0} - Y_b}{(UB_0)^2}\right](dX_u) + \left[\frac{X_b - X_{u_0}}{(UB_0)^2}\right](dY_u)$$

$$= \theta_2 - \left\{\tan^{-1}\left(\frac{X_{u_0} - X_b}{Y_{u_0} - Y_b}\right)\right.$$

$$\left. -\tan^{-1}\left(\frac{X_a - X_b}{Y_a - Y_b}\right) + 0°\right\} + v_2$$

$$\times\left[\frac{Y_b - Y_{u_0}}{(BU_0)^2}\right](dX_u) + \left[\frac{X_{u_0} - X_b}{(BU_0)^2}\right](dY_u)$$

$$= \theta_3 - \left\{\tan^{-1}\left(\frac{X_c - X_b}{Y_c - Y_b}\right)\right.$$

$$\left. -\tan^{-1}\left(\frac{X_{u_0} - X_b}{Y_{u_0} - Y_b}\right) + 180°\right\} + v_3$$

$$\times\left[\frac{Y_{u_0} - Y_c}{(UC)^2}\right](dX_u) + \left[\frac{X_c - X_{u_0}}{(UC_0)^2}\right]$$

$$= \theta_4 - \left\{\tan^{-1}\left(\frac{X_{u_0} - X_c}{Y_{u_0} - Y_c}\right)\right.$$

$$\left. +\tan^{-1}\left(\frac{X_b - X_c}{Y_b - Y_c}\right) + 0°\right\} + v_4$$

3. Substituting control point coordinates and approximate coordinates X_{u_0} and Y_{u_0} into the linearized observation equations and multiplying by ρ, we can form the following **A** and **K** matrices. [Note that Equation (16-19) is used to calculate their **K** values.]

$$\mathbf{A} = \rho\begin{bmatrix} \dfrac{4527.15 - 3727.59}{6049.00^2} & \dfrac{6861.35 - 865.40}{6049.00^2} \\[2mm] \dfrac{3727.59 - 2047.25}{4736.83^2} & \dfrac{2432.55 - 6861.35}{4736.83^2} \\[2mm] \dfrac{2047.25 - 3727.59}{4736.83^2} & \dfrac{6861.35 - 24.32.55}{4736.83^2} \\[2mm] \dfrac{3727.59 - 27.15}{5446.29^2} & \dfrac{2865.22 - 6861.35}{5446.29^2} \end{bmatrix}$$

$$= \begin{bmatrix} 4.507 & 33.800 \\ 15.447 & -40.713 \\ -15.447 & 40.713 \\ 25.732 & -27.788 \end{bmatrix}$$

$$\mathbf{K} = \begin{bmatrix} 50°06'50'' - \left\{ \tan^{-1}\left(\dfrac{2432.55 - 865.40}{2047.25 - 4527.15} \right) - \tan^{-1}\left(\dfrac{6861.35 - 865.40}{3727.59 - 4527.15} \right) + 0° \right\} \\[10pt] 101°30'47'' - \left\{ \tan^{-1}\left(\dfrac{6861.35 - 2432.55}{3727.59 - 047.25} \right) - \tan^{-1}\left(\dfrac{865.40 - 2432.55}{4527.15 - 2047.25} \right) + 0° \right\} \\[10pt] 98°41'17'' - \left\{ \tan^{-1}\left(\dfrac{2865.22 - 2432.55}{27.15 - 047.25} \right) - \tan^{-1}\left(\dfrac{6861.35 - 2432.55}{3727.59 - 2047.25} \right) + 180° \right\} \\[10pt] 59°17'01'' - \left\{ \tan^{-1}\left(\dfrac{6861.35 - 2865.22}{3727.59 - 27.15} \right) - \tan^{-1}\left(\dfrac{2432.55 - 2865.22}{2047.25 - 27.15} \right) + 0° \right\} \end{bmatrix} = \begin{bmatrix} 0.00'' \\ 0.00'' \\ -0.69'' \\ -20.23'' \end{bmatrix}$$

Note that values in the $\mathbf{K}$ matrix for angles θ_1 and θ_2 are exactly equal to zero for the first iteration because initial coordinates X_{u_0} and Y_{u_0} were calculated using these two angles.

4. Matrix solution of Equation (16-7) for adjusted coordinates X_u and Y_u.

$$\mathbf{A}^T\mathbf{A} = \begin{bmatrix} 1159.7 & -1820.5 \\ -1820.5 & 5229.7 \end{bmatrix},$$

$$\mathbf{Q} = (\mathbf{A}^T\mathbf{A})^{-1} = \begin{bmatrix} 0.001901 & 0.000662 \\ 0.000662 & 0.000422 \end{bmatrix}$$

$$\mathbf{A}^T\mathbf{K} = \begin{bmatrix} -509.9 \\ 534.1 \end{bmatrix}$$

$$\mathbf{X} = (\mathbf{A}^T\mathbf{A})^{-1}(\mathbf{A}^T\mathbf{K})$$

$$= \begin{bmatrix} 0.001901 & 0.000662 \\ 0.000662 & 0.000422 \end{bmatrix} \begin{bmatrix} -509.0 \\ 534.1 \end{bmatrix}$$

$$= \begin{bmatrix} dX_u \\ dY_u \end{bmatrix}$$

and

$$dX_u = -0.62 \text{ ft}$$

$$dY_u = -0.11 \text{ ft}$$

$$X_u = X_{u_0} + dX_u = 6861.35 - 0.62 = 6860.73$$

$$Y_u = Y_{u_0} + dY_u = 3727.59 - 0.11 = 3727.48$$

Note that a second iteration produced negligible-sized values for dX_u and dY_u, thus, the solution converged after one iteration. The second iteration is not shown.

5. Matrix solutions for residuals and precisions by Equation (16-10) with $\mathbf{L}$ replaced by $\mathbf{K}$ is as follows:

$$\mathbf{V} = \mathbf{AX} - \mathbf{K}$$

$$= \begin{bmatrix} 4.507 & 33.80 \\ 15.447 & -40.713 \\ -15.447 & 40.713 \\ 25.732 & -27.788 \end{bmatrix} \begin{bmatrix} -0.62 \\ -0.11 \end{bmatrix}$$

$$\times \begin{bmatrix} 0.00'' \\ 0.00'' \\ -0.69'' \\ -20.23'' \end{bmatrix}$$

$$= \begin{bmatrix} -6.5'' \\ -5.1'' \\ +5.8'' \\ +7.3'' \end{bmatrix}$$

By Equation (16-11),

$$\mathbf{V}^T\mathbf{V} = \begin{bmatrix} -6.5 & -5.1 & 5.8 & 7.3 \end{bmatrix}$$

$$\times \begin{bmatrix} -6.5 \\ -5.1 \\ 5.8 \\ 7.3 \end{bmatrix} = 155.2 \text{ sec}^2$$

$$S_0 = \sqrt{\dfrac{\mathbf{V}^T\mathbf{V}}{m - n}}$$

$$= \sqrt{\dfrac{155.2}{4 - 2}} = \pm 8.8''$$

By Equation (16-13),

$$S_{X_u} = S_0\sqrt{Q_{X_uX_u}} = 8.8\sqrt{0.001901} = \pm 0.38 \text{ ft}$$

$$S_{Y_u} = S_0\sqrt{Q_{Y_uY_u}} = 8.8\sqrt{0.000422} = \pm 0.18 \text{ ft}$$

Finally, the standard deviation in the position of U, S_u, is the square root of the sum of the squares of and S_{X_u} *and* S_{Y_u} or as follows:

$$S_u = \sqrt{S_{X_u}^2 + S_{Y_u}^2} - \sqrt{0.38^2 + 0.18^2}$$

$$= \pm 0.42 \text{ ft}$$

16-12-2. Adjustment of Resections

Resection is a method that may be employed for locating the unknown horizontal position of an occupied theodolite station by measuring a minimum of two horizontal angles to a minimum of three stations whose horizontal positions are known. If more than three stations are available, redundant observations may be obtained and the position of the unknown occupied station can be computed by employing the least-squares procedure. Like intersection, this method is suitable for locating an occasional isolated point and is especially well-adapted for use over inaccessible terrain.

Consider the resection position computation for the occupied point U of Figure 16-9, having observed the three horizontal angles shown. Points A, B, C, and D are fixed control stations. Angles θ_1, θ_2, and θ_3 are observed from station U, whose position is unknown.

Utilizing prototype Equation (16-22), we may write a linearized observation equation for each angle. These observation equations can be expressed in matrix notation as follows:

$$_3\mathbf{A}_2 \ _2\mathbf{X}_1 = {}_3\mathbf{K}_1 + {}_3\mathbf{V}_1$$

Application of the least-squares routine yields corrections dX_u and dY_u, the most probable coordinates X_u and Y_u, residuals, and the estimated standard deviation of the position of point U.

16-12-3. Adjustment of More Complex Triangulation Networks

The basic figure for triangulation is generally considered to be the quadrilateral, but frequently other geometrical figures, such as chains of quadrilaterals, central-point figures, etc., are used. Regardless of the geometrical shape of the triangulated figure, the basic least-squares approach is simply an extension of principles already discussed, which involves writing one observation equation for each measured angle in terms of the most probable coordinates of the points involved. The following example (see Figure 16-10) illustrates the adjustment of a quadrilateral.

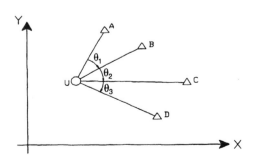

Figure 16-9. Resection example.

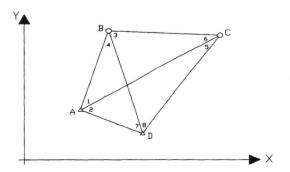

Figure 6-10. Quadrilateral of Example 16-8.

Example 16-8. Adjust, by least squares, the quadrilateral of Figure 16-10, given the following observations and known data:

Observed angles (assume equal weights):

$$1 = 42°35'29.0'', \quad 5 = 21°29'23.9''$$
$$2 = 87°35'10.6'', \quad 6 = 39°01'35.4''$$
$$3 = 79°54'42.1'', \quad 7 = 31°20'45.8''$$
$$4 = 18°28'22.4'', \quad 8 = 39°34'27.9''$$

Fixed coordinates:

$$X_a = 4270.33, \quad X_d = 7610.58$$
$$Y_a = 8448.90, \quad Y_d = 4568.75$$

A computer program has been used to form the matrices and solve the problem. Equal weights were used. In the program, the angles were input in order 1 through 8. The **X** matrix in the solution is

$$\mathbf{X} = \begin{bmatrix} dX_b \\ dY_b \\ dX_c \\ dY_c \end{bmatrix}$$

The following computer-output listing, which is self-explanatory, gives the solution for Example 16-8. As shown, one iteration was satisfactory to achieve convergence because for the second iteration, the unknowns—corrections to the initial coordinates—are all zeros. Residuals, adjusted coordinates, their standard deviation, and adjusted angles are all given at the end of the listing.

Triangulation Example

Number of Observed Distances	≫ 0
Number of Observed Angles	≫ 8
Number of Observed Azimuths	≫ 0
Number of Unknown Stations	≫ 2
Number of Control Stations	≫ 2

Control Stations

Station	Northing	Easting
A	8448.900	4270.330
D	4568.750	7610.580

Initial Approximations of Unknown Station Coordinates

Station	Northing	Easting
B	16,749.769	5599.549
C	16,636.185	14,633.027

Angle Observations

Backsighted Station	Occupied Station	Foresighted Station	Angle
B	A	C	42°35'29.0''
C	A	D	87°35'10.6''
C	B	D	79°54'42.1''
D	B	A	18°28'22.4''
D	C	A	21°29'23.9''
A	C	B	39°1'35.4''
A	D	B	31°20'45.8''
B	D	C	39°34'27.9''

Weight Matrix
Dimensions: 8 × 8

1	0	0	0	0	0	0	0
0	1	0	0	0	0	0	0
0	0	1	0	0	0	0	0
0	0	0	1	0	0	0	0
0	0	0	0	1	0	0	0
0	0	0	0	0	1	0	0
0	0	0	0	0	0	1	0
0	0	0	0	0	0	0	1

Iteration No. 1
A Matrix
Dimensions: 8 × 4

−24.2301	3.8804	9.6822	−12.2549
0.0000	0.0000	−9.6822	12.2549
16.2008	−20.1080	0.2845	22.8298
7.7448	−6.6023	0.0000	0.0000
0.0000	0.0000	−3.0864	−4.8244
0.2845	22.8298	−9.9668	−10.5750
16.4853	2.7219	0.0000	0.0000
−16.4853	−2.7219	12.7686	−7.4305

K Matrix
Dimensions: 8 × 1

4.7351
−4.7406
4.3168
−4.3190
−7.7651
4.1671
−7.8755
8.5812

Covariance Matrix (Q_{xx})
Dimensions: 4 × 4

0.0016	0.0008	0.0018	−0.0003
0.0008	0.0079	0.0077	0.0066
0.0018	0.0077	0.0106	0.0063
−0.0003	0.0066	0.0063	0.0073

Covariance Matrix (Q_{xx})
Dimensions: 4 × 4

0.0016	0.0008	0.0018	−0.0003
0.0008	0.0079	0.0077	0.0066
0.0018	0.0077	0.0106	0.0063
−0.0003	0.0066	0.0063	0.0073

Unknowns (X) / Residuals (V)
Dimensions: 4 × 1 / Dimensions: 8 × 1

Unknowns (X)	Residuals (V)
0.0001	−2.1010
−0.0001	−5.0324
0.0002	4.1834
−0.0003	1.4172
	−1.7581
	5.4004
	−6.4838
	1.4744

Unknowns (X) / Residuals (V)

Unknowns (X)	Residuals (V)
−0.1909	−2.1010
0.6449	−5.0324
0.8542	4.1834
0.6987	1.4172
	−1.7582
	5.4004
	−6.4839
	1.4744

STANDARD ERROR OF UNIT WEIGHT = 5.606

Standard Deviations

SXB = ±0.226
SYB = ±0.498
SXC = ±0.577
SYC = ±0.478

Iteration No. 2 — A Matrix
Dimensions: 8 × 4

−24.2285	3.8793	9.6815	−12.2538
0.0000	0.0000	−9.6815	12.2538
16.2000	−20.1054	0.2843	22.8272
7.7441	−6.6012	0.0000	0.0000
0.0000	0.0000	−3.0860	−4.8235
0.2843	22.8272	−9.9658	−10.5734
16.4844	2.7218	0.0000	0.0000
−16.4844	−2.7218	12.7675	−7.4303

Adjusted Angles

Backsighted Station	Occupied Station	Foresighted Station	Angle
B	A	C	42°35′31.1″
C	A	D	87°35′15.6″
C	B	D	79°54′37.9″
D	B	A	18°28′21.0″
D	C	A	21°29′25.7″
A	C	B	39°01′30.0″
A	D	B	31°20′52.3″
B	D	C	39°34′26.4″

K Matrix
Dimensions: 8 × 1

−2.0977
−5.0381
4.1798
1.4184
−1.7575
5.3996
−6.4825
1.4781

16-13. TRAVERSE ADJUSTMENT

Of the many methods for traverse adjustment, the characteristic that distinguishes traverse adjustment by least squares from approximate methods is that distance and direction observations are adjusted simultaneously in the least-squares approach. The result is an adjustment in which all geometrical conditions are satisfied; but more important, the values of the adjusted quantities are more probable with least-squares adjustment than with any other method because the sum of the weights times their corresponding squared residuals is minimized. In addition, least squares enables the assignment of relative weights to the observations based on their expected relative reliabilities.

In this section, least-squares traverse adjustment by the method of observation equations is presented in which an observation equation is written for each measured distance and angle. The basic observation equations for distance and angle measurements are Equations (16-18) and (16-22), respectively.

The apparent inconsistency arising from the fact that distance and angle observations with differing units are combined into one adjustment is resolved through the use of appropriate weights. In this procedure, a relative weight is assigned to each observation in accordance with the inverse of its variance (square of standard deviation). Based on variances, relative weights of distance and angle observations are given by

$$\text{Distance: } P_{L_{ij}} = \frac{1}{(S_{L_{ij}})^2} \qquad (16\text{-}24)$$

$$\text{Angle: } P_{\theta_{jik}} = \frac{1}{(S_{\theta_{jik}})^2} \qquad (16\text{-}25)$$

In Equations (16-24) and (16-25), $P_{L_{ij}}$ and $P_{\theta_{jik}}$ are weights, respectively, for distance and angle observations, and $S_{L_{ij}}$ and $S_{\theta_{jik}}$ are their respective standard deviations. Since standard deviations are generally not available before adjustment, they are usually estimated and referred to as a priori values. They are important because they significantly influence the

adjustment. The fundamental condition that is enforced in least-squares adjustment is the minimization of the pv^2 terms. Thus, it follows that if observations are weighted in accordance with the inverse of variance and the same units are used for the residual and standard deviation, the pv^2 terms will all be dimensionless and, therefore, compatible for simultaneous adjustment. The procedure is demonstrated in the example problem given in this section.

For an n-sided closed traverse, there are n distances and $n + 1$ angles, if we consider one angle for orientation. In Figures 16-11a and b, e.g., each closed traverse has four sides, with four distances, and five angles measured. Each new traverse station introduces two unknowns, their X- and Y-coordinates; thus, $2(n - 1)$ unknowns exist for any completely surveyed closed traverse. Therefore, for such a traverse, regardless of its number of sides, the number of redundant equations r (number of observations minus number of unknowns) is equal to $(2n + 1) - 2(n - 1) = 3$. Specifically, for the example traverses of Figure 16-11, $r = \{2(4) + 1\} - 2(4 - 1) = 3$.

Example 16-9. Adjust by least squares the simple traverse shown in Figure 16-12 from

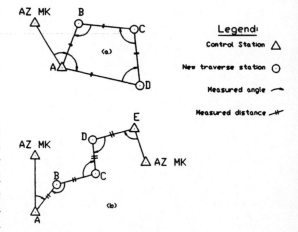

Figure 16-11. Closed traverse examples.

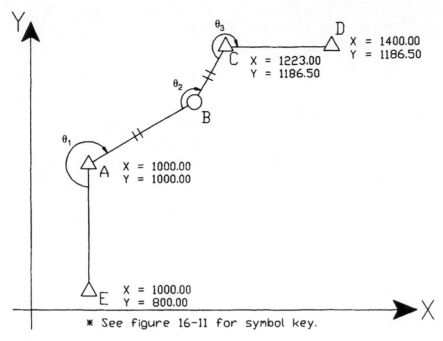

Figure 16-12. Traverse of Example 16-9.

the following measured survey data (coordinates of the control points are shown in the figure):

Observations	A priori Standard Deviations
$AB = 200.00$	± 0.05 ft
$BC = 100.00$	± 0.08 ft
$\theta_1 = 240°00'$	± 30 sec
$\theta_2 = 150°00'$	± 30 sec
$\theta_3 = 240°01'$	± 30 sec

1. Calculate initial approximate coordinates for station B.

$$X_{b_0} = 1000.00 + 200.00 \sin 60° = 1173.20$$

$$Y_{b_0} = 1000.00 + 200.00 \cos 60° = 1100.00$$

2. Formulate the **X** and **K** matrices.

$$\mathbf{X} = \begin{bmatrix} dX_b \\ dY_b \end{bmatrix}, \qquad \mathbf{K} = \begin{bmatrix} k_{L_{ab}} \\ k_{L_{bc}} \\ k_{\theta_1} \\ k_{\theta_2} \\ k_{\theta_3} \end{bmatrix}$$

$$\mathbf{K} = \begin{bmatrix} 200.00 - 200.00 \text{ ft} \\ 100.00 - 99.81 \text{ ft} \\ 120°00'00'' - 120°00'00'' \\ 150°00'00'' - 149°55'51'' \\ 119°59'00'' - 119°55'48'' \end{bmatrix} = \begin{bmatrix} 0.00 \\ 0.19 \\ 0'' \\ 249'' \\ 192'' \end{bmatrix}$$

The values in the **K** matrix are derived by subtracting computed quantities, based on itial coordinates, from their respective observed quantities.

3. Calculate the **A** matrix. The **A** matrix is formed using prototype Equation (16-18) for the distances and Equation (16-22) for the angles. Note that angle coefficients are multiplied by $\rho(206'264.8''/\text{rad})$ to convert to seconds and be compatible with the values given in the **K** matrix.

$$
A = \begin{bmatrix}
\left(\dfrac{1173.20 - 1000.00}{200.00}\right) & \left(\dfrac{1100.00 - 1000.00}{200.00}\right) \\[2mm]
\left(\dfrac{1173.20 - 1223.00}{100.00}\right) & \left(\dfrac{1100.00 - 1186.50}{100.00}\right) \\[2mm]
\left(\dfrac{1000.00 - 1100.00}{200.00^2}\right)\rho & \left(\dfrac{1173.20 - 1000.00}{200.00^2}\right)\rho \\[2mm]
\left(\dfrac{1000.00 - 1100.00}{200.00^2} - \dfrac{1186.50 - 1100.00}{100.00^2}\right)\rho & \left(\dfrac{1173.20 - 1000.00}{200.00^2} - \dfrac{1173.20 - 1223.00}{100.00^2}\right)\rho \\[2mm]
\left(\dfrac{1100.00 - 1186.50}{100.00^2}\right)\rho & \left(\dfrac{1223.00 - 1173.20}{100.00^2}\right)\rho
\end{bmatrix}
$$

$$
A = \begin{bmatrix}
0.866 & 0.500 \\
-0.498 & -0.865 \\
-515.7 & 893.1 \\
-2299.8 & 1920.3 \\
-1784.2 & 1027.2
\end{bmatrix}
$$

4. Formulate the **P** matrix. Based on the given a priori standard deviations and if we use Equations (16-24) and (16-25), the **P** matrix is

$$
P = \begin{bmatrix}
\dfrac{1}{0.05^2} & & & & \text{zeros} \\[2mm]
& \dfrac{1}{0.08^2} & & & \\[2mm]
& & \dfrac{1}{30^2} & & \\[2mm]
& & & \dfrac{1}{30^2} & \\[2mm]
\text{zeros} & & & & \dfrac{1}{30^2}
\end{bmatrix}
$$

$$
= \begin{bmatrix}
400.00 & & & & \text{zeros} \\
& 156.2 & & & \\
& & 0.0011 & & \\
& & & 0.0011 & \\
\text{zeros} & & & & 0.0011
\end{bmatrix}
$$

5. Solving Equation (16-9) using these matrices yields the following values for the corrections to the initial coordinates (a second iteration, not shown, produced zeros for dX_b and dY_b):

$$dX_b = -0.11 \text{ ft}$$

$$dY_b = -0.01 \text{ ft}$$

The residuals, computed by Equation (16-10), are

$$v_{ab} = -0.10 \text{ ft}$$

$$v_{bc} = -0.12 \text{ ft}$$

$$v_{\theta_1} = 47.0 \text{ sec}$$

$$v_{\theta_2} = -17.4 \text{ sec}$$

$$v_{\theta_3} = -6.4 \text{ sec}$$

By Equations (16-11), (16-12), and (16-13), the standard deviation of unit weight and standard deviations of the adjusted coordinates are

$$S_0 = \pm 1.74$$

$$S_{X_b} = \pm 0.04 \text{ ft}$$

$$S_{Y_b} = \pm 0.05 \text{ ft}$$

6. The adjusted coordinates are

$$X_b = 1173.20 - 0.11 = 1173.09$$

$$Y_b = 1100.00 - 0.01 = 1099.99$$

7. Finally, the adjusted observations obtained by adding their residuals are

$$AB = 200.00 - 0.10 = 199.90$$

$$BC = 100.00 - 0.12 = 99.88$$

$$\theta_1 = 360° - (120°00'49'') = 239°59'13''$$

$$\theta_2 = 150°00'00'' - 0°00'17'' = 149°59'43''$$

$$\theta_3 = 360° - (119°59'00'' - 0°00'06'')$$

$$= 240°01'06''$$

16-14. CORRELATION AND THE STANDARD ERROR ELLIPSE

As given in Section 16-6, Equation (16-13) can be used to determine the standard deviations in a station's adjusted coordinates. These uncertainties are a reflection of the geometry of the problem and inexactness of the measurements used to determine the station's position. Standard deviations computed by Equation (16-13) are parallel to the X-Y adjustment axis. However, the station's largest positional uncertainties will not generally be aligned with these axes. For example, consider the uncertainties in the position of station U shown in Figure 16-13. Obviously, the uncertainties in this station's position due to the distance inexactness will not be aligned with the **X**- and **Y**-coordinate system, but rather they will vary

along the direction of the line itself. For any horizontal adjustment problem, there is a correlation between the station's coordinate uncertainties. This correlation can be determined from elements of the covariance matrix by the equation

$$r_{xy} = \frac{Q_{xy}}{\sqrt{Q_{xx}}\sqrt{Q_{yy}}} \tag{16-26}$$

where r_{xy} is the correlation coefficient between two unknown parameters; Q_{xy} the off-diagonal element in the xth row and yth column; Q_{xx} the diagonal element in the xth row and column; and Q_{yy} the diagonal element in the yth row and column. It should be noted that r_{xy} is always between zero and 1. When r_{xy} equals 1, a change in one of the unknowns directly influences the value of the other unknown. Zero indicates no correlation between the two unknowns, and thus changes in one unknown would not affect the other unknown. With Equation (16-26), the x and y coordinates of station C in Example 16-8 have a correlation coefficient of

$$r_{c_x c_y} = \frac{Q_{c_x c_y}}{\sqrt{Q_{c_x c_x}}\sqrt{Q_{c_y c_y}}}$$

$$= \frac{0.0063}{\sqrt{0.0106}\sqrt{0.0073}} = 0.716$$

From the standard deviations of a station's coordinates, the *standard error rectangle* can be drawn as shown in Figure 16-14. This rectangle has half-dimensions of S_x and S_y and encloses the so-called *standard error ellipse*. The semimajor and semiminor axes of this ellipse (the *U-V* axis) exist in the directions of the largest and smallest uncertainties of the point, respectively. The proper amount of angular rotation required to produce these axes is given by t in the equation

$$2t = \tan^{-1}\left(\frac{2Q_{xy}}{Q_{yy} - Q_{xx}}\right) \tag{16-27}$$

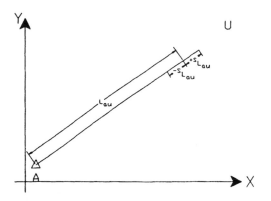

Figure 16-13. Uncertainty in station coordinates due to distance uncertainty.

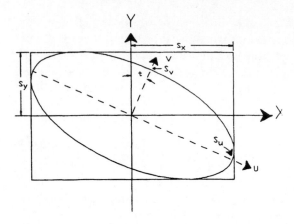

Figure 16-14. The standard error rectangle and ellipse.

where t is the rotation angle from the Y-axis to the semiminor axis of the ellipse, the V-axis. Note that $2t$ must be located in its appropriate quadrant according to the signs of Q_{xy} and $(Q_{yy} - Q_{xx})$ using the sign convention given in Table 16-2 before t is determined.

As noted previously, in this rotated system, the maximum positional uncertainty occurs along the rotated U-axis, and the minimum along the V-axis. The covariance matrix elements for the point rotated into these axes are

$$Q_{uu} = Q_{xx} \sin^2 t + 2Q_{xy} \cos t \sin t$$
$$+ Q_{yy} \cos^2 t \qquad (16\text{-}28)$$

$$Q_{vv} = Q_{xx} \cos^2 t - 2Q_{xy} \sin t \cos t$$
$$+ Q_{yy} \sin^2 t \qquad (16\text{-}29)$$

From these Q_{uu} and Q_{vv} elements, the resulting uncertainties in the U- and V-axis system are

$$S_u = S_0 \sqrt{Q_{uu}} \qquad (16\text{-}30)$$

$$S_v = S_0 \sqrt{Q_{vv}} \qquad (16\text{-}31)$$

where S_u is the maximum positional uncertainty of the station—at a confidence level of 39.4%— and S_v the axis perpendicular to the U-axis.

Equations (16-30) and (16-31) produce the semimajor and semiminor axes, for the *stan-*

Table 16-2. Sign conventions to determine the proper quadrant of $2t$

Sign Q_{xy}	Sign $(Q_{yy} - Q_{xx})$	Quadrant
+	+	I (0–90%)
+	−	II (90–180°)
−	−	III (180–270°)
−	+	IV (270–360°)

dard error ellipse. This ellipse can be modified to produce an error ellipse at any confidence level, or percent probability. The magnification factor is derived from F-statistics. These modifiers allow changing the standard error ellipse to an ellipse of any probability level. They give ratios of variances for varying degrees of freedom, for as the degrees of freedom increase, precision also can be expected to increase. The $F_{(2,\text{ degrees of freedom}, \alpha)}$ statistics are shown in Table 16-3. It can be seen that for small degrees of freedom the F-statistics' modifier rapidly decreases and stabilizes for larger degrees of freedom. The confidence level of the error ellipse may be increased to any confidence level by the multiplier $c = \sqrt{2F_{\text{statistic}}}$. The resulting new uncertainties are

$$S_{u_\%} = S_u c = S_u \sqrt{2F_{\text{statistic}}}$$
$$S_{v_\%} = S_v c = S_v \sqrt{2F_{\text{statistic}}} \qquad (16\text{-}32)$$

Table 16-3. $F_{2,\text{ degree of freedom}, \alpha}$ statistics for selected certainty levels

Degrees of freedom	90%	95%	99%
3	5.46	9.55	30.82
4	4.32	6.94	18.00
5	3.78	5.79	13.27
6	3.46	5.14	10.92
7	3.26	4.74	9.55
8	3.11	4.46	8.65
9	3.01	4.26	8.02
10	2.92	4.10	7.56
15	2.70	3.68	6.36
20	2.59	3.49	5.85
25	2.53	3.39	5.57
30	2.49	3.32	5.39
60	2.39	3.15	4.98

Example 16-10. Compute the standard and 95% error ellipses for the station C of Example 16-8.

1. Recall from Example 16-8 that $S_0 = 5.626$ with four degrees of freedom. Recall also that station C's covariance matrix elements were

$$Q_c = \begin{bmatrix} Q_{xx} & Q_{xy} \\ Q_{xy} & Q_{yy} \end{bmatrix} = \begin{bmatrix} 0.0106 & 0.0063 \\ 0.0063 & 0.0073 \end{bmatrix}$$

2. From Equation (16-27), two times the rotation angle t is

$$2t = \tan^{-1}\left[\frac{2(0.0063)}{0.0073 - 0.0106}\right] = -75°20'$$

Since $2t$ is negative with a positive numerator and negative denominator, it lies in quadrant II. Therefore, 180° must be added to $2t$ to obtain the proper value for t. Thus, $t = \frac{1}{2}(180° - 75°20') = 52°20'$.

3. From Equations (16-28) and (16-29), the station's rotated covariance matrix elements are

$$Q_{uu} = 0.0106\sin^2(t)$$
$$+ 2(0.0063)\cos(t)\sin(t)$$
$$+ 0.0073\cos^2(t) = 0.015$$
$$Q_{vv} = 0.0106\cos^2(t)$$
$$- 2(0.0063)\sin(t)\cos(t)$$
$$+ 0.0073\sin^2(t) = 0.002$$

4. From Equations (16-30) and (16-31), the semimajor and semiminor axes for the standard error ellipse are

$$S_u = S_0\sqrt{Q_{uu}} = 5.606\sqrt{0.015} = \pm 0.70 \text{ ft}$$
$$S_v = S_0\sqrt{Q_{vv}} = 5.606\sqrt{0.002} = \pm 0.29 \text{ ft}$$

5. If we use Equations (16-32) and Table 16-3, at a 95% level of confidence, the semimajor and semiminor axes are

$$S_{u95\%} = \pm 0.70\sqrt{2 \times 6.94} = \pm 2.59 \text{ ft}$$
$$S_{v95\%} = \pm 0.29\sqrt{2 \times 6.94} = \pm 1.08 \text{ ft}$$

16-15. SUMMARY

In this chapter, only a brief treatment of the subject of survey adjustments by least squares was presented, and emphasis has been placed on methods of adjusting the most commonly employed types of traditional ground surveys. These include level nets, trilateration, triangulation, and traverses. Example problems of each were given to clarify computational procedures. Many other more advanced topics in least squares are useful and important, but space limitations do not allow their discussion here. References cited pursue the subject in greater depth and should be consulted by those interested in further study.

REFERENCES

HIRVONEN, R. A. 1965. *Adjustment by Least Squares in Geodesy and Photogrammetry.* New York: Frederick Ungar Publishing Company.

LEICK, A. 1990. *GPS Satellite Surveying.* New York: John Wiley & Sons.

MIKHAIL, E. M. 1976. *Observations and Least Squares.* New York: Dun-Donnelly.

MIKHAIL, E. M., and G. GRACIE. 1981. *Analysis and Adjustment of Survey Measurements.* New York: Van Nostrand Reinhold.

RAINSFORD, H. F. 1957. *Survey Adjustments and Least Squares.* London: Constable.

WOLF, P. R. 1980. *Adjustment Computations: Practical Least Squares for Surveyors,* 2nd ed. Rancho Cordova, CA: Landmark Enterprises.

17

Field Astronomy for Azimuth Determinations

Richard L. Elgin, David R. Knowles, and Joseph H. Senne

17-1. INTRODUCTION

Surveying has been defined as the science of determining positions of points on the earth's surface. The four components of surveying measurements are: (1) vertical (elevations), (2) horizontal (distances), (3) relative direction (angles), and (4) absolute direction (azimuths). Due to recent developments in technology, the accuracy and efficiency of measuring these first three components have increased dramatically. This has resulted in accurate determination of the size and shape of figures. Unfortunately, determination of the orientation of figures, the fourth component, has not kept pace, even though inexpensive technology and equipment exist, such as precise timepieces, portable time signal receivers, ephemerides, programmable calculators, and computers. The purpose of this chapter is to provide sufficient theory, calculations, and field procedures so surveys in both the northern and southern hemispheres can be accurately oriented without significant increases in time and expense, in both the northern and southern hemispheres.

17-2. CELESTIAL SPHERE AND DEFINITIONS

To better visualize positions and movements of the sun, stars, and celestial coordinate circles, they are projected onto a sphere of infinite radius surrounding the earth. This sphere conforms to all the various motions of the earth as the earth rotates on its axis and revolves around the sun. Figure 17-1 illustrates the celestial sphere and principal circles necessary to understand celestial observations and calculations. Definitions of important points and circles on the sphere follow:

Great circle. A circle described on the sphere's surface by a plane that includes the sphere's center.

Zenith (Zn). The point directly overhead or where an observer's vertical line pierces the celestial sphere. Opposite zenith is the nadir.

Equator. A great circle on the celestial sphere defined by a plane that is perpendicular to the poles.

Horizon. A great circle on the celestial sphere defined by a plane that is perpendicular to an observer's vertical.

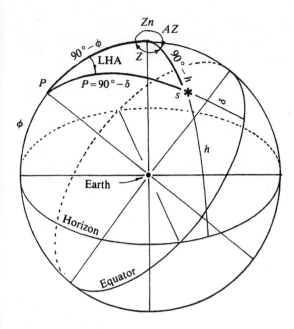

Figure 17-1. Celestial sphere.

Hour circle. A great circle that includes the poles of the celestial sphere. It is analogous to a line of longitude and perpendicular to the equator.

Vertical circle. A great circle that includes the zenith and nadir. It is perpendicular to the horizon.

Meridian. The hour circle that includes an observer's zenith. it represents north-south at an observer's location.

Altitude (h). The angle measured from the horizon upward along a vertical circle. It is the vertical angle to a celestial body.

Declination (δ). The angular distance measured along an hour circle north (positive) or south (negative) from the equator to a celestial body. It is analogous to latitude.

Prime meridian. Reference line (zero degrees longitude) from which longitude is measured. It passes through the Royal Naval Observatory in Greenwich, England; hence, it is also known as Greenwich meridian.

Longitude (λ). Angle measured at the pole, east or west from the prime meridian. Varies from zero degrees to 180°E and 180°W.

Latitude (ϕ). Angle measured along a meridian north (positive) and south (negative) from the equator. It varies from zero to 90°.

Greenwich hour angle (*GHA*). Angle at the pole measured westward from the prime (Greenwich) meridian to the hour circle through a celestial body. It is measured in a plane parallel to the equator and varies from zero to 360°. The GHA of a celestial body is always increasing—moving westward—with time.

Local hour angle (*LHA*). Angle at the pole measured westward from an observer's meridian to the hour circle through a celestial body. It differs from GHA by the observer's longitude.

Meridian angle (t). Equivalent to LHA, except it is measured either eastward or westward and is always less than 180°.

Astronomical triangle (*PZS triangle*) Spherical triangle formed by the three points: (1) celestial pole P, (2) observer's zenith Zn, and (3) celestial body S.

Polar distance (*PS or p*). Angular distance from the celestial pole P to a celestial body S. Known as codeclination: $p = 90° - \delta$.

Zenith distance (*ZS or z*). Angular distance along a vertical circle from an observer's zenith Zn to a celestial body S. Known as coaltitude: $z = 90° - h$.

Slide PZ. Angular distance from the pole P to an observer's zenith Zn. Known as coaltitude: $PZ = 90° - \phi$.

Angle Z. Angle measured at the zenith Zn, in a plane parallel with the observer's horizon, from the pole to a celestial body. Angle Z is denoted as AZ if it is measured as an azimuth, clockwise from zero to 360°.

Angle S. Angle at a celestial body between the pole P and observer's zenith Zn. Known as the parallactic angle.

Astronomical refraction. As light from a celestial body penetrates the earth's atmosphere, direction of the light ray is bent. Astronomical refraction is the angular difference between the direction of a light ray when it enters the atmosphere and its direction at the point of observation. Refraction causes celestial objects to appear higher than they actually are. Refraction corrections are required in some observation methods (altitude method).

Parallax. Apparent displacement of a point with respect to the reference system, caused by a shift in observation location. Celestial observations are considered to be made at the earth's center instead of on the earth's surface (a distance of approximately 3963 mi, or 6378 kn). Parallax corrections are required in some sun observation methods (altitude method).

Mean solar time. Uniform time based on a mean or fictional sun position. The mean sun is a point that moves at a uniform rate around the earth, making one revolution in exactly 24 hr.

Apparent sun time. Nonuniform time based on the apparent sun's position. Because the earth's orbit is eccentric and inclined to the equator, the apparent sun does not cross the observer's meridian exactly every 24 hr throughout the year.

Equation of time. Difference between apparent time and mean time.

Coordinated universal time (*UTC*). Uniform time based on mean-time at Greenwich. Since the earth's rotation is gradually slowing, approximately one leap second is added per year. UTC is broadcast by radio station WWV.

*UT*1 *time.* Mean universal time at the prime meridian obtained directly from the stars. It contains all the irregular motions of the earth and is corrected for polar wandering. UT1 is the time required for celestial observations.

DUT correction. Difference between UT1 time and UTC time.

17-3. AZIMUTH OF A LINE

Azimuth is defined as an angle measured clockwise from a reference meridian (north-south direction) to a line. Several types of meridians exist: astronomic, geodetic, grid, magnetic, record, and assumed. For field measurements, the most convenient, accurate, and retraceable reference is astronomic north. Once obtained, astronomic north can be converted to geodetic and grid north. Astronomic north is based on the direction of gravity (vertical) and axis of rotation of the earth. Geodetic north is based on a mathematical approximation of the earth's shape. The difference between astronomic and geodetic north is the LaPlace correction.

The term true north has frequently been used to indicate either astronomic or geodetic north. In central and eastern portions of the United States, LaPlace corrections tend to be small. Corrections in western states, however, may approach 20 arc-sec. Given present angle-measuring accuracies, LaPlace corrections may be significant, and use of the term true north should be discontinued. A direction determined from celestial observations results in an astronomic north reference meridian.

The azimuth of a line can be determined by measuring an angle from the line to a reference of known azimuth and computed from the following equation (see Figure 17-2).

$$AZL = AZ - \text{ang rt} \qquad (17\text{-}1)$$

This is a general equation. If *AZL* computes to be negative, 360° is added to normalize the azimuth.

For astronomic observations, a celestial body becomes the reference direction. If we know geographic location (latitude and longitude), ephemeris data, and either time or altitude, the azimuth of a celestial body can be computed. If time is used, the procedure is known as the *hour-angle method*. Likewise, if altitude is measured, the procedure is termed the *altitude method*. The basic difference between them is that the altitude method requires approximate

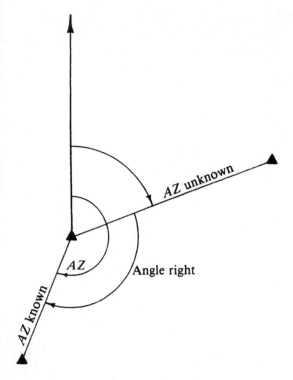

Figure 17-2. Azimuth/angle relationships.

time and an accurate vertical angle corrected for parallax and refraction, whereas the hour-angle method requires accurate time and no vertical angle.

In the past, the altitude method has been more popular for sun observations primarily due to the difficulty of obtaining and maintaining accurate time in the field (time accuracy requirements of the hour-angle method for the sun are greater than for Polaris). Recent developments of time receivers and accurate timepieces, particularly digital watches with split-time features, and time modules for calculators have eliminated this obstacle. The hour-angle method is more accurate, faster, requires shorter training for proficiency, has fewer restrictions on time of day and geographic location, is more versatile, and is applicable to the sun, Polaris, and other stars. Consequently, the hour-angle method is emphasized, and its use by surveyors is encouraged.

17-4. AZIMUTH OF A CELESTIAL BODY BY THE HOUR-ANGLE METHOD

Applying the law of sines to the *PZS* triangle shown in Figure 17-1, we obtain

$$\frac{\sin(90° - h)}{\sin LHA} = \frac{\sin(90° - \delta)}{\sin Z}$$

$$\sin(90° - h)\sin Z = \sin(90° - \delta)\sin LHA$$

$$\cos h \sin Z = \cos \delta \sin LHA \quad (17\text{-}2)$$

From the five-part formula for spherical triangles,

$$\sin(90° - h)\cos Z = \sin(90° - \phi)\cos(90° - \delta)$$
$$- \cos(90° - \phi)$$
$$\times \sin(90° - \delta)\cos LHA \quad (17\text{-}3)$$
$$\cos h \cos Z = \cos \phi \sin \delta$$
$$- \sin \phi \cos \delta \cos LHA$$

Dividing Equation (17-2) by (17-3) yields

$$\frac{\cos h \sin Z}{\cos h \cos Z} = \frac{\sin LHA \cos \delta}{\cos \phi \sin \delta - \sin \phi \cos \delta \cos LHA}$$

Then

$$\tan Z = \frac{\sin LHA}{\cos \phi \tan \delta - \sin \phi \cos LHA} \quad (17\text{-}4)$$

Since

$$Z = 360° - AZ \quad \text{and} \quad \tan(360° - AZ) = -\tan AZ$$

then

$$\tan AZ = \frac{-\sin LHA}{\cos \phi \tan \delta - \sin \phi \cos LHA}$$

or $\qquad (17\text{-}5)$

$$AZ = \tan^{-1}\frac{-\sin LHA}{\cos \phi \tan \delta - \sin \phi \cos LHA}$$

Solving Equation (17-5) using the arctangent function on a typical calculator will result in a value between $-90°$ and $+90°$ and must be

Table 17-1.   Correction to normalize azimuth

When LHA Is	Correction	
	If AZ Is Positive	If AZ Is Negative
0–180°	180°	360°
180–360°	0°	180°

normalized to between 0 and 360° by adding algebraically a correction from Table 17-1.

Rather than arctangent function, if the calculator has rectangular to polar conversion ($R \rightarrow P$), AZ can be directly computed without referring to Table 17-1. If we use an HP calculator, e.g., the numerator and denominator are reduced, placing the numerator value in the Y register and denominator number in the X register (display). Executing $R \rightarrow P$ will yield AZ (decimal degrees) in the Y register. Pressing the XY interchange key ($X \langle \rangle Y$) displays the azimuth. For most calculators, azimuths between 180° and 360° are displayed as minus values. Consequently, for a negative result simply add 360°. If the observation has been made in the southern hemisphere, the sign of ϕ should be negative. In this case, the azimuth is still measured clockwise from north.

Rotation of the celestial sphere increases the local hour angle (LHA) of a celestial body by approximately 15° per hour. (The average sun increases by exactly 15° per hour). Therefore, to calculate the LHA at the instant of observation, accurate time is required. In the United States, coordinated universal time (UTC) is broadcast by the National Bureau of Standards radio station WWV (WWVH in Hawaii) on 2.5, 5, 10, 15, and 20 MHz. Inexpensive receivers pretuned to WWV are available. Also, the signal can be received by calling (303) 499-7111. Calling 1-900-410-TIME gives accurate time, but does not provide DUT corrections. In Canada, Eastern standard time (EST) is broadcast on radio station CHU (3.33, 7.335, and 14.67 MHz). This can be converted to UTC by adding 5 hr to EST. GPS receivers are another source of precise time.

Time based on the actual rotation of the earth (UT1) is obtained by adding a correction

(DUT) to coordinated universal time (UT1 = UTC + DUT). DUT is obtained from WWV (WWVH and CHU) by counting the number of double ticks following any minute tone. These double ticks are not obvious and must be carefully listened for. Each double tick represents one-tenth of a second and is positive for the first 7 sec (ticks). Beginning with the ninth second, each double tick is a negative correction. The total correction, either positive or negative, will not exceed 0.7 sec. Although this DUT correction is very small, it is easy to apply and increases azimuth accuracy.

A stopwatch with a split (or lap) time feature is excellent for obtaining times of pointings. The stopwatch is set by starting on a WWV minute tone and then checked 1 min later with a split time. If a significant difference is observed, restart the stopwatch or apply this difference as a correction. Split times are taken for each pointing on the celestial body and added to the beginning UT1 time (beginning UTC corrected for DUT).

A calculator with a time module, such as the HP-41CX or HP48SX, is ideal for obtaining times. The module can be accurately set to the UTC, and the DUT applied using the $T + X$ function. UT1 will now be displayed by the calculator. Time (UT1) of each pointing on the celestial body can be stored and then recalled for subsequent calculations. To ensure that the timepiece has not gained or lost a significant amount of time, it should be rechecked with WWV after the observations. The key to accurate azimuths by the hour-angle method is obtaining accurate time. Surveyors should develop skilled techniques for synchronizing starting time and obtaining split times on the celestial body.

To utilize ephemeris tables, the Greenwich date as well as the time of observation must be known. For observations in the western hemisphere, if UT1 is greater than local time, the Greenwich date is the same as the local date. If UT1 is less than local time, the Greenwich date is the local date plus 1 day. For the eastern hemisphere, if UT1 is less than local time (24-hr basis), Greenwich date is the same as the local date. If UT1 is greater than local time, the Greenwich date is the local date minus 1 day.

Both the observer's latitude and longitude are required for the hour-angle method. Usually, these values are readily obtained by scaling from a map such as a USGS 7.5-min quadrangle sheet. In general, to achieve equivalent azimuth accuracies, latitude and longitude must be more accurately determined for observations on celestial bodies close to the equator—e.g., the sun—than for bodies near the pole—e.g., Polaris.

Declination of celestial bodies—the sun, Polaris, and selected stars—is tabulated in most ephemeris tables for 0 hr UT1 of each day (Greenwich date). Linear interpolation for declination at the UT1 time of observation can be performed using the following equation:

$$Decl = decl\,0^h + (decl\,24^h$$
$$- decl\,0^h)\left(\frac{UT1}{24}\right) \qquad (17\text{-}6)$$

where decl 0^h is the declination at 0 hr for the Greenwich date of observation, and decl 24^h the declination at 0 hr for the next Greenwich day.

Linear interpolation of a nonlinear function results in an error. Except for the sun's declination, this error is insignificant. On sun observations, the contribution to total error in azimuth depends on numerous factors and usually is negligible. It can be eliminated, however, by using three-point interpolation or the special two-point nonlinear interpolation equation as follows:

$$Decl = decl\,0^h$$
$$+ (decl\,24^h - Decl\,0^h)\left(\frac{UT1}{24}\right)$$
$$+ (0.0000395)(decl\,0^h)\sin(7.5\,UT1)$$
$$(17.7)$$

where declination is expressed in decimal degrees.

This equation is unique for the sun's declination. It should not be used for the declina-

tion of Polaris or any other star. A negative declination indicates that the celestial body is south of the equator and must be a negative value in Equations (17-5), (17-6), and (17-7).

The LHA at UT1 time of observation is necessary to compute the azimuth of a celestial body. It is defined as an angle measured westward at the north celestial pole, from the observer's meridian, to the celestial body's hour circle. hence, as can be seen from Figure 17-3, an equation for the LHA is

$$LHA = GHA - W\lambda \quad \text{(west longitude)} \quad (17\text{-}8)$$

or

$$LHA = GHA + E\lambda \quad \text{(east longitude)} \quad (17\text{-}9)$$

LHA should be normalized to between 0° and 360° by adding or subtracting 360°, if necessary.

Greenwich hour angle (GHA) of celestial bodies—the sun, Polaris, and selected stars—is

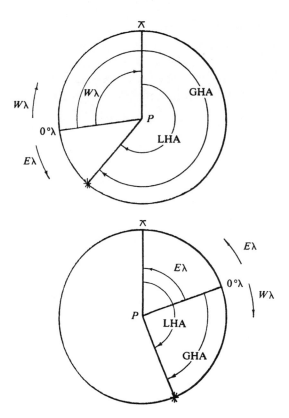

Figure 17-3. Relationship between LHA, GHA, and longitude as viewed from top of celestial sphere.

tabulated in some ephemeris tables for 0 hr UT1 time of each day (Greenwich date). Interpolation for GHA at the time of observation is required. Due to the earth's rotation, the GHA increases by approximately 360° during a 24-hr period, rather than changing by a degree or less as the ephemeris tables appear to indicate. Consequently, interpolation for GHA at UT1 time of observation can be performed using the following equation:

$$HGA = GHA\,0^h + (GHA\,24^h - GHA\,0^h + 360°)$$
$$\times \left(\frac{UT1}{24}\right) \qquad (17\text{-}10)$$

where GHA 0^h = GHA at 0 hr for the Greenwich date of observation, and GHA 24^h = GHA at 0 hr for the next day.

Except for the sun, once each year the tabulated GHA will increase past 360° from one day to the next. For observations on this one day only, 720° rather than 360° should be used in Equation (17-10).

Rather than listing the sun's GHA, some ephemeris tables provide the equation of time E. E is usually defined as apparent time minus mean time and can be converted to GHA using the following equation:

$$HGA = 180° + 15\,E \qquad (17\text{-}11)$$

where E is in decimal hours.

In those cases where E is listed as mean time minus apparent time, the algebraic sign of E should be reversed.

17-5. AZIMUTH OF A CELESTIAL BODY BY THE ALTITUDE METHOD

Applying the law of cosines to the *PZS* triangle in Figure 17-1 results in the following:

$$\cos(90° - \delta) = \cos(90° - h)\cos(90° - \phi)$$
$$+ \sin(90° - h)$$
$$\times \sin(90° - \phi)\cos Z$$
$$\cos Z = \frac{\sin \delta - \sin h \sin \phi}{\cos h \cos \phi}$$

Since $Z = 360° - AZ$ and $\cos(360° - AZ) = \cos AZ$, then

$$\cos AZ = \frac{\sin \delta - \sin h \sin \phi}{\cos h \cos \phi} \qquad \text{or}$$

$$AZ = \cos^{-1} \frac{\sin \delta - \sin h \sin \phi}{\cos h \cos \phi} \qquad (17\text{-}12)$$

Solving Equation (17-12) using the arccosine function on a typical calculator will result in a value between 0 and 180° and must be normalized to between 0 and 360°. A celestial body will be at equal altitudes with approximately equal angles from the meridian twice during a 12-hr period. Consequently, if the celestial body is east of the meridian (morning observation on the sun), AZ found from Equation (17-12) is the correct azimuth. If the body is west of the meridian (afternoon observation on the sun), AZ must be subtracted from 360° to obtain the correct azimuth. Normalizing the azimuth for celestial bodies other than the sun may be difficult or confusing.

Latitude is normally scaled from a topographic map as for the hour-angle method. Due to inaccuracies resulting from errors associated with altitude, declination at the precise time of observation is not so critical as in the hour-angle method. For sun observations, the zone time announced on radio or television corrected to Greenwich is sufficient when interpolating to obtain declination. For other celestial bodies, using declination at 0 hr without interpolating is usually sufficient.

Altitude h is a theoretical angle measured at the earth's center and assumes that light from a celestial body passes straight through the atmosphere (see Figure 17-4). It is normally determined by measuring a vertical angle near the earth's surface and correcting for parallax and refraction. In general, the earth's radius is insignificant compared with distances to celestial bodies, and parallax is considered to be zero. For the sun, however, parallax is significant and computed using the

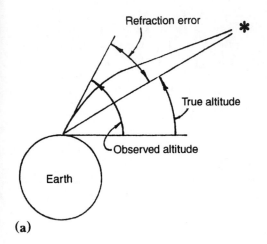

(a)

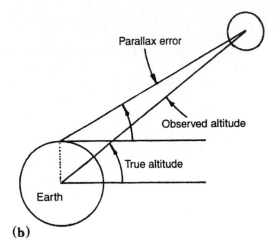

(b)

Figure 17-4. (a) Refraction. (b) Parallax.

following equation:

$$C_p = 0.002443 \cos V \qquad (17\text{-}13)$$

where C_p is in decimal degrees, and V the measured vertical angle.

Refraction is computed using the following equation:

$$C_r = \frac{0.273\,b}{(460 + F)\tan V} \qquad (17\text{-}14)$$

where C_r is in decimal degrees, and

b = absolute barometric pressure (not corrected to sea level) in inches of mercury

F = temperature in degrees Fahrenheit

V = measured vertical angle

If b is unknown, it can be calculated from the following:

$$b = \text{inverse log} \frac{92{,}670 - \text{elevation}}{62{,}737} \qquad (17\text{-}15)$$

where elevation is in feet.

Refraction corrections are large for celestial bodies close to the horizon owing to the amount of atmosphere through which the light must pass. Since pressure and temperature are usually known only at the observation station, errors in refraction may become quite large and have a significant adverse effect on azimuth accuracies.

Applying parallax and refraction corrections to the measured vertical angle results in the following:

$$h = V + C_p - C_r \qquad (17\text{-}16)$$

17-6. SUN OBSERVATIONS (HOUR-ANGLE METHOD)

Sun observations, as compared with those on stars, provide surveyors with a more convenient and economical method of determining an accurate astronomic azimuth. A sun observation can be readily incorporated into a regular work schedule; requires little additional field time; and with reasonable care and proper equipment, an accuracy to within 6 arc-sec can be obtained. In the hour-angle method, a horizontal angle from a line to the sun is measured. Knowing accurate time of the observation and geographic position, we can compute the sun's azimuth. This azimuth and horizontal angle are combined to yield the line's azimuth.

Horizontal angles from a line to the sun are obtained from direct and reverse (face left and face right or face I and face II) pointings taken on the backsight mark and sun. It is suggested that repeating theodolites be used as directional instruments, with one of two general measuring procedures being followed: (1) a single foresight pointing on the sun for each pointing on the backsight mark or (2) multiple foresight pointings on the sun for each pointing on the backsight mark.

For the single foresight procedure, the sighting sequence is (1) direct on mark, (2) direct on sun, (3) reverse on sun, and (4) reverse on the mark—with times being recorded for each sun pointing. The two times and four horizontal circle readings constitute one data set. An observation consists of one or more sets, and a minimum of three sets is recommended. This procedure is similar to measuring an angle at a traverse station using a directional theodolite.

For the multiple foresight procedure, the sighting sequence is (1) direct on mark, (2) several direct on the sun, (3) an equal number of reverse on sun, and (4) reverse on mark—with times being recorded for each pointing on the sun. A minimum of 6 pointings (3D and 3R) on the sun is recommended. The multiple times, multiple horizontal circle readings on the sun, and two horizontal circle readings on the backsight mark constitute one observation.

In general, which measuring procedure to use is a matter of preference. The single foresight procedure is based on an assumption that sun pointings are of approximately equal precision as backsighting pointings. In turn, the multiple foresight procedure is based on an assumption that sun pointings are less precise than backsight pointings. Errors in accurately setting or synchronizing the timepiece have the same effect on both methods. The single foresight system lends itself to proper procedure for incrementing horizontal circle and micrometer settings on the backsight. The multiple foresight method permits a greater number of pointings on the sun in a shorter time span.

Since a large difference usually exists between vertical angles to the backsight mark and sun, it is imperative that an equal number of both direct and reverse pointings be taken. When reducing data to compute horizontal angles, direct readings on the backsight mark should always be subtracted from direct foresight readings on the sun and likewise for reverse readings. Add 360° if the resulting angle is negative.

Vertical angles to the sun are usually larger than those in typical surveying work, thereby increasing the importance of accurately leveling instruments. For a vertical angle of 5°, a leveling error of 10 arc-sec perpendicular to the direction pointed will result in smaller than a 1-arc-sec error in the horizontal circle reading. For a vertical angle of 45°, however, this error would be 10 arc-sec. Because of this and other errors, it is recommended that observations not be made when the sun's altitude is greater than 45°.

The sun cannot be observed directly through the telescope without using either an eyepiece or objective lens filter. In lieu of a filter, the sun's image and cross hair can be projected onto a white surface held approximately 1 ft behind the eyepiece. Both eyepiece and telescope focus must be adjusted to obtain a sharp image. Usually, only that portion of the cross-hair system situated within the sun's image is clearly visible. Observations with a filter are more convenient and slightly improve pointing accuracies. For total stations, an objective lens filter is mandatory to protect EDMI components. For the same reason, a telescope-mounted EDMI should be removed or covered with a lens cap before making observations.

The sun's image is large in diameter—approximately 32 min of arc—making accurate pointings on the center impractical. In lieu of pointing the center, both direct and reverse pointings may be taken on only one edge of the sun—usually the trailing edge (see Figure 17-5). The sun's trailing edge is pointed by allowing it to move onto the vertical cross hair. The leading edge is pointed by moving the vertical cross hair forward, until it becomes tangent to the sun's image. A correction to the sun's center is calculated from semidiameter and altitude. It is applied to the measured horizontal angle from a backsight mark to the

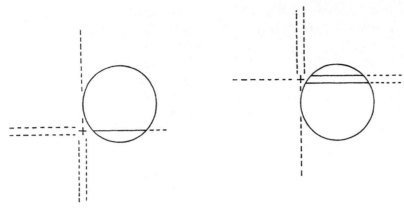

Figure 17-5. Pointing the sun.

sun's edge. This correction *dH* is computed using Equations (17-17) and (17-18). The semidiameter of the sun is tabulated in most ephemeris tables.

$$h = \sin^{-1}(\sin \phi \sin \delta$$
$$+ \cos \phi \cos \delta \cos \text{LHA}) \qquad (17\text{-}17)$$
$$dH = (\text{sun's semidiameter})/\cos h \quad (17\text{-}18)$$

When pointing the left edge (left when facing the sun), *dH* is added to an angle right. When pointing the right edge, subtract. The left edge is always the trailing edge at latitudes greater than 23.5° N and the leading edge at latitudes greater than 23.5° S. An azimuth of the line *AZL* should be computed for each pointings on the sun.

After azimuths of the line have been computed, they are compared and, if found within acceptable limits, averaged. Azimuths computed with telescope direct and telescope reversed should be compared independently. Systematic instrument errors and use of an objective lens filter can cause a significant difference between the two. An equal number of direct and reverse azimuths should be averaged.

An alternate calculation procedure averages times and angles, or points on opposite edges of the sun and averages to eliminate semidiameter corrections. Due to the sun traveling on an apparent curved path and its changing semidiameter correction with altitude, this procedure usually introduces a sig-

nificant error in azimuth. Also, it does not provide a good check on the final azimuth. Averaging times and angles reduces the number of calculations; however, since modern calculators and software are readily available, accuracy should not be sacrificed by shortcutting calculations.

Errors in determining a line's azimuth can be divided into two categories: (1) measuring horizontal angles from a line to the sun and (2) errors in determining the sun's azimuth. Except when pointing on the sun, errors in horizontal angles are similar to any other field angle. The width of a theodolite cross hair is approximately 2 or 3 arc-sec. With practice, the sun's edge, particularly the trailing one, can be pointed to within this width. In many instances, pointing the sun may introduce a smaller error than pointing the backsight mark.

Total error in the sun's azimuth is a function of errors in obtaining UT1 time, and scaling latitude and longitude. The magnitude these errors contribute to total error is a function of the observer's latitude, declination of the sun, and time from local noon. The relationship of these parameters for selected latitudes and declinations is shown in Table 17-2, which illustrates the importance of input data accuracies at different times of day and year. It should be noted that 10 arc-sec of latitude are equivalent to approximately 1000 ft (300 m) on the earth's surface. Ten arc-sec of longitude are equivalent to approximately 880 ft (270 m) at 30° latitude and 500 ft (150 m) at 60° latitude.

As an example, for an early morning observation (time of 4 hr from local noon), during late fall (declination of −23°), at latitude of 30° N, assume that UT1 time is accurate within 0.3 sec, and scaling of latitude and longitude within 5″. From Table 17-2, the error due to time is 3″, those due to latitude and longitude are 1″ and 3″, respectively, and the total error is smaller than 5″ ($\sqrt{3^2 + 1^2 + 3^2}$). For this same example, if the observation was taken close to local noon, time and scaling longitude would be more critical.

Since errors in scaling latitude and longitude are constant for all data sets of an obser-

vation, each computed azimuth of the sun contains a constant error. Errors in time affect azimuth in a similar manner. Consequently, increasing the number of data sets does not appreciably reduce the sun's azimuth-error. The increase can, however, improve horizontal angle accuracy and therefore have a desirable effect.

17-7. EXAMPLE SUN OBSERVATION (HOUR-ANGLE METHOD)

Calculations are shown for the example field notes in Figure 17-6. An azimuth is computed for each pointing; however, in order to elimi-

Table 17-2. Azimuth errors related to time, latitude, longitude, and declination (Sun)

Latitude = 30° N		Declination = +23°		
Data Error		Time from Local Noon		
	0^h	2^h	6^h	6^h
1^s of Time	1′53″	11″	6″	6″
10″ of Latitude	0″	19″	7″	2″
10″ of Longitude	1′15″	7″	4″	4″
Latitude = 30° N		Declination = −23°		
Data Error		Time from Local Noon		
	0^h	2^h	4^h	6^h
1^s of Time	17″	14″	9″	7″
10″ of Latitude	0″	3″	2″	2″
10″ of Longitude	12″	9″	6″	4″
Latitude = 60° N		Declination = +23°		
Data Error		Time from Local Noon		
	0^h	2^h	4^h	6^h
1^s of Time	23″	19″	14″	12″
10″ of Latitude	0″	7″	7″	4″
10″ of Longitude	15″	13″	10″	8″
Latitude = 60° N		Declination = −23°		
Data Error		Time from Local Noon		
		0^h	2^h	4^h
1^s of Time		14″	13″	13″
10″ of Latitude		0″	3″	1″
10″ of Longitude		9″	9″	8″

For southern latitudes, similar errors are obtained by reversing the sign of the declination.

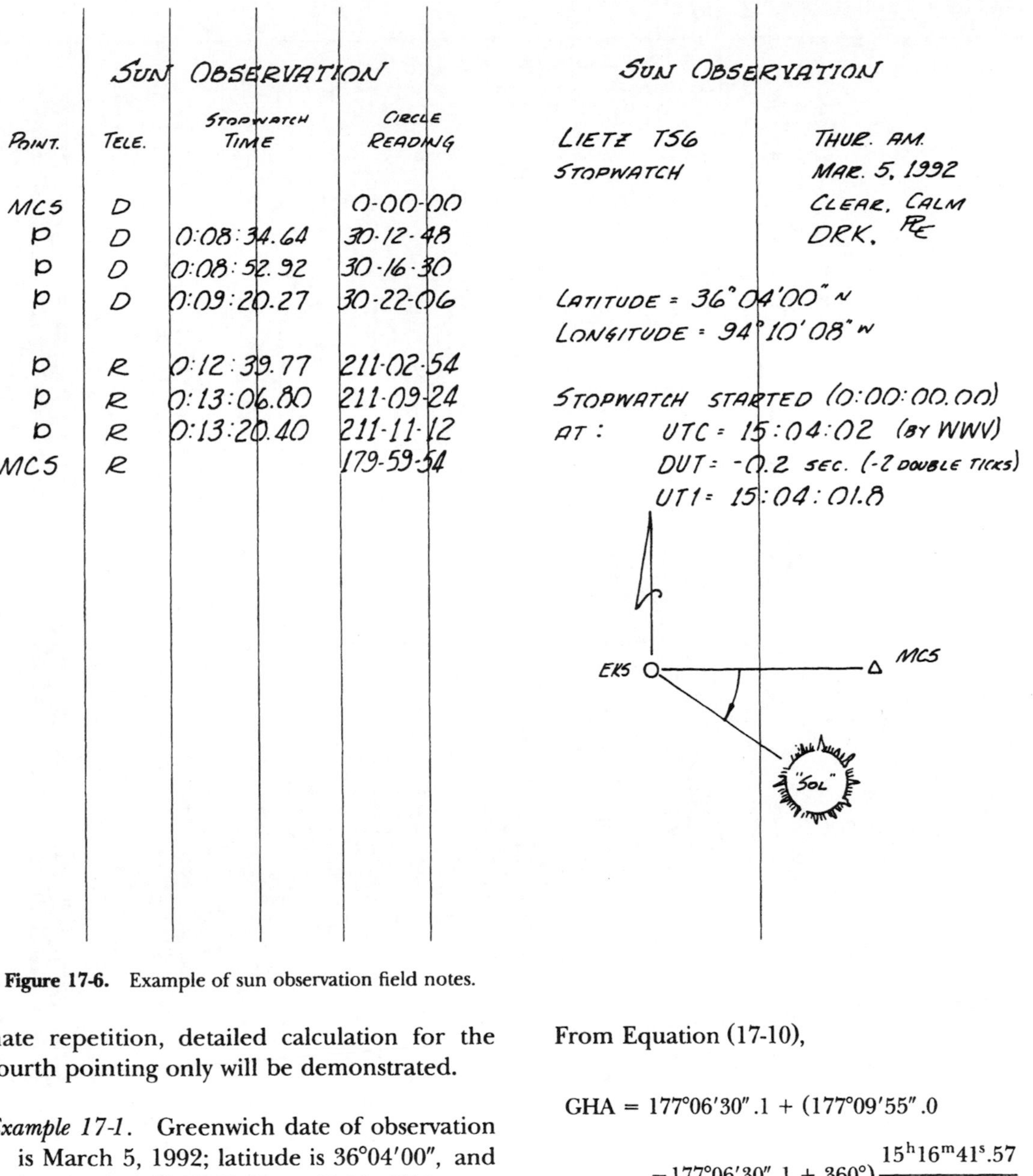

SUN OBSERVATION

POINT.	TELE.	STOPWATCH TIME	CIRCLE READING
MCS	D		0-00-00
p	D	0:08:34.64	30-12-48
p	D	0:08:52.92	30-16-30
p	D	0:09:20.27	30-22-06
p	R	0:12:39.77	211-02-54
p	R	0:13:06.80	211-09-24
p	R	0:13:20.40	211-11-12
MCS	R		179-59-54

SUN OBSERVATION

LIETZ T56
STOPWATCH

THUR. AM.
MAR. 5, 1992
CLEAR. CALM
DRK. $\mathcal{R}_E$

LATITUDE = 36°04'00" N
LONGITUDE = 94°10'08" N

STOPWATCH STARTED (0:00:00.00)
AT : UTC = 15:04:02 (BY WWV)
  DUT = -0.2 SEC. (-2 DOUBLE TICKS)
  UT1 = 15:04:01.8

Figure 17-6. Example of sun observation field notes.

nate repetition, detailed calculation for the fourth pointing only will be demonstrated.

Example 17-1. Greenwich date of observation is March 5, 1992; latitude is 36°04'00", and longitude 94°10'08". From Table 17-3,

$$\text{GHA}\, 0^h = 177°06'30.1$$
$$\text{GHA}\, 24^h = 177°08'06''.3$$
$$\text{Decl}\, 0^h = -6°15'05''.9$$
$$\text{Decl}\, 24^h = -5°38'56''.7$$
$$\text{Semidiameter} = 0°16'09''.0$$

Fourth pointing (telescope reversed),

$$\text{UT1} = 15^h04^m01^s.8 + 0:12:39.77$$
$$= 15^h16^m4^s.57$$

From Equation (17-10),

$$\text{GHA} = 177°06'30''.1 + (177°09'55''.0$$
$$-177°06'30''.1 + 360°)\frac{15^h16^m41^s.57}{24^h}$$
$$= 406.31780°$$
$$= 46.31780°$$

From Equation (17-8),

$$\text{LHA} = 46.28788° - 94°10'08''$$
$$= -47.85109°$$
$$= 312.14891°$$

Table 17-3. Ephemeris tables

MARCH 1992

GREENWICH HOUR ANGLE FOR THE SUN AND POLARIS FOR 0 HOUR UNIVERSAL TIME

DAY	GHA (SUN)	DECLINATION	EQ. OF TIME APPT-MEAN	SEMI-DIAM.	GHA (POLARIS)	DECLINATION	GREENWICH TRANSIT
	o ' "	o ' "	M S	' "	o ' "	o ' "	H M S
1 SU	176 53 57.7	- 7 34 13.4	-12 24.15	16 10.0	123 20 35.8	89 14 08.10	15 44 03.
2 M	176 56 55.2	- 7 11 21.6	-12 12.32	16 09.7	124 20 06.5	89 14 07.94	15 40 05.
3 TU	176 59 59.9	- 6 48 23.6	-12 00.00	16 09.5	125 19 37.4	89 14 07.76	15 36 08.
4 W	177 03 11.7	- 6 25 19.8	-11 47.22	16 09.2	126 19 08.1	89 14 07.57	15 32 10.
5 TH	177 06 30.1	- 6 02 10.7	-11 33.99	16 09.0	127 18 38.0	89 14 07.35	15 28 13.
6 F	177 09 55.0	- 5 38 56.7	-11 20.33	16 08.7	128 18 06.7	89 14 07.11	15 24 16.
7 SA	177 13 26.2	- 5 15 38.1	-11 06.25	16 08.5	129 17 33.9	89 14 06.86	15 20 19.
8 SU	177 17 03.5	- 4 52 15.4	-10 51.77	16 08.2	130 16 59.2	89 14 06.61	15 16 22.
9 M	177 20 46.4	- 4 28 49.0	-10 36.90	16 08.0	131 16 22.9	89 14 06.36	15 12 25.
10 TU	177 24 34.9	- 4 05 19.3	-10 21.67	16 07.7	132 15 45.2	89 14 06.12	15 08 28.
11 W	177 28 28.7	- 3 41 46.6	-10 06.09	16 07.5	133 15 06.8	89 14 05.90	15 04 31.
12 TH	177 32 27.4	- 3 18 11.4	-09 50.17	16 07.2	134 14 28.4	89 14 05.69	15 00 34.
13 F	177 36 30.8	- 2 54 34.0	-09 33.95	16 07.0	135 13 50.8	89 14 05.50	14 56 37.
14 SA	177 40 38.6	- 2 30 54.8	-09 17.43	16 06.7	136 13 14.4	89 14 05.32	14 52 40.
15 SU	177 44 50.4	- 2 07 14.2	-09 00.64	16 06.5	137 12 39.6	89 14 05.13	14 48 43.
16 M	177 49 06.0	- 1 43 32.6	-08 43.60	16 06.2	138 12 05.8	89 14 04.92	14 44 46.
17 TU	177 53 25.0	- 1 19 50.3	-08 26.33	16 05.9	139 11 32.2	89 14 04.69	14 40 49.
18 W	177 57 47.0	- 0 56 07.6	-08 08.86	16 05.7	140 10 57.9	89 14 04.42	14 36 52.
19 TH	178 02 11.7	- 0 32 24.8	-07 51.22	16 05.4	141 10 21.8	89 14 04.13	14 32 55.
20 F	178 06 38.6	- 0 08 42.3	-07 33.43	16 05.1	142 09 43.7	89 14 03.83	14 28 58.
21 SA	178 11 07.4	0 14 59.6	-07 15.51	16 04.8	143 09 03.4	89 14 03.52	14 25 02.
22 SU	178 15 37.7	0 38 40.6	-06 57.48	16 04.6	144 08 21.7	89 14 03.22	14 21 05.
23 M	178 20 09.2	1 02 20.4	-06 39.38	16 04.3	145 07 39.1	89 14 02.94	14 17 09.
24 TU	178 24 41.6	1 25 58.4	-06 21.23	16 04.0	146 06 56.4	89 14 02.68	14 13 12.
25 W	178 29 14.4	1 49 34.6	-06 03.04	16 03.7	147 06 14.1	89 14 02.42	14 09 16.
26 TH	178 33 47.5	2 13 08.3	-05 44.84	16 03.5	148 05 32.6	89 14 02.18	14 05 19.
27 F	178 38 20.4	2 36 39.4	-05 26.64	16 03.2	149 04 51.9	89 14 01.93	14 01 22.
28 SA	178 42 52.9	3 00 07.3	-05 08.47	16 02.9	150 04 12.0	89 14 01.68	13 57 26.
29 SU	178 47 24.7	3 23 31.8	-04 50.35	16 02.6	151 03 32.6	89 14 01.41	13 53 29.
30 M	178 51 55.6	3 46 52.5	-04 32.29	16 02.3	152 02 53.5	89 14 01.14	13 49 32.
31 TU	178 56 25.2	4 10 09.0	-04 14.32	16 02.0	153 02 14.1	89 14 00.84	13 45 35.

APRIL 1992

GREENWICH HOUR ANGLE FOR THE SUN AND POLARIS FOR 0 HOUR UNIVERSAL TIME

DAY	GHA (SUN)	DECLINATION	EQ. OF TIME APPT-MEAN	SEMI-DIAM.	GHA (POLARIS)	DECLINATION	GREENWICH TRANSIT
	o ' "	o ' "	M S	' "	o ' "	o ' "	H M S
1 W	179 00 53.3	4 33 20.9	-03 56.44	16 01.8	154 01 33.9	89 14 00.53	13 41 39.
2 TH	179 05 19.7	4 56 27.9	-03 38.69	16 01.5	155 00 52.6	89 14 00.19	13 37 42.
3 F	179 09 44.1	5 19 29.7	-03 21.06	16 01.2	156 00 09.5	89 13 59.85	13 33 46.
4 SA	179 14 06.2	5 42 25.8	-03 03.59	16 00.9	156 59 24.6	89 13 59.51	13 29 49.
5 SU	179 18 25.9	6 05 15.8	-02 46.27	16 00.7	157 58 37.7	89 13 59.16	13 25 53.
6 M	179 22 42.9	6 27 59.6	-02 29.14	16 00.4	158 57 49.3	89 13 58.83	13 21 57.
7 TU	179 26 57.1	6 50 36.6	-02 12.20	16 00.1	159 57 00.0	89 13 58.52	13 18 01.
8 W	179 31 08.1	7 13 06.5	-01 55.46	15 59.8	160 56 10.4	89 13 58.23	13 14 05.
9 TH	179 35 15.8	7 35 28.9	-01 38.95	15 59.6	161 55 21.3	89 13 57.96	13 10 09.
10 F	179 39 19.9	7 57 43.6	-01 22.67	15 59.3	162 54 33.3	89 13 57.70	13 06 13.
11 SA	179 43 20.3	8 19 50.1	-01 06.65	15 59.0	163 53 46.6	89 13 57.44	13 02 16.
12 SU	179 47 16.7	8 41 48.2	-00 50.89	15 58.8	164 53 00.9	89 13 57.17	12 58 20.
13 M	179 51 08.7	9 03 37.4	-00 35.42	15 58.5	165 52 15.6	89 13 56.87	12 54 24.
14 TU	179 54 56.2	9 25 17.5	-00 20.25	15 58.2	166 51 29.7	89 13 56.56	12 50 27.
15 W	179 58 38.8	9 46 48.1	-00 05.41	15 58.0	167 50 42.5	89 13 56.21	12 46 31.
16 TH	180 02 16.4	10 08 09.0	00 09.09	15 57.7	168 49 53.2	89 13 55.86	12 42 35.
17 F	180 05 48.4	10 29 19.9	00 23.23	15 57.4	169 49 01.7	89 13 55.50	12 38 39.
18 SA	180 09 14.8	10 50 20.4	00 36.99	15 57.2	170 48 08.5	89 13 55.15	12 34 43.
19 SU	180 12 35.2	11 11 10.2	00 50.35	15 56.9	171 47 13.9	89 13 54.81	12 30 48.
20 M	180 15 49.3	11 31 49.1	01 03.29	15 56.7	172 46 19.0	89 13 54.50	12 26 52.
21 TU	180 18 56.9	11 52 16.7	01 15.80	15 56.4	173 45 24.2	89 13 54.20	12 22 56.
22 W	180 21 57.8	12 12 32.7	01 27.85	15 56.1	174 44 30.0	89 13 53.92	12 19 01.
23 TH	180 24 51.7	12 32 36.7	01 39.45	15 55.9	175 43 36.6	89 13 53.64	12 15 05.
24 F	180 27 38.5	12 52 28.4	01 50.57	15 55.6	176 42 44.1	89 13 53.36	12 11 09.
25 SA	180 30 17.9	13 12 07.5	02 01.20	15 55.3	177 41 52.3	89 13 53.07	12 07 13.
26 SU	180 32 49.9	13 31 33.6	02 11.32	15 55.1	178 41 00.7	89 13 52.78	12 03 17.
27 M	180 35 14.1	13 50 46.4	02 20.94	15 54.8	179 40 09.0	89 13 52.47	11 59 21.
28 TU	180 37 30.6	14 09 45.6	02 30.04	15 54.6	180 39 16.7	89 13 52.14	11 55 25.
29 W	180 39 39.1	14 28 30.9	02 38.61	15 54.3	181 38 23.3	89 13 51.80	11 51 30.
30 TH	180 41 39.6	14 47 01.8	02 46.64	15 54.1	182 37 28.3	89 13 51.45	11 47 34.

SUN, POLARIS

TABLE A

SOKKIA

From Equation (17-7),

$$\text{Decl} = -6°02'10''.7$$
$$+ (-5°38'56''.7 + 6°02'10''.7)$$
$$\times \frac{15^{h}16^{m}41^{s}.57}{24^{h}}$$
$$+ (0.0000395)(-6°02'10''.7)$$
$$\times \sin(7.5 \times 15^{h}16^{m}41^{s}.57)$$
$$= -5.78980° - 0.00022°$$
$$= -5.79002°$$

From Equation (17-5),

$$AZ = \tan^{-1}\frac{-\sin(312.14891°)}{\begin{aligned}&\cos(36°04'00'') \tan(-5.79002°)\\ &-\sin(36°04'00'') \cos(312.14891°)\end{aligned}}$$

$$= \tan^{-1}\frac{0.74140323}{-0.47703565}$$

Using $R \to P$, we obtain $AZ = 122.75815°$, or using $\tan^{-1}$, $AZ = -57.24185°$.

Since LHA is between 180° and 360° and *AZ* negative, the normalize correction from Table 17-1 equals 180°.

$$AZ = -57.24185° + 180°$$
$$= 122.75815°$$
$$\text{Field R Ang Rt} = 211°02'54'' - 179°59'54°$$
$$= 31.05000°$$

From Equation (17-17),

$$h = \sin^{-1}[\sin(36°04'00'')\ln(-5.79002°)$$
$$+ \cos(36°04'00'')\cos(-5.79002°)$$
$$\times \cos(312.14891°)]$$
$$= 28.7037°$$

From Equation (17-18),

$$dH = \frac{0°16'09''.0}{\cos(28.7037°)}$$
$$= 0.30688°$$

The left edge is pointed D & R; therefore, *dH* is positive.

$$\text{Cor R Ang Rt} = 31.05000° + 0.30688°$$
$$= 31.35688°$$

From Equation (17-1),

$$AZL = 122.75815° - 31.35688°$$
$$= 91.40127°$$
$$= 91°24'04''.6$$

Using the same calculation procedure for remaining pointings yields the following:

Direct		Reverse	
91°24'03''.6	(1)	91°24'04''.6	(4)
91°24'04''.8	(2)	91°23'09''.6	(5)
91°24'03''.1	(3)	91°24'10''.3	(6)

A comparison of direct and reverse azimuths indicates that the fifth value contains excessive error. Throw out

this azimuth and a direct azimuth (third) before averaging. (Average = 91°24'06'')

17-8. SUN OBSERVATIONS (ALTITUDE METHOD)

Outdated and less accurate (approximately 1 min of arc, depending on numerous factors), the altitude method should only be considered when accurate time cannot be determined. Simultaneous vertical and horizontal angles to the sun's center are required. Therefore, a special sighting accessory must be used or both edges of the sun pointed simultaneously (quadrant method, see Figure 17-7). Due to the suns's large diameter, both edges cannot be accurately observed simultaneously using a filter. Since total stations require an objective lens filter, this essentially eliminates their use for this method.

The altitude method requires very accurate vertical angles that must be corrected for parallax and refraction. This is particularly critical when the sun is close to local noon because of rapid changes in azimuth, with little or no change in altitude. Therefore, observations should not be made within 2 to 3 hr of local noon. Refraction corrections are large and more difficult to accurately determine when the sun is close to the horizon. This restricts observations during the first hour or two after sunrise and before sunset.

Due to problems involved in obtaining horizontal and vertical pointings at the same instant, and the importance of vertical angle accuracy, a set of data should consist of several foresights on the sun for each backsight. A recommended sighting procedure is (1) direct on backsight mark, (2) three direct on sun, (3) three reverse on sun, and (4) reverse on backsight mark. The three direct and three reverse angles (horizontal and vertical) are averaged, with a single azimuth computed for each set. In order to minimize errors due to curvature of the sun's apparent path, time spans from first direct to last reverse pointings should be kept as short as possible.

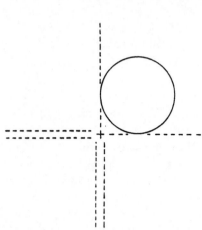

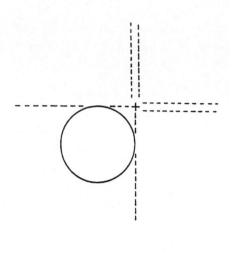

Figure 17-7.   Quadrant method.

17-9.   EXAMPLE SUN CALCULATION (ALTITUDE METHOD)

Example 17-2.

Local date = April 10, 1992
Avg. ang rt to sun's center = 326°47′30″
Avg. zenith angle to sun's center = 65°20′42″
Avg. time = 4:40 PM CST
Elevation = 1400 ft
Temperature = 70°F
Latitude = 36°04′00″
Greenwich date = April 10, 1986
From Table 17-3,

$$\text{Decl } 0^h = 7°57′43″.6$$
$$\text{Decl } 24^h = 8°19′50″.1$$
$$\text{UT} = 4^h40^m + 12^h + 6^h$$
$$= 22^h40^m$$

From Equation (17-6),

$$\text{Decl} = 7°57′43″.6$$
$$+ (8°19′50″.1 - 7°57′43″.6)\frac{22^h40^m}{24^h}$$
$$= 8.3101°$$
$$\text{Vert ang} = 90° - 65°20′42″$$
$$= 24.6550°$$

From Equation (17-13),

$$C_p = 0.002443 \cos(24.6550°)$$
$$= 0.0022°$$

From Equation (17-15),

$$b = \text{inverse log} \frac{92{,}670 - 1400}{62{,}737}$$
$$= 28.50 \text{ in. of Hg}$$

From Equation (17-14),

$$C_r = \frac{0.273(28.50)}{(460 + 70)\tan(24.6550°)}$$
$$= 0.0320°$$

From Equation (17-16),

$$h = 24.6550° + 0.0022° - 0.0320°$$
$$= 24.6252°$$

From Equation (17-12),

$$AZ = \cos^{-1}\frac{\sin(8.3101°)}{\cos(24.6252°)\cos(36°04′00″)}$$
$$= 97.8830°$$

For an afternoon observation (sun west of meridian),

$$AZ = 360° - 97.8830°$$
$$= 262.1170°$$

From Equation (17-1),

$$AZL = 262.1170° - 326°47'30''$$
$$= -64.6746°$$
$$= 295.3254°$$
$$= 295°20'$$

17-10. POLARIS OBSERVATIONS (HOUR-ANGLE METHOD)

In most land surveying situations, determination of the astronomic azimuth by sun observations is satisfactory. However, for direction accuracy requirements of approximately 6 arcsec or fewer, a star observation is required. At middle latitudes of the northern hemisphere, Polaris is preferred. It moves very slowly as seen by an observer on earth and is easily located. At near-pole and near-equator latitudes, a star other than Polaris should be selected. If close to the equator, Polaris may not be visible, and horizontal refraction can be a problem. When near the pole, time and leveling become very critical in azimuth determinations.

Several observation methods and calculation procedures can be applied to determine astronomic azimuth from Polaris. For all practical purposes, however, the hour-angle method is the only one that should be considered. Figure 17-8 depicts the apparent motion of Polaris to an observer on earth. The relationship between the north celestial pole, azimuth of Polaris, and horizontal angle from line *AB* to Polaris is shown.

Four important positions of Polaris during its daily rotation around the pole are (1) upper culmination (UC), (2) western elongation (WE), (3) lower culmination (LC), and (4) eastern elongation (EE). If an observer sights Po-

laris exactly at upper or lower culmination, its azimuth is zero, which of course simplifies computing the line's azimuth. This method of observation is not practical in that culmination occurs for only an instant, and movement is most rapid in either an east or west direction. Consequently, accurate direct and reverse pointings on Polaris are not possible. Likewise, observation procedures and computations can be simplified by making observations at either eastern or western elongation. These also have distinct disadvantages, since elongation normally occurs only once each day during hours of darkness and possibly at an inconvenient time of night. Instead of observing Polaris at culmination or elongation, surveyors should be prepared to make observations at any time (hour angle) and perform the necessary calculations.

Figure 17-9 can assist in locating Polaris. It is the end star in the handle of constellation Ursa Minor (Little Dipper) located between the constellation Ursa Major (Big Dipper) and Cassiopeia. Two stars of the Big Dipper's cup point to Polaris. Finding Polaris in the instrument's field of view for the first time can be exasperating, but it need not be if approached in a systematic manner. The instrument can be prefocused on some distant object, such as a bright star or the moon. Horizontally, Polaris can be located with the telescope sight. Approximate vertical angles to Polaris can be computed by estimating expected UT1 times of observation and applying the following equation:

$$h = \phi + (90° - \delta)$$
$$\times \cos[\text{GHA } 0^h - W\lambda + (\text{UT1})(15°02')]$$
$$(17\text{-}19)$$

As an example, assume a Polaris observation is to be taken at 7:15 PM CST, December 7, 1992 (local date), at longitude 91°46' W and latitude 37°57' N. UT1 time and Greenwich date of observation would be $1^h 15^m$ December

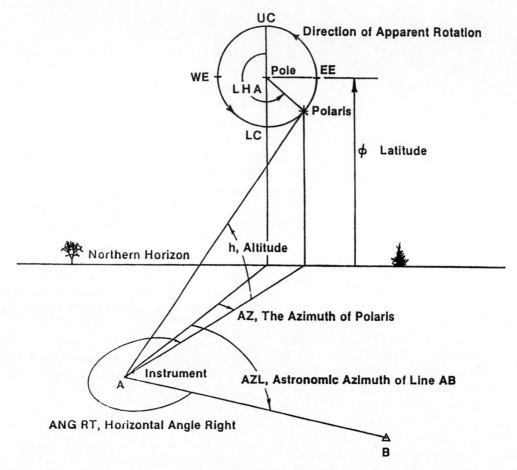

Figure 17-8. Polaris movement and relationship of horizontal angle, azimuth, and local hour angle.

8. GHA 0^h on this date is $40°31'$ and declination $89°15'$.

$$h = 37°57' + (90° - 89°15')$$
$$\times \cos[40°31' - 91°46' + (1.25^h)(15°02')]$$
$$= 38°35'$$

Converting this approximate vertical angle to zenith angle yields $51°25'$. Since the altitude of Polaris is approximately equal to the observer's latitude, precise leveling of the theodolite is extremely critical for mid and high northern latitudes.

Polaris observations made at night present problems not encountered in survey work performed during daylight hours. In particular, illumination is necessary to see the cross hairs against the night sky. Theodolites usually have lighting systems as accessories. If a lighting system is not available, point a light into the hole located at the theodolite's reflecting mirror to illuminate both horizontal circle and cross hairs, or use a flashlight to illuminate the cross hairs by reflecting light at an angle into the objective lens.

Horizontal angles from a line to Polaris are obtained from direct and reverse pointings taken on the backsight mark and star. The single foresight for each backsight procedure (as discussed under sun observations) is recommended with at least three sets being taken. For high-order surveys, horizontal circle and micrometer settings should be incremented between sets.

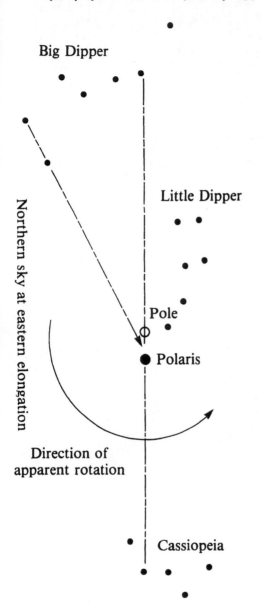

Northern sky at upper culmination

Big Dipper

Little Dipper

Northern sky at eastern elongation

Northern sky at western elongation

Pole

Polaris

Direction of
apparent rotation

Cassiopeia

Northern sky at lower culmination

Figure 17-9. Chart to locate Polaris.

for hand computations, averaging angles and times for each set is normally sufficient. Azimuths for each set should be compared and, if found within acceptable limits, averaged. A slightly more accurate procedure is computing an azimuth for each pointing as recommended for sun observations.

Excluding instrument and personal errors, accuracy of an azimuth determination is affected by errors in obtaining UT1 time of pointing Polaris and scaling latitude and longitude. Determining specifications for time, latitude, and longitude in order to meet a prescribed azimuth accuracy is somewhat complicated since they vary with the observer's latitude and LHA at time of pointing. Table 17-4 lists maximum azimuth errors resulting from inaccuracies in time, latitude, and longitude at several latitudes. In general, for any observations made on Polaris in the continental United States, if pointing times are obtained to 1 sec and latitude and longitude are scaled to 1000 ft (300 m), the resulting azimuth accuracy is within 0.5 arc-sec (if we disregard errors in the horizontal angle from mark to star).

17-11. EXAMPLE POLARIS CALCULATION

Calculations are detailed for the example field notes in Figure 17-10. These notes illustrate standard procedures to increment horizontal circle and micrometer settings for a directional theodolite. An azimuth is computed for

If the time span between direct and reverse pointings on Polaris is kept below 4 min, an azimuth computed from averaged horizontal angles and UT1 times would contain a maximum error of only 0.2 arc-sec at 60° latitude and smaller at lower latitudes. Consequently,

Table 17-4. Maximum azimuth error related to time, latitude, and longitude (polaris)

| Data | Latitude | | |
Error	20°	40°	60°
1ˢ of Time	0″.23	0″.28	0″.42
10″ of Latitude	0″.05	0″.15	0″.49
10″ of Longitude	0″.15	0″.18	0″.28

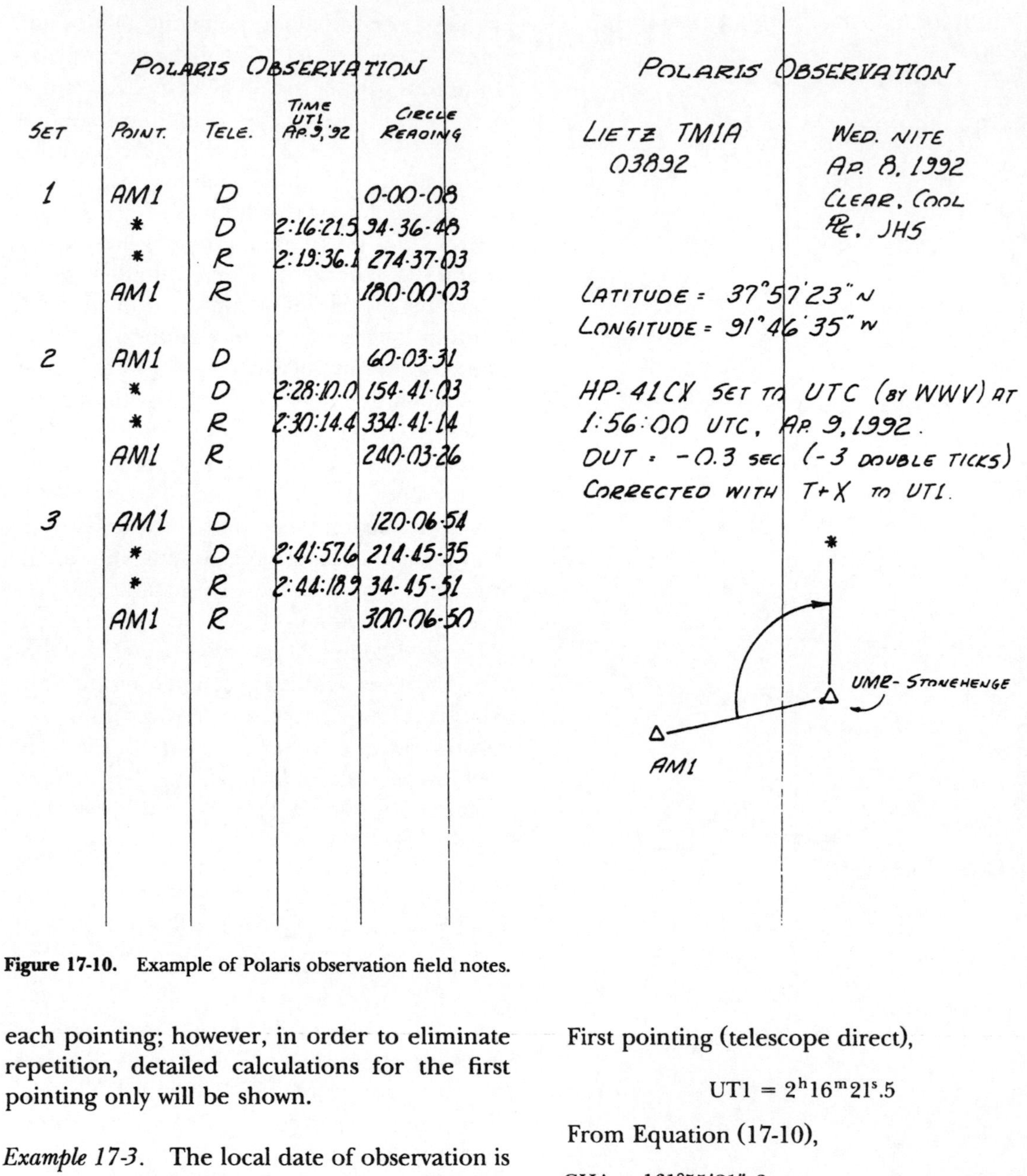

POLARIS OBSERVATION

SET	POINT.	TELE.	TIME UT1 Ap.9,'92	CIRCLE READING
1	AM1	D		0-00-08
	*	D	2:16:21.5	94-36-48
	*	R	2:19:36.1	274-37-03
	AM1	R		180-00-03
2	AM1	D		60-03-31
	*	D	2:28:10.0	154-41-03
	*	R	2:30:14.4	334-41-14
	AM1	R		240-03-26
3	AM1	D		120-06-54
	*	D	2:41:57.6	214-45-35
	*	R	2:44:18.9	34-45-51
	AM1	R		300-06-50

POLARIS OBSERVATION

LIETZ TM1A
03892

WED. NITE
Ap. 8, 1992
CLEAR, COOL
P.E. JHS

LATITUDE = 37°57'23" N
LONGITUDE = 91°46'35" W

HP. 41CX SET TO UTC (BY WWV) AT
1:56:00 UTC, AP. 9, 1992.
DUT = -0.3 SEC (-3 DOUBLE TICKS)
CORRECTED WITH T+X TO UT1.

UM2- STONEHENGE

AM1

Figure 17-10.   Example of Polaris observation field notes.

each pointing; however, in order to eliminate repetition, detailed calculations for the first pointing only will be shown.

Example 17-3.   The local date of observation is April 8, 1992. Greenwich date of observation is April 9, 1992.
From Table 17-3,

$$\text{GHA}\,0^h = 161°55'21''.3$$

$$\text{GHA}\,24^h = 162°54'33''.3$$

$$\text{Decl}\,0^h = 89°13'57''.96$$

$$\text{Decl}\,24^h = 89°13'57''.70$$

First pointing (telescope direct),

$$\text{UT1} = 2^h16^m21^s.5$$

From Equation (17-10),

$$\text{GHA} = 161°55'21''.3$$
$$+ (162°54'33''.3 - 161°55'21''.3 + 360°)$$
$$\times \frac{2^h16^m21^s.5}{24^h}$$
$$= 196.10560°$$

From Equation (17-8),

$$\text{LHA} = 196.10560° - 91°46'35''$$
$$= 104.32921°$$

From Equation (17-6),

$$Decl = 89°13'57''.96$$
$$+ (89°13'57''.70 - 89°13'57''.96)$$
$$\times \frac{2^h16^m21^s.5}{24^h}$$
$$= 89.232760°$$

From Equation (17-5),

$$AZ = \tan^{-1} \frac{-\sin(104.32921°)}{\cos(37°57'23'') \tan(89.232760°) - \sin(37°57'23'') \cos(104.32921°)}$$

$$= \tan^{-1} \frac{-0.96888969}{59.03056203}$$

Using $R \to P$, we obtain

$$AZ = -0.94033°$$
$$= 359.05967°$$

or using $\tan^{-1}$,

$$AZ = -0.94033°.$$

Since LHA is between 0 and 180° and AZ is negative, the normalize correction from Table 17-1 is 360°.

$$AZ = -0.94033° + 360°$$
$$= 359.05967°$$
$$Field\ D\ ang\ rt = 94°36'48'' - 0°00'08''$$
$$= 94.61111°$$

From Equation (17-1),

$$AZL = 359.05967° - 94.61111°$$
$$= 264.44856°$$
$$= 264°26'54''8$$

Using the same calculation procedure for remaining pointings yields the following:

Set	Direct	Reverse
1	264°26'54''.8	264°26'47''.9
2	264°26'53''.7	264°26'47''.5
3	264°26'55''.3	264°26'48''.5

Avg. $AZL = 264°26'51''.3$.

17-12. STARS OTHER THAN POLARIS

At times, neither Polaris nor the sun is suitable for azimuth observations. Polaris obviously cannot be observed south of the equator and may not produce accurate results at high northern latitudes. In some cases, sun observations may not produce sufficient accuracy for the work desired, or a cloudy day can require a night observation if a time schedule is to be met. Also, Polaris may be cloud covered while the southern sky is clear. There is a fifth magnitude star, Sigma Octantis, within one degree of the south celestial pole that can be observed in the southern hemisphere using the same procedures as Polaris. Of course, both latitude and declination are negative, and the azimuth will, as usual, be measured clockwise from north. Since the star is nearly invisible to the naked eye, accurate star maps or precomputation are necessary for location.

In general, pointings to stars can be made more accurately than those on the sun since the image is precisely defined, atmospheric turbulence is lower, and thermal expansions of the theodolite and tripod are eliminated. In addition, bright stars are visible shortly after sundown, thus permitting observations to be made in twilight.

If we assume that GHA and declination ephemeris data are available for other stars, calculations are identical to those for Polaris. However, their apparent motion is rapid, and for stars close to the equator, the movement is similar to the sun's. Consequently, observations should be avoided when the vertical angle is above 45°—i.e., their directions change rapidly with time, particularly if near the observer's meridian.

Some stars are not visible during the entire night, and their locations change throughout the year. Before attempting any observations, it is necessary to become familiar with a star's position. This involves determining when it will be above the horizon and identifying its location. Many publications, such as Sokkia's Celestial Observation Handbook and Ephemerist, provide visibility charts to aid in this procedure.

17-13. SUMMARY

Astronomic azimuth provides an accurate and efficient means of orienting field surveys. Textbooks and surveying courses have covered celestial observations for many years, but until recently this aspect of surveying measurements has been essentially academic and not employed in surveying practice. This is due in part to emphasis on the altitude method for sun observations and the inherent associated problems. Surveyors are encouraged to take advantage of available technology and use the hour-angle method for all azimuth determinations, whether the sun, Polaris, or selected stars are observed.

Ample software is available to perform all necessary calculations for the hour-angle method. Included are modules for hand-held calculators that not only serve as timepieces, data collectors, and computers, but also generate ephemeris data. As professionals, however, surveyors must have a basic understanding of the underlying theory of celestial

observations and be able to test and verify accuracies of the software used.

In the near future, it is hoped that *direction*, the fourth component of surveying measurements, will no longer be a stepchild to the other three. As a result, all field surveys will be accurately oriented to a retraceable reference.

REFERENCES

BUCKNER, R. B. 1984. *A Manual on Astronomic and Grid Azimuth*. Rancho Cordora: Landmark Enterprises.

ELGIN, R. L., D. R. KNOWLES, and J. H. SENNE. *Celestial Observation Handbook and Ephemeris.* 1985–1993 Overland Park: Sokkia/Lietz.

ELGIN, R. L., D. R. KNOWLES, and J. H. SENNE. 1989. *Practical Surveying for Observations*. Canton: P.O.B. Publishing.

ELGIN, R. L., D. R. KNOWLES, and J. H. SENNE. 1986–1989. The tech on celestial observations. *P.O.B. Magazine* 12 1 (Oct.–Nov. 1986 through June–July 1990).

MACKIE, J. B. 1978. *The Elements of Astronomy for Surveyors*, 8th ed. London: Charles Griffin.

MUELLER, I. I. 1969. *Spherical and Practical Astronomy as Applied to Geodesy*. New York: Frederick Ungar Publishing Company.

NASSAU, J. J. 1948. *Practical Astronomy*. New York: McGraw-Hill.

18

Map Projections

Porter W. McDonnell

18-1. INTRODUCTION

The earth is round; maps are flat. If a particular map is to show only a very small portion of the earth, such as a few city blocks, the roundness of the earth is insignificant. On the other hand, if a map is to show the western hemisphere, the roundness presents a major problem—i.e., some kind of deformation will be necessary. To illustrate, a large section of orange peel can only be flattened if it is stretched and torn.

A map depicting only a small area is often called a *plan*. A map showing a large portion (or all) of the earth, where curvature of the surface becomes a factor, is called a *map* or *chart*, the latter term being used for a map designed for navigational purposes. Preparation of a plan involves a simple rectangular grid, whereas a map or chart commonly requires the selection of a suitable map projection to deal with the earth's shape.

Although the roundness of the earth is not a factor in drawing a plan—a map of limited area—the topic of map projections is nevertheless important to land surveyors. Increasingly, land surveyors are making use of plane coordinate systems (Chapter 19) that extend over hundreds of kilometers (or miles) even though the job at hand is small. When a plan shows a limited land area and is drawn as if the earth were flat, the data shown may be so precise that a knowledge of map projections is needed in the survey computations.

For the purposes of this chapter, and for small-scale mapping generally, the earth can be considered a sphere with a radius of 6370 km (or about 3960 mi). Actually, the dimension is greater across the equator than from top to bottom (pole to pole).

For large-scale mapping and in geodetic surveying, the earth's true shape has to be considered (see Chapter 19). In these cases, the earth is assumed to be a spheroid instead of a sphere. Figure 18-1 is a flowchart in which the choice of a datum (sphere or spheroid) is shown as the first step in evolving a map projection.

This chapter is based on the author's textbook, *Introduction to Map Projections* (New York: Marcel Dekker, 1979), with permission from the publisher. (Second edition available from Landmark Enterprises, Rancho Cordova, CA, © 1992.)

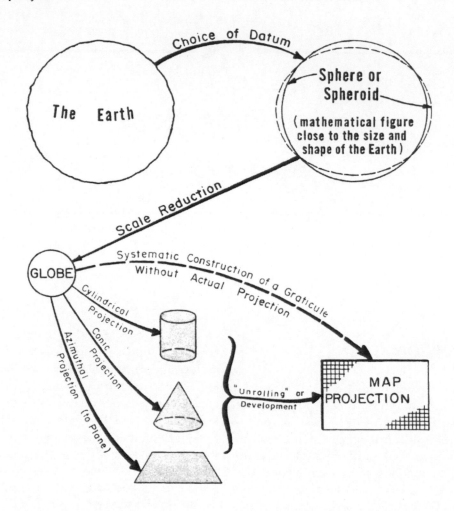

Figure 18-1. Evolution of a map projection. In some cases, a geometric projection to a developable surface is involved, but usually the term *cylindrical, conic,* or *azimuthal* is used to classify a projection that only resembles such a case. The dashed arrow shows this possibility. (From P. W. McDonnell, Jr., 1979, New York: Marcel Dekker, with permission.)

18-2.   PROJECTION

If all points within some large portion of the earth—e.g., the western hemisphere—are to be represented on a flat map, *two* transformations of the sphere or spheroid are necessary. First, there must be a scale reduction to make sure a huge area fit into the limits of a sheet. Second, there needs to be a systematic way of deforming the rounded surface of the sphere or spheroid to make it flat (see Figure 18-1).

It is very useful to think of these operations as always being done in two steps, in the order mentioned. First, the full-sized sphere is greatly reduced to an exact model called a *globe*. Second, a map projection is generated to convert all or part of the globe into a flat map. There are an infinite number of ways, literally, of accomplishing this second step.

Certain reference lines and points have been established on the earth. The equator and two poles are known to all (see Figure 18-2). Lines running north and south, from pole to pole, are *meridians*. One of them, passing through Greenwich, England, has been chosen arbitrarily to be the *prime meridian*. It is assigned an angular value of 0°. Each other meridian is identified by its angular distance

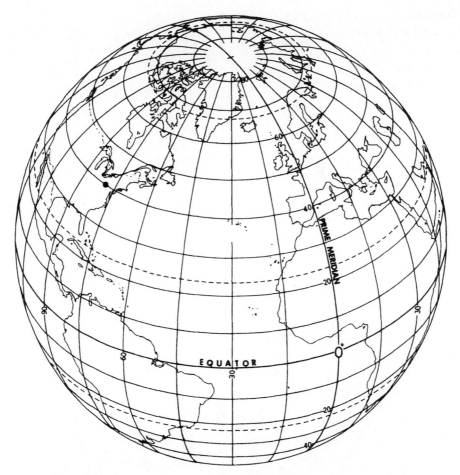

Figure 18-2. Network of meridians and parallels. Pittsburgh is located at $\phi = 40°$ N and $\lambda = 80°$ W.

east or west of the prime meridian. The meridian through Pittsburgh, e.g., is 80° west of Greenwich and said to have a *longitude* of 80° W. (The line itself is a meridian, and its spherical coordinate the longitude). The Greek letter lambda (λ) is used for longitude. Lines crossing all meridians at right angles and running parallel to the equator are called *parallels*. Each parallel is identified by its angular distance north or south of the equator, known as its *latitude*. The parallel passing through Madrid, Pittsburgh, and Peking is at 40° N and is often called the 40th parallel if there is no chance of confusing it with the parallel at 40° S. The Greek letter phi (ϕ) is used for latitude.

A network of meridians and parallels is called a *graticule*. When the sphere or spheroid is reduced in size, as shown in Figure 18-1, the graticule becomes, of course, a reference network for all points on the *globe* just as it is on the earth itself.

The map being produced is two-dimensional. Points on the map sheet have x and y positions based on some rectangular system of reference. The map projection process is the systematic transformation of all spherical coordinates ϕ and λ of the globe into corresponding rectangular coordinates x and y on the map. Mathematically, $x = f_1(\phi, \lambda)$ and $y = f_2(\phi, \lambda)$, meaning that x and y positions on the map are functions of ϕ and λ. These functions must be: (1) unique, so a particular point will appear at only one position on the map; (2) finite, so a particular point will not appear at infinity and be unplottable; and (3) continuous, so although stretching or shrinking of features may occur, there will be no gaps. Projections do exist in which the functions are not finite for the entire globe.

In some cases, the x and y positions may be obtained by imagining an intermediate step involving a cylinder, cone, or plane as shown in Figures 18-1 and 18-3. To illustrate this type of projection, imagine a ray of light projected radially from the globe's center to a tangent surface (Figure 18-3). A point on the globe having a certain ϕ and λ can be transferred to the surrounding surface, which then is "unrolled" or *developed* to form a plane map. More commonly, the functions used to get x and y positions are purely mathematical and do not involve a developable surface. Many of these mathematical concoctions bear some resemblance to the geometrically projected cases shown in Figure 18-3. The cylinder, cone, and plane thus provide a convenient basis for classifying a large number of projections. A projected graticule is classified as (1) *cylindrical* if it takes on a rectangular appearance (Figure 18-3a), (2) *conic* if it looks fan-shaped (Figure 18-3b), and (3) *azimuthal* if its resembles a map projected directly to a plane (Figure 18-3c). The term azimuthal refers to the property that azimuths (or bearings or directions) from the central point to other points are not deformed during the projection process. This term is discussed further in Sections 18-9 and 18-12. An example of the cylindrical group is the well-known Mercator projection.

If a projected graticule has only a slight similarity to geometrically projected cases, it may be classified as pseudocylindrical, pseudoconic, or pseudoazimuthal. The pseudocylindrical projection is not rectangular in appearance, but the parallels are all horizontal, suggesting a relationship to the cylindrical group (see Section 18-12).

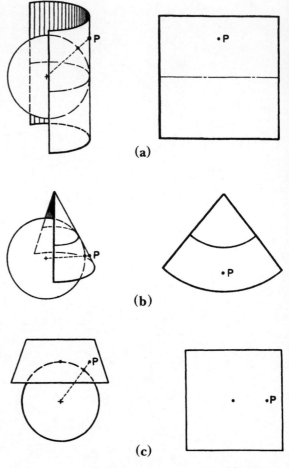

Figure 8-3. Developable projection surfaces. (a) Cylinder tangent to a globe at the equator and developed or unrolled map. (b) Cone tangent to a globe along a parallel and developed or unrolled map. (c) Plane tangent to a globe at the North Pole and part of resulting map.

18-3. SCALE

When a large area, such as the western hemisphere, is shown on a small sheet of paper, the result is said to be a small-scale map. A large-scale map, of course, is the opposite; an example would be the map of a few city blocks on a large sheet. The large-scale map is generated from a relatively large globe. For a plane coordinate system to be used in surveying computations, a full-size globe is used (see Chapter 19).

The usual way of expressing scale in numerical terms is by a dimensionless *ratio* or *representa ive fraction* (RF).

$$\text{RF scale} = \frac{\text{globe distance}}{\text{earth distance}} \quad (18\text{-}1)$$

If 200 km (124.3 mi) are represented on a globe as 1 cm, the RF scale is

$$\frac{1.00 \text{ cm}}{200 \text{ km}} = \frac{0.0100 \text{ m}}{200{,}000 \text{ m}} = \frac{1}{20{,}000{,}000}$$

Distances on the globe are 20,000,000 times smaller than on the earth itself. (Two-dimensional surfaces or areas are reduced in both dimensions and thus are smaller by a factor of $20{,}000{,}000^2$ or 4.00×10^{14}, but only the linear scale is usually stated.) This scale, often called the principal scale, may be written for convenience as 1 : 20,000,000. If the radius of the generating globe R is known, the RF scale is equal to $R/6370$ km converted to a dimensionless ratio with a numerator of unity.

Scale also may be expressed in *unit equiva lents*. In the case mentioned, 1 cm represents 200 km or 1 cm = 200 km, in which it is understood that the smaller unit is a globe distance, and the larger unit an earth distance. It is standard practice to assign unity to the smaller unit rather than to say 1 km = 0.005 cm or 1 km = 0.05 mm. Other examples of unit equivalent scales are 1 in. = 300 mi and 1 in. = 2000 ft.

Scale may be shown graphically, with convenient multiples of earth distances marked off along a bar. In the first example (1 cm = 200 km), scale divisions equivalent to 100 or 500 km might be used. The size of a 500-km division would be 500,000 m/20,000,000 = 0.025 m or 2.5 cm. Graphic scales, being pictorial, are very helpful to a map user.

18-4. SCALE FACTOR

All dimensions of the earth are reduced proportionately when it is reduced to a globe.

Some dimensions do not undergo any further change as the surface of the generating globe is projected to become a map. Figure 18-3a shows the case of a cylinder tangent to the globe at the equator. As the cylinder is unrolled, or developed, the equator maintains its original length. Such a line is called a *standard* line or line of exact scale. It is said to have a *scale factor* or "particular scale" of 1.000. If a certain line is doubled in length during the projection process, it is said to have a scalar factor of 2.000. In equation form,

$$\text{Scale factor} = \frac{\text{map distance}}{\text{globe distance}} \quad (18\text{-}2)$$

The scale factor on any map will vary from point to point and may vary in different directions at the same point, being 1.000 along only standard lines or at a standard point. No map has a uniform scale. An RF, such as 1 : 20,000,000 applies to the generating globe itself and is correct for the map only when the scale factor is 1.000.

18-5. MATHEMATICS OF THE SPHERE

Before proceeding with this discussion of map projections, it is important to review the geometry of spheres.

If the radius of the globe is R, the circumference is $2\pi R$. That, of course, is the length of the equator and any meridian circle. The various parallels are shorter in circumference than the equator. The North and South poles are really the 90th parallels in the northern and southern hemispheres, but they have zero lengths.

The length of a particular parallel can be calculated by multiplying length of the equator by cosine of latitude. If we use ϕ for latitude, the relationship is

Length of parallel

$$= (\text{length of equator})(\cos \phi) \quad (18\text{-}3)$$

Figure 18-4 shows why Equation (18-3) is correct. The figure depicts a cross section of a sphere in which the radius of any parallel, in its own plane, varies with cos ϕ. The circumference of any parallel, in turn, is $2\pi R \cos \phi$. Obviously, partial lengths of parallels, falling between any two meridians, also vary with cos ϕ, becoming zero at the poles where cos ϕ is zero.

To illustrate this relationship, consider the meridians through Pittsburgh (80° W) and Denver (105° W). The distance between these two meridians is a maximum at the equator —namely,

$$\frac{25}{360} 2\pi R = 2779 \text{ km}$$

using 6370 km as the radius of the earth. Between the two cities, as measured along the 40th parallel, the distance is

$$\frac{25}{360} 2\pi R \cos 40° = 2129 \text{ km}$$

If a map having an RF scale of 1 : 20,000,000 shows the 40th parallel as a standard line, the distance between the cities, measured along the parallel, will be

$$\frac{25}{360} \left(\frac{2\pi R \cos 40°}{20,000,000} \right) = 0.000106 \text{ km} \quad \text{or} \quad 10.6 \text{ cm}$$

If another map of the same scale has a scale factor along the 40th parallel of 1.15, the distance will be (10.6) (1.15) = 12.2 cm. The ratio of $2\pi/360$ can be viewed as a conversion factor for degrees to radians.

The surface area of a sphere is $4\pi R^2$, exactly equal to that of a cylinder having the same diameter and height. The circumference

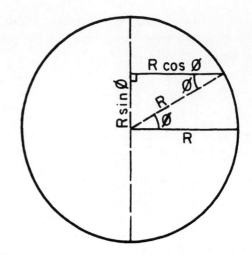

Figure 18-4. Cross section of a globe, showing that the radius of any parallel of latitude is $R \cos \phi$ and the distance between its plane and that of the equator $R \sin \phi$.

of such a cylinder is $2\pi R$ and the height $2R$. The area is $(2\pi R)(2R) = 4\pi R^2$, as stated for the sphere.

The surface area of a zone between any two parallels on a sphere may be found in a similar manner. It is equal to a strip on the surrounding cylinder having the same height (see Figure 18-5). The height of the strip shown is $R \sin \phi$ and the area, therefore, is

Area, equator to parallel

$$= (2\pi R)R \sin \phi = 2\pi R^2 \sin \phi \quad (18\text{-}4)^*$$

This formula may be used, e.g., to find what fraction of the earth's surface lies within the Arctic Circle (66.5° N). The plane of the Arctic Circle is parallel to the equator. The distance between the two planes is $R \sin 66.5°$. This is a height like that of the gray strip shown in Figure 18-5. The area within the Arctic Circle

---

*Equation (18-4), for the area of the zone between the equator and any parallel, also may be found by integration. A narrow zone located at any latitude will have a width $R\,d\phi$ and radius $R \cos \phi$. Its area is $(2\pi R \cos \phi)(R'd\phi)$ and the total area desired therefore is

$$\int_0^{\leq} 2\pi R^2 \cos \phi \, d\phi = 2\pi R^2 \sin \phi$$

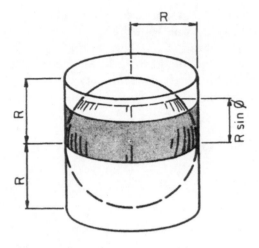

Figure 18-5. Globe surrounded by a cylinder of same height $2R$. Surface area of shaded zone on the sphere is equal to the gray strip or band on a cylinder having the same height.

may be found by subtraction from the area of a hemisphere.

$$2\pi R^2 - 2\pi R^2 \sin 66.5° = 2\pi R^2(1 - \sin 66.5°)$$

This may be seen as a circumference $2\pi R$ times a strip height of $R - R \sin 66.5°$. The fraction of surface area within the Arctic Circle is found by dividing this by the area of a sphere. The answer is independent of R, as follows:

$$\frac{2\pi R^2(1 - \sin 66.5°)}{4\pi R^2} = \tfrac{1}{2}(1 - \sin 66.5°)$$

$$= 0.0415 \quad \text{or} \quad 4.15\%$$

The term *great circle* refers to any arc on the earth, or globe, formed by a plane containing the center of the sphere. Each meridian is a great circle (or actually half of one, running only from pole to pole); the equator is another example. The shortest distance between any two points on the earth's surface is a great circle route. The shortest route between Pittsburgh and Denver is not the one discussed earlier, but rather a great circle route running slightly above the 40th parallel.

The shortest distance between two points may be found by first calculating the central angle subtended by the two points—measured in the plane of the great circle—using an expression from spherical trigonometry. In Figure 18-6, if D is the central angle between points A and B, ϕ_a the latitude of A, ϕ_b the latitude of B, and $\Delta\lambda$ the difference of longitude between A and B, the expression is

$$\cos D = \sin \phi_a \sin \phi_b$$
$$+ \cos \phi_a \cos \phi_b \cos \Delta\lambda \quad (18\text{-}5)$$

Note that ϕ_a and ϕ_b must be expressed as plus or minus (north or south of the equator), but $\Delta\lambda$ may be the longitudinal difference in either direction—not necessarily smaller than 180°. The latter statement is correct because $\cos \Delta\lambda$ is the same either way, e.g., $\cos 20° = \cos 340°$ and $\cos 200° = \cos 160°$. Central angle D can be converted to a surface distance or arc distance by assuming that the earth is spherical. The length of each degree of a great circle is, of course, $2\pi R/360°$.

The shortest distance between Pittsburgh and Denver is found as follows, using latitudes

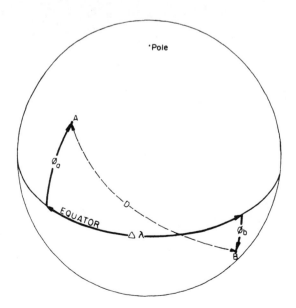

Figure 18-6. Terms in Equation (18-5) include desired central angle D between points A and B, latitudes ϕ_a and ϕ_b of the two points, and difference in longitude $\Delta\lambda$.

and longitudes from the previous example:

$$\cos D = \sin 40° \sin 40°$$
$$+ \cos 40° \cos 40° \cos 25°$$
$$D = 19.09°$$
$$\text{Surface distance} = (2\pi R/360°)(19.09°)$$
$$= 2122 \text{ km}$$

This figure should be rounded to 2120 km.

Although a great circle provides the shortest possible route between two points, it may be a difficult one to follow if navigating manually by compass. In flying from Pittsburgh to Peking, it is simpler to go over Denver, due west all the way, than to follow a route with a constantly changing bearing—i.e., the great circle route would be northwest at first and southwest later. The route of constant bearing, or constant azimuth (in this case, the 40th parallel), is called a *loxodrome* or *rhumb line*. Of course, parallels and meridians are loxodromes, but in general a loxodrome is a route that crosses every meridian at the same angle (see Figure 18-7). The trip from Pittsburgh to Peking could follow a series of loxodromes that together approximate a great circle route. The pilot would change his or her bearing a

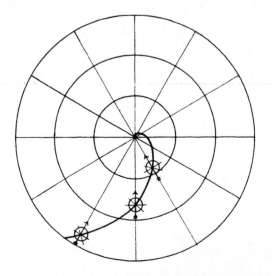

Figure 18-7. Loxodrome crossing meridians at constant angle. (From E. Raisz, 1962, New York: McGraw-Hill, *Principles of Cartography*, with permission.)

number of times, but not continually. There is a map projection with the valuable characteristic that straight lines drawn on it are great circle routes (the gnomonic, see Figure 18-24) and another map projection on which straight lines are loxodromes (the Mercator, see Figure 18-19).

Another useful bit of geometry relates to the trapezoid. Its area is

$$\text{Area} = \frac{b_1 + b_2}{2} h \qquad (18\text{-}6)$$

A 1° quadrangle on a globe (bounded by parallels and meridians 1° apart) resembles a trapezoid but is on a curved surface instead of a plane. The two bases and height are slightly curved. A 1-*min* quadrangle is even more like a trapezoid because there is much less curvature involved; this idea is used in Section 18-10.

18-6. CONSTANT OF THE CONE

On some map projections, including one used in the state plane coordinate system (Chapter 19), a parallel of latitude is drawn as part of a circle. Longitudinal coverage $\Delta\lambda$ represented by the circular arc may be a full 360° or any smaller number, such as 16° for a map of France.

Figure 18-8 shows a central angle L that is related to $\Delta\lambda$ but generally not equal to it. For example, a map may show all 360° of the 40th parallel as a circular arc in which $L = 231°$. Central angle L is related to $\Delta\lambda$ by a constant k as follows:

$$L = k \Delta\lambda \qquad (18\text{-}7)$$

In the case cited, $L = (0.643)(360°) = 231°$. For reasons explained in Section 18-12 and Chapter 19, k is called the *constant of the cone*.

In dealing with circles, it is useful to remember that a chord length can be calculated as follows, referring again to Figure 18-8:

$$\text{Chord} = 2r \sin \tfrac{1}{2}L \qquad (18\text{-}8)$$

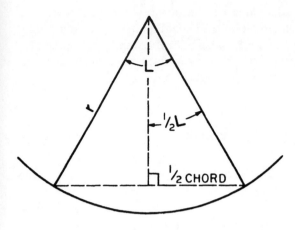

Figure 18-8. Central angle L in conic projections is not equal to $\Delta\lambda$ on globe, and radius r must be derived. (It is not equal to that of globe or of parallel itself, as given in Figure 18-4).

This expression is applied in route surveying as well as the study of map projections.

Often, there is a need to calculate x- and y-coordinates (or tangent offsets) of several points along the circular arc (see Figure 18-9). This computation is useful in plotting, route surveying, and state plane coordinate computations. Assume that a particular parallel has a radius of 100.0 cm and the meridians will cross it at intervals of 5.60 cm. (Figure 18-9 is not drawn to scale.) The central angle for each 5.60-cm arc is

$$L = \frac{5.60}{100.0}\ \text{rad}$$

or

$$\left(\frac{5.60}{100.0}\right)\left(\frac{360}{2\pi}\right) = 3.2086°$$

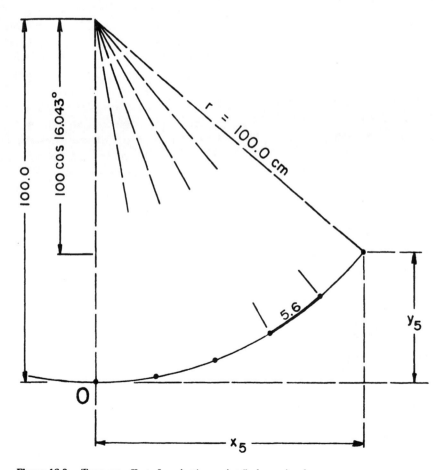

Figure 18-9. Tangent offsets for plotting point 5 along circular arc.

A plotting table may be prepared giving x- and y-coordinates from the point 0, where the central meridian meets the parallel. If five graticule points are needed, the last one will be as follows:

$$x_5 = 100.0 \sin(5 \times 3.2086°)$$
$$= 27.64 \text{ cm}$$
$$y_5 = 100.0 - 100.0 \cos(5 \times 3.2086°)$$
$$= 3.89 \text{ cm}$$

If the meridians are spaced 4° apart in longitude, the constant of the cone must be $3.2086/4.00 = 0.8022$.

The next five sections describe some characteristics found in various map projections.

18-7. STANDARD LINES

In the discussion of scale factor (Section 18-4), standard lines were defined as lines that do not change length when projected from a generating globe to a map. Some projections have only one standard line, such as the equator in Figure 18-3a, but others have many. Several conic projections, e.g., have two standard parallels.

18-8. EQUIDISTANT PROJECTIONS

Although no map projection can offer a uniform scale, some have it in one direction. A projection may have a scale factor of 1.000 in the north-south direction (all meridians are standard; see Figure 18-15), or in the east-west direction (all parallels are standard; see Figure 18-14). Such projections may be described as equidistant.

18-9. AZIMUTHAL PROJECTIONS

Section 18-2 mentioned the characteristic that all azimuthal projections share, namely, that directions to all points with respect to a central point are not deformed during projection from globe to map. (The direction to a distant point is important in the operation of airports, seismographs, radio stations, etc.) An example is shown in Figure 18-18.

18-10. EQUAL-AREA PROJECTIONS

If the *relative size* of all features on a generating globe is maintained during the process of projection to a map, the projection is said to be *equal-area* (also equivalent or equiareal).

It has been pointed out that no map has a scale factor of 1.000 everywhere. If area is to be preserved but scale cannot be, then a given feature on the globe, such as a state, will have to be plotted with a scale factor greater than 1.000 in one direction and smaller than 1.000 in another. It can be shown that such "compensatory scale factors" on an equal-area projection occur in perpendicular directions, often called the *principal directions*. If a circle with a radius of 1 cm is drawn on the surface of a large generating globe, it appears as an *ellipse* when projected to an equal-area map (see Figure 16-10). If one semiaxis is reduced to 0.5 cm, the other will be increased to 2.0 cm to make the ellipse and circle contain the same area.

18-11. CONFORMAL PROJECTIONS

Although equal-area projections have compensatory scale factors and allow a tiny circle, as just described, to be distorted in order to avoid a change in its area, *conformal* projections have equal scale factors in all directions at any one point. A tiny circle is not distorted at all but becomes simply a larger or smaller circle, depending on its location on the map. Instead of preserving size, conformal projections preserve *shape* (see Figure 18-11). They also are called orthomorphic projections.

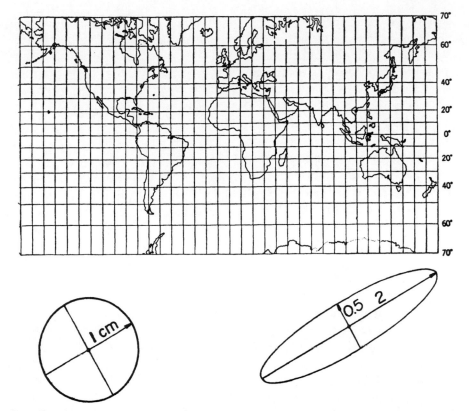

Figure 18-10. Circle on a globe, shown at left, projects as an ellipse on a map. Example from an equal-area projection is shown.

If the shape of everything on a globe could be preserved on a map, as for the tiny circle in the preceding paragraph, the map would have a uniform scale, which is impossible. Thus, it is correct to say that shapes of small features will be preserved in the course of projection to the conformal map. Conformal projections are ideal for setting up plane coordinate systems for use in surveying, because a surveyor's transverse is small in comparison to the portion of the earth on a particular system. Its angles will be the same when placed on the

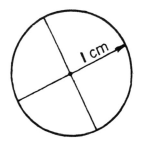

Figure 18-11. In conformal projections, a circle on a globe, at left, will project as another circle (a special kind of ellipse). A scale factor of 0.6 is illustrated.

coordinate system as when they were measured in the field. A single scale factor (generally not 1.000) will apply to all distance measurements, unless the survey is of unusually great size of precision (see Chapter 19).

A map certainly cannot be both equal-area and conformal. It cannot have minimum and maximum scale factors at a point that are compensatory and yet equal. Some projections are neither equal-area nor conformal. The tiny circle referred to earlier is then distorted in both size and shape (see Figure 18-12).

It can be shown that for any projection, such a circle is invariably transformed into an ellipse of some size and has a pair of axes (the principal directions) that remained perpendicular during projection. Unless the projection is conformal, other angular relationships at the point are disturbed. Angular deformation at a point is zero for the principal directions and reaches some maximum value for another pair of lines (see Figure 18-13). Figure 18-14 shows how this maximum angular deformation varies over an equal-area projection of the world. Clearly, the meridians and parallels do not always meet at right angles as they did on the globe. It is evident in the figure that the pair of perpendiculars that remain perpendicular after projection is not necessarily the pair of graticule lines at a point.

The mathematics of how the tiny circle is deformed into an ellipse was developed by M. A. Tissot in 1881. Further discussion appears in *Introduction to Map Projections*.[1]

18-12.   SIMPLE EQUIDISTANT PROJECTIONS

If we use the three basic projection surfaces —cylinder, cone, and plane—it is possible to generate three very simple equidistant projections. They are equidistant in the sense that all meridians are standard (the scale factor is 1.000 in the north-south direction). In each case, there is one standard parallel.

All three projections are examples of an idea mentioned in Section 18-2; they are not *literally* projected to a cylinder, cone, or plane but rather are designed mathematically to have a desirable property. They may be thought of as true projections on which the spacing of parallels has been later modified to match their spacing on the globe, making them equidistant.

The first two in the group—cylindrical and conic—are not very important in themselves, but serve to introduce more complex projections that have great value to surveyors.

Cylindrical Equidistant Projection

This projection is also called *plane chart*, *plate carrée*, *simple cylindrical*, or the *cylindrical equal-spaced projection*.

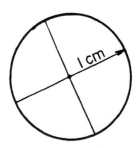

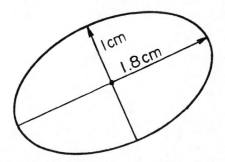

Figure 18-12.   In a projection that is neither equal-area nor conformal, a circle on a globe, at left, will project as a nonlinear ellipse of different size.

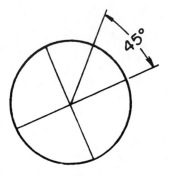

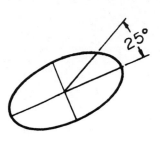

Figure 18-13. When a circle is projected as a noncircular ellipse, the angle between two radial lines will be deformed, except for the two perpendicular that become the ellipse axes.

A cylinder wrapped all the way around the generating globe touching the equator will have the same circumference as the sphere —namely, $2\pi R$. If the whole world is shown on this projection, construction is begun by drawing the equator as a straight line of this length (the equator is a standard line). The meridians are standard also and drawn as straight vertical lines with a length πR. Figure 18-15 shows the resulting graticule, consisting of perfect squares. They are standard in their north-south dimension but, except at the equator, are wider than the corresponding "squares" or quadrangles on the globe. The

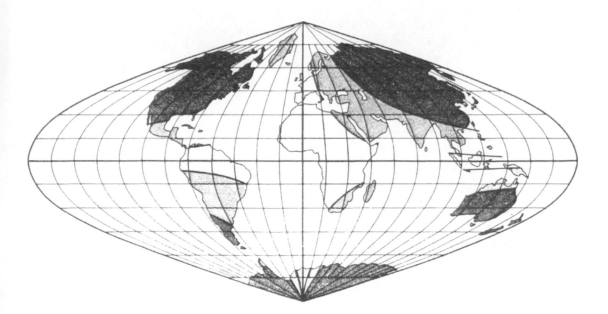

Figure 18-14. World map on a sinusoidal projection, showing lines of equal maximum angular deformation (10 and 40°). Projection is equal-area. Standard lines include the central meridian and all parallels. (From A. H. Robinson and R. D. Sale, 1984, *Elements of Cartography*, 5th ed., New York: John Wiley & Sons, with permission.)

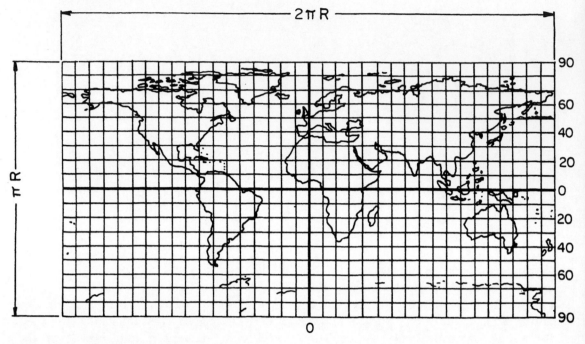

Figure 18-15. Cylindrical equidistant projection.

meridians fail to converge, resulting in the north and south poles appearing as lines the length of the equator instead of points.

By definition, the scale factor is 1.000 along the standard lines. It is greater than 1.000 along the parallels. The 60th parallel, e.g., has a "globe distance," or length, of $2\pi R \cos 60°$ (see Section 18-5) but a "map distance" equal to $2\pi R$—the same as the equator. The scale factor is the ratio of these lengths.

$$\text{Scale factor} = \frac{\text{map distance}}{\text{globe distance}}$$

$$= \frac{2\pi R}{2\pi R \cos 60°} = \sec 60° = 2.000$$

The east-west scale factor varies with $\sec \phi$, being 1.000 on the equator and infinity at the poles.

This projection is so easy to construct, there is little need to think in terms of x- and y-coordinates being functions of ϕ and λ. The relationship, however, is $x = C\lambda$ and $y = C\phi$, meaning λ and ϕ are plotted to some scale as if they were rectangular coordinates.

Conical Equidistant Projection

The projection just discussed was classified as cylindrical, even though spacing of the parallels was determined by the requirement that it be equidistant rather than by any actual geometric projection to a cylinder. The *conical equidistant* is designed in exactly the same way.

A conical equidistant projection is best suited for mapping areas in the vicinity of one standard parallel just as the cylindrical equidistant is appropriate for areas near the equator. This projection, as well as the other conics, is generally chosen for an area lying entirely on one side of the equator.

A cross-sectional view of a globe and tangent cone is depicted in Figure 18-16. The apex is at A and the point of tangency at T. The angle at the apex between globe axis and cone element AT is seen to be equal to the latitude of the standard parallel. In triangle ATO, the tangent of ϕ is R/AT and

$$AT = \frac{R}{\tan \phi} = R \tan \phi$$

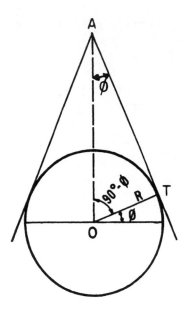

Figure 18-16. Cross-sectional view of a globe and tangent cone.

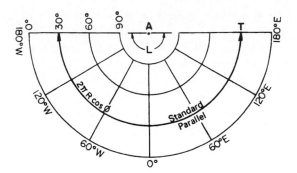

Figure 18-17. Developed cone for the northern hemisphere, conical equidistant projection, with standard parallel at 30°.

Distance AT may be used as a compass setting r for drawing the standard parallel on a map.

The radius of the standard parallel *on the globe is $R \cos \phi$*, as illustrated in Figure 18-4. Its length, of course, is $2\pi R \cos \phi$. *On the map*, after the cone has been "unrolled," the parallel has the same length but the radius used in drawing it, AT or $R \cot \phi$, makes it appear as less than a full circle (see Figure 18-17). The central angle at A called L, in radians, is equal to the arc length divided by the radius.

$$L = \frac{2\pi R \cos \phi}{R \cot \phi} = 2 \cos \phi \tan \phi$$

$$= 2\pi \cos \phi \frac{\sin \phi}{\cos \phi} = 2\pi \sin \phi$$

This angle may be converted to degrees (if we multiply by $360°/2\pi$).

$$L \text{ (in degrees)} = 2\pi \sin \phi \frac{360°}{2\pi} = 360° \sin \phi$$

The L angle is independent of scale (R was cancelled out of the expression). If $\phi = 30°$, then $L = 180°$, as shown in Figure 18-17. A full 360° of longitude is shown within a semicircle. The "constant of the cone" k, defined in Section 18-6, is $\sin \phi$ in the case of a conical equidistant projection.

The standard parallel is divided into equal parts and the meridians drawn as straight radial lines of standard length. As in the cylindrical equidistant, the North Pole will be a line instead of a point. For $\phi = 30°$, distance AT is $1.732R$, while the distance from T to the pole is $(60/180)\pi R$ or only $1.04R$.

The cylindrical equidistant projection, covered previously, is really only a special case of the conical equidistant in which the standard parallel is at $\phi = 0°$, radius $AT = $ infinity, the constant of the cone $k = \sin \phi = 0$, and central angle $L = 0°$ (the meridians being parallel). In other words, a cylinder is merely a special kind of a cone having its apex at infinity.

The conical equidistant projection is known also as the "simple conic." It could reasonably be chosen for a map covering only a few degrees of latitude, such as a tourist map of the Trans-Canada Highway. In that case, the standard parallel might be 50°N; however, better conic projections are available.

Azimuthal Equidistant Projection

It has been pointed out that a cylinder is really a special cone with its apex at infinity. A plane that is tangent at the pole may be viewed as a special cone also. It has an altitude equal to zero and its standard parallel is at 90° N or S. (It is just a little bit flatter than a cone made tangent to 80° N, e.g.) Distance AT is zero or $R \cot 90°$. The constant of the cone $k = \sin 90° = 1.000$, meaning that the central angle L, in degrees, for a full 360° of longitude is 360°. A graticule has a fan-shaped appearance like regular conics if the fan is thought of as being wide open (see Figure 18-18, p. 450). Meridians radiate like spokes of a wheel and are separated by the same angles as on the globe. The projection is called the polar azimuthal equidistant. *Introduction to Map Projections* discusses the nonpolar, or oblique, case where the plane may be tangent to any selected point.

In the polar case, the azimuthal property requires meridians to be drawn with their actual differences in longitude. If the projection is to be equidistant, all of them will be standard lines; parallels will be equally spaced concentric circles. The opposite pole will be a large circle drawn with a radius of πR.

The oblique case is often centered at an airport, radio station, or seismograph because

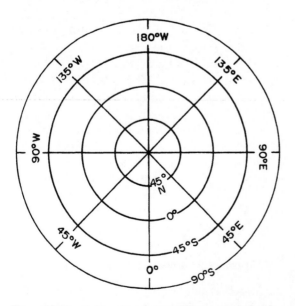

Figure 18-18. Polar azimuthal equidistant projection for the entire world.

it correctly presents both directions and distances from that point.

18-13. PROJECTIONS FOR PLANE COORDINATE SYSTEMS

The three simple projections described in Section 18-12 are neither equal-area nor conformal. Variations exist that do have one or the other of these properties. For surveying purposes, the three projections discussed are modified in the following ways:

1. They are made conformal by sacrificing the equidistant property—i.e., the scale factor is allowed to vary along the line formerly held standard.

2. A spheroid is adopted instead of a sphere because there is to be no scale reduction; field work and computations are done with an RF scale of 1:1 (full size). The dimensions are not purely for plotting purposes, as in cartography.

3. In the cases of the cylindrical and conic projections, *two* standard parallels are used instead of one. This serves to keep the scale factor closer to 1.000 over a wide region. Instead of having all scale factors equal to or greater than 1.000 at all points, as was true in the simple cases of the previous section, projections with two standard parallels include values slightly smaller than 1.000 between the parallels and greater than 1.000 beyond them.

4. For the cylindrical projection, the supposed cylinder is turned 90°, running transversely to the earth's axis. The two standard lines thus are not parallels of latitude but parallel lines adjacent to a selected central meridian. Only a limited area—a "zone" near the central meridian—is included in the coordinate system.

Among cylindrical projections, the conformal one is the Mercator, shown in Figure 18-19 (p. 451). The meridians fail to converge just as they did in Figure 18-15; therefore, the east-west scale factor again varies with sec ϕ.

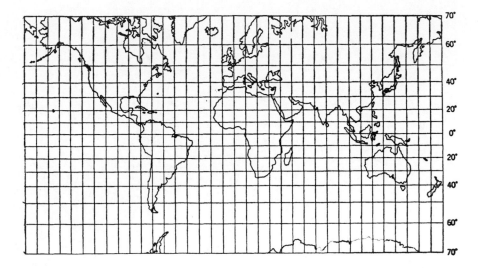

Figure 18-19. Mercator projection. Scale factor in all directions increases with sec ϕ, becoming 2.00 at 60° and infinity at the poles. (From A. H. Robinson and R. D. Sale, 1984, *Elements of Map Projections*, 5th ed., New York: John Wiley & Sons, with permission.)

Because the Mercator is conformal, the scale factor along the meridians must vary in the same way. The two poles, therefore, fall at infinity. The *transverse* Mercator is shown in several figures of Chapter 19.

Among conic projections, the conformal one is the Lambert conformal conic. It is used in the state plane coordinate system for more than half of the states. In a particular state, such as Connecticut, a best-fitting "cone" was selected having its central meridian about midway across the state and its standard parallels just inside the north and south borders (see Chapter 19).

Among azimuthal projections, the conformal one is the stereographic. When based on a sphere, the stereographic is formed as a true projection to a tangent plane from a diametrically opposite point (see Figure 18-23). The polar case is used in the military grid systems (Chapter 19). The oblique case has also been chosen for plane coordinate systems in a few places, including New Brunswick and Prince Edward Island.

Figures 18-20, 18-21, 18-22, 18-23, and 18-24 illustrate five map projections and show how a particular triangle projects on each. The an-

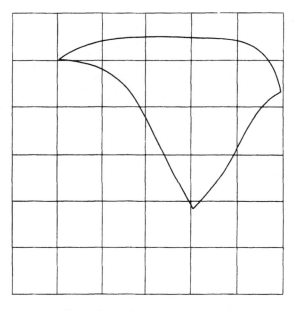

Figure 18-20. Cylindrical equidistant projection of the western hemisphere. Great circle routes joining Anchorage, Madrid, and Buenos Aires are plotted on this and the following four figures; in each case, the central meridian is 90° W. (From P. W. McDonnell, Jr., 1979, *Introduction to Map Projections*, New York: Marcel Dekker, with permission.)

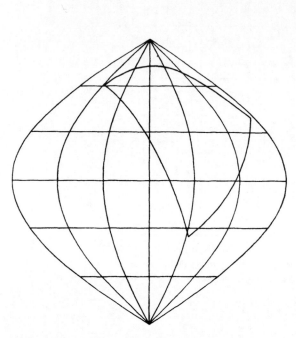

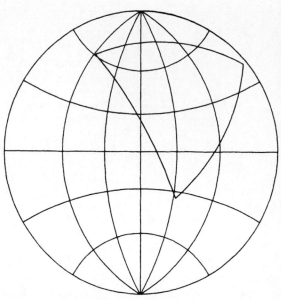

Figure 18-23. Conformal stereographic projection centered on the equator; meridians and parallels are perpendicular.

Figure 18-21. Equal-area sinusoidal projection of the western hemisphere.

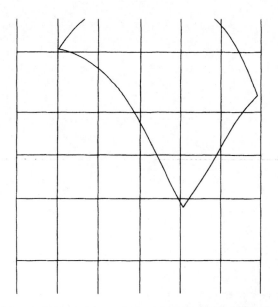

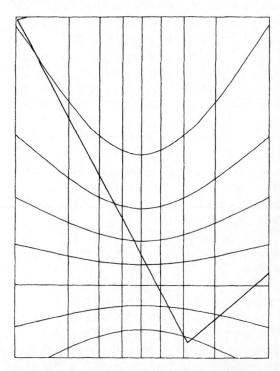

Figure 18-22. Conformal Mercator projection. Angles formed where great circles meet are equal to corresponding angles on earth.

Figure 18-24. Gnomonic projection centered on the equator and 90° W, with graticule lines at 15° intervals. The projection extends to infinity in all directions, but it offers the unique advantage of displaying any great circle as a straight line.

gles at each apex are the same (not deformed) in the conformal projections shown in Figures 18-22 and 18-23.

NOTE

1. McDonnell, P. W., Jr. 1992. *Introduction to Map Projections*, 2nd ed. Rancho Cordova, CA: Landmark Enterprises.

REFERENCES

DEETZ, C. H., AND O. S. ADAMS. 1944 *Elements of Map Projection*. Special Publication No. 68, U.S. Coast & Geodetic Survey.

ROBINSON, A. H., AND R. D. SALE. 1984 *Elements of Cartography*, 5th ed. New York: John Wiley & Sons.

Printed by Books on Demand, Germany